DISSOLVED AIR FLOTATION FOR WATER CLARIFICATION

ABOUT THE AMERICAN WATER WORKS ASSOCIATION

The **American Water Works Association** is the authoritative resource for knowledge, information, and advocacy to improve the quality and supply of water in North America and beyond. AWWA is the largest organization of water professionals in the world. AWWA advances public health, safety, and welfare by uniting the efforts of the full spectrum of the entire water community. Through our collective strength we become better stewards of water for the greatest good of the people and the environment.

American Water Works Association
6666 W. Quincy Ave.
Denver, CO 80235
303.794.7711
www.awwa.org

Executive Director: David LaFrance
Director of Publishing: Liz Haigh
Publications Manager: Gay Porter De Nileon

DISSOLVED AIR FLOTATION FOR WATER CLARIFICATION

James K. Edzwald, Ph.D.

Johannes Haarhoff, Ph.D.

American Water Works Association

McGraw Hill

New York Chicago San Francisco Lisbon London Madrid
Mexico City Milan New Delhi San Juan Seoul
Singapore Sydney Toronto

The McGraw-Hill Companies

McGraw-Hill books are available at special quantity discounts to use as premiums and sales promotions, or for use in corporate training programs. To contact a representative, please e-mail us at bulksales@mcgraw-hill.com.

Dissolved Air Flotation for Water Clarification

1 2 3 4 5 6 7 8 9 0 LPI/LPI 1 7 6 5 4 3 2 1

ISBN 978-0-07-174562-8
MHID 0-07-174562-9

This book is printed on acid-free paper.

Sponsoring Editor
Larry S. Hager

Editing Supervisor
Stephen M. Smith

Production Supervisor
Pamela A. Pelton

Acquisitions Coordinator
Michael Mulcahy

Project Manager
Aloysius Raj, Newgen Publishing and Data Services

Copy Editor
Anupama Gopinath, Newgen Publishing and Data Services

Proofreader
Helen Mules

Indexer
Doreen McLaughlin

Art Director, Cover
Jeff Weeks

Composition
Newgen Publishing and Data Services

ABOUT THE AUTHORS

James K. Edzwald is Professor Emeritus in the Department of Civil and Environmental Engineering at the University of Massachusetts, Amherst. He earned his B.S. and M.S. degrees in Civil Engineering and Environmental Health Engineering at the University of Maryland, and a Ph.D. in Water Resources Engineering at the University of North Carolina, Chapel Hill. He also held faculty positions at the University of Missouri, Clarkson University, and Rensselaer Polytechnic Institute. His research interests include water supply, drinking water treatment, and aquatic chemistry. Dr. Edzwald has authored or coauthored over 150 publications on water quality and treatment. He is the editor of the sixth edition of *Water Quality & Treatment: A Handbook on Drinking Water*, published by AWWA and McGraw-Hill. He received the 2004 A.P. Black Award from AWWA for his contributions in water supply research and the 2009 Founders' Award from the Association of Environmental Engineering and Science Professors for his contributions to environmental engineering education and practice. He is a registered professional engineer in New York.

Johannes Haarhoff is Professor in the Department of Civil Engineering Science at the University of Johannesburg (UJ), South Africa. He earned an Honors B.Eng. in Civil Engineering at the University of Stellenbosch, South Africa, followed by an M.Eng. in Water Engineering at the same university. He spent eight years working in municipal engineering and civil construction before enrolling at Iowa State University, where he earned a Ph.D. in Sanitary Engineering. After a few years as a water treatment specialist at a prominent South African consultancy, Dr. Haarhoff was appointed a Professor at the Rand Afrikaans University (since 2005 the University of Johannesburg). He established the UJ Water Research Group, which focuses on practical aspects of water supply and drinking water treatment. He is a registered professional engineer in South Africa, a Fellow of the South African Institution of Civil Engineering, and a Senior Fellow of the Water Institute of Southern Africa.

Cover photo: Segaliud Water Treatment Plant (Sandakan, Sabah, East Malaysia). Photo courtesy of Purac, Sweden.

CONTENTS

Preface xi
Acknowledgments xiii

Chapter 1. Introduction 1-1

1-1 Types of Flotation Methods / *1-1*
1-2 Overview Description of DAF for Drinking Water Clarification / *1-6*
References / *1-9*

Chapter 2. A History of Dissolved Air Flotation 2-1

2-1 The Initial Use of Flotation for Mineral Separation / *2-1*
2-2 DAF Development for the Paper Industry / *2-2*
2-3 DAF Adapted for Industrial Wastewater Treatment / *2-3*
2-4 The Theoretical Basis for DAF is Consolidated / *2-4*
2-5 DAF for Potable Water Treatment in the 1960s / *2-5*
2-6 The Spread of DAF Since 1970 / *2-6*
2-7 International DAF Conferences / *2-7*
2-8 The Application and Acceptance of DAF / *2-8*
References / *2-10*
Records of International Conferences / *2-12*

Chapter 3. Air Saturation 3-1

3-1 Air Requirements for DAF / *3-2*
3-2 The Principles of Air Saturation / *3-7*
3-3 Open-End Saturation / *3-9*
3-4 Dead-End Saturation / *3-11*
3-5 Saturator Efficiency / *3-25*
3-6 Air Flow and Energy Requirements / *3-28*
3-7 Design Example / *3-33*
References / *3-35*

Chapter 4. Air Precipitation 4-1

4-1 Distribution and Adjustment of Saturator Flow / *4-1*
4-2 Bubble Suspensions / *4-2*

4-3 Bubble Formation / *4-6*
4-4 Injection Nozzles / *4-9*
References / *4-14*

Chapter 5. Air Bubbles and Particles in Water **5-1**

5-1 Air Bubbles in Water / *5-1*
5-2 Particles in Water / *5-6*
5-3 Particle-Bubble Interactions and Forces / *5-15*
References / *5-25*

Chapter 6. Pretreatment Coagulation and Flocculation **6-1**

6-1 Introduction to Coagulation / *6-2*
6-2 Contaminants / *6-4*
6-3 Bulk Water Chemistry and Temperature / *6-9*
6-4 Rapid Mixing / *6-12*
6-5 Coagulation Chemistry and Mechanisms / *6-14*
6-6 Guidance on Coagulant Dosing for DAF / *6-33*
6-7 Introduction to Flocculation / *6-41*
6-8 Flocculation Fundamentals / *6-41*
6-9 Flocculation Tanks / *6-46*
6-10 Flocculant, Flotation, and Filter Aids / *6-47*
6-11 Guidance on Flocculation for DAF / *6-48*
References / *6-49*

Chapter 7. Contact Zone **7-1**

7-1 Contact Zone Hydraulics / *7-2*
7-2 Contact Zone Modelling / *7-7*
7-3 White Water Bubble-Blanket Model / *7-10*
7-4 Practical Applications of the Contact Zone Model / *7-29*
References / *7-34*
Appendix: Derivation of the White Water Bubble-Blanket Contact Zone Model / *7-35*

Chapter 8. Separation Zone **8-1**

8-1 The Boundary between Contact and Separation / *8-1*
8-2 The Rise Rate of Air Bubbles / *8-2*
8-3 The Nature of Floc-Bubble Aggregates / *8-5*
8-4 The Rise Rate of Floc-Bubble Aggregates / *8-6*
8-5 An Over-Simplified View of DAF Separation / *8-9*
8-6 Separation-Zone Measurements and Modelling / *8-10*
8-7 Inlet and Outlet Considerations / *8-12*
8-8 Separation-Zone Flow Patterns / *8-14*
References / *8-17*

Chapter 9. Float Layer Removal 9-1

9-1 The Nature of the Float Layer / *9-1*
9-2 Hydraulic or Mechanical Float Layer Removal? / *9-2*
9-3 Hydraulic Float Layer Removal / *9-4*
9-4 Mechanical Float Layer Removal / *9-6*
References / *9-9*

Chapter 10. Process Selection and Design 10-1

10-1 DAF Selection Based on Raw Water Quality Considerations / *10-1*
10-2 Bench-Scale Testing / *10-5*
10-3 Pilot-Scale Testing / *10-9*
10-4 Supplemental Measurements / *10-12*
References / *10-16*

Chapter 11. Conventional Applications for Drinking Water Treatment 11-1

11-1 DAF Configurations / *11-1*
11-2 DAF Treatment of Various Water Quality Types / *11-6*
References / *11-20*

Chapter 12. Additional Applications 12-1

12-1 DAF Pretreatment for Membrane Processes / *12-1*
12-2 Water Reuse / *12-7*
12-3 Spent Filter Backwash Water / *12-9*
References / *12-14*

Chapter 13. Dissolved Air Flotation for Desalination Pretreatment 13-1

13-1 Seawater Chemistry / *13-2*
13-2 Contaminants / *13-8*
13-3 Coagulants and Coagulation / *13-11*
13-4 Physical Properties of Seawater / *13-16*
13-5 Air Saturation in Seawater / *13-18*
13-6 Examples of DAF Pretreatment / *13-19*
References / *13-21*

Appendix A. Abbreviations and Equation Symbols A-1

Appendix B. Useful Conversions B-1

Appendix C. Useful Constants C-1

PREFACE

The dissolved air flotation (DAF) process has now been in use for more than 40 years as an important process for the clarification of drinking water. During the past 20 years it became a widely adopted alternative for sedimentation in all parts of the world, for plants large and small. A rich body of experience and fundamental understanding has steadily accumulated among researchers, designers, and manufacturers and yet there is no professional book on this subject. The authors over the last 25 years have published scholarly journal papers, conducted research, have participated as keynote speakers at international conferences on DAF, and consulted worldwide to engineering companies and water utilities. We know from our own experience and from numerous discussions with professionals in the water field that there is a great need for this book. Our primary purpose with this book is to consolidate and to interpret this knowledge, for the first time, in a single volume entirely devoted to DAF for water clarification.

The book should be of particular interest to those sectors of the water field involved with DAF including academic researchers, design engineers from consulting practices, water plant manufacturers and suppliers, and water plant managers and operators. Others who should consider this book include:

- University professors, upper-level undergraduate and graduate students who study water treatment processes. This book will serve as a comprehensive reference in courses on Water Treatment Processes, Advanced Water Treatment, Membrane Processes and Desalination, and Design of Water Treatment Systems.
- Engineers and scientists who design water treatment plants and advanced water treatment plants for water reuse and water desalination.
- Water utility managers, engineers, scientists, and operators.
- Engineers and scientists involved with industrial water and wastewater treatment processes.

The material in the book is developed along three main objectives. The first is to develop a fundamental basis for understanding how DAF works. The fundamental principles developed in the book apply to all fields that use DAF. However, discussion and evaluation of DAF for the following drinking water applications are addressed in detail: (1) clarification in conventional treatment, (2) clarification pretreatment in low-pressure and high-pressure membrane plants—for the latter especially in desalination applications, (3) clarification in water reuse, and (4) treatment of spent filter backwash waters. Therefore, a second objective is to provide guidance for process engineers and water managers toward these applications where DAF might be incorporated in the various treatment schemes. The third objective is to develop the necessary DAF design concepts and to illustrate them by descriptions of practical applications.

Chapter 1 provides background material on various methods of flotation so the reader gains an appreciation of DAF versus other methods and gives a brief description of DAF as used within a conventional drinking water treatment plant. Chapter 2 traces the evolution of DAF technology from its mining roots a hundred years ago up to its present status as a widely used and accepted clarification process.

The next six chapters develop fundamentals of the process, Microbubbles, essential for successful DAF treatment, are covered in Chaps. 3 and 4. Chapter 3 considers the options for saturating the recycle stream with air under pressure and develops a complete set of equations for the design of common saturation systems, while Chap. 4 analyzes the manner in which the recycle flow is injected into the DAF tank and how the bubbles are formed. Chapter 5 considers breifly some general properties of air bubbles in water, and then emphasizes the forces between particles, bubbles, and particles-bubbles. The crucially important roles of appropriate pretreatment coagulation and flocculation are addressed in Chap. 6. Chapter 7 continues with our fundamental approach dealing with the contact zone of the DAF tank where bubbles and floc particles collide and attach to each other. Finally, we reach the separation zone in Chap. 8, where the floc-bubble aggregates are separated before the clarified water flows out of the DAF tank. We stress in Chaps. 3 to 8 fundamental principles pertaining to DAF, but we also provide practical and illustrative examples, provide ranges or typical values of design and operating variables, and relate the concepts to drinking water treatment practice.

The book closes with five practical chapters. Chapter 9 is devoted to methods for removing the float layer from the surface of the DAF tank. Chapter 10 addresses whether DAF is a process that should be considered for a water treatment plant. It starts with the presentation of a clarification process selection guide based on raw water quality, moves to what information bench-scale DAF studies provide, and then considers pilot-scale DAF studies and the kind of design and operating criteria generated from these studies. Supplementary and complementary measurements and tests are also addressed. Chapter 11 discusses and evaluates DAF in conventional drinking water plants. We begin with the presentation of different DAF configurations including discussion of conventional-rate and high-rate plants. DAF treatment experience for various type raw water qualities are presented and evaluated. The discussion according to raw water quality is supplemented with several case studies, each demonstrating different aspects of DAF. Chapter 12 explores additional applications of DAF pretreatment for membrane systems, especially low-pressure ones. DAF is also considered in the fields of water reuse and in treating spent filter backwash water. Case studies are presented throughout the chaptaer. The last chapter (Chap. 13) examines DAF as a seawater-pretreatment process in reverse osmosis desalination plants. The chapter starts with summarizing seawater chemistry, discussing contaminants of concern in pretreatment, and discussing and contrasting seawater coagulation to freshwater. The physical properties of seawater are examined and how they might affect DAF performance. We also consider the effects of seawater on dissolving air and available air in the saturator. Finally, we present examples of DAF in full-scale desalination plants.

James K. Edzwald, Ph.D.
Professor Emeritus
University of Massachusetts

Johannes Haarhoff, Ph.D.
Professor
University of Johannesburg

ACKNOWLEDGMENTS

This professional book, *Dissolved Air Flotation for Water Clarification*, is a tribute to the many water process engineers, scientists, consultants, water utility managers, and academics who have for many years worked on this subject. Without their research, development, and water plant experience, we would not have a book like this that lays out the principles and practice of this water treatment technology. There are several individuals who were most helpful in discussing and providing information on dissolved air flotation, providing plant visits, providing "gray" literature, and in reviewing sections of the book. We are most grateful to Tony Amato (Enpure Ltd.), Ian Crossley, (Hazen and Sawyer), Jan Dahlquist (formerly with Lackeby Water, Purac), John Dyksen (United Water New Jersey), Gary Kaminiski (Aquarion Water Company of Connecticut), Matthew Valade (Hazen and Sawyer), and David Pernitsky (CH2M Hill).

Chapter 2 required the patience and generosity of numerous individuals who shared their DAF experiences of many years ago—these individuals are listed in the references of that chapter. John Tobiason (University of Massachusetts) and John Gregory (University College, London) reviewed portions of Chap. 5 and made invaluable suggestions for improvement. Kerry Howe (University of New Mexico), Steven Duranceau (University of Central Florida), and John Tobiason made valuable suggestions and comments on parts of Chap. 12. In preparing Chaps. 11 and 12, we obtained information and data on several applications of DAF, and several of the individuals who provided assistance are acknowledged above. We also thank the following people who generously helped with this effort: Keith Cartnick (United Water New Jersey), Laurel Passantino (Malcolm Pirnie), Peter Tymkiw (Malcolm Pirnie), and Ben van der Merwe (Environmental Engineering Services, Windhoek). Finally, we are grateful for the comments and suggestions made by the book reviewers: John Bratby (Brown and Caldwell), Ross Gregory (Consultant), David Pernitsky, and David Russell (Global Environmental Operations). They were most helpful in making revisions and in improving the book.

DISSOLVED AIR FLOTATION FOR WATER CLARIFICATION

CHAPTER 1
INTRODUCTION

1-1 TYPES OF FLOTATION METHODS 1-1
Dispersed Air Flotation 1-2
Electrolytic Flotation 1-3
Combination of Ozone and Flotation 1-3
Dissolved Air Flotation 1-3
1-2 OVERVIEW DESCRIPTION OF DAF FOR DRINKING WATER CLARIFICATION 1-6
Pretreatment 1-6
DAF Tank and Recycle Flow 1-8
REFERENCES 1-9

Dissolved air flotation (DAF) is a widely used particle separation (clarification) process. It has many applications in several fields, as summarized in Table 1-1, including drinking water, municipal and industrial wastewaters, industrial water supply, and other fields such as in situ treatment of lakes for algae and seawaters for algae and oil spills. DAF removes particles from water by using air bubbles that collide and attach to particles, and the particle-bubble aggregates are then separated by rising to the surface of the tank. The most common method of producing bubbles, and the primary method discussed in the book, is the recycle-flow pressurization method. In this method, a recycle flow of treated water is pumped to a saturator vessel where air is added under pressure and dissolved. The pressurized-recycle flow from the saturator is then introduced into the front portion (contact zone) of the DAF tank where it is injected typically through nozzles. The pressure change across the nozzles causes the dissolved air to come out of solution in the form of small air bubbles. There are other methods of dissolving air that are explained briefly later in this chapter and where appropriate in other chapters, but our main focus is the recycle-flow pressurization method.

In this chapter we provide background material on various methods of flotation, so the reader gains an appreciation of DAF versus other methods. We also describe briefly the application of DAF for a conventional drinking water treatment plant. We do this for several reasons: because this type of plant is a widely used application, because it provides an opportunity to define various terms, and because we can introduce some typical design and operating variables for DAF.

1-1 TYPES OF FLOTATION METHODS

There are different ways of forming bubbles for flotation processes. Although DAF is the subject of this book, we describe briefly the three main ways of bubble formation: electrolytic flotation, dispersed air flotation, and dissolved air flotation. We also include the special case of combination of ozone and flotation in which bubbles are produced by dispersed or dissolved air methods.

TABLE 1-1 Dissolved Air Flotation Clarification Applications

Drinking Water Treatment
Clarification in a conventional water treatment plant
Clarification in low-pressure membrane treatment plants and nanofiltration-membrane plants
Clarification in reverse osmosis desalination plants
Clarification in water reclamation/water reuse
Treatment of spent filter backwash water
Municipal Wastewater Treatment
Primary clarification
Secondary clarification
Tertiary treatment: Suspended solids removal, phosphorus removal following chemical precipitation
Combined sewer water and storm water treatment
Wastewater reclamation
Thickening* of waste suspensions
Industrial Water Supply and Industrial Wastewater Treatment
Chemical industry
Food wastes: Vegetable wastes, dairies, meat packing, poultry processing, vegetable oil production
Oil production and refineries
Pharmaceutical plants
Pulp and paper mills
Steel mills
Soap manufacturing
Other
Separation of minerals from ores
Fiber separation in the internal process water recovery in pulp and paper industry
Removal of PCBs at hazardous waste sites
In situ treatment of lakes for algae and seawaters for algae and oil spills

*Not a clarification process; thickening concentrates the solids concentrations of wastes.

Dispersed Air Flotation

In dispersed air flotation, bubbles are produced by mechanical means either by passing air through porous diffusers or by aeration using turbines or impellers. The bubbles formed are much larger in size, 0.5 to 2 mm (Matis and Zouboulis, 1995), than those produced by other flotation methods. The main application of dispersed air flotation is in mineral processing using these large bubbles to float high-density mineral particles.

For several reasons, dispersed air flotation is not used for drinking water treatment and wastewater clarification. First, the large bubbles produced in dispersed air flotation systems yield lower collision efficiencies with floc particles and lower process efficiency compared to the smaller bubbles in DAF. Second, large bubbles are not necessary to float the low-density floc particles usually found in drinking waters and wastewaters. Third, mechanical aeration with turbines creates turbulence and a high degree of mixing that shear flocs. Fourth, surface active chemicals (surfactants) are sometimes used to reduce bubble size, and the addition of these chemicals is prohibited in drinking water treatment.

Electrolytic Flotation

In electrolytic flotation or electroflotation, hydrogen and oxygen bubbles are produced by applying a direct current between two electrodes. An advantage of electrolytic flotation is that small bubbles of 20 to 40 μm are formed (Burns et al., 1997). Electrode materials include steel, aluminum, lead oxide deposited on titanium, and stainless steel. There are several reasons why electrolytic flotation is not used in drinking water applications. First, power consumption and operating costs are high for treating low-conductivity drinking waters. Second, the hydraulic loadings are low compared to DAF, hence larger tank footprints. There is also the problem of potential contamination of the floated water with metals originating from the electrodes. Other problems are electrode fouling, maintenance, and replacement. There has been some use of electroflotation for sludge thickening in small municipal wastewater plants (Gregory and Edzwald, 2010). Other applications are the treatment of animal wastes, textile wastewaters, and industrial wastewaters containing emulsions and heavy metals.

Combination of Ozone and Flotation

In France, in the 1990s, there was interest in the development of the combination of ozone and flotation for treating water supplies with high algae concentrations. Interest in these drinking water technologies by Veolia Water may have waned since then as they no longer appear on the company's website; therefore, we describe them only briefly.

One combination is a dispersed air flotation method called Ozoflot®. This combination was tested in at least two pilot studies (Bourbigot et al., 1991; Wilson et al., 1993) in which the supplies were high in algae. Ozonation takes place in the first chamber where ozone in air is fed through porous plate diffusers. A flow of water is maintained across the diffusers to break up the air bubbles and carry the water with floc particles and bubbles to the second chamber for flotation. The other is the use of dissolved air flotation in which the recycle water is saturated with ozone-rich air and is called Flottazone®. A full-scale plant was built at Agen (southern France) as reported by Baron et al. (1997).

Dissolved Air Flotation

In DAF processes, bubbles are produced by reducing the pressure of water saturated with air. There are three ways to achieve this: (1) vacuum flotation, (2) direct injection of air into the water, followed by pressurization by either pump or hydraulic head, and (3) pumping of water through a pressurized air vessel, or saturator. Each method is described briefly in the following sections.

Vacuum DAF. The water undergoing treatment is saturated with air under atmospheric pressure. Air bubbles are then produced by applying a vacuum to the flotation tank. This process has several disadvantages for continuous flow clarification processes. The amount of air released is small. The vacuum drawn is limited practically to, say, half an atmosphere, while the other DAF methods use saturation pressures of four to six atmospheres. The equipment required to produce a vacuum is fairly sophisticated and cumbersome to operate. The process lends itself to batch-type operation. A major disadvantage for drinking water applications is that the treated water, after exposed to a partial vacuum, would contain little or no dissolved oxygen. Hence, reducing conditions would occur that can increase the dissolved iron and manganese in water leaving the tank or from coatings of downstream granular media filters. Vacuum flotation was used in the pulp and paper industry, but has been replaced by pressure DAF. In the past, it was used in the thickening of waste sludges, but it has been replaced mostly by gravity and pressure belt filters.

Pressure DAF with Direct Air Injection. Air is added directly to the water to be treated where either the entire water flow (full-stream saturation) or a part of the flow (part-stream saturation) is pressurized to dissolve the added air. This air-saturated stream is then injected into the contact zone.

The most common method for raising the pressure after air addition is a specially adapted centrifugal pump (see Fig. 1-1a), which is discussed further in Chap. 3. Another method is to add air to water under a large elevation hydraulic head, as depicted in

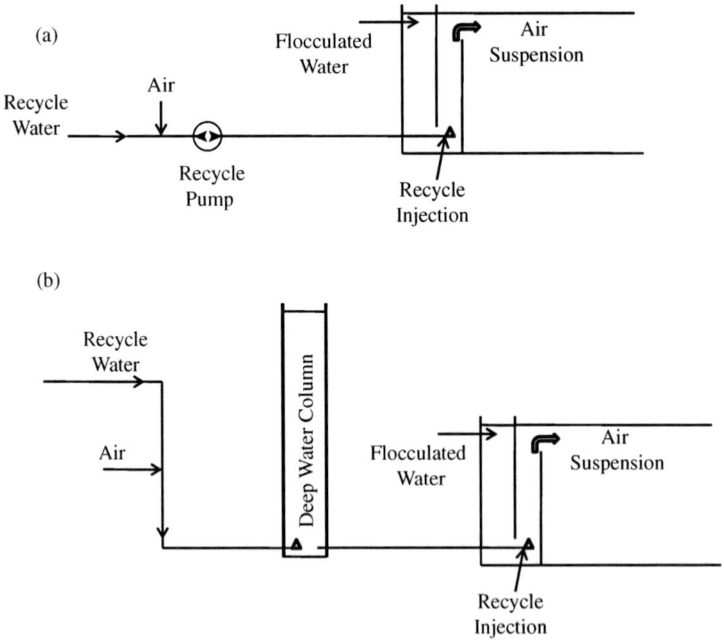

FIGURE 1-1 Pressure DAF with direct air injection: (a) air addition with centrifugal pump, (b) induced air dissolution by hydraulic head of water.

Fig.1-1b. The elevation head from the influent water surface in the chamber to the point of air injection causes the air to dissolve into the water under pressure. The water with dissolved air flows upward and to the DAF tank where the reduction in water pressure releases air bubbles. Induced air saturation has seen limited use. It has been applied in wastewater treatment applications for oxygen transfer where the process is referred to as "deep shaft" flotation. The authors are not aware of any drinking water applications of this method.

Pressure DAF with Saturators. The water is pressurized in the presence of air until the water is close to saturation and is then injected into the DAF tank using pressure-reducing valves or nozzles producing small bubbles of 10 to 100 μm. Three flow schemes are possible as illustrated in Fig. 1-2: (1) full-flow pressurization, (2) split-flow pressurization, and (3) recycle-flow pressurization. All three schemes are used in industrial water and wastewater treatment applications. The recycle-flow pressurization scheme is used in drinking water applications. The most common method of air saturation is by using saturation vessels, or simply called saturators, at pressures between 350 and 700 kPa, more commonly at pressures of 400 to 600 kPa.

Chapter 3 is directed primarily at recycle-flow pressurization in saturators, but it also covers the fundamentals of DAF by direct air injection with a comparison of their respective energy requirements.

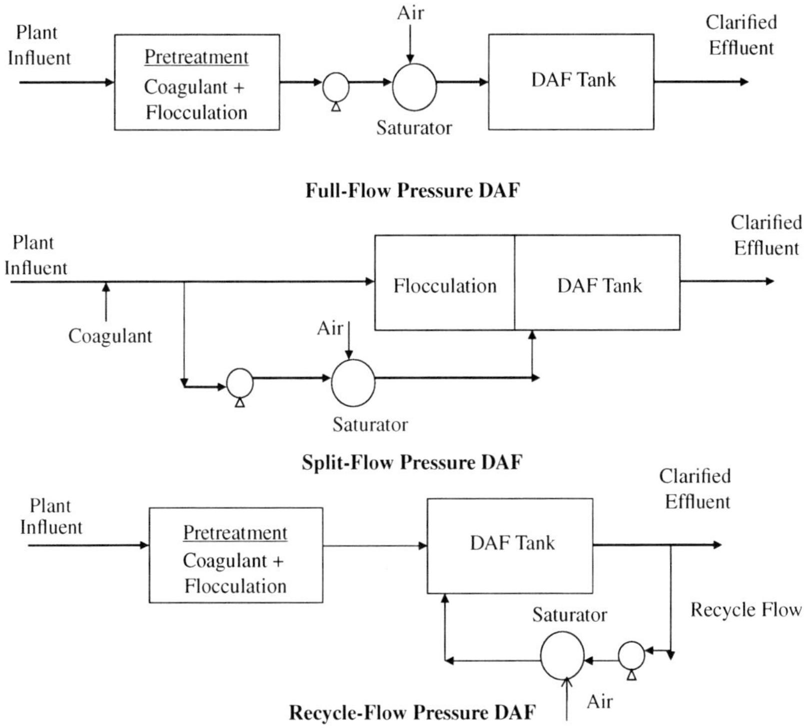

FIGURE 1-2 Flow schemes for pressure DAF systems.

1-2 OVERVIEW DESCRIPTION OF DAF FOR DRINKING WATER CLARIFICATION

Figure 1-3 shows a process schematic of a conventional-type drinking water treatment plant with DAF as the clarification step instead of sedimentation. This is the most widely used application of DAF for drinking water and thus gets special attention. It is often referred to in many of the chapters. Note that the plant is divided into the pretreatment processes of coagulation and flocculation followed by DAF and, then, granular media filtration. An overview is given to provide the reader with information of the general features of the plant, some terminology, and some typical design and operating conditions. To aid our description, Fig. 1-4 shows a picture of a pilot-scale DAF unit. In the same figure are pictures of the baffled contact zone of an empty DAF tank in a full-scale plant showing the recycle manifold and nozzles, and a saturator in a full-scale plant.

Pretreatment

Coagulation and flocculation are necessary pretreatment processes. Coagulation is a chemical addition step. The coagulants used most often in water treatment are aluminum or ferric salts. Coagulation has two purposes: to destabilize particles present in the raw water and to convert dissolved natural organic matter (NOM) into particles. The latter purpose, over the last 25 years, has become an essential part of water treatment used to remove as much dissolved organic carbon (DOC) as feasible through coagulation. The chemistry of coagulation involves selecting proper dosing and pH conditions to produce particles—originally in the raw water and new particles from metal coagulant precipitation—that can form flocs in the flocculation tank and, ultimately, to attach bubbles to these flocs in the DAF tank.

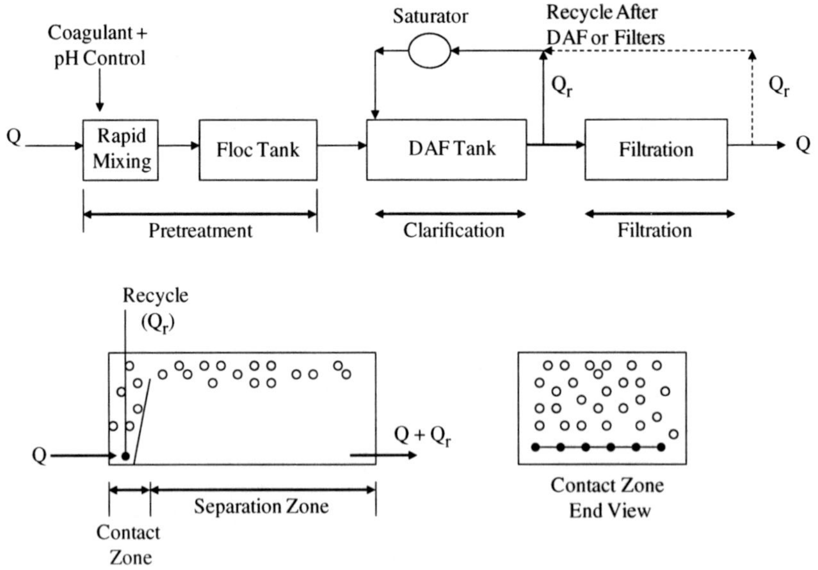

FIGURE 1-3 Schematic diagram of DAF drinking water treatment plant.

(a) (b) (c)

FIGURE 1-4 (a) DAF pilot plant (Courtesy: David J. Pernitsky, CH2M Hill, Inc.), (b) contact zone of empty DAF tank showing recycle manifold and nozzles (West Nyack, New York, United States), and (c) DAF saturator (Cambridge, Massachusetts, United States).

The goal of coagulation chemistry, then, is to produce particles with little or no net electrical surface charge and with a relatively hydrophobic character (Chap. 5). We shall see in Chaps. 6 and 7 that good coagulation chemistry is essential for good performance of the DAF process.

The flocculation step involves mixing of the water to induce collisions among particles and formation of floc particles or simply flocs. The sizes of flocs in the influent to the DAF tank affect collisions between flocs and bubbles in the DAF tank contact zone and consequently the rise velocities of flocs with attached bubbles in the DAF tank separation zone. The flocs with attached bubbles are called floc-bubble aggregates. Flocculation is used in both sedimentation and DAF plants, but the objectives differ. When settling follows flocculation, the goal is to produce large flocs capable of settling at a fairly rapid rate. Flocs with sizes of 100 μm and greater are required to produce reasonable settling rates. DAF differs: we do not want floc settling to take place. Rather, we desire to attach bubbles to smaller flocs to form floc-bubble aggregates with densities less than the density of water. We can do this efficiently with flocs of sizes less than 100 μm, more like tens of micrometers. This does not mean that we cannot float larger flocs. It means that the objective of the floc tank in a DAF plant is to produce smaller flocs than those produced in a settling plant. Fundamental material regarding flocculation is covered in Chap. 6. In Chap. 7 the effect of floc size on the collection of flocs by bubbles is explained, and in Chap. 8, the effect of floc-bubble aggregate size on the rise velocities and removal of aggregates is explained.

DAF Tank and Recycle Flow

The DAF tank is divided into two different sections, as illustrated at the bottom of Fig. 1-3. The first section, separated by a baffle from the second, is called the contact zone, where floc particles are introduced and contacted with air bubbles. In some older literature the contact zone is referred to as the reaction zone; throughout this book the commonly accepted term contact zone is used. Here, collisions occur among bubbles and floc particles. If the floc particles are prepared properly via coagulation with respect to their surface chemistry, then bubbles colliding with the flocs may attach yielding floc-bubble aggregates. The water carrying the suspension of bubbles, flocs, and aggregates flows to the second part of the tank, called the separation zone. Here, bubbles not attached to flocs and floc-bubble aggregates may rise to the surface of the tank. The float layer at the surface of the tank consists of a mixture of bubbles and floc particles attached to bubbles. Over time, this float layer is concentrated producing a sludge that is collected and removed from the tank as described in Chap. 9. Clarified water, often referred to as the subnatant, is withdrawn from the bottom of the tank. In a standard-type DAF water plant, granular media filtration follows DAF. In some applications, DAF is placed on top of the filters. Both applications are discussed in Chap. 11. The more traditional standard plant is the horizontal layout with filtration following DAF. This allows for the two processes to be designed and operated independently at different hydraulic loadings.

The most common method of providing air bubbles, as described briefly above, is recycle-flow pressurization with a saturator. A recycle flow (Q_r) is pumped to a saturator where air is added under pressure. The recycle fraction, also called the recycle ratio and a rate (although the latter term is a misnomer), is a surrogate measure of the amount of air supplied to DAF tank. It is an important design and operating variable. The recycle ratio (r) is defined as the recycle flow divided by the plant throughput flow (Q) as shown in Eq. (1-1).

$$r = \frac{Q_r}{Q} \tag{1-1}$$

Typical ranges for recycle ratios and saturator pressures in drinking water are 8 to 12 percent and 400 to 600 kPa, respectively. The recycle flow is pumped to a saturator vessel (Fig. 1-4c) where air is added under pressure. For the standard plant with filtration separated from DAF, the recycle flow is taken either directly following DAF or after filtration. For plants with DAF over filtration, the recycle flow is taken after the filters. Chapter 3 deals with saturators and the addition and dissolution of air to the recycle flow. The recycle flow containing the dissolved air under pressure is injected into the contact zone, as illustrated at the bottom of Fig. 1-3 and discussed in detail in Chap. 4. Briefly, the recycle flow is injected with nozzles or valves to reduce the pressure to the hydrostatic head in the contact zone. The pressure drop through the nozzles or valves causes precipitation of the air from the water. A distinctive characteristic for these large pressure drops is to produce very small air bubbles in the range of 10 to 100 μm. These small air bubbles scatter light producing a milky or white appearance. The name *white water* is given to this air bubble suspension that occurs in the contact zone and that carries into the separation zone.

Finally, some comments are made about the hydraulic loadings of DAF tanks. The hydraulic loading sets the size of the tank. Usually, the nominal hydraulic loading (HL_{nom}) is used to describe DAF. It is presented in Eq. (1-2) and uses the plant throughput flow and gross plan area (A) of the contact and separation zones.

$$HL_{nom} = \frac{Q}{A} \tag{1-2}$$

The removal of bubbles and floc-bubble aggregates occurs in the separation zone. The separation of bubbles and aggregates depends on their rise velocities relative to the hydraulic loading of the separation zone; in principle, the rise velocities must be greater, but it is shown in Chap. 8 that the flow patterns affect greatly the performance. The separation-zone hydraulic loading (HL_{sz}) should be described using the sum of the plant throughput flow and the recycle flow divided by the separation-zone plan area (A_{sz}) as presented in Eq. (1-3).

$$HL_{sz} = \frac{Q + Q_r}{A_{sz}} \tag{1-3}$$

Drinking water DAF plants designed in the 1960s through the 1980s had nominal hydraulic loadings of 5 to 10 m^3/m^2-h (2 to 4 gpm/ft^2). The units show flow rate divided by area, which of course is a velocity, so it is common to show the hydraulic loading as simply with units of m/h. These rates were increased in the 1990s, so it is now common to refer to conventional-rate DAF loadings in the range of 10 to 15 m/h (4 to 6 gpm/ft^2). There have been innovations in the hydraulic design of DAF tanks that have led to the development of high-rate DAF systems. These are discussed in Chaps. 8 and 11. The high-rate systems are designed in the nominal loading range of 20 to 30 m/h (8 to 12 gpm/ft^2) or greater and with separation zone loadings of 20 to 40 m/h (8 to 16 gpm/ft^2).

REFERENCES

Baron, J., Ionesco, N. M., and Bacquet, G. (1997), "Combining flotation and ozonation—the Flottazone® process," in *Dissolved Air Flotation*, Chartered Institution of Water and Environmental Management, London.

Bourbigot, M. -M., Martin, N., Faivre, M., Le Corre, K., and Quennell, S. (1991), Efficiency of an ozoflotation-filtration process for the treatment of the River Thames at Walton Works, *Journal of Water Supply: Research and Technology – Aqua*, 40 (2), 88–96.

Burns, S. E., Yiacoumi, S., and Tsouris, C. (1997), Microbubble generation for environmental and industrial separations. *Separation and Purification Technology*, 11, 221–232.

Gregory, R., and Edzwald, J. K. (2010), "Sedimentation and flotation," in J. K. Edzwald (ed.), *Water Quality and Treatment: A Handbook on Drinking Water*, AWWA, 6th ed., McGraw-Hill, New York.

Matis, K. A., and Zouboulis, A. I. (1995), "An overview of the process," in K. A. Matis (ed.), *Flotation Science and Engineering*, Marcel Dekker, Inc., New York.

Wilson, D., Lewis, J., Nogueria, F., Faivre, M., and Boisdon, V. (1993), The use of ozoflotation for the removal of algae and pesticides from a stored lowland water, *Ozone Science & Engineering*, 15, 481–496.

CHAPTER 2

A HISTORY OF DISSOLVED AIR FLOTATION

2-1 THE INITIAL USE OF FLOTATION FOR MINERAL SEPARATION 2-1
2-2 DAF DEVELOPMENT FOR THE PAPER INDUSTRY 2-2
2-3 DAF ADAPTED FOR INDUSTRIAL WASTEWATER TREATMENT 2-3
2-4 THE THEORETICAL BASIS FOR DAF IS CONSOLIDATED 2-4
2-5 DAF FOR POTABLE WATER TREATMENT IN THE 1960s 2-5
2-6 THE SPREAD OF DAF SINCE 1970 2-6
2-7 INTERNATIONAL DAF CONFERENCES 2-7
2-8 THE APPLICATION AND ACCEPTANCE OF DAF 2-8
REFERENCES 2-10
RECORDS OF INTERNATIONAL CONFERENCES 2-12

2-1 THE INITIAL USE OF FLOTATION FOR MINERAL SEPARATION

Up to the Middle Ages, minerals were separated by hand. Mechanical stamps and crushers introduced since, can mill the ore into much finer particles, which allows the separation of minerals on the basis of different densities in a liquid medium. This brought a vast improvement in recovery, but these wet mechanical separation methods were effective only for particles larger than about 0.3 mm—the smaller particles ran to waste. In 1860, William Haynes, in England, discovered the affinity of very small particles to adhere to oil droplets, which could concentrate the particles from a watery suspension in a surface layer as the oil rose to the top (Glembockaja, 1977). The principle of *bulk oil flotation* was thus born. In 1877, the brothers Adolph and August Bessel, who owned a tile manufacturing firm in Dresden, Germany, employed this principle to separate graphite particles. They found that faster separation was achieved if air was added to the suspension. At first, they produced the bubbles by boiling the suspension, a principle patented in 1877. The 1877 patent of the Bessel brothers was commemorated with a centenary symposium in Freiberg, Germany (Bergakademie Freiberg, 1977). In 1886, another Bessel patent followed, which was for the method of producing gas bubbles by the reaction of carbonates with hydrochloric acid (Glembockaja, 1977). After the Bessel patents, a flurry of patents followed in Britain, the United States, and Australia to improve the process. By 1912 no less than 140 different patents, by 57 different individuals, had been registered that had some connection with flotation, with oil alone or aided by air (Kitchener, 1984). A number of detailed reviews on the early patent history has been published by Megraw (1916), Rickard and Ralston (no date), Wark (1938), Gaudin (1957), Crabtree and Vincent (1962), and Glembockaja (1977).

In the first decade of the 20th century, more methods of introducing air into the flotation tank were proposed. In 1902 Potter in the United States suggested that flotation could be achieved by the reaction of sulfuric acid with carbonates alone, without any oil whatsoever. In the same year Delprat proposed that sodium bisulphate could be used as an alternative to the more expensive sulfuric acid (Glembockaja, 1977). In 1904 Francis Elmore obtained a British patent for bubble formation by the electrolysis of water, as well as vacuum flotation (Crabtree and Vincent, 1962). The direct injection of air and its dispersion by an impeller followed (United States patents by Sulman, Picard and Ballot in 1906; Hoover in 1910), as well as the direct dispersion of air through a porous bottom (United States patent by Callow in 1914) (Gaudin, 1957). The mining industry was quick to capitalize on these inventions. The first full-scale application of froth flotation (where the float layer accumulates to a few centimeters thick) was made during 1905 in Broken Hill, Australia, using the Potter-Delprat method. This event was commemorated with a centenary symposium in Brisbane in 2005 (Australasian Institute of Mining and Metallurgy, 2005). The first use of froth flotation in the United States was in Basin Mill, Montana, in 1911, also commemorated 50 years later with an anniversary volume of papers (American Institute of Mining Metallurgy and Petroleum Engineers, 1962).

After experimentation with the numerous technical possibilities of introducing air into the flotation tank, the applications for mineral separation soon converged on the direct injection of air, or *dispersed air flotation,* which remains the method of choice in the mining industry to this day. Mineral particles are heavy and require bubbles with diameters of about 2 mm, such as those formed in dispersed air flotation, to attain adequate buoyancy. These larger bubbles will remain attached to the particles only if the mineral surfaces are more hydrophobic. The further development of flotation for mineral separation therefore focused on finding more efficient "collectors"—chemicals that were added to render mineral surfaces sufficiently hydrophobic. Furthermore, the suspensions had to be stabilized to avoid the collapse of bubbles and the settling of particles, which led to the search for better chemical "frothers." The refinements of flotation subsequent to the first full-scale applications for mineral separation mainly centered on the development of more efficient organic and inorganic chemical collectors and frothers to render flotation more effective. The typical particles encountered in water and wastewater engineering have vastly different properties than dense mineral particles, and these particles had to await alternative flotation technologies more suited to their removal.

2-2 DAF DEVELOPMENT FOR THE PAPER INDUSTRY

The paper industry in Scandinavia, at the start of the 1920s, faced a two-pronged problem. The wastewater from the paper mills was rich in paper fibers, which were recovered in large sedimentation basins. Fiber recovery, however, was incomplete and losses of more than 5 percent were common. In addition, the liquid waste from the paper mills was a foul mixture of paper fibers, excess machine lubricant, bacterial slime, and other debris, which caused severe environmental damage to the rivers into which these wastes were discharged. A solution was desperately sought, both for commercial and environmental reasons. Flotation brought a substantial improvement in fiber recovery. The Norwegian engineer Nils Petersen patented the concept of full-stream pressure flotation. He experimented with the process without using any chemicals at first, but the process brought the desired results only after a mixture of chemicals (consisting of aluminum sulfate and formaldehyde, among others) was added. The chemical mixture was developed by the Swede, Karl Sveen. This method was subsequently known as the Sveen-Petersen system.

Shortly thereafter, in 1928, the Swede Adolf Karlström developed the Adka (after the first letters of his name and surname) vacuum flotation system. Both the Sveen-Petersen and Adka systems required only 10 percent of the reactor volume needed by the earlier settling systems and brought huge improvements to the paper industry (Hanisch, 1960). The technology was soon used at paper mills in other regions, including Finland and Germany (Suutarinen, 2008).

A summary and comparison of these systems, which were classified by Kiuru (2001) as the first-generation DAF systems, are instructive (Hanisch, 1960). In both systems, the paper waste was mixed with treatment chemicals first, before air was drawn into a Venturi on the suction side of the pump feeding the flotation system. Provision was made in both systems for coarse air bubbles to escape. For the Sveen-Petersen system, the tank was a pressure vessel from which the water flowed to the flotation tank. For the Adka system, the tank was at atmospheric pressure from which the water was drawn into a vacuum chamber. The surface loading in the Sveen-Petersen system was between 4 and 6 m/h, while it was between 5 and 8 m/h for the Adka system. The electrical consumption of the Sveen-Petersen system was 0.25 to 0.30 kWh/m^3, while the Adka system used less energy.

2-3 DAF ADAPTED FOR INDUSTRIAL WASTEWATER TREATMENT

Despite the flotation successes in the paper industry in the 1930s, there was some hesitation to apply the process to the wastewaters of other industries. The first adaptation of DAF for industrial wastewater treatment was reported from the United States in the 1940s. Before the 1940s, flotation in the United States (other than for mineral separation) was limited to the removal of fat from water by the direct injection of air bubbles with diameters of 1 mm or larger. In 1942, the Dorr Company introduced the "Vacuator"—a vacuum DAF process for the separation of fat and other particles from abattoir wastewater. The technology closely followed the earlier Adka system from the paper industry, using a two-stage process where air was dissolved in a first chamber, followed by separation in a second chamber under vacuum. The air was introduced by sparging compressed air through porous underdrains and the large bubbles were vented off before the suspension entered the vacuum chamber. The pressure in the vacuum chamber was typically 30 to 35 kPa below atmospheric pressure (Hanisch, 1960). In 1950, the Gibbs Company constructed the first pressure DAF system in the United States for industrial water. This process was an adaptation of the Sveen-Petersen system, but with the important advance that part-stream aeration was introduced, as opposed to the original full-stream aeration. The air was introduced on the suction side of the recycle pump. A pressure vessel was placed downstream of the pump in which the water had a retention time of about 60 seconds, to allow further air dissolution and to vent off the large bubbles. This plant was soon followed by others. Their hydraulic loading was 6 to 10 m/h for round tanks and up to 12 m/h for rectangular tanks. The recycle stream was pressurized between 100 and 500 kPa, depending on the particle concentration. The energy demand varied from 0.05 kWh/m^3 for large plants with capacity of 200 m^3/h to 0.25 kWh/m^3 for small plants with capacity of 10 m^3/h (Hanisch, 1960).

During the 1950s, pressure DAF was also successfully used in the United States for the thickening of activated sludge. Attempts to use DAF for the separation of sludge from domestic wastewater, dating from the 1940s, were successful only in 1943 after the addition of detergents—an analogy with the frothers developed earlier for mineral separation

(Hanisch, 1960). By 1960, despite the technological development of part-stream pressure DAF and its successful use in industry, it had not been applied for the clarification of potable water.

2-4 THE THEORETICAL BASIS FOR DAF IS CONSOLIDATED

The experiences with DAF in industry, largely empirical at first, steadily contributed to a better understanding of the process. By the start of the 1960s, it was evidenced by the publication of a number of papers and books that consolidated the practical and theoretical understanding at the time. Three of the most important publications are discussed as illustration.

In 1959, Edward Vrablik, an engineer from the Eimco Corporation in Illinois, presented a paper *Fundamental Principles of Dissolved-Air Flotation for Industrial Wastes* (Vrablik, 1959) at the 14th Industrial Waste Conference at Purdue University in the United States. This paper presented a theoretical basis for the design of DAF systems in general. Typical bubble sizes were measured, theoretical bubble rise rates were calculated, and practical suggestions were made on how the critical design parameters could be determined with simple bench-scale tests. The paper concluded with a proposed design procedure, along with some design guidelines. Although Vrablik recognized 12 different industrial DAF applications and also discussed the benefits and different types of chemical collectors and frothers, the main contribution of Vrablik was to present a general theory of DAF, which was not tied to any specific industry or technology—it could conceivably be used for any suspension. He reviewed, for example, different methods of air dissolution, part-stream DAF as well as full-stream DAF, high versus low-saturation pressure, and other options without prejudice. As long as the fundamental requirements of DAF were met, the process should work. In his own words: "DAF requires the study of three phases: gas, liquid and solid."

In the same year as the Vrablik paper, Baldefrid Hanisch obtained his doctorate at the University of Stuttgart with a thesis on the application of DAF to wastewater. Hanisch graduated as an engineer in 1950 at the Technical University of Stuttgart. During 1952–53 he obtained a masters degree in sanitary engineering at the University of Michigan, after which he returned to Stuttgart to eventually complete his doctoral thesis entitled *The Efficient Application of Flotation with Very Small Bubbles to the Clarification of Waste Water* (Stuttgarter Unikurier, 2000). Being in the United States in the early 1950s provided Hanisch an opportunity to follow the DAF developments described in the previous section. In the first part of his thesis, Hanisch reviewed the development of DAF up to that point. The second and technical part of his thesis systematically reviewed the physical principles of air solubility and surface tension, and the behavior of bubbles in suspension before reporting the results of his experimental studies (Hanisch, 1960).

The book *An Introduction to the Theory of Flotation* was published in the USSR by Klassen and Mokrousov of the Institute of Mining in Moscow. The importance of this book is evidenced by the translation of its second edition into English by Leja and Poling at the University of Alberta, published in 1963 (Klassen and Mokrousov, 1963). Although the book was aimed at the mining industry, its approach was extremely rigorous and fundamental. Its value for DAF, among others, lies in a chapter of 24 pages entitled *The Formation of Air Bubbles*, where the formation of bubbles during pressure release and bubble coalescence are discussed in detail—matters of great relevance to this day.

At the start of the 1960s, there existed a substantial body of knowledge on the theory and practice of DAF, jointly accumulated by inventors, practitioners, and researchers from all over the world. The time was ripe for the application of DAF to potable water treatment.

2-5 DAF FOR POTABLE WATER TREATMENT IN THE 1960S

In the middle of the 1950s, the Swede Ruben Rausing (who started the international packaging company Tetrapak in 1951) bought a paper mill with a water supply containing natural color, which was detrimental to its paper quality. He called upon Adolf Karlström, the developer of the Adka process, to improve the quality. Karlström solved the problem with a part-stream pressure DAF process, which was patented in 1957 in Sweden under the joint names of Karlström and Rausing. The two inventors were sufficiently enthused by the process that they formed the company Purac (derived from "pure aqua") in 1956 with the specific purpose of further developing DAF for all applications other than the paper industry. The paper industry remained the business domain of Karlström. His wife was German and he used his regular visits to Germany to do brisk business with many small German paper mills, under the name of Deutsche Adka. His German visits also brought him into contact with Baldefrid Hanisch in Stuttgart, whom he regularly visited (Dahlquist et al., 2008).

Purac, formed as a specialist DAF company to exploit the opportunities awaiting DAF in any field other than the paper industry, had a slow start. Adolf Karlström's son, Bengt Karlström, joined Purac in 1960 to provide more momentum. Their first industrial DAF application was the treatment of colored water at the Östana (Sweden) paper mill in 1960. Their first DAF applications for potable water, both plants using vacuum DAF, were in the towns of Flen and Strängnås, Sweden, in 1964 (Rosén, 2010).

One reason for the long delay before DAF made its first breakthrough in water treatment was the careful scrutiny of the new process by the generally conservative drinking water industry. A good example is provided by the introduction of the process in Finland, documented by Tanhuala (1994). In 1962, representatives of Purac visited Finland to market DAF as a water treatment process. Some clients were interested, but first referred the matter to the Technical Committee of the Finnish Association of Cities, with the request that they should investigate whether the process would be suitable for Finnish water conditions. The Technical Committee, in turn, posed the question to their counterparts at the Swedish Association of Municipal Engineering. By this time, DAF was already in use at two wastewater treatment plants and three industrial water treatment plants in Sweden, and it was confirmed that DAF would be suitable for Finland.

Purac built their first Finnish DAF plant in the town of Kokkola in 1964 to provide water to the Outokumpu industrial plant. Purac provided a vacuum DAF system, while a new Finnish company Rictor was responsible for the filters, pipes, pumps, and the rest of the equipment (Suutarinen, 2010). The Rictor company (the name was adapted from "rego" in Latin, meaning to guide or to direct) was founded in 1964 by Oiva Suutarinen and a partner, and its first contracts entailed the installation of wastewater treatment equipment. After Rictor's first potable water treatment contract in the city of Seinäjoki (the upgrade of a conventional settling/filtration plant that did not include DAF), it got permission to experiment in one of the abandoned tanks that had become too small. By building a complete wooden pilot plant inside the tank, they could establish and improve their own design criteria for DAF (Suutarinen, 2008). The first DAF plant for water treatment in Finland was constructed by a joint venture between Purac and Rictor in the city Kemijärvi in Lapland during 1965 (Tanhuala, 1994). This was followed by another DAF plant for potable water in Kusaankoski by Rictor, commissioned in 1967 (Suutarinen, 2010). The rapid acceptance of DAF in Finland was driven by the cold and often colored raw water supplies of Finland. To clarify these waters with sedimentation, long flocculation times had to be combined with low sedimentation rates—an expensive option. The lower cost and higher efficiency of DAF gave it an immediate advantage over sedimentation. By 1970, an estimated 20 DAF plants were already in use in Finland (Tanhuala, 1994). After 1970, practically no more sedimentation plants were built in Finland as the process had been displaced by DAF (Kiuru, 2001).

On the other side of the world, scientists and engineers in Southern Africa were grappling with an entirely different water quality problem. The city of Windhoek, capital of Namibia, was forced to augment the water supply of the city by the reclamation of its domestic wastewater. From 1963 to 1968, an intensive research program was conducted to verify that the proposed water quality criteria, the treatment processes, and their safety barriers were adequate. The first treatment barrier was a conventional biological wastewater treatment plant, followed by a retention period of about 14 days in a series of shallow-maturation ponds to exploit the viricidal properties of the harsh sunlight in this semi-desert region. The retention in the ponds caused algal growth that had to be removed. The scientists soon found that if water samples from the ponds were poured into measuring cylinders prefilled with some liquid alum, the turbulence mixed the water with the alum, and also led to the precipitation of supersaturated oxygen formed by photosynthesis. The result was the rapid formation of a float layer with exceptionally clear water below. This phenomenon was successfully verified at pilot scale, with some additional air added through a Venturi to the full inlet stream in the event of low photosynthetic activity (Van Vuuren et al., 1965). This process was thus a DAF system, with some dispersed air added when required. The full-scale reclamation plant, commissioned in October 1968, first recarbonated the maturation pond effluent to lower the pH. After recarbonation, the excess carbon dioxide provided sufficient air for the subsequent flotation step (Van Vuuren et al., 1970). The current plant at Windhoek is different from the original and is a case study reviewed in Chap. 12.

The Windhoek experience influenced the further development of DAF in two ways. First, it inspired a South African research program over the next decade that provided the basis for numerous successful DAF applications for industry, water, and wastewater (reviewed by Offringa, 1995). Second, Dr. Ron Packham of the British Water Research Association (now the Water Research Centre or WRc) visited the Windhoek site in 1969 and witnessed the work on flotation. This was the origin of the WRc's own comprehensive, systematic investigation into the use of DAF for potable water treatment. At first, the objective was to demonstrate its cost efficiency in comparison with sedimentation, with the emphasis later shifting toward optimization and the effects on the subsequent filtration step. The outcomes of the WRc program have been summarized by Gregory (1997).

2-6 THE SPREAD OF DAF SINCE 1970

Hercules Powder, a chemical company based in Delaware in the United States, bought Purac in 1971. Hercules Powder already owned an English company called Water Management, which hereby became a sister company to Purac. The Purac technology thus became readily available in the United Kingdom, soon after Ron Packham returned from Windhoek. The WRc made themselves fully aware of the available DAF technologies in the marketplace and visited numerous industrial plants using DAF and dispersed air flotation. Ron Packham and Bob Hyde of the WRc, for example, visited Purac's new DAF plant in Arvika, Sweden in 1972. While the WRc pursued a substantial demonstration program to the water suppliers in the United Kingdom, the 1975–76 drought in the United Kingdom resulted in Water Management providing two DAF package plants to Wessex Water in 1976 to help overcome localized supply problems that helped to draw attention to the viability of DAF (Gregory, 2010).

In 1975, Hercules Powder withdrew from the water industry. Purac in Sweden and Water Management in England were released to operate independently as before. The good cooperation between these two companies led to Purac buying Water Management in 1981, which eventually became Purac England. Before the takeover by Purac, Water Management built a few more DAF plants in England (Dahlquist et al., 2008).

With the proof that DAF could be successfully applied for potable water clarification, the emphasis shifted to making DAF more cost effective. An important result was that DAF was combined with rapid gravity filtration, as the typical surface loadings of the two processes were approximately the same. The process, originally named Flofilter™, could thus be stacked into the same reactor—DAF in the headspace above the filter bed. The first Flofilter™ was commissioned in Kristinestad, Finland in 1967 (Rosén, 2010), while at least five such Finnish plants were operational by 1970 (Suutarinen, 2010). The first Flofilter™ in Sweden was commissioned at Gullspång in 1968 (Rosén, 2010). The Flofilter™ at Lysekil in Sweden, commissioned in 1974, is described in Chap. 11. The Flofilter™ configuration is also referred to as "In-Filter DAF" or "flotation over filtration" or "DAFF."

DAF was unknown in the Netherlands up to 1974. In this year, a number of initiatives started. A Dutch delegation visited a number of flotation plants in Sweden, a pilot testing program was started in Amsterdam, and a Dutch flotation working group, coordinated by KIWA (a Dutch research institute), had been set up with representatives from the waterworks of Amsterdam, Rotterdam, Brabant, and South West Holland, and the Technical University of Delft. By 1979, a full-scale DAF plant was commissioned at Zevenbergen in Brabant, the Netherlands, followed closely by five more in the Netherlands and one in Belgium (Van Puffelen, 1990). During the 1980s and 1990s, the projected water demand in the Netherlands did not materialize and the perceived need to utilize surface water supplies was not realized. The lack of new plants and the preference for groundwater sources led to reduced interest for DAF and no new plants were built in the 1990s in the Netherlands (Van Dijk, 2009).

The practical development of DAF in the 1960s and early 1970s was followed by an academic interest into DAF, covering diverse areas such as bubble behavior in dilute suspensions, the effect of bubble size on separation, and the interaction between bubbles and flocs. By the 1980s, numerous full-scale and pilot-scale applications provided much empirical data to stimulate further research. This book draws heavily on the many contributions emanating from these and later researchers, and these contributions are reviewed in the following chapters. By 1990, DAF had attracted the attention of many interested water professionals in different parts of the world. Moreover, DAF expertise was not confined to small isolated patches any longer, as communication among these groups improved.

2-7 INTERNATIONAL DAF CONFERENCES

The developments of the 1960s and early 1970s stimulated a great interest in DAF. Technical reports and papers appeared more regularly, and a number of full-scale and pilot-scale applications had been ongoing or completed in different parts of the world. All these developments paved the way for the ground-breaking conference held during June 1976 in Felixstowe (England), organized by the WRc. At this conference, despite only 13 papers being presented, the international DAF community shared a common platform and discussion forum for the first time with speakers from Sweden, France, the United States, the Netherlands, and England. Among the 270 delegates, 80 were from countries other than the United Kingdom (Gregory, 1997). This large attendance, at a time when specialist conferences were a rarity, for a relatively new, untried process, remains remarkable. The four sessions at the conference were devoted to sludge thickening, industrial treatment, potable water treatment (mostly the WRc's own studies), and practical experience with commercial DAF plants (with speakers from Sweden, the United Kingdom, and the Netherlands). The numerous new plants built in the period following Felixstowe allowed a critical comparison of design parameters in England (Longhurst and Graham, 1987), Finland (Heinänen, 1988), and South Africa (Haarhoff and Van Vuuren, 1993).

The valuable international experience gained with DAF since Felixstowe precipitated a workshop held in Antwerp in 1991, attended by a small invited group. Seven speakers from Belgium, France, Germany, the Netherlands, Scandinavia, the United Kingdom, and the United States reported on the great progress that DAF had made in their respective countries since Felixstowe. This was followed by the first full-fledged DAF conference after Felixstowe during April 1994 in Orlando, Florida. The breadth and depth of the papers presented at the important Orlando Conference provided clear evidence that DAF had gathered a critical mass of researchers and practitioners. A record 28 papers were presented from 10 different countries, of which a large number were subsequently published as peer-reviewed papers in *Water Science and Technology*. Orlando decisively demonstrated the value of academic and professional interaction among the DAF community and set the standard and format for the conferences to follow. The third conference was held in 1997 in London with 36 papers from 12 countries, and with a hard-bound book of the proceedings being published. The fourth conference was held in 2000 in Helsinki, Finland with 58 papers and posters from 16 countries, with a selection of papers subsequently published in *Water Science and Technology*. The fifth conference was held in Seoul, Korea in 2007, with 41 papers and 12 posters from 13 countries, with a hard-bound book of the proceedings being published. (The references to the conference records are provided in a separate section of the reference list.) At the time of writing, the sixth conference is being planned for 2012 in New York City. A remarkable feature of all the past international DAF conferences was the approximately equal participation of academics, consultants, and specialist contractors.

2-8 THE APPLICATION AND ACCEPTANCE OF DAF

Hundreds of DAF water treatment installations have been built in many parts of the world. In the early years, many utilities experimented with DAF as a pilot or on a small scale. Of special importance is the number of large installations. Water utilities will not invest large amounts of capital and become dependent on DAF if they do not have complete faith in the process. Using an arbitrary yardstick of 50 ML/d or more as a measure of a "large" plant, an international survey was recently conducted to gauge the extent to which DAF had been accepted by the water community. The cumulative treatment capacity of plants larger than 50 ML/d is shown in Fig. 2-1. There are 70 such plants reflected in Fig. 2-1, with a total installed capacity exceeding the 10 000 ML/d mark. By the year 2000, in comparison, there were only 12 plants in this category, while records of about 240 DAF plants of all sizes have been reported (Schofield, 2001).

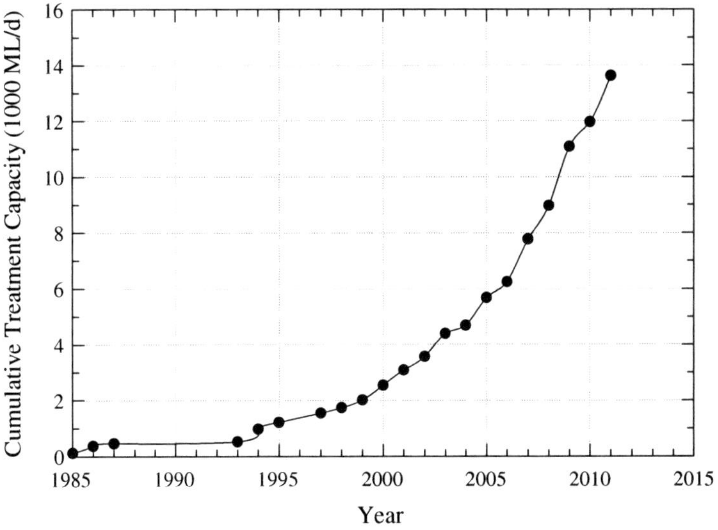

FIGURE 2-1 Cumulative installed DAF capacity in the world, expressed in 1000's of ML/d, counting only treatment plants with capacity of 50 ML/d or larger.

At this time, can DAF be considered a mature, robust technology alongside much older processes such as sedimentation and filtration? In 1976, DAF was perceived to be an "art" rather than a "science." The 1990 edition of the authoritative *Water Quality and Treatment* published by the American Water Works Association recognized the merits of DAF by including it alongside sedimentation in the same chapter. At the London Conference in 1997, the carefully worded hope was expressed that "most of the perception should have been dispelled" by the advances between 1976 and 1997 (Gregory, 1997). At the Helsinki Conference in 2000 a bolder opinion was voiced that DAF "may be considered a mature technology," although "there is still considerable scope for research and development to make the process even more efficient" (Schofield, 2001).

The progress of DAF is evaluated here in terms of a structured model of technology acceptance, such as the one proposed by Moore (1991). This model, depicted in Fig. 2-2, was found to hold true for many technologies (Sroufe et al., 2000). The natural progression is the adoption by a few "innovators" at first, followed by a slightly larger group of "early adopters." The bulk of the technology users are in the "early majority" and "late majority" categories, followed by a much smaller group of "laggards." The single strongest point of resistance in this progression is moving from the "early adopters" to the "early majority"—a resistance point that is labeled as the "chasm." For technology developers, the principal challenge is that of "crossing the chasm." The end of the cycle is reached when the technology is fully accepted, or displaced by another.

Water treatment plants require long lead times to plan, test, design, and construct, and have life spans of decades once they are built. The life cycle of a DAF plant is measured in decades compared to electronic products that have life cycles of a few years at most. The first point to note is that the product cycle in Fig. 2-2 will run over a few decades. Moreover, water treatment plants are only built when increased demand warrants the investment. Even if water treatment professionals fully endorse DAF as the preferred option, it may take

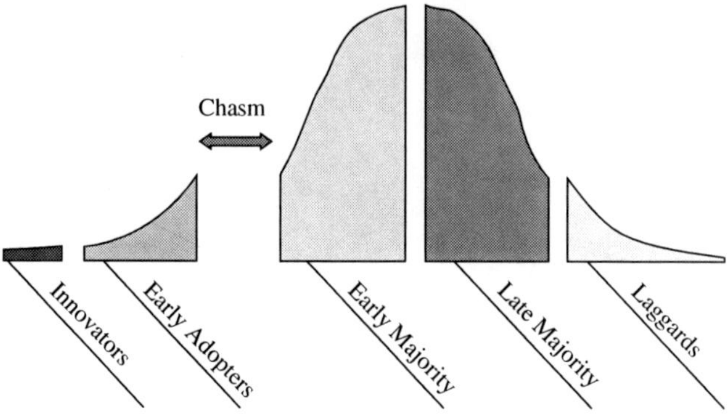

FIGURE 2-2 Phases in the adoption of a new product or technology (after Moore, 1991) (*Source: Sroufe et al., 2000, The new product design process and design for environment—"Crossing the chasm," International Journal of Operations and Production Management, 20 (2), 267–291, with permission from the Emerald Group Publishing Limited*).

many years before the opportunity arises to exercise this option. In terms of the Moore model, the following periodization of DAF development is offered (Haarhoff, 2008):

- The DAF "innovators" were those who experimented and started to apply DAF as a new option in the 1960s and early 1970s, before there was the body of supporting knowledge of today.
- The "early adopters" were those who applied the process at full scale between 1975 and 1985. The 1976 Conference may have been a stimulant in this regard, having demonstrated that DAF is indeed a credible technology embraced by many others all over the world.
- The "chasm" seems to have been the period between 1985 and 1995. Although many smaller plants were commissioned all over the world, the technology was stagnant and utilities, except possibly those in Sweden and Finland, seemed hesitant to fully endorse the process at larger scale. The 1994 Orlando Conference, with its impressive display of new studies, ideas, and consolidated understanding, may have been instrumental in dispelling many doubts about DAF to bring the chasm to an end.
- The era of the "early majority" dawned in the early 1990s when a number of larger utilities turned their serious attention to DAF, clearly evident from Fig. 2-1. With so much more at stake in large cities like New York and Antwerp, for example, much more stringent pilot testing had to be performed beforehand.

Having bridged the critical chasm of the Moore model, DAF is firmly established as a viable process option with a solid theoretical base and many examples of successful application. It is the objective of this book to present a consolidated summary in the following chapters.

REFERENCES

American Institute of Mining Metallurgy and Petroleum Engineers (1962), D. W. Fuerstenau (ed.), *Froth Flotation: 50th Anniversary Volume*, New York.

Australasian Institute of Mining and Metallurgy (2005), *Centenary of Flotation Symposium*, Brisbane, Queensland. Proceedings on CD-ROM.

Bergakademie Freiberg (1977), *100 Jahre Flotation Teil I*, Freiberger Forschungsheft A593, VEB Deutscher Verlag für Grundstoffindustrie, Leipzig.

Crabtree, E. H., and Vincent, J. D. (1962), "Historical outline of major flotation developments," in D. W. Fuerstenau (ed.), *Froth Flotation 50th Anniversary Volume*, New York: American Institute of Mining, Metallurgical and Petroleum Engineers.

Dahlquist, J., Hilmer, A., and Rosén, B. (2008), Interviewed by J. Haarhoff on May 28, 2008, in Lund, Sweden.

Gaudin, A. M. (1957), *Flotation*, New York: McGraw-Hill.

Glembockaja, T. V. (1977), Die Entwicklung der modernen Schaumflotation und die Rolle der Erfindungen der Gebrüder Bessel (The development of modern foam flotation and the role of the discoveries of the brothers Bessel), in *100 Jahre Flotation Teil I*, VEB Deutscher Verlag für Grundstoffindustrie, Leipzig.

Gregory, R. (1997), "Summary of general developments in DAF for water treatment since 1976," in *Dissolved Air Flotation*, London: The Chartered Institution of Water and Environmental Management.

Gregory, R. (2010), Email communication J. Haarhoff dated September 13, 2010.

Haarhoff, J., and Van Vuuren, L. R. J. (1993), *A South African Design Guide for Dissolved Air Flotation*, Report TT60/03, Pretoria, South Africa: Water Research Commission.

Haarhoff, J. (2008), Dissolved air flotation: progress and prospects for drinking water treatment, *Journal of Water Supply: Research and Technology–Aqua*, 57 (8), 555–567.

Hanisch, B. (1960), *Die wirtshaftliche Anwendung der Flotation met sehr kleinen Luftblasen zur Reinigung von Abwasser* (The Efficient Application of Flotation with Very Small Bubbles to the Clarification of Waste Water), Stuttgarter Berichte zur Siedlungswasserwirtschaft 8, Kommissionsverlag R. Oldenbourg, München.

Heinänen, J. (1988), Use of dissolved air flotation in potable water treatment in Finland, *Aqua Fennica*, 18 (2), 113–123.

Kitchener, J. (1984), The froth flotation process: past, present and future—in brief, in *The Scientific Basis of Flotation*, The Hague, Netherlands: Martinus Nijhoff Publishers.

Kiuru, H. J. (2001), Development of dissolved air flotation technology from the 1st generation to the newest or 3rd one (very thick micro-bubble bed) with high flow-rates (DAF in turbulent flow conditions), *Water Science & Technology*, 43 (8) 1–7.

Klassen, V. I., and Mokrousov, V. A. (1963), *An Introduction to the Theory of Flotation*, London: Butterworths.

Longhurst, S. J. and Graham, N. J. D. (1987), Dissolved air flotation for potable water treatment: a survey of operational units in Great Britain, *Public Health Engineer*, 14 (6), 71–76.

Megraw, H. A. (1916), *The Flotation Process*, New York: McGraw-Hill.

Moore, G. (1991), *Crossing the Chasm*, New York: Harper Business.

Offringa, G. (1995), Dissolved air flotation in Southern Africa, *Water Science & Technology*, 31 (3–4), 159–172.

Rickard, T. A., and Ralston, O. C. (no date, circa 1917), *Flotation*, San Francisco: Mining and Scientific Press.

Rosén, B. (2010), Email communication to J. Haarhoff dated September 23, 2010.

Schofield, T. (2001), Dissolved air flotation in drinking water production, *Water Science & Technology*, 43 (8), 9–18.

Sroufe, R., Curkovic, S., Montabon, F., and Melnyk, S. A. (2000), The new product design process and design for environment—"crossing the chasm," *International Journal of Operations and Production Management*, 20 (2), 267–291.

Stuttgarter Unikurier 84/85 (2000), Doppelfeier: Karl-Heinz Hunken und Baldefrid Hanisch 80 (Double Celebration: Karl-Heinz Hunken and Baldefrid Hanisch 80.) Issue 1/2000.

Suutarinen, O. (2008), Interviewed by J. Haarhoff on June 2, 2008, in Helsinki, Finland.

Suutarinen, O. (2010), Email communication to J. Haarhoff dated September 17, 2010.

Tanhuala, T. (1994), *Development of Water Treatment in Finland* (in Finnish), Master's Thesis, Tampere University of Technology, August 1994. (Partial translation by Tapio Katko, Tampere University of Technology.)

Van Dijk, J. C. (2009), Interviewed by J. Haarhoff on June 13, 2009, in Delft, the Netherlands.

Van Puffelen, J. (1990), Flotatie: een topper in de waterzuivering (Flotation: a winner for water treatment), *H_2O*, 23 (10), 256–262.

Van Vuuren, L. R. J., Meiring, P. G. J., Henzen, M. R., and Kolbe, F. F. (1965), The floatation of algae in water reclamation, *International Journal of Air and Water Pollution*, 9, 823–832.

Van Vuuren, L. R. J., Henzen, M. R., Stander, G. J., and Clayton, A. J. (1970), The full-scale reclamation of purified sewage effluent for the augmentation of the domestic supplies of the City of Windhoek, *Presented at the 5th International Water Pollution Research Conference*, San Francisco, July 1970, 1–32.

Vrablik, E. R. (1959), Fundamental principles of dissolved-air flotation for industrial wastes, *Proceedings 14th Industrial Waste Conference, Purdue University*, 743–779.

Wark, I. C. (1938), *Principles of Flotation*, Melbourne: Australasian Institute of Mining and Metallurgy.

RECORDS OF INTERNATIONAL CONFERENCES

1976: The proceedings were compiled by the WRc in 1977 as *Flotation for Water and Waste-Water Treatment, Water Research Centre Conference*, Felixstowe.

1994: A number of the papers were published in 1995 in *Water Science & Technology*, 31 (3–4).

1997: The proceedings were published as *Dissolved Air Flotation*, London: The Chartered Institution of Water and Environmental Management.

2000: A number of the papers were published in 2001 in *Water Science & Technology*, 43 (8).

2007: The proceedings were published as *Proceedings of the 5th International Conference of Flotation in Water and Wastewater Systems*, September 11–14, Seoul, South Korea.

CHAPTER 3
AIR SATURATION

3-1 AIR REQUIREMENTS FOR DAF 3-2
Properties of Atmospheric Air 3-2
How Much Air Is Required for DAF? 3-4
Specification of Recycle Flow Rate and Saturation Pressure 3-5
3-2 THE PRINCIPLES OF AIR SATURATION 3-7
Water in Equilibrium with Atmospheric Air 3-7
The Difference between Open-End and Dead-End Saturation 3-9
3-3 OPEN-END SATURATION 3-9
Principle of Operation 3-9
Specification of Open-End Saturation Systems 3-10
3-4 DEAD-END SATURATION 3-11
Principle of Operation 3-11
Composition of Saturator Air 3-13
The Air Available for DAF 3-15
Modelling the Mass Transfer in a Saturator 3-17
Application of the Mass Transfer Model 3-20
End Effects 3-24
Saturator Control 3-25
Saturator Start-up 3-25
3-5 SATURATOR EFFICIENCY 3-25
The Problematic Definition of Saturation Efficiency 3-25
Experimental Measurement of Air Transfer 3-26
3-6 AIR FLOW AND ENERGY REQUIREMENTS 3-28
Maximum Air Flow for Open-End Saturation 3-28
Minimum Air Flow for Dead-End Saturation 3-29
Energy Requirements 3-30
3-7 DESIGN EXAMPLE 3-33
REFERENCES 3-35

The success of DAF depends on the dissolution of air into water with a saturation system, followed by the release of the air in the DAF contact zone. It is therefore appropriate to start with a fundamental discussion of air, water, and the dissolution of air into water. After dealing with how much air is required by DAF, and with different ways of expressing the air requirement, the solubility of nitrogen, oxygen, and argon—the principal three gases in atmospheric air—is discussed. Two practical methods of air dissolution are considered next, with a major part devoted to the prediction of the air transfer efficiency of a packed saturator, which is the most common air saturation method used currently. This is followed by an examination of the energy and air flow requirements of the different air saturation methods before a final design example is presented. The material is complemented with the development of rigorous mathematical models to allow designers to analyze and optimize saturation systems from first principles.

3-1 AIR REQUIREMENTS FOR DAF

Properties of Atmospheric Air

In dry atmospheric air, the gases nitrogen, oxygen, and argon account for 99.96 percent of the total volume. DAF saturation systems employ either atmospheric air (open-end saturation) or air in saturators (dead-end saturation) with higher nitrogen content—an important difference covered in more detail further on. The gas fractions in the standard atmosphere $f_{atm,X}$, as well as those in typical saturator air $f_{sat,X}$, are shown in Table 3-1.

Throughout this chapter, air concentrations are considered for two situations, namely, either air at atmospheric conditions or air at saturator conditions, that prevail when air is trapped within a pressure vessel. In both cases, the pressure is expressed as the absolute pressure p_{abs}. For atmospheric air, the absolute pressure of atmospheric air p_{atm} is provided in Table E-1. For air at saturator conditions, the absolute pressure has to be calculated:

$$p_{abs} = p_{atm} + p_{sat} - p_{vap} \tag{3-1}$$

The absolute dry air pressure is p_{abs}; p_{atm} the atmospheric pressure; p_{sat} the gauge pressure in the air vessel or saturator; p_{vap} the vapor pressure of water (all expressed as kPa). Tables D-1 and D-2 provide values for p_{vap}. The high degree of turbulence and large interfacial area in DAF saturation devices ensure full saturation of the air with water vapor.

Air concentration is expressed in different ways, depending on the need of the application. For purposes of contact zone modelling (Chap. 7) and separation zone modelling (Chap. 8) it is more practical to express the air concentration as the number of bubbles of a certain size in a unit volume, as demonstrated in those chapters. For the purpose of expressing the total air transferred during saturation, the air concentration required for DAF is expressed in one of three other ways—volumetric air concentration (mL air/L of water, or ppm by volume), molar concentration (mol/m^3), or mass concentration (mg/L). To convert between these different air expressions, consider the molar concentration of a specific gas in air (knowing that 1 m^3 of dry air contains 44.6 moles at a temperature of 0°C and pressure of 101.3 kPa):

$$\text{molar gas concentration } G_X = f_X \times 44.6\left(\frac{273.2}{273.2+t}\right)\left(\frac{p_{abs}}{101.3}\right) \tag{3-2}$$

TABLE 3-1 Gas Properties at 20°C for Atmospheric Air and Typical Saturator Air

		Nitrogen	Oxygen	Argon	Total
Molecular mass, *mw*	g/mol	28.0	32.0	39.9	—
Molar fraction of atmosphere, f	(–)	0.7808	0.2095	0.0093	0.9996
Product of *mw* and f_{atm}	g/mol	21.86	6.70	0.37	28.94
Molar fraction of saturator air, f	(–)	0.8600	0.1345	0.0055	1.0000
Product of *mw* and f_{sat}	g/mol	24.08	4.30	0.22	28.60

Values for typical saturator air are taken at sea level and saturator gauge pressure of 500 kPa.

The molar gas concentration in air is G_X (mol/m^3); the molar fraction f_X is the molar or volumetric fraction of gas X (dimensionless); t is the air temperature (°C); p_{abs} is the absolute dry atmospheric pressure (kPa). The absolute dry pressure p_{abs} for atmospheric air is simply equal to p_{atm}.

To find the cumulative molar concentration of gases in a mixture of nitrogen, oxygen, and argon where $\Sigma f_x = 1$:

$$\text{molar air concentration } \sum G_X = 44.6\left(\frac{273.2}{273.2+t}\right)\left(\frac{p_{abs}}{101.3}\right) \tag{3-3}$$

To obtain the mass concentration of dry air, the molar concentration in Eq. (3-3) is multiplied with the molecular weights of the gases, provided in Table 3-1:

$$\text{mass concentration of dry air} = 44.6\left(\frac{273.2}{273.2+t}\right)\times\left(\frac{p_{atm}}{101.3}\right)\sum(mw_X \times f_X) \tag{3-4}$$

For consistent DAF performance, the volumetric concentration of air bubbles should be kept constant. It is the volume of air bubbles that determines the efficiency of contact (Chap. 7), the bubble rise rate, and separation efficiency (Chap. 8). Although it is customary to express the air requirement of DAF as a mass concentration, the same mass of air can occupy different volumes, depending on temperature and pressure. It is considered to be fundamentally more correct to use the volumetric air concentration as a design parameter, but molar, mass, and volumetric air concentrations are used interchangeably in this book. Table 3-2 shows the relationship between volumetric, molar, and mass concentrations for different temperatures and altitudes. More detailed values are provided in Tables E-2 and E-3.

TABLE 3-2 Number of Moles or Milligrams in 1 L of Air for Different Temperatures and Altitudes

Altitude	0 m	0 m	1000 m	1000 m	0 m	0 m	1000 m	1000 m
Concentration	Molar (mol)	Mass (mg)	Molar (mol)	Mass (mg)	Molar (mol)	Mass (mg)	Molar (mol)	Mass (mg)
Temperature (°C)	Atmospheric air				Typical saturator air			
0	0.0446	1.291	0.0395	1.144	0.0446	1.275	0.0395	1.131
5	0.0438	1.267	0.0388	1.124	0.0438	1.252	0.0388	1.111
10	0.0430	1.245	0.0381	1.104	0.0430	1.230	0.0381	1.091
15	0.0423	1.223	0.0375	1.085	0.0423	1.209	0.0375	1.072
20	0.0416	1.202	0.0368	1.066	0.0416	1.188	0.0368	1.054
25	0.0409	1.182	0.0362	1.048	0.0409	1.168	0.0362	1.036
30	0.0402	1.163	0.0356	1.031	0.0402	1.149	0.0356	1.019
35	0.0395	1.144	0.0351	1.014	0.0395	1.131	0.0351	1.003
40	0.0389	1.126	0.0345	0.998	0.0389	1.112	0.0345	0.987

Values for typical saturator air are taken at saturator gauge pressure of 500 kPa.

EXAMPLE 3-1 Conversion of Air Concentrations An air volume concentration of 8 mL/L (8000 ppm) in the DAF reactor is specified for an application at an altitude of 1000 m and water temperature 10°C. Convert the specification to its molar and mass equivalents.

Solution The dry atmospheric pressure at 1000 m is 89.8 kPa (Table E-1). As we are interested in the air volume after saturation, when the water is at atmospheric pressure, the saturation and vapor pressures are irrelevant to this problem. Calculate the molar concentration of saturator air:

Eq. (3-3): $G = 0.0381$ mol/L

This number can be either calculated with Eq. (3-3) or looked up in Table 3-1 or Table E-2.

Convert to molar concentration:

8 mL of saturator air = 0.008 × 0.0381 = 0.000305 mol

The molar concentration corresponding to the specification of 8 mL/L is 0.305 mol/m^3.

Convert to mass concentration:

The air mass in 1 L of saturator air is obtained from Table 3-1 or Table E-2:

1 L of saturator air = 1.091 g

8 mL of saturator air = 0.008 × 1.091 = 0.00873 g

The mass concentration corresponding to the specification of 8 mL/L is 8.73 mg/L. ▲

How Much Air Is Required for DAF?

As noted before, the success of DAF is dependent on the available air volume. The distribution of the air volume among a multitude of bubbles is, of course, also important—this aspect is discussed in Chap. 4. The emphasis of this chapter is on ensuring that enough air is dissolved in the saturation system.

The air requirement can be determined from the contact and separation models presented in Chaps. 7 and 8. In Chap. 7, the number of collisions for efficient contact is used to calculate the required number of bubbles, and thus the air volume. In Chap. 8, the actual volume of air attached to the flocs is used to calculate the required air volume. In practice, however, much more air is added than that indicated by these theoretical requirements, on two counts. First, there must be adequate air to ensure spatial coverage. Even if there were only one floc to be floated, bubbles must be present everywhere to ensure the availability of a bubble in the vicinity of that floc. Second, a surplus of air is required to compensate for air losses. Air dissolution is not 100 percent efficient, air precipitation is not complete, and macro-bubbles that contribute nothing to floc removal represent a large fraction of the air volume—factors that are elaborated on in this and the following chapter. In addition, in the case of raw water that is not fully saturated with respect to oxygen, more air has to be added to satisfy the oxygen demand of the raw water, which is discussed in the following text. The practical air requirement to compensate for all the above factors was determined empirically with pilot and full-scale experimentation during the early DAF development of the 1960s and 1970s.

The body of experience accumulated over the years provides reliable estimates of the air requirement for the clarification of drinking water by DAF. An early review of international practice suggested an air requirement of 6 to 8 mg/L for South African DAF plants (Haarhoff and Van Vuuren, 1995), which typically treat warm water (20°C) at high altitudes (higher than 1000 m) and low hydraulic loading rates. Using the conversion factors in Table 3-2, this early guideline translates into 5.7 to 7.6 mL/L (5700 to 7600 ppm). A modern, updated guideline suggests an air volume of 7.6 mL/L (Gregory and Edzwald, 2010). Edzwald (2010), after considering typical saturation pressures and recycle ratios, found that the volumetric air concentration in successful plants ranges between 5.6 and 8.6 mL/L (5600 to 8600 ppm). This agrees with an air concentration of 8400 ppm suggested in Chap.7, based on consideration of collisions among bubbles and flocs. As explained in Chap. 8, modern high-rate DAF reactors may require even greater air concentrations. The volumetric air requirement for current DAF design practice therefore converges on 7 to 9 mL/L, with larger values for high-rate applications—the latter has to be determined empirically. Figure 3-1 provides this range in molar and mass equivalents.

The guideline above is only valid if the raw water is saturated with oxygen. If there is an oxygen deficit in the raw water, a part of the air added during saturation will be dissolved and consumed by the raw water, leaving less air for bubble formation. The equilibrium concentration of oxygen is of the same order of magnitude as the total DAF air requirement. Figure 3-1 shows typical air mass concentrations of 7 to 11 mg/L; water at atmospheric pressure could dissolve up to 14 mg/L of oxygen (Table F-3). If the raw water is only slightly undersaturated in oxygen, the oxygen in the air going into the saturation system could dissolve completely and will not be released in the contact zone.

Designers should endeavor to eliminate an oxygen deficit in the raw water before the DAF process, by mechanical or splash (hydraulic) aeration. The elimination of an oxygen deficit with the DAF saturation system is costly. For typical DAF applications, about 10 percent of the flow is pressurized to 500 kPa, which is equivalent to pumping the entire plant flow through a head of 5 m. Splash aeration typically requires 2 m of head. If there is no other way, the recycle flow rate has to be increased, illustrated as follows:

If atmospheric air is dissolved into water, then air with the same atmospheric composition will be released in the DAF contact zone. Assume that an atmospheric air concentration of 9 mL/L is released in the contact zone. This is equivalent to releasing $0.2095 \times 9 = 1.89$ mL/L of oxygen, or $1.89 \times 0.0416 = 0.078$ mol/m^3, if an average temperature of 20°C is assumed (the conversion factor of 0.0416 is taken from Table E-2.) The solubility of oxygen at this temperature is 0.285 mol/m^3, which is much greater. All the oxygen that would have been released as air bubbles would therefore be consumed by the oxygen demand of the raw water. To compensate, the recycle flow rate has to be increased to allow nitrogen and argon, without oxygen, to form an adequate bubble suspension. The recycle flow rate has to be increased to $100/(1 - 0.2095) = 127$ percent of what it would have been if the water was fully saturated with oxygen. For air supplied from a saturator, the oxygen fraction is about 13 percent, which means that the recycle flow rate from a saturator has to be increased to $100/(1 - 0.13) = 115$ percent of what it would have been.

Specification of Recycle Flow Rate and Saturation Pressure

Once the air requirement for DAF is determined, the sizing of the saturation system proceeds with Eq. (3-5):

$$M_{av} = \left(\frac{1+r}{r}\right) M_{req} \tag{3-5}$$

M_{av} is the molar air concentration released in the DAF reactor by the recycle flow (mol/m^3); r is the recycle ratio as earlier defined in Chap. 1; M_{req} is the required air concentration available in the DAF reactor after blending (mol/m^3). The air requirement can be met by any combination of recycle ratio r and recycle air concentration M_{av}.

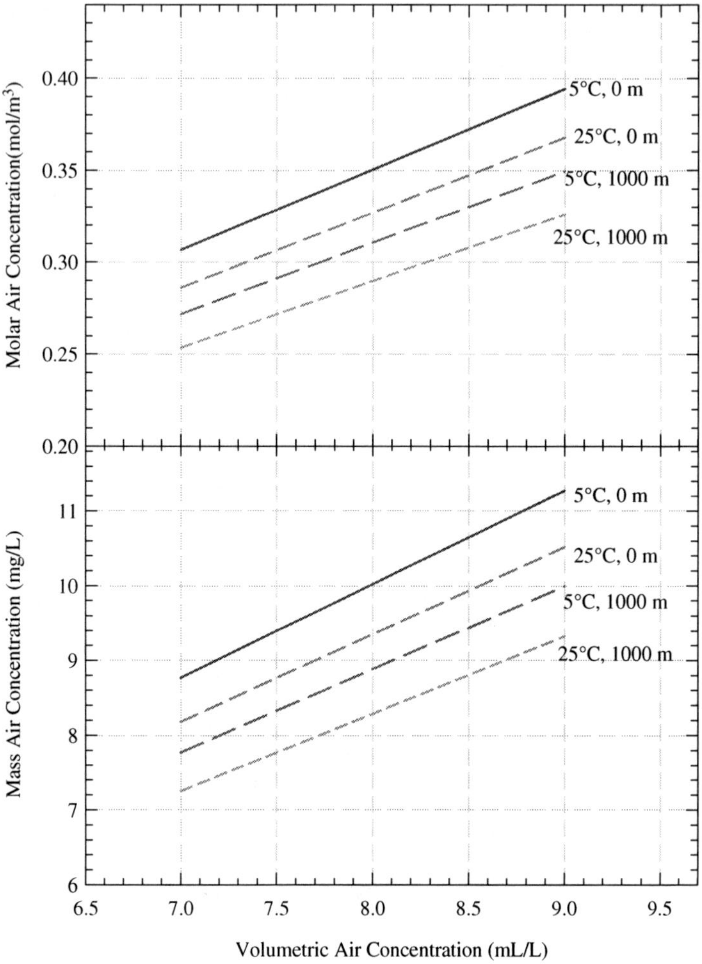

FIGURE 3-1 The molar (top) and mass (bottom) equivalents of the volumetric air concentration at different temperature and altitude.

EXAMPLE 3-2 Required Air Concentrations for Different Recycle Flow Rates A proposed DAF plant, which treats water from an impoundment with occasional dissolved oxygen deficits, is to be built at an altitude of 1000 m. The water temperatures (minimum; average; maximum) are 5, 10, and 15°C, respectively. The volumetric air specification is 8 mL/L. It is not possible to aerate the raw water before treatment and the oxygen deficit has to be satisfied through the saturation system using an air saturator. Specify the molar air requirement in terms of a relationship between the recycle flow rate and air concentration in the water leaving the saturator.

Solution The molar equivalent of 8 mL/L is 0.310, 0.300, and 0.290 mol/m^3 at 5, 15, and 25°C, respectively (calculated from Tables E-2 and E-3). To compensate for the undersaturation with respect to oxygen, these values should be increased by 15 percent to 0.357, 0.345, and 0.333 mol/m^3, respectively.

Obtain M_{atm} = 1.150, 0.921, and 0.768 mol/m^3 from Table F-3 for the three temperatures. These values are at sea level and have to be multiplied by 89.8/101.3 to compensate for the altitude, leading to M_{atm} = 1.019, 0.816 and 0.681 mol/m^3, respectively. Use Eq. (3-5) to calculate the air concentration leaving the saturator for different recycle ratios:

Recycle ratio (%)	6%	8%	10%	12%	14%
5°C: Air from saturator (mol/m^3)	6.31	4.82	3.93	3.33	2.91
15°C: Air from saturator (mol/m^3)	6.101	4.66	3.80	3.22	2.81
25°C: Air from saturator (mol/m^3)	5.88	4.50	3.66	3.11	2.71

This table provides 15 different combinations of recycle ratio and temperature, all exactly meeting the volumetric air requirement of 8 mL/L or 8000 ppm. In other words, if the recycle flow rate is high, the air concentration in the recycle flow is low, maintaining the air dosing at 8 mL/L. If the raw water is fully saturated with oxygen, then the recycle flow rate could be reduced by 15 percent. It is reiterated that raw water oxygen deficits should be eliminated prior to DAF where possible. ▲

Having found a general relationship between M_{sat} and r (as in Example 3-2), one can select a specific point on the curve. This decision is influenced by other factors. M_{sat} is important from the perspective of bubble size (discussed in Chap. 4), as well as a practical matter dependent on the preferred duty point of the recycle pumps, safety regulations for pressure vessels, and costs. Also, the air transfer efficiency determines how much air will actually be transferred due to kinetic constraints. The transfer efficiency of saturators is an important topic discussed later in this chapter.

3-2 THE PRINCIPLES OF AIR SATURATION

Water in Equilibrium with Atmospheric Air

The design of the recycle saturation system hinges on Henry's law. At equilibrium conditions, Henry's law relates the molar concentration of a gas in air to the molar concentration of the gas in water:

$$G_X = H_X \times \left(M_X^*\right) \tag{3-6}$$

G_X is the molar concentration of the gas in air (mol/m^3); H_X is the dimensionless Henry's constant of gas X (dimensionless); M_X is the molar concentration of gas X in the water (mol/m^3); superscript * refers to equilibrium. Henry's constant is affected by both temperature and ionic strength. Tables F-1 and F-2 provide a detailed listing of Henry's constant for nitrogen, oxygen, and argon—the principal air gases. For normal drinking water applications, the values in Table F-1 are appropriate. For detailed work, the correlations in App. F can be used to calculate precise values. A set of air solubility curves in water, based on the Henry's constants in Tables F-3 and F-4, is shown in Fig. 3-2.

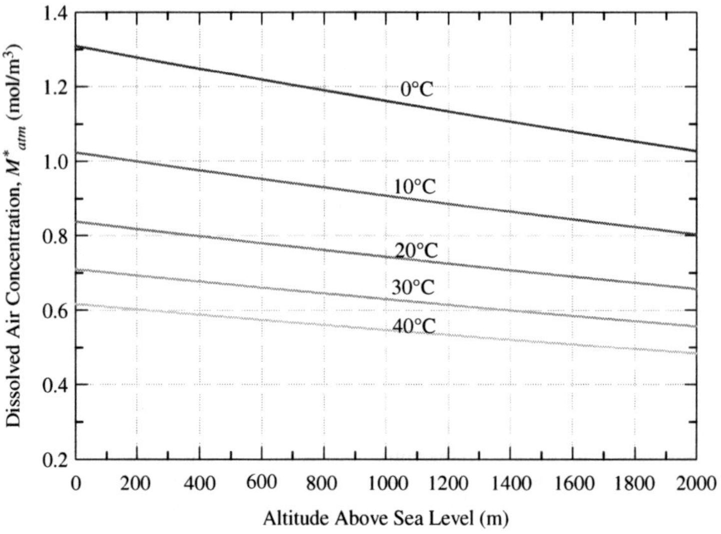

FIGURE 3-2 Equilibrium concentration of atmospheric air in water at different temperatures and altitudes.

EXAMPLE 3-3 Precipitation of Air due to Depressurization A water volume is at equilibrium with atmospheric air at sea level and 20°C (Condition 1). The volume is then pressurized to 500 kPa in the presence of atmospheric air and left for a while to attain equilibrium with water, while the water temperature stays the same (Condition 2). The pressure is then released back to atmospheric pressure and the water left until a new equilibrium is reached (Condition 3). How much air (volumetric, molar and mass concentrations) is released as bubbles between Conditions 2 and 3?

Solution The absolute dry air pressure during Conditions 1 and 3 is 101.3 kPa (Table E-1). The absolute dry air pressure during Condition 2 is 101.3 + 500.0 – 2.3 (vapor pressure) = 599.0 kPa. The amount of air dissolved into the water between Conditions 1 and 2 is exactly the same as the amount released between Conditions 2 and 3. Calculate the molar concentration of dissolved air at Conditions 1 and 3:

Eq. (3-2): $G_N = 0.03245$; $G_O = 0.00871$; $G_{Ar} = 0.00039$ mol/L

Table F-1: $H_N = 60.27$; $H_O = 30.56$; $H_{Ar} = 27.66$ mol/mol

Eq. (3-6): $M_N^* = 0.000538$; $M_O^* = 0.000285$; $M_{Ar}^* = 0.000014$ mol/L

$M_{total}^* = 0.000837$ mol/L

Alternatively, M_{total}^* can be read from Table F-3.

Calculate the molar concentration of dissolved air at Condition 2:

Eq. (3-2): $G_N = 0.1919$; $G_O = 0.0515$; $G_{Ar} = 0.0023$ mol/L

Table F-1: The same Henry's constants as before

Eq. (3-6): $M_N^* = 0.003184$; $M_O^* = 0.001685$; $M_{Ar}^* = 0.00083$ mol/L

$M_{total}^* = 0.004951$ mol/L

The molar concentration of air released between Conditions 2 and 3 is:

Air released = 0.004951 − 0.000837 = 0.004114 mol/L.

Its volumetric equivalent is:

Table E-2: 0.004114/0.0416 = 0.0989 L air/L water or 98.9 mL/L

Its mass equivalent is:

Table E-3: 0.0989 × 1.202 = 0.1189 g/L or 118.9 mg/L ▲

The Difference between Open-End and Dead-End Saturation

There are two practical methods to saturate water with air. The first method is called open-end saturation, where a small air flow rate is continuously introduced into the water, followed by high turbulence to rapidly dissolve all the air. In other words, the air enters the saturation device at the same rate at which it leaves as dissolved air. Two practical technologies, listed in Chap. 1, belong to this category. In the first, air is injected directly into the suction side of a recycle pump, or air is injected before or at a point where the static water head is high—the induced-air technology. The second method is dead-end saturation, where the air is pumped into a pressure vessel to form an air cushion, not immediately leaving with the water. As the water flows through the vessel, it dissolves away a part of the air, with the rest of the air trapped in the vessel.

Henry's law is essential for the design of both open-end and dead-end saturation. For open-end saturation, it is used to ensure that no undissolved air leaves the saturation device. For dead-end saturation, it is used to calculate the solubility of the saturator air, that has a composition different from atmospheric air.

At present, the vast majority of DAF installations for drinking water treatment make use of dead-end saturators, which are mostly filled with plastic packing to improve transfer efficiency, but some are also used without packing. The analysis and design of packed saturators therefore take up the major part of the chapter. Open-end saturator pumps, however, offer a simple, elegant solution that deserves more rigorous attention than afforded in the past. The remainder of this chapter develops the necessary design equations for both open-end and dead-end saturation. All the material is based on recycle-flow pressure DAF, but can be easily adapted for full-flow and split-flow pressure DAF, the differences illustrated in Fig. 1-2.

3-3 OPEN-END SATURATION

Principle of Operation

Open-end saturation takes place when a small quantity of air is injected into a pressurized volume of water with the intention of completely dissolving the air. The air does not accumulate in the contactor and has to be continuously replenished—hence "open-end" saturation.

The most common open-end method is to inject air into the low-pressure suction side of a pump. The turbulence within the pump volute disperses the air into small bubbles. As the water leaves the pump, the high water pressure and large interfacial area of the bubbles encourage the dissolution of the air into the water. Special centrifugal DAF pumps are available for this purpose, which look similar to regular centrifugal pumps but with important small internal changes. The pump impellers are adapted to deal with two-phase flow and to disperse the incoming air flow as finely as possible. The DAF pumps have to meet two other technical challenges—that of preventing cavitation once air is injected, and of maintaining the pump efficiency, which tends to drop when air is introduced.

Alternative open-end saturation systems do not use the pump housing as the contacting chamber, but employ a separate proprietary static mixing device downstream of the pump. Air and water enter the mixing device under conditions of high turbulence, which serves the same purpose as the DAF pump housing. To inject the air after the pump, the air has to be pressurized to the delivery pressure of the pump, or more. This requires a little more energy than injecting the air on the low-pressure suction side of the pump. With a Venturi on the suction line, the air can be drawn into the pump without any pressurization whatsoever.

The air flow into an open-end saturation system is a compromise between injecting as little air as possible (to achieve complete dissolution of the air quickly) and as much air as possible (to minimize the required water flow rate and thus the pumping energy). For many industrial thickening and industrial separation applications, it does not matter that much if some air bubbles remain, in which case a larger flow of air into the pump can be tolerated. For drinking water clarification, large air bubbles interfere with the delicate collision and attachment processes in the DAF tank and should be avoided. For this reason, some DAF pump manufacturers provide a small air separation tank immediately downstream of the pump to allow undissolved air to vent off. Such an air release device, however, makes the system more complicated and is unnecessary if the air flow is controlled properly.

Specification of Open-End Saturation Systems

Some general comments and suggestions for the specification of open-end systems are appropriate. The literature on commercially available DAF pumps often makes exaggerated claims that do more harm than good. However, DAF pumps do provide a viable option for air saturation and deserve serious consideration, based on sound technical grounds.

The key air specification for DAF is the air volume precipitated after the release of the recycle stream in the DAF tank. It can be stated explicitly in volume, molar, or mass concentrations, as shown earlier, or implicitly as combinations of saturation pressure, water flow rate, air flow rate, and transfer efficiency. Later in Sec. 3-6, a relationship is derived to make a precise estimate of the air requirements. None of the DAF pump literature surveyed by the authors over the years provided enough technical information to make such estimates.

The commercial DAF pump literature, on the other hand, abound with claims on relative energy savings, presumably in relation to dead-end saturators. What would be more useful is the actual energy consumption per unit of dissolved air, backed by guarantees and performance tests, to allow DAF designers to make rational comparisons between different saturation options.

For drinking water clarification, complete dissolution of the air is important, as noted before. This may not be easy to measure, but reputable suppliers of DAF pumps should be able to demonstrate complete air dissolution by an absence of air collected in a separator vessel immediately downstream of the pump. Getting as much of the air as possible into the recycle stream but ensuring complete air dissolution at the same time requires modulation of the air flow rate in conjunction with the recycle flow rate. Drinking water plants have to produce water at a variable rate to match the demand pattern, creating a need for operational changes to the recycle flow rate, and thus also the air flow rate. This could compromise the claimed simplicity of open-end saturation and requires special attention.

Some DAF pump literature include claims of the favorable bubble sizes being produced (mostly between 20 and 30 μm in diameter) when the recycle is discharged into the DAF tank. Although the claim may be true, the bubble size is independent of the DAF pump. The bubble size of the air suspension depends on the saturation pressure and on the details of the air injection nozzles used in the DAF tank. The means of air dissolution has nothing to do with the means of air precipitation. A specific set of air injection nozzles in a DAF tank should produce the same bubble size, regardless of the air saturation method.

3-4 DEAD-END SATURATION

Principle of Operation

Dead-end saturation requires a pressure vessel into which air and water are pumped. These pressure vessels, manufactured of steel, are known as saturators and are used in practically all large drinking water DAF plants. The air forms a cushion within the saturator, from which the water draws its increased dissolved air concentration before leaving the saturator. Like all industrial pressure vessels, saturators are subject to rigorous industrial safety regulations regarding structural design, manufacture, and periodic inspections.

Saturators can be either filled with packing material (packed saturators) or left empty (unpacked saturators). This choice is a compromise between potential clogging of the packing material, and energy cost (unpacked saturators are less efficient). Drinking water treatment plants need a recycle stream, taken from either the outflow of the DAF tank or the filtered water supply (recycle-flow pressure DAF). The usual good quality from either stream practically eliminates the potential for clogging of the packing, and packed saturators are therefore often used, sometimes with a screen at the water inlet to trap coarse material. The recycle stream can also be taken from the incoming raw water (split-flow pressure DAF), but this is an unusual case and not considered here. Unpacked saturators are more common for wastewater thickening by DAF, which requires much larger recycle streams usually taken from unfiltered effluent of lower quality.

Unpacked saturators are positioned either lying on their side or upright. The incoming water has to be sprayed into the saturator at the top to increase the surface area by dispersion for greater air transfer efficiency. Some systems have an addition internal recirculation system to improve their air transfer efficiency. The energy demand of unpacked saturators is greater than that of packed saturators. An early study showed that unpacked saturators had to be operated at higher pressure than packed saturators to achieve equivalent air transfer (Zabel and Hyde, 1977). This may be due partly to the need for unpacked saturators to have an energy-consuming spray nozzle inside the saturator to disperse the flow into the open saturator volume. Packed saturators work well with a simpler distribution manifold or plate that consumes less energy, as the water is dispersed by the packing material. It should be noted that the saturation pressure, in the case of a spray nozzle being used, is the pressure downstream of the nozzle, and not the pressure at the inlet to the saturator. Some major equipment suppliers, which serve both the drinking water and industrial markets, nevertheless prefer to use unpacked saturators only (e.g., Degremont, 1991). The internal details of unpacked saturators determine their efficiency and, unlike packed saturators, no general theory for predicting their efficiency has yet been proposed. Unpacked saturators are therefore not covered further in this book. Figure 3-3 shows three examples of typical air saturators with the top two examples of packed saturators, and a set of unpacked saturators at the bottom. Figure 3-4 shows two examples of popular packing types.

FIGURE 3-3 Examples of saturators at full-scale drinking water plants: (a) packed saturator, Stamford, CT (U.S.); (b) packed saturators, Windhoek (Namibia)—see case study in Chap. 12; and (c) unpacked saturators, Haworth, NJ (U.S.)—see case study in Chap. 11.

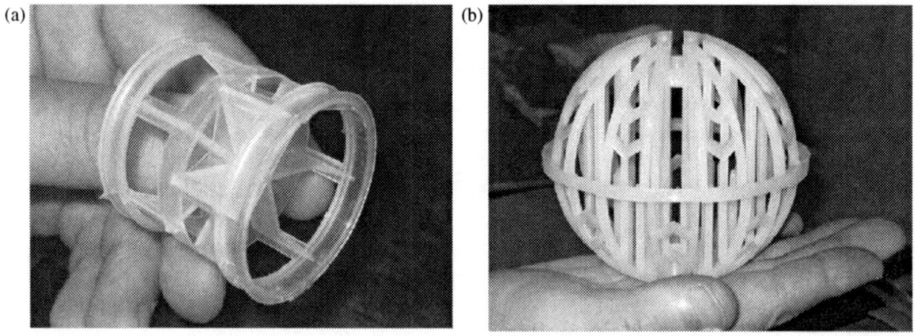

FIGURE 3-4 Examples of plastic packing used in packed saturators: (a) Pall rings, nominal size 38 mm; (b) Jaeger Tri-Pack®, nominal size 76 mm.

A schematic cross-section of a packed saturator is given in Fig. 3-5. Water is introduced at the top of the saturator, is distributed evenly over its cross-sectional area, and then drops onto the packing material. The packing material splits the water flow into a multitude of minute flow paths, thus providing large interfacial area to improve air transfer. The water is collected in a pool at the bottom of the saturator, from where the saturated water is

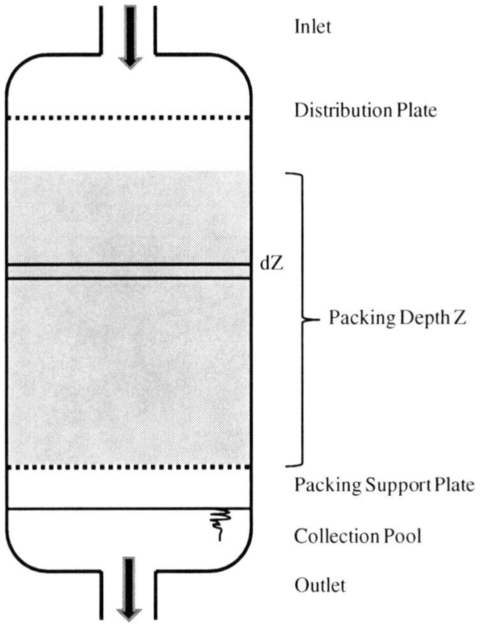

FIGURE 3-5 Schematic cross-section of a packed saturator.

distributed to the DAF tank(s). The packing support plate is often replaced with a basket screen before the outlet, allowing the saturator to be filled with packing from the bottom. Compressed air is pumped into the tank to maintain the air cushion in the saturator, keeping the water level a safe distance above the outlet. Typical design parameters for packed saturators are given in Table 3-3. Readers should note that the hydraulic loading of the saturator is expressed as a mass loading for the purposes of the mass transfer model that follows. To convert to volumetric loading, the mass loading has to be divided by the water density, obtained from Tables D-1 and D-2.

Composition of Saturator Air

The air entering a saturator is trapped in an air cushion, thus forming a dead end for the air flow. The only way for the air to leave the saturator is by dissolution into the water. Because the three atmospheric gases have different Henry's constants, they do not dissolve to the same extent. The different depletion rates of the gases in the air cushion eventually lead to an air composition that is different from atmospheric air. In this section, a method for calculating the composition of saturator air is developed.

Assuming equilibrium conditions, a molar balance for gas X across the saturator takes the form:

$$\left[Q_r \times M_X^* + Q_{air} \times G_X\right]_{in} = \left[Q_r \times M_X^*\right]_{out} \tag{3-7}$$

TABLE 3–3 Typical Design Parameters for Packed Saturators

		Range	Typical
Hydraulic loading*	kg/m²-s	20–60	40
Gauge pressure	kPa	350–600	500
Packing depth	mm	800–2000	1400
Packing size	mm	25–90	50

Right column shows typical values, which are used for illustration of saturator performance in Figs. 3-8 to 3-11.

*Expressed as a mass loading.

Q_r is the recycle flow rate and Q_{air} is the volumetric air flow rate into the saturator (both in m³/s). As the water flowing into and out of the saturator is at equilibrium with air, Henry's law applies:

$$Q_r\left[\frac{G_X}{H_X}\right]_{in} + Q_{air}\left[G_X\right]_{in} = Q_r\left[\frac{G_X}{H_X}\right]_{out} \tag{3-8}$$

The gas concentrations are determined from the gas fractions and the pressures in atmospheric and saturator air, respectively. With constant temperature throughout, Eq. (3-8) is reformulated:

$$\left(\frac{1}{H_X}\right)Q_r \times f_{X,atm} \times p_{atm} + Q_{air} \times f_{X,atm} \times p_{atm} = \left(\frac{1}{H_X}\right)Q_r \times f_{X,sat} \times \left(p_{atm} + p_{sat} - p_{vap}\right) \tag{3-9}$$

Equation (3-9) simplifies to:

$$\left(\frac{Q_{air}}{Q_r}\right) = \frac{f_{sat,X}}{H_x \times f_{atm,X}}\left(\frac{p_{atm} + p_{sat} - p_{vap}}{p_{atm}}\right) - \frac{1}{H_X} \tag{3-10}$$

Application of Eq. (3-10) to nitrogen, oxygen, and argon in turn allows the simultaneous solution of each gas fraction as well as the volumetric air-water flow rate ratio required to maintain the air volume in the saturator. If nitrogen, oxygen, and argon are considered, there are four unknowns—the three saturator gas fractions and the volumetric air-water flow rate ratio. The fourth equation is provided by $\Sigma f_x = 1$. The nitrogen fraction in saturator air at equilibrium is solved first:

$$f_{sat,N} = \frac{f_{atm,N}\left[H_N\left(S - f_{atm,Ar} - f_{atm,O}\right) + f_{atm,Ar} \times H_{Ar} + f_{atm,O} \times H_O\right]}{S\left(f_{atm,O} \times H_O + f_{atm,N} \times H_N + f_{atm,Ar} \times H_{Ar}\right)} \tag{3-11}$$

The pressure ratio S in Eq. (3-11) is provided by:

$$S = \frac{p_{atm} + p_{sat} - p_{vap}}{p_{atm}} \tag{3-12}$$

From here, the other fractions follow:

$$f_{sat,O} = \frac{f_{atm,O}}{S} + \left(\frac{H_O}{H_N}\right)\left(\frac{f_{sat,N} \times f_{atm,O}}{f_{atm,N}}\right) - \left(\frac{H_O}{H_N}\right)\frac{f_{atm,O}}{S} \tag{3-13}$$

$$f_{sat,Ar} = 1 - f_{sat,N} - f_{sat,O} \tag{3-14}$$

Figure 3-6 was generated using Eq. (3-11), which shows that the air in the saturator is enriched with nitrogen as the saturator pressure is increased. Although Fig. 3-6 demonstrates differences due to temperature and saturator pressure, it should be noted that the differences are small, with the nitrogen fraction mostly between 0.85 and 0.87, but still distinctly greater than its value of 0.78 in atmospheric air.

The Air Available for DAF

When the pressure of the saturated water is released in the DAF reactor, not all the dissolved air precipitates from the water. The air precipitates only to the point when a new equilibrium with the water in the DAF reactor is reached, again dictated by Henry's law.

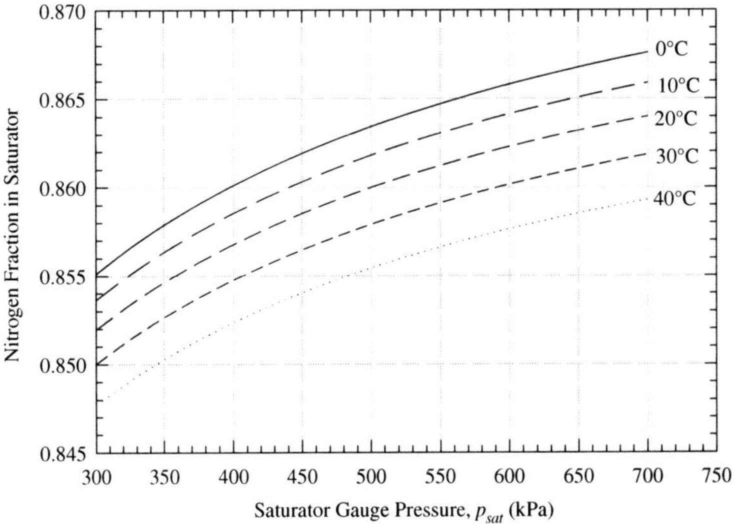

FIGURE 3-6 Volumetric or molar nitrogen fraction in saturator air as a function of temperature and saturator gauge pressure, at sea level.

The air actually precipitated, which forms bubbles and makes DAF possible, is called *available air*. The equilibrium between water and air in the DAF reactor is dependent on the "atmosphere" within the millions of tiny bubbles inside the tank, and not on the normal atmosphere above the water surface of the DAF tank, as often assumed. The concentration of dissolved air remaining in the saturated water, in the first few minutes after pressure release, is thus dictated by the air composition of the bubbles, which is the same as the air composition in the saturator.

Assuming complete saturation, the total available air is calculated with the saturator air fractions determined earlier:

$$M_{av}^{\ *} = \left(p_{sat} - p_{vap}\right)\left(\frac{44.6}{101.3}\right)\left(\frac{273.2}{273.2+t}\right)\left(\frac{f_{sat,N}}{H_N} + \frac{f_{sat,O}}{H_O} + \frac{f_{sat,Ar}}{H_{Ar}}\right) \tag{3-15}$$

$M_{av}^{\ *}$ is the total available air after complete saturation (mol/m^3). Figure 3-7 shows a general solution to Eq. (3-15) to indicate the effects of water temperature and saturator pressure.

Equations (3-7) to (3-15) assume that the recycle water flowing into the saturator is fully saturated with air. For most practical applications, this is a valid assumption as the water into the saturator is recycled from a point after the DAF reactor. In the unusual case where the water into the saturator is taken from another source that might be deficient in dissolved oxygen, the saturator gas fractions are slightly different with an even larger nitrogen fraction as more oxygen is depleted. With the oxygen deficit being known, we can estimate the effect on available air using the same approach as discussed earlier.

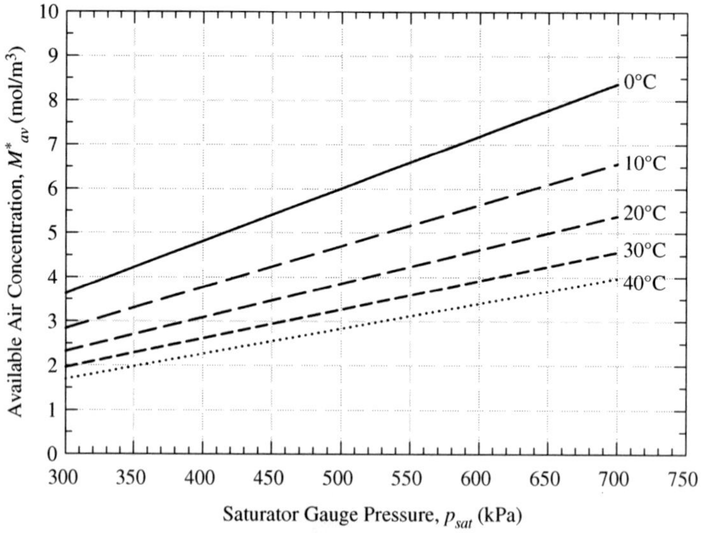

FIGURE 3-7 Available air concentration (M^*_{av}) from a saturator at complete saturation, as a function of temperature and saturator gauge pressure, at sea level.

EXAMPLE 3-4 Available Air from a Saturator at Equilibrium Calculate the available air M^*_{av} at equilibrium from a saturator with a total gauge pressure of 500 kPa, installed at an altitude of 700 m above sea level. The water temperature is 20°C.

Solution At 20°C the Henry's constants are 60.27, 30.56, and 27.66 for nitrogen, oxygen, and argon, respectively, as determined earlier in Example 3-3. The gas fractions in atmospheric air are 0.7808, 0.2095, and 0.0093 for nitrogen, oxygen, and argon, respectively. At an altitude of 700 m, the dry atmospheric pressure is 93.2 kPa (Table E-1) and the vapor pressure 2.3 kPa (Table D-1).

Calculate the pressure ratio S:

Eq. (3-12): $S = (93.2 + 500 - 2.3)/93.2 = 6.340$

Calculate the saturator gas fractions:

Eq. (3-11): $f_{sat,N} = 0.8611$

Eq. (3-13): $f_{sat,O} = 0.1334$

Eq. (3-14): $f_{sat,Ar} = 0.0055$

The available air at full saturation is:

Eq. (3-15): $M_{av}^* = 3.85 \text{ mol/m}^3$

To get the volume and mass equivalents, the appropriate conversion factors are found at an altitude of 700 m, by interpolation from Tables E-2 and E-3:

Table E-2: Conversion factor = 0.0383 mol/L

Volume equivalent = 3.85/0.0383 = 100.6 mL/L

Table E-3: Conversion factor = 1.093 mg/L

Mass equivalent = 100.6 × 1.093 = 110.0 mg/L ▲

Modelling the Mass Transfer in a Saturator

The available air concentration provided by Eq. (3-15) is the theoretical maximum, corresponding to complete air transfer in the saturator. In reality, the retention time of the recycle water in the saturator is less than a minute and equilibrium is not reached. For design and optimization, it is important to predict the transfer efficiency. In this section, a mass transfer model is developed, broadly following the one proposed by Haarhoff and Rykaart (1995).

For the infinitesimal element dZ shown in Fig. 3-5, the gas transfer is described in terms of the increase of the gas concentration in water:

$$J = \frac{Q}{A}\left(\frac{dM_X}{dZ}\right) \tag{3-16}$$

J is the mass transfer rate per unit volume of packing [mol/m^3-s]; dM the concentration difference in the water entering and leaving the element dZ (mol/m^3); A the saturator cross-sectional area (m²); subscript X refers to the air gas under consideration. J can also be formulated in terms of the interfacial mass transfer:

$$J = K_L \times a_w \left(M_X^* - M_X \right) \tag{3-17}$$

K_L is the mass transfer rate constant (m/s); a_w is the wetted packing area (m²/m³); superscript * refers to the equilibrium gas concentration in water. Elimination of J and substitution of Q/A with HL_{mass}/ρ yields:

$$dZ = \left(\frac{HL_{mass}}{\rho \times K_L \times a_w} \right) \frac{dM_X}{\left(M_X^* - M_X \right)} \tag{3-18}$$

HL_{mass} is the liquid mass loading rate [kg/m²-s]; ρ is the density of water (kg/m³). With the assumption of constant gas concentration throughout the packing depth, we integrate Eq. (3-17) from top to bottom to provide a mass transfer model:

$$Z = \left(\frac{HL_{mass}}{\rho \times K_L \times a_w} \right) \ln \left(\frac{M_X^* - M_{X,out}}{M_X^* - M_{X,in}} \right) \tag{3-19}$$

Subscripts in and out refer to the gas concentrations in the water entering and leaving the saturator (mol/m³).

Expressions are required for the mass transfer rate constant K_L and the wetted packing area a_w. The modelling of the mass transfer across the air-water interface follows the Lewis-Whitman two-film approach, which postulates the interface to consist of two laminar layers—one air and the other water. In this case, the laminar air film offers practically no resistance due to the low Henry's constants of the air gases. The resistance is only dependent on the water layer. The Onda correlation (Onda et al., 1968) provides a means of estimating the transfer rate constant:

$$K_L = 0.0051 \left(\frac{\mu \times g}{\rho} \right)^{1/3} \left(\frac{HL_{mass}}{a_w \times \mu} \right)^{2/3} \left(\frac{\rho \times D}{\mu} \right)^{1/2} \left(a \times d_p \right)^{2/5} \tag{3-20}$$

The dynamic viscosity of water is μ (kg/m-s); g is gravitational acceleration (m/s²); D is the molecular diffusivity of the gas in water (m²/s); a is the specific (dry) packing area (m²/m³); d_p is the nominal packing size (m).

Onda et al. (1968) also provide a correlation for the wetted packing area:

$$a_w = a \left(1 - \exp \left(-1.45 \left(\frac{\sigma_c}{\sigma} \right)^{3/4} \left(\frac{HL_{mass}}{a \times \mu} \right)^{1/10} \left(\frac{HL_{mass}^2 \cdot a}{\rho^2 \times g} \right)^{1/20} \left(\frac{HL_{mass}^2}{\rho \times \sigma \times a} \right)^{1/5} \right) \right) \tag{3-21}$$

The critical surface tension of the packing material is σ_c (kg/s²); σ is the surface tension of water (kg/s²). It is customary, especially when the transfer rate constant K_L cannot be

TABLE 3-4 Specific Area for Commercial Plastic Packing (m^2/m^3)

	Packing size d_p				
Packing name	25 mm	38 mm	51 mm	76 mm	90 mm
Ballast rings	213	13120	105		85
Ballast saddles	213		112	92	
Flexirings	213	131	115		92
Hiflow rings	214		150		76
Hollow ball	460	325	236	150	
Nor-Pac	180	180	102		
Novalox	256	256	121	105	
Pall rings	220	220	108	85	79
Super Intalox	207		108	89	
Tellerettes	180		125		
Tripacks	278		157		125
Critical surface tension of all plastic packing is 0.033 kg/s^2					

Source: Partly from Haarhoff, J., and Cleasby, J. L. (1990), Evaluation of Air Stripping for the Removal of Organic Drinking-Water Contaminants, Water SA 16 (1) 13–22, with permission from the Water Research Commission, Pretoria, South Africa.

estimated separately from the interfacial area a_w (such as in bubble aeration systems), to measure and report them as a single mass transfer constant $K_L a_w$.

Some properties of water and the packing material are affected by temperature. The physical properties of water are provided in Tables D-1 and D-2. The surface tension of plastic packing is 0.033 kg/s^2. Table 3-4 shows the specific area of a number of commercial packings. As a general guideline, the specific area of plastic packing, with the packing nominal size in mm, can be approximated as:

$$a = 2500 \times d_p^{-0.75} \tag{3-22}$$

EXAMPLE 3-5 Calculate the Mass Transfer Constants K_{Law} Estimate the mass transfer constant for a packed saturator filled with 50 mm plastic packing, hydraulic loading of 40 kg/m^2-s, and water temperature of 20°C.

Solution At 20°C, the water density is 998.2 kg/m^3; the dynamic viscosity is 0.001002 kg/m-s; the surface tension is 0.07275 kg/s^2 (from Table D-1).

If the packing type is known, the exact specific area can be obtained from the manufacturer. If not, the specific area of the plastic packing can be estimated:

Eq. (3-22): $a = 120\ m^2/m^3$

The wetted packing area is:

Eq. (3-21): $a_w = 68.1\ m^2/m^3$

The transfer rate constant is different for the three air gases due to their different molecular diffusivities in water. From Table F-1, the values are 1.66×10^{-9}, 2.19×10^{-9}, and

2.28 × 10^{-9} m^2/s for nitrogen, oxygen, and argon, respectively. Calculate the values for the transfer rate constant, and the overall mass transfer constants:

Eq. (3-20): Nitrogen $K_L = 0.000637$ m/s; $K_La_w = 0.0434$ /s

Eq. (3-20): Oxygen $K_L = 0.000732$ m/s; $K_La_w = 0.0499$ /s

Eq. (3-20): Argon $K_L = 0.000747$ m/s; $K_La_w = 0.0509$ /s ▲

Application of the Mass Transfer Model

The mass transfer model in Eq. (3-19) is rearranged to solve for the air concentration leaving the saturator:

$$M_{sat,X} = M^*_{sat,X} - \left(M^*_{sat,X} - M^*_{atm,X}\right)e^{\left(-\frac{\rho \times Z \times K_L \times a_w}{HL_{mass}}\right)} \tag{3-23}$$

Equation (3-23) has to be applied to nitrogen, oxygen, and argon in turn to yield the molar concentration of each gas leaving the saturator. The concentrations are added to get the total concentration of air in the water leaving the saturator. To obtain the available air concentration, the concentration in equilibrium with the DAF reactor (earlier shown to have the same composition as saturator air) has to be subtracted:

$$M_{av} = \left(M_{sat,N} + M_{sat,O} + M_{sat,Ar}\right) - 44.6\left(\frac{273.2}{273.2+t}\right)\left(\frac{p_{atm}}{101.3}\right)\left(\frac{f_{sat,N}}{H_N} + \frac{f_{sat,O}}{H_O} + \frac{f_{sat,Ar}}{H_{Ar}}\right) \tag{3-24}$$

M_{av} is the available air that precipitates as bubbles in the DAF reactor (mol/m^3).

EXAMPLE 3-6 Available Air from a Saturator with the Mass Transfer Model Assume the same conditions as for Example 3-5, namely, a saturator gauge pressure of 500 kPa, at altitude 700 m and temperature 20°C. If the liquid loading on the saturator is 40 kg/m^2-s and 50 mm plastic packing, 1000 mm deep, is used, calculate the available air M_{av} actually transferred.

Solution At 20°C the Henry's constants are 60.27, 30.56, and 27.66 for nitrogen, oxygen, and argon, respectively (earlier determined in Example 3-3). The mass transfer constants for nitrogen, oxygen, and argon are 0.0434, 0.0499 and 0.0509 /s, respectively (from Example 3-5). The gas fractions in saturator air are 0.8611, 0.1334, and 0.0055 for nitrogen, oxygen, and argon (from Example 3-4). At an altitude of 700 m, the dry atmospheric pressure is 93.2 kPa and the vapor pressure 2.3 kPa (from Example 3-4).

The molar concentrations M^*_{atm} of the gases in the water entering the saturator are calculated first:

Eqs. (3-3) and (3-6): Nitrogen 0.495 mol/m^3

Oxygen 0.262 mol/m^3

Argon 0.013 mol/m^3

The molar concentrations M^*_{sat} of the water at equilibrium in the saturator are calculated next:

Eqs. (3-3) and (3-6):	Nitrogen	3.463 mol/m^3
	Oxygen	1.058 mol/m^3
	Argon	0.048 mol/m^3

The molar concentrations M_{sat} of the water leaving the saturator are:

Eq. (3-23):	Nitrogen	2.459 mol/m^3
	Oxygen	0.829 mol/m^3
	Argon	0.038 mol/m^3

The available air concentration in the water from the saturator is finally obtained:

Eq. (3-24): $M_{av} = 2.61$ mol/m^3

The actual available air is less, as one would expect, than the previously calculated available air concentration of $M^*_{av} = 3.85$ mol/m^3 (Example 3-4), which was calculated for perfectly efficient air transfer.

The volume and mass equivalents of M_{av} can be calculated with the same conversion constants used in Example 3-4. ▲

A saturator with typical design parameters is defined in Table 3-3. The mass transfer model is applied to this saturator to demonstrate the relative importance of the four saturator design variables. The results are shown in four figures; Fig. 3-8 shows the effect of saturator pressure, Fig. 3-9 the effect of the mass hydraulic loading, Fig. 3-10 the effect of packing depth, and Fig. 3-11 the effect of packing size. In all cases, the available air transferred

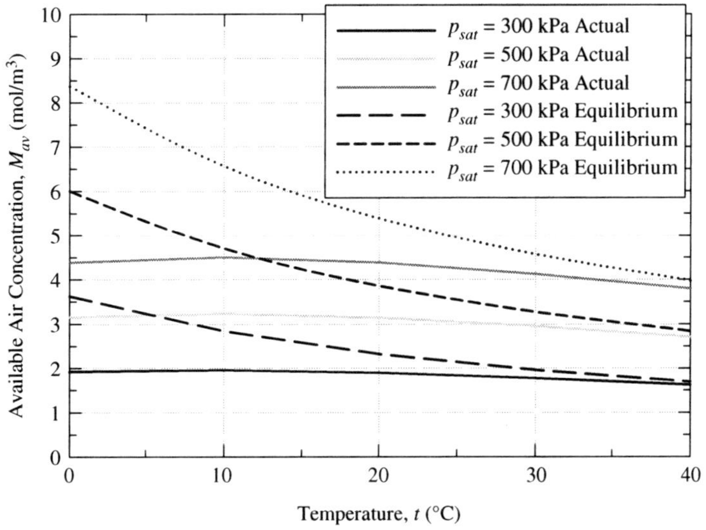

FIGURE 3–8 Available air transferred by a typical saturator showing the effects of saturator pressure and water temperature (saturator parameters are shown in Table 3-3).

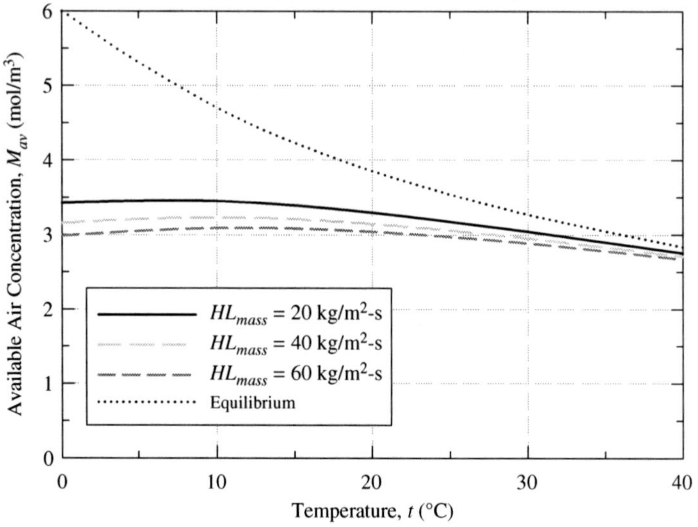

FIGURE 3–9 Available air transferred by a typical saturator showing the effects of hydraulic loading and water temperature (saturator parameters are shown in Table 3-3).

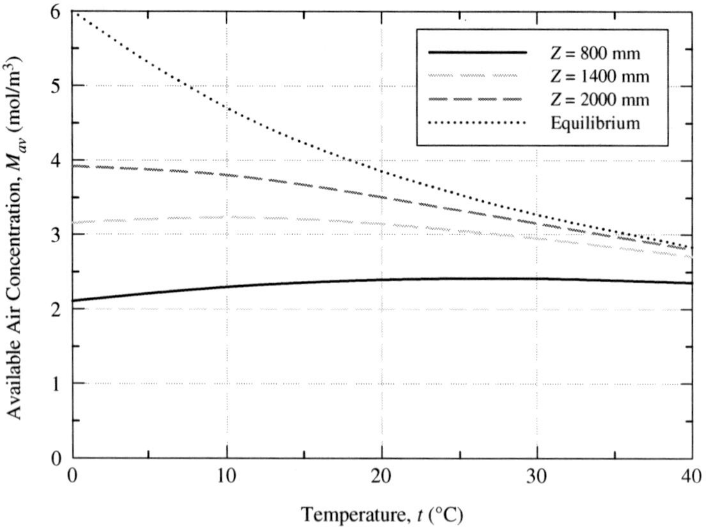

FIGURE 3–10 Available air transferred by a typical saturator showing the effects of packing depth and water temperature (saturator parameters are shown in Table 3-3).

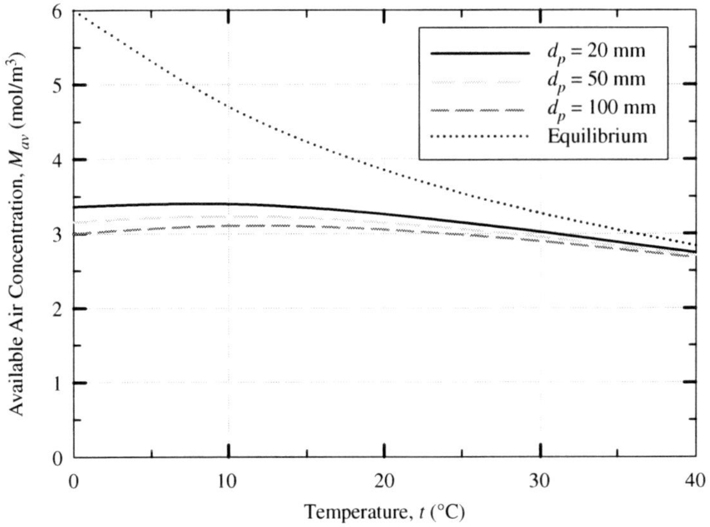

FIGURE 3–11 Available air transferred by a typical saturator showing the effects of nominal packing size and water temperature (saturator parameters are shown in Table 3-3).

is shown as a function of water temperature, with the equilibrium value (for a perfectly efficient saturator) indicated for comparison.

Over the typical ranges of the design parameters considered, a number of remarks follow from Figs. 3-8 to 3-11:

- Saturation pressure, over the range indicated, has the largest effect on the production of available air.
- Packing depth is the next most important design variable. By increasing the packing depth from 800 to 2000 mm, the production of available air is roughly doubled at low temperatures.
- Hydraulic loading has a small effect on the production of available air. The poorer transfer intuitively expected at higher hydraulic loading is largely offset by the increase in wetted packing area. Higher loadings cause the water to spread more over the plastic packing surfaces to increase the interfacial area.
- Packing size has a small effect with a slight advantage for smaller packing sizes.
- For all the design variables, the effect of water temperature is surprisingly small. This is contrary to the equilibrium conditions indicated in the figures, which do show large differences between cold and warm water. Why does the transfer efficiency drop off so rapidly at colder temperatures? The answer lies in the opposing effects of air solubility and molecular diffusivity. At cold temperatures, the lower molecular diffusivity limits the air transfer rate, despite the greater air solubility. This phenomenon is illustrated in Fig. 3-12, where air solubility is quantified by the inverse of Henry's constant and the air transfer efficiency by the mass transfer coefficient $K_L a_w$. For all three gases, the conclusion is the same—the greater solubility is almost exactly paralleled by a lower mass transfer coefficient.

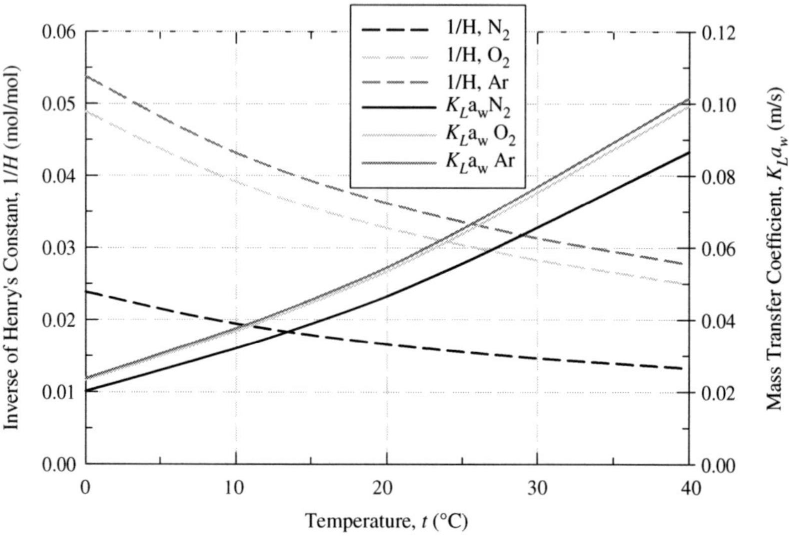

FIGURE 3–12 Opposing effects of water temperature on the mass transfer coefficient and the capacity for air transfer, demonstrated for nitrogen, oxygen, and argon.

The mass transfer model has been validated by a number of studies (Haarhoff and Rykaart, 1995; Steinbach and Haarhoff, 1997, Valade et al., 2001). In general, the predictions of the model agree with their experimental results, and those of previous studies. With the saturator pressure about 500 kPa and larger, the predictions were good, while a larger scatter of experimental results was evident for lower pressures.

End Effects

The mass transfer model describes the mass transfer as the water flows through the packing. This is not the only opportunity for air to be transferred to the water. As the water is distributed at the top of the saturator by a spray nozzle or a perforated plate, air transfer immediately starts to take place. As the water is collected at the bottom, more air is transferred before the water flows out of the saturator. These air transfer steps before and after the packing are known as the end effects. The end effects have not been modelled as few saturators have identical internal details, and are conservatively ignored because their contributions are relatively small because time of contact and interfacial area for contact are substantially greater in the packing.

Some general conclusions about end effects were drawn by Steinbach (1998), who performed tests on a laboratory saturator that closely resembled a full-scale saturator. The vertical dimensions of the saturator were similar to those used at full scale, while the inner diameter was 263 mm. The incoming water was pumped onto a distribution plate, with 4-mm-diameter holes at a density of about 3700 /m^2. From the plate, the water dropped about 400 mm before it reached the top of the packing bed. The packing bed of 800 mm rested on a support plate from where the water dropped a short distance of about 40 mm into a pool of saturated water about 200 mm deep. By removing the saturator section with the packing material, the end effects could be measured directly from the shortened saturator.

The air transfer at the top and bottom ends was significantly ($\alpha = 0.01$) reduced by increasing hydraulic loading. The saturator pressure, although also statistically significant, had a much smaller effect. The air transfer at the ends was reduced as the saturator pressure increased. The air transfer at the top and bottom ends, for the details of that particular saturator, increased the available air by 18 to 20 percent beyond the concentration predicted by the mass transfer model for the transfer of air within the packing. These experiments were done within the temperature range 15 to 20°C.

Saturator Control

In order to maintain a relatively constant air cushion in the saturator, the water and air flows into the saturator have to be carefully balanced to maintain a constant water level in the saturator. It is advantageous to set the minimum level just above the bottom of the packing, to always leave the bottom end of the packing submerged. The submerged packing effectively counters any swirling of the water in the bottom pool and thus prevents vortex formation and air entrainment.

Many applications rely on the adjustment of the water flow into the saturator. The other simpler option is to leave the water flow at its set rate and to only control the air flow into the saturator. Some systems open and close the air-inlet line according to the maximum and minimum water levels allowed in the saturator, and feed the air from an independent, pressurized air vessel. Other systems leave the air-inlet line open all the time, controlling the level with a release valve on the inlet line, which opens and closes according to the minimum and maximum level allowed in the saturator. The latter system has the practical advantage of the air compressors operating continuously. There is no need for the traditional air-vessel downstream of conventional compressors—the air volume in the saturator is sufficiently large to double as an air balancing reservoir.

Saturator Start-up

The air in the saturator, as shown in the previous sections, is eventually enriched with nitrogen due to the different solubilities of the air gases. When the saturator is started, however, the air volume has normal atmospheric composition. How long does it take before the saturator air composition reaches equilibrium? This question does not have much relevance for full-scale applications, where the saturators run for days and weeks on end, but is an important consideration for pilot- and bench-scale studies. In these cases, little or no attention is given to optimize the air saturation system and it often turns out to be excessively large in relation to the small saturator air flow required. Moreover, the saturation system is operated intermittently for each experimental run and the conversion from atmospheric to saturator air composition has to start from scratch each time.

Haarhoff and Steinbach (1996) derived a kinetic model that shows that it takes about 3 h at a mass liquid loading of $HL_{mass} = 15$ kg/m^2-s to attain saturator air equilibrium. As HL_{mass} is lowered, the time increases and reaches 12 h at a liquid loading of 4 kg/m^2-s.

3-5 SATURATOR EFFICIENCY

The Problematic Definition of Saturation Efficiency

It is customary to characterize a mass transfer process in terms of its transfer efficiency, taken as the ratio of the mass that is actually transferred to the mass that could be transferred under equilibrium conditions. The application of this definition to the air saturation systems described is ambiguous for a number of reasons. First, the efficiency can be expressed either in terms of the total air concentration or in terms of the available air after saturation. Second, the equilibrium condition can be expressed in terms of the atmospheric air composition (in which case a saturator could never be 100 percent efficient) or in terms of the saturator air composition. Third, in open-end systems the efficiency is expressed in terms of how much of the supplied air is actually dissolved—this is a completely different concept and should always be 100 percent or close to it. Different studies use different definitions, often not rigorously recorded. To avoid similar confusion, efficiency ratios are not used in this book.

Experimental Measurement of Air Transfer

The available air produced by the saturation system should be measured—either as a plant commissioning test or as a routine test to ensure optimal operation of the saturation system. A number of methods were used by different researchers in the past and these were reviewed by Haarhoff and Steinbach (1997). Some methods have sampled the saturated water under pressure and measured the dissolved air directly (Leininger and Wall, 1974; Packham and Richards, 1975; Henry and Gehr, 1981), a technically difficult option not further pursued. The majority of studies measured the air volume after the saturated water had been depressurized to atmospheric conditions. For measurement of full-scale systems with large saturator flow rates, continuous measurement with an air flow meter is best (e.g., Bratby and Marais, 1975; Robert et al., 1978; Rees et al., 1980; Shannon and Buisson, 1980); for smaller pilot or bench-scale studies, the available air is more conveniently measured by liquid displacement (Conway et al., 1981; Lovett et al., 1984; Vosloo et al., 1986; Casey and Naoum, 1986; Smits, 1990).

The elements of the batch and continuous measurement systems are shown schematically in Fig. 3-13. The water flow rate, in both cases, is measured volumetrically at the water outlet. For the batch system, the available air volume is read from the calibrated air collection tube; for the continuous system the available air flow rate is recorded by the air flow meter.

The air injection nozzle in the measuring bottle is at the heart of both systems. Analogous to the difficulty of reaching saturation close to equilibrium values in the saturator, it is equally difficult to precipitate all the available air upon pressure release. Air injection is therefore characterized by its air precipitation efficiency. Steinbach (1998) performed a comprehensive set of experiments to evaluate the design parameters of the air injection nozzle. The nozzle with the best air precipitation efficiency had an orifice at the end of the pipe, shown in Fig. 3-14 along with the rest of the equipment. The exit orifice had a diameter of 1 mm and length of 5 mm, with an impinging screen between 5 mm and 13 mm away from the mouth of the orifice. At a saturation gauge pressure of 500 kPa the air precipitation efficiency was always about 95 percent. It is pointed out that this nozzle guideline was developed by only looking at its precipitation efficiency. It should not be used for the design of the nozzles in the DAF tank, as discussed in the next chapter.

From such measurements, the volumetric air concentration in the water is directly determined. If we assume that the air precipitation efficiency of the air injection nozzle is 95 percent, the molar concentration of air in the saturated water is estimated as:

$$M_{measured} = \left(\frac{\text{air volume}}{\text{water volume}}\right)\left(\frac{44.6}{0.95}\right)\left(\frac{273.2}{273.2+t}\right)\left(\frac{p_{atm}}{101.3}\right) \qquad (3\text{-}25)$$

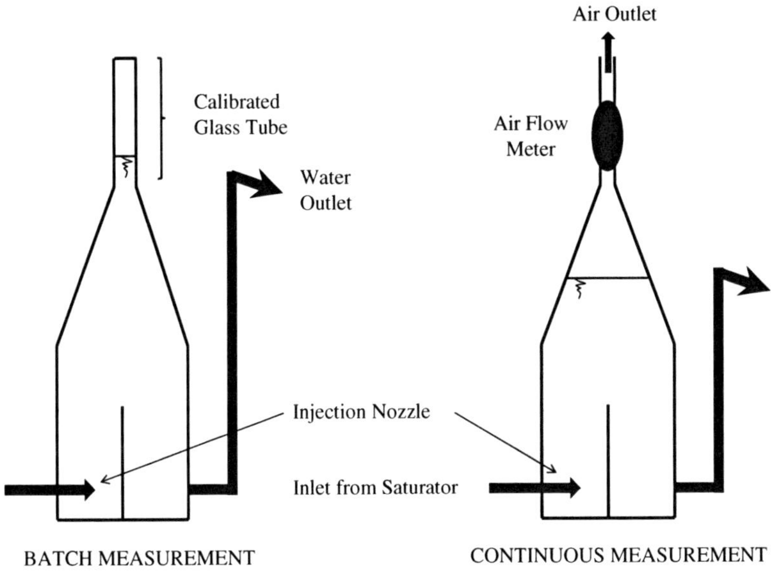

FIGURE 3-13 Illustration of different approaches to the measurement of available air after saturation.

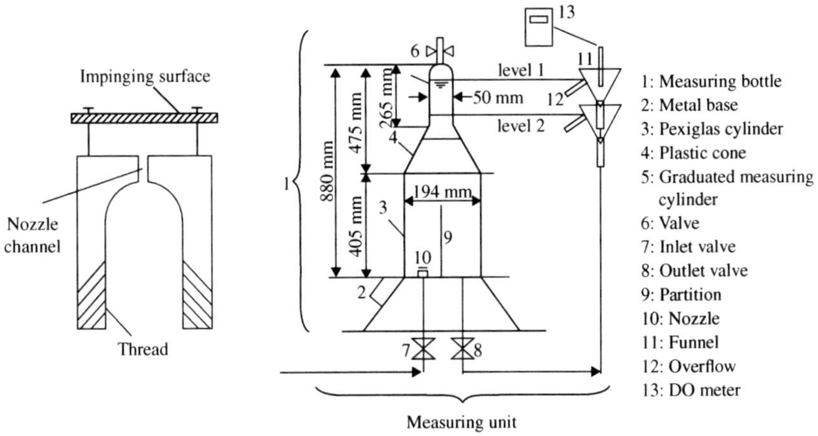

FIGURE 3-14 Details of a batch air measuring vessel, showing the discharge nozzle on the left. For high air precipitation efficiency, the nozzle channel should have a length of 5 mm and diameter of 1 mm, while the impinging surface should be between 5 mm and 10 mm from the channel exit (details from Steinbach, 1998).

The $M_{measured}$ must be compared with the M_{av} predicted from the mass transfer model. It should be slightly greater than M_{av} because of the end effects not accounted for. If it is lower than predicted, a problem is indicated due to blocked packing, faulty instrumentation, or some other cause.

For most practical applications, Eq. (3-25) is adequate. For more precise analysis of the measured data, a detailed mathematical model was developed by Haarhoff and Steinbach

(1997). This model provides a direct measure of the air precipitation efficiency and also takes dissolved oxygen deficits into account. It requires extensive computation with sophisticated software and is not covered here.

EXAMPLE 3-7 Measured Molar Concentration in Saturated Water A batch air measurement vessel is used to evaluate a saturation system at a water temperature of 17°C and an altitude of 400 m above sea level. Over a period of 30 s, an air volume of 94 mL is trapped while 1415 mL of water is collected at the water outlet. Calculate the molar concentration of available air in the saturated water.

Solution The dry atmospheric pressure at 400 m above sea level is 96.6 kPa (Table E-1).

Eq. (3-25): $M_{\text{measured}} = 2.800\ \text{mol/m}^3$ ▲

3-6 AIR FLOW AND ENERGY REQUIREMENTS

The concerns about the air flow rate into the saturation device are different for open- and dead-end systems. For open-end systems, it is of utmost importance to limit the air flow rate below the maximum that could be dissolved into the water, to prevent the presence of undissolved air in the recycle flow toward the saturator. For dead-end systems, it is equally important to guarantee the air flow rate above the maximum that could be dissolved into the water, to ensure that the air cushion within the saturator stays adequately replenished. It is therefore necessary to consider the required air flow rate for open-end saturation at the highest anticipated temperature, where air solubility is lowest. For dead-end saturation, it was shown that the available air is fairly insensitive to water temperature—the minimum air flow rate should be checked at both temperature extremes. The following discussion considers basic open-end saturation, without refinements such as air separation vessels to trap undissolved air and systems that adjust the air flow rate according to temperature.

Maximum Air Flow for Open-End Saturation

Consider a molar balance of gas X across an open-end saturation system. Express the volumetric air flow rate at atmospheric conditions:

$$Q_{air} \times G_{X,atm} + Q_r \times M^*_{X,atm} = Q_r \times M^*_{X,sat} \tag{3-26}$$

With the recycle flow at equilibrium with atmospheric air before and within the saturation device, we apply Henry's law, which leads to:

$$\frac{Q_{air}}{Q_r} = \frac{1}{H_X}\left(\frac{p_{sat} - p_{vap}}{p_{atm}}\right) \tag{3-27}$$

The volumetric flow rate is Q (m^3/s) with the subscripts *air* and r denoting air flow rate and recycle flow rate, respectively. To make sure that no air remains undissolved after the saturation device, the Henry's constant of the least soluble air gas, namely, nitrogen, has to be used.

Equation (3-27) provides the absolute maximum air flow rate at atmospheric conditions and leaves no margin of error in air dosing. To account for kinetic transfer limitations, inaccuracies in air dosing, and small variations in water flow rate, it is prudent to allow some safety margin—a design value of 80 percent of the theoretical maximum is suggested. The maximum air flow should be determined for the highest anticipated water temperature, where Henry's constant is a maximum.

Equation (3-27) provides the actual volumetric air flow rate Q_{air} at atmospheric conditions. To convert the volumetric air flow rate to a standardized free air delivery (FAD) rate, the effects of pressure and temperature have to be accounted for:

$$Q_{standard} = Q_{air}\left(\frac{p_{atm}}{p_{standard}}\right)\left(\frac{273.2 + t_{standard}}{273.2 + t_{sat}}\right) \quad (3\text{-}28)$$

There is no universally accepted definition of the standard air flow rate. Different authorities and manufacturers use different standard pressures, standard temperatures, and even different degrees of humidity. Equation (3-28) allows the expression of the volumetric air flow rate for any definition of "standard" air.

EXAMPLE 3-8 Air Flow Limitation for Open-End Saturation Calculate the required air flow rate (as a ratio of the recycle flow rate) for an open-end system at a saturation gauge pressure of 500 kPa, absolute dry atmospheric pressure of 95.0 kPa, and maximum water temperature of 20°C. To prevent the possibility of undissolved air bubbles in the recycle, restrict the air flow to 80 percent of the maximum. If the recycle flow rate is 40 L/s, express the air flow for standard conditions of 101.3 kPa and 15°C.

Solution The vapor pressure p_{vap} is 2.3 kPa (Table D-1). The Henry's constant for nitrogen is 60.27 (Table F-1).

Calculate Q_{air}/Q_r with the air flow rate at atmospheric conditions, and reduce by 20 percent:

Eq. (3-27): $Q_{air}/Q_r = 0.0695$

For a recycle flow rate of 40 L/s:

$Q_{air} = 0.0695 \times 40 = 2.782$ L/s

The above rate is at saturation pressure and the actual water temperature. At the standard conditions specified in this particular instance:

Eq. (3-28): $Q_{standard} = 2.564$ L/s = 153.8 L/min = 9231 L/h at standard conditions ▲

Minimum Air Flow for Dead-End Saturation

The required air flow rate into a dead-end saturator would be at a maximum if the air transfer was perfectly efficient. The mass transfer model showed that well-designed saturators can come fairly close to perfect efficiency, especially at higher temperatures, and the uncertain end effects can bring it even closer. It is therefore suggested that the minimum air flow rate is determined by assuming that the water leaving the saturator is at equilibrium with the saturator air. In fact, in practice one would specify an air flow rate, say, 20 percent more than equilibrium to ensure that the air cushion is not depleted.

The molar balance across the saturator, presented earlier as Eq. (3-10), provides a convenient expression for estimating the average air flow rate into the saturator, repeated here:

$$\left(\frac{Q_{air}}{Q_r}\right) = \frac{f_{sat,X}}{H_x \times f_{atm,X}}\left(\frac{p_{atm} + p_{sat} - p_{vap}}{p_{atm}}\right) - \frac{1}{H_X}$$

Once the saturator gas fractions are known, Eq. (3-10) is applied to any of the air gases to obtain the ratio between air and water flow into the saturator. Unlike open-end systems, there is no danger of undissolved air in the water leaving the saturator.

EXAMPLE 3-9 Air Flow Requirement for Dead-End Saturation Calculate the required air flow rate for a dead-end system at a saturation gauge pressure of 500 kPa, absolute dry atmospheric pressure of 95.0 kPa, water temperature of 20°C, and recycle water flow rate of 40 L/s. Allow for an air flow rate 20 percent more than what would be dissolved at perfect transfer efficiency.

Solution Use the molar balance for nitrogen to calculate the required air flow rate. (Oxygen or argon could be used just as well.) The Henry's constant for nitrogen is 60.27, oxygen 30.56, and argon 27.66 (Table F-1).

At the pressure conditions and water temperature given, calculate the nitrogen fraction in the saturator air:

Eq. (3-11): $f_{sat,N} = 0.8608$

Calculate the required flow rate ratio with 20 percent surplus for safety:

Eq. (3-10): $Q_{air}/Q_r = 0.1204 = 12.04\%$

Calculate the required air flow rate:

$Q_{air} = 0.1204 \times 40 = 4.81$ L/s at atmospheric conditions. ▲

Energy Requirements

There are two energy components to consider for air-water saturator systems, namely, the energy required to pressurize the air and the energy required to pressurize the water. In comparison, the energy required for air pumping is negligible and ignored. The net energy e imparted to every cubic meter of water pumped through the saturation:

$$e = p_{sat} \tag{3-29}$$

The energy e is expressed per cubic meter (kJ/m^3). Equation (3-29) considers the net pumping head only, which is the head necessary to reach saturation pressure without allowance for pump inefficiency, and frictional and additional static head. The actual energy will always be higher, as the water has to be pumped through some static head due to the hydraulic head at the withdrawal point always being lower than the water surface in the DAF reactor. Also, the frictional and secondary losses in the saturation system pipes need to be added. These additional energy components differ from site to site and are ignored in this section, that only considers the net energy required for saturation. Designers should add these energy components for each specific application.

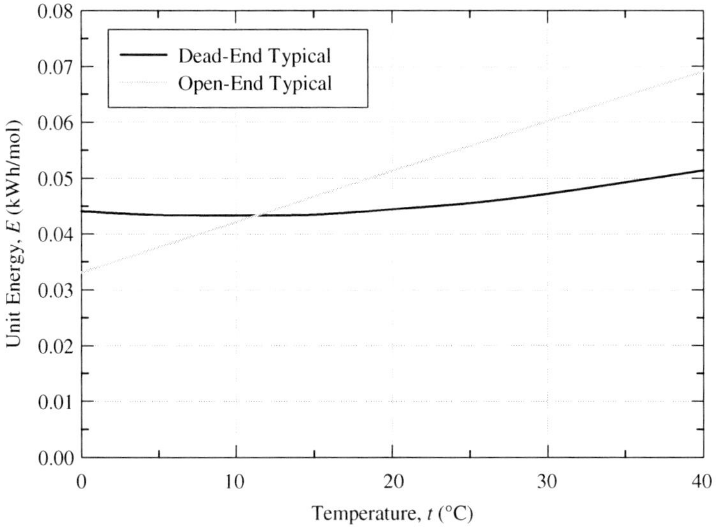

FIGURE 3-15 Comparison of unit energy required for open-end and dead-end saturation (typical dead-end saturator is defined in Table 3-3; the typical open-end system is shown for an air flow rate of 80% of the theoretical maximum).

To get the unit energy consumption E, the energy in Eq. (3-29) has to be divided by the moles of available air produced by the saturation system:

$$E = \frac{e}{M_{av}} \tag{3-30}$$

E is the unit energy consumption (kJ/mol), e is the energy input per volume (kJ/m^3), and M_{av} the available air as before. The energy consumption in kilojoules can be converted to the more customary kilowatt-hour by knowing that 1 kWh = 3600 kJ.

In the case of open-end saturation, the available air is given by:

$$M_{av} = \left(\frac{Q_{air}}{Q_r}\right) 44.6 \left(\frac{273.2}{273.2+t}\right)\left(\frac{p_{atm}}{101.3}\right) \tag{3-31}$$

In the case of dead-end saturation, the available air depends on the transfer efficiency of the saturator, which has to be calculated for every application. As a general guide, the energy requirements for both open-end and dead-end saturation are shown in Fig. 3-15.

EXAMPLE 3-10 Comparison of Energy Requirements Compare the energy requirements for open-end and dead-end saturation at 20°C. For open-end saturation, assume 80 percent of the maximum air flow rate and for the saturator assume the typical saturator described in Table 3-3. The atmospheric pressure is 101.3 kPa and the saturation gauge pressure 500 kPa.

Solution The vapor pressure is 2.3 kPa (Table D-1). H_N is 60.27 mol/mol.

The energy necessary to generate the saturation pressure:

Eq. (3-29): $e = 500$ kJ/m^3

For open-end saturation, calculate the available air concentration:

Eq. (3-27): $Q_{air}/Q_r = 0.0815$

Restrict the air flow to 80 percent of the theoretical maximum: $Q_{air}/Q_r = 0.0815 \times 0.80 = 0.0652$

Eq. (3-31): $M_{av} = 2.71$ mol/m^3

The unit energy consumption is:

Eq. (3-30): $E = 500/2.71 = 184$ kJ/mol $= 0.0512$ kWh/mol

For dead-end saturation, the energy required is the same as above, namely, $e = 500$ kJ/m^3. To get the available air concentration, read it off the middle line of any of Figs. 3-9 to 3-11 for 20°C:

Fig. (3-11): $M_{av} = 3.15$ mol/m^3

The unit energy consumption is:

Eq. (3-30): $E = 500/3.15 = 159$ kJ/mol $= 0.0441$ kWh/mol ▲

For preliminary design and assessment, engineers desire rules of thumb for quick estimation of the capital costs and running costs associated with DAF. It is therefore useful to convert the unit energy consumption E to an estimate for installed pump power in kW (which determines the capital cost) and for energy consumption in kWh/m^3 (which determines the running cost). The conversions are illustrated in Example 3-11.

EXAMPLE 3-11 Different Ways of Expressing Energy Requirements A DAF plant is planned at sea level for a raw water flow rate of 40 ML/d. The air requirement is 8 mL/L (8000 ppm) at a temperature of 10°C, and the saturation pressure is 500 kPa. Estimate the net required pump power and the energy consumed for every cubic meter treated. Use the performance of a typical saturator, as defined in Table 3-3 for preliminary assessment.

Solution At 10°C, there is 0.043 mol/L of air (Table E-1). The required air concentration of 8 mL/L is $0.043 \times 0.008 = 0.000344$ mol/L $= 0.344$ mol/m^3. At the saturation pressure of 500 kPa and 10°C, the actual air transferred by a typical saturator is 3.2 mol/m^3 (read off Fig. 3-8).

Calculate the recycle ratio r for $M_{req} = 0.344$ mol/m^3 and $M_{av} = 3.2$ mol/m^3:

Eq. (3-5): $r = 0.1204 = 12.04$ percent of the raw water flow rate.

From Fig. 3-15, the unit energy E is read off as 0.043 kWh/mol. The required air concentration is 0.344 mol/m^3. The energy per cubic meter of raw water is:

$0.043 \times 0.344 \times (1 + 0.1204) = 0.0166$ kWh/m^3

or $40000 \times 0.0166 = 663$ kWh/day at full production

The installed pumping power for the recycle flow is:

663 kWh/d for 40 ML/d = 16.6 kW-d/ML

or 16.6/24 = 0.691 kW-d/ML.

It is emphasized that this example has the primary aim to demonstrate the energy conversions, and not to generate typical design values. These numbers are for the net energy use (to overcome saturator pressure only), without allowance for static head, and frictional and secondary losses. ▲

3-7 DESIGN EXAMPLE

Problem Make a preliminary design proposal for a DAF recycle system, using a packed saturator and fresh water, at the following conditions:

GIVEN	Elevation	300 m
GIVEN	Water temperature	23°C
GIVEN	Volumetric air requirement	7 mL/L

Step One: Assemble the physical constants There are no design decisions to make in this step. All the values are dependent on the altitude and water temperature. The values can all be directly taken from the tables in the appendices, or calculated with the correlation equations provided in the appendices. The equation numbers for the correlations are provided below. As an alternative, the values can be found in the tables in the like-numbered appendices:

Eq. (D-1)	Water density	997.5 kg/m^3
Eq. (D-1)	Water viscosity	0.000933 kg/m-s
Eq. (D-1)	Surface tension	0.07243 kg/s^2
Eq. (E-2)	Atmospheric pressure	97.8 kPa
Eq. (D-1)	Vapor pressure	2.8 kPa
Eq. (F-3)	Henry's constant (nitrogen)	62.75 mol/mol
Eq. (F-3)	Henry's constant (oxygen)	32.02 mol/mol
Eq. (F-3)	Henry's constant (argon)	28.97 mol/mol
Table F-1	Diffusivity (nitrogen)	1.80E-9 m^2/s
Table F-1	Diffusivity (oxygen)	2.37E-9 m^2/s
Table F-1	Diffusivity (argon)	2.48E-9 m^2/s

Step Two: Conditions at equilibrium with atmospheric air There are no design decisions to make in this step. All the values are dependent on the original input parameters and the values calculated in Step One:

Table 3-1	Nitrogen fraction in atmosphere	0.7808
Table 3-1	Oxygen fraction in atmosphere	0.2095
Table 3-1	Argon fraction in atmosphere	0.0093

Eq. (3-2)	Nitrogen concentration in air	31.00 mol/m³
Eq. (3-2)	Oxygen concentration in air	8.32 mol/m³
Eq. (3-2)	Argon concentration in air	0.37 mol/m³
Eq. (3-6)	Nitrogen concentration in water	0.494 mol/m³
Eq. (3-6)	Oxygen concentration in water	0.260 mol/m³
Eq. (3-6)	Argon concentration in water	0.013 mol/m³
	Total concentration in water	0.767 mol/m³

Step Three: Conditions at equilibrium with saturator air At this point, the saturation gauge pressure has to be selected, one of the most important variables of the design process. Once this is selected in the first line, the rest of Step Three follows automatically:

SELECT	Saturation gauge pressure	500 kPa
Eq. (3-1)	Absolute dry saturator pressure	595 kPa
Eq. (3-12)	Pressure ratio	6.085
Eq. (3-11)	Nitrogen fraction in saturator	0.8599
Eq. (3-13)	Oxygen fraction in saturator	0.1346
Eq. (3-14)	Argon fraction in saturator	0.0056
Eq. (3-2)	Nitrogen concentration in air	207.75 mol/m³
Eq. (3-2)	Oxygen concentration in air	32.52 mol/m³
Eq. (3-2)	Argon concentration in air	1.34 mol/m³
Eq. (3-6)	Nitrogen concentration in water	3.311 mol/m³
Eq. (3-6)	Oxygen concentration in water	1.015 mol/m³
Eq. (3-6)	Argon concentration in water	0.046 mol/m³
	Total concentration in water	4.372 mol/m³

Step Four: Saturator design Step Four requires the selection of three saturator parameters:

SELECT	Mass hydraulic loading	55 kg/m²-s
SELECT	Plastic packing depth	1400 mm
SELECT	Plastic packing size	32 mm
Eq. (3-22)	Specific packing area	186 m²/m³
Eq. (3-21)	Wetted packing area	112 m²/m³
Eq. (3-20)	Nitrogen transfer rate constant	6.24 m/s
Eq. (3-20)	Oxygen transfer rate constant	7.16 m/s
Eq. (3-20)	Argon transfer rate constant	7.32 m/s
Eq. (3-23)	Nitrogen leaving saturator	2.832 mol/m³
Eq. (3-23)	Oxygen leaving saturator	0.917 mol/m³
Eq. (3-23)	Argon leaving saturator	0.042 mol/m³
Eq. (3-24)	Available air in recycle	3.072 mol/m³

Step Five: Recycle rate Step Five requires the specification of the volumetric air concentration of the available air in the contact zone:

SELECT	Required volumetric air concentration	7.0 mL/L
Eq. (3-3)	Required molar air concentration	0.278 mol/m³
Eq. (3-4)	Required mass air concentration	7.95 mg/L
Eq. (3-5)	Required recycle ratio	9.95 percent of raw water flow

By specifying the mass hydraulic loading as an areal unit rate, and the recycle ratio as a percentage of the raw water flow, the raw water flow rate is not required. If the designer wishes to assess the diameter(s) of the saturator and the actual volumetric flow rate of the recycle stream, it can be calculated at this point by using the raw water flow rate.

Step Six: Air flow and energy requirements for dead-end saturation All the design decisions have been made at this point. In Step Six, the air flow and energy consequences of the chosen design parameters are quantified:

Eq. (3-10)	Air flow rate (atmospheric conditions)	9.09 percent of recycle flow rate (perfect transfer)
Eq. (3-29)	Energy consumption	500 kJ/ m³ of recycle flow
Eq. (3-30)	Unit energy consumption	162.8 kJ/mol of available air
	Energy consumption	45.2 kJ/m³ of raw water flow or 0.0126 kWh/m³

REFERENCES

Bratby, J., and Marais, G. (1975), Saturator performance in dissolved-air (pressure) flotation, *Water Research*, 9 (11), 929–936.

Casey, T. J., and Naoum, I. E. (1986), Air saturators for use in dissolved air flotation processes, *Water Supply*, 4, 69–84.

Conway, R. A., Nelson, R. F., and Young, B. A. (1981), High-solubility gas flotation, *Journal WPCF*, 53 (11), 1198–1205.

Degremont (1991) *Water Treatment Handbook*, 6th ed., Volume 2. Paris: Lavoisier Publishing.

Edzwald, J. K. (2010), Dissolved air flotation and me, *Water Research*, 44 (7), 2077–2106.

Gregory, R., and Edzwald, J. K. (2010), "Sedimentation and flotation," in J. K. Edzwald, ed., *Water Quality and Treatment: A Handbook on Drinking Water*, 6th ed., New York: AWWA and McGraw-Hill.

Haarhoff, J., and Cleasby, J. L. (1990), Evaluation of air stripping for the removal of organic drinking-water contaminants, *Water SA*, 16 (1), 13–22.

Haarhoff, J., and Rykaart, E. M. (1995), Rational design of packed saturators, *Water Science and Technology*, 31 (3–4), 179–190.

Haarhoff, J., and Van Vuuren, L. R. J. (1995), Design parameters for dissolved air flotation in South Africa, *Water Science and Technology*, 31 (3–4), 203–212.

Haarhoff, J., and Steinbach, S. (1996), A model for the prediction of the air composition in pressure saturators, *Water Research*, 30 (12), 3074–3082.

Haarhoff, J., and Steinbach, S. (1997), A comprehensive method for measuring the air transfer efficiency of pressure saturators, *Water Research*, 31 (5), 981–990.

Henry, J. G., and Gehr, R. (1981), *Dissolved Air Flotation for Primary and Secondary Clarification*, Sewage Collection and Treatment Report SCAT-9, Canada.

Leininger, K. V., and Wall, D. J. (1974), Available air measurements applied to flotation thickener evaluations, *Deeds and Data, WPCF*, 11, 3–7.

Lovett, D. A., Travers, S. M., and Maas, R. L. (1984), Treatment of Abattoir Wastewater by Dissolved Air Flotation, Part 1: Wastewater not Pretreated. Meat Research Report No 9, CSIRO Meat Research Laboratory, Cannon Hill, Australia.

Onda, K., Takeuchi, H., and Okumoto, Y. (1968), Mass transfer coefficients between gas and liquid phases in packed columns, *Journal of Chemical Engineering in Japan*, 1 (1), 56–62.

Packham, R. F. and Richards, W. N. (1975), *The Determination of Dissolved Air in Water*, Technical Memorandum *TM106*, Water Research Centre, Medmenham, England.

Rees, A. J., Rodman, D. J., and Zabel, T. F. (1980), Evaluation of Dissolved-Air Flotation Saturator Performance. *WRC Technical Report TR143*, Medmenham Laboratory, UK.

Robert, K. L., Weeter, D. W., and Ball, R. O. (1978), Dissolved air flotation performance, *Proceedings 33rd Industrial Waste Conference*, Purdue University, 194–199.

Shannon, W. T., and Buisson, D. H. (1980), Dissolved air flotation in hot water, *Water Research*, 14 (7), 759–765.

Smits, J. (1990), Bepaling van de Versadigingsgraad van Gesatureerd Water bij Flotatie (Determination of the Degree of Saturation of Saturated Water during Flotation). Thesis submitted to the High Technical School, The Hague.

Steinbach, S. (1998), *Dissolution and Precipitation of Air in Dissolved Air Flotation*, D.Ing Thesis, Rand Afrikaans University, Johannesburg, South Africa.

Steinbach, S., and Haarhoff, J. (1997), Assessment of air transfer efficiency in packed saturators used in dissolved air flotation, *Journal of the American Water Works Association,* 89 (12), 71–82.

Valade, M., Nickols, D., Haarhoff, J., Barret, K., and Dunn, H. (2001), Evaluation of the performance of full-scale packed saturators, *Water Science and Technology,* 43 (8), 67–74.

Vosloo, P. B. B., Williams, P. G., and Rademan, R. G. (1986), Pilot and full-scale investigations on the use of combined dissolved-air flotation and filtration (DAFF) for water treatment, *Water Pollution Control*, 85, 114–121.

Zabel, T. F., and Hyde, R. A. (1977), Factors influencing dissolved-air flotation as applied to water clarification, in J. D. Melbourne & T. F. Zabel, eds., *Papers and Proceedings of Water Research Centre Conference on Flotation for Water and Waste Treatment*, Medmenham, UK: Water Research Centre.

CHAPTER 4
AIR PRECIPITATION

4-1 DISTRIBUTION AND ADJUSTMENT OF SATURATOR FLOW 4-1
4-2 BUBBLE SUSPENSIONS 4-2
Measurement of Bubble Size Distribution 4-2
Size Distribution by Ranges 4-3
Size Distribution by Numbers 4-4
Size Distribution by Volume 4-4
Macro-Bubbles 4-5
4-3 BUBBLE FORMATION 4-6
A Conceptual Bubble Formation Model 4-6
Nucleation 4-6
Bubble Growth 4-8
Bubble Coalescence 4-8
4-4 INJECTION NOZZLES 4-9
Number and Spacing 4-9
Nozzle Types 4-9
Nozzle Design Features and Their Effects 4-13
REFERENCES 4-14

4-1 DISTRIBUTION AND ADJUSTMENT OF SATURATOR FLOW

At smaller treatment plants, it is common to provide one central saturator, from where the saturated water is distributed first to each individual DAF tank, and then to the individual injection nozzles within each tank. At larger plants, it is necessary to use more than one saturator to limit the size of individual saturators and the length of the distribution manifolds from the saturators to the DAF tanks. Saturators are usually provided without standby units, which makes the latter option more flexible for saturator shutdown and maintenance. The air compressors and recycle pumps, in both cases, are installed with 50 or 100 percent standby.

The primary design concern with saturator flow distribution is the prevention of premature air precipitation. Bubble formation within the pipe leading to the injection nozzles leads to the release of large bubbles into the DAF tank. Premature air precipitation has two possible causes. First, the saturation point can be located too low in relation to the highest points of the distribution piping, which could happen if the saturators are placed next to the recycle pumps in a deep pumping station. By the time the water reaches the DAF tanks, the reduction in static height can initiate premature air precipitation. Second, the friction and secondary losses in the distribution piping can be large enough to allow air release before the nozzles are reached. It is therefore sound design practice to put all flow meters, valves, and other fittings upstream of the saturator, and to limit the friction and secondary losses as far as possible in the pipes downstream of the saturator.

The concern about premature air precipitation is more important for very efficient saturators. Consider a saturator operating at a typical gauge pressure of 500 kPa or about, say, 50 m of hydraulic head. If the saturator dissolves 94 percent of the maximum available air (readily achievable by packed saturators), it leaves a margin of only 3 m of head for the sum of static height difference, friction, and secondary losses. If a less efficient system would dissolve, say, 80 percent of the maximum available air, the margin would increase to 10 m.

The pumping energy for the saturator flow is the major operational DAF expense. As the flow rate is variable at almost all treatment plants, there is a cost incentive to reduce the saturator flow to the minimum that will still provide good DAF separation. Different strategies are used. An obvious option is to control the saturator flow rate, which is readily and efficiently achieved with variable-speed drives on the recycle pumps. Unless special self-adjusting nozzles are used (discussed in more detail in Sec. 4-4), the usual fixed-orifice injection nozzles cause the saturation pressure to drop at the same time, which may be undesirable if the properties of the air suspension are affected. Another option is to provide more than one manifold of injection nozzles in each DAF tank, which allows step-wise adjustment—if the flow drops below a certain level, a complete manifold is shut down. Up to three manifolds per DAF tank, varying in their flow proportions, are used in practice.

4-2 BUBBLE SUSPENSIONS

Measurement of Bubble Size Distribution

The milky bubble suspensions (*white* water) in DAF have incredibly high bubble concentrations. Even after the saturator flow is blended with the raw water flow, the concentration is about 10^{10} to 10^{11} bubbles per cubic meter. The counting and measuring of a representative number of bubbles at these concentrations pose experimental problems that have not been entirely resolved.

The earliest approach was to capture enough visual images of the bubble suspension to allow the bubbles to be individually measured and counted. The principal problem with this method is visual penetration of a dense suspension. Bubbles close to the observation window can be readily imaged, but the wall effects are uncertain. Focusing the plane of observation farther away from the observation window leads to the problem of partially obscured imaging. The studies that used side observation windows employed two methods to improve the resolution of the method. In the first method, the bubble suspension could be illuminated by a strong, thin sheet of light running parallel to the window, which amplified the clarity of the bubbles within the illuminated sheet (Moruzzi and Reali, 2007). The other option was to divert the bubbles near the wall and only allow the bubbles in the center to continue rising into the observation cell, thereby permitting much deeper visual penetration into the cell (Rykaart and Haarhoff, 1995).

The problem with nonintrusive measurement techniques, of not seeing far enough into the bubble suspension, is overcome by inserting a borescope into the bubble suspension (Jönsson and Lundh, 2000). The system works best if the suspension is first transferred to a laboratory beaker where it can be strongly illuminated. Direct use of the borescope in a flotation tank is less successful because of the low light level. A borescope is a thin observation tube with the imaging optics embedded at its tip, capable of focusing a few millimeters from the tip. From the perspective of a small bubble, it represents an enormously large obstacle in the flow path. This raises a concern about the possible disruption of the natural flow pattern around the borescope, which may deflect the smaller bubbles selectively from the field of view.

More recently, the measurement of bubbles by conventional particle counters has been compared with the results obtained by visual imaging. In the absence of flocs, the methods gave comparable results with average number-based sizes within 10 percent of each other. It was noted that on-line measurement of bubbles in this way could be done 300 times faster than with computer-based image analysis (Han et al., 2002a).

Size Distribution by Ranges

Most DAF literature cites the bubble size distribution as a range between minimum and maximum sizes. The sizes quoted cover a large range from about 10 to 150 μm. A summary by Edzwald (2010) is presented as Table 4-1. Most bubbles in the contact zone have sizes of 40 to 80 μm. There is some growth in the sizes of bubbles as they rise from the bottom of the tank to near the top, due to less hydrostatic pressure, but the growth is small. Bubble size measurements made at full scale show that bubble sizes grow as the suspension moves from the contact zone into the separation zone (Leppinen and Dalziel, 2004), presumably due to bubble coalescence. Although bubbles of about 40 to 80 μm were found over the depth of the contact zone, the bubbles in the separation zone were mostly in the range of 50 to 150 μm. Based on these findings, the typical size for contact zone modelling is taken as 60 μm (Chap. 7) and the typical size for separation zone modelling is taken as 100 μm (Chap. 8).

TABLE 4-1 Bubble Sizes Reported from Different Flotation Studies

Bubble sizes (μm)	Conditions or effects	Reference
10–120	Needle valve: most bubbles 40–90 μm WRC nozzle: most bubbles 20–50 μm	Zabel (1984)
33–75 (median sizes)	Hague nozzle; larger bubbles at a pressure of 350 kPa compared to 500 and 620 kPa	De Rijk et al. (1994)
200 kPa pressure: 82 and 22 (mean and standard deviation)	Wide variety of nozzles studied; Percentages of large bubbles (>150 μm) were 7.7% for 200 kPa and 3.4% for 500 kPa	Study by Rykaart reported in Haarhoff and Edzwald (2004)
500 kPa pressure: 62 and 22 (mean and standard deviation)		
15–85	Mean sizes of ~30 μm for pressures of 350–608 kPa Increasing size for pressures <350 kPa	Han et al. (2002b)
Albert Plant 70–84 (median: contact zone) 72–145 (median: separation zone)	Full-scale plants observed bubble clusters (large group of bubbles attached to a floc particle)	Leppinen and Dalziel (2004)
Graincliffe Plant 40–60 (contact zone) 50–150 (separation zone)		

Source: Edzwald (2010), Dissolved air flotation and me, Water Research, 44 (7), 2077–2106, with permission from Elsevier, copyright 2010.

Size Distribution by Numbers

Rather than just stating a range, a more precise description of the bubble size distribution is given by the frequency or cumulative distribution functions, expressed in terms of bubble numbers. After an adequate number of bubbles have been measured, the results are summarized in a graph from which the median or any percentile bubble size can be obtained. Fig. 4-1 shows a typical graph of this kind.

Size Distribution by Volume

The shape of bubbles in water is determined by the rise rate of the bubbles. Provided that the bubble size is below about 500 μm, as found in DAF applications, the bubbles retain their shape as rigid spheres as they rise through water. Large bubbles with sizes of about 1 to 10 mm, typically of use in dispersed air flotation applications, deform as they rise and take the shape of ellipsoids. Very larger bubbles with sizes greater than 10 mm have the shape of spherical caps (Clift et al., 1978). Bubbles in DAF therefore are spherical and this shape is used to predict bubble volumes, to model contact zone performance in Chap. 7, and to predict bubble rise velocities in Chap. 8.

The frequency distribution function in terms of numbers can therefore be converted to volume by treating the bubbles as spheres. When expressed as volumes, the cumulative distribution function is pushed toward larger bubble sizes, as demonstrated in Fig. 4-1. The frequency distribution by volume is useful for detecting excessive bubble coalescence. The larger bubbles are fewer in number and may not show up on a distribution diagram based on numbers. Once converted to volume, a bimodal distribution may be evident because of bubble coalescence, as shown for a typical bubble suspension in Fig. 4-2.

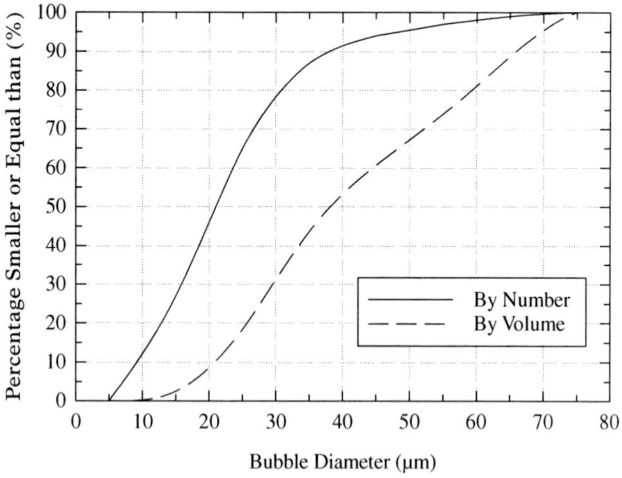

FIGURE 4-1 Cumulative distribution functions for the same bubble suspension, shown in terms of bubble numbers and bubble volumes (*Data for a typical DAF bubble suspension, taken from Han et al., 2002a*).

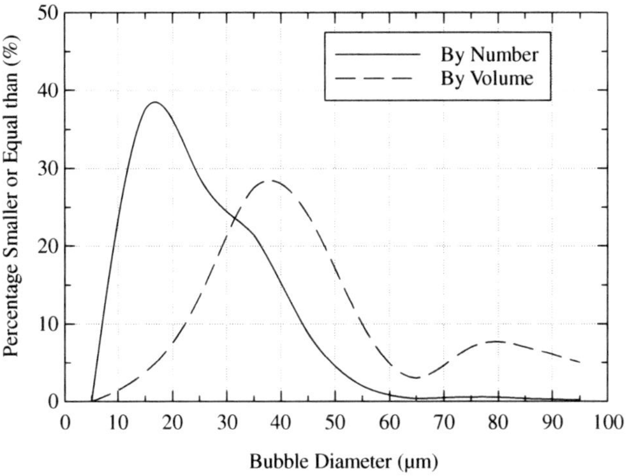

FIGURE 4-2 Frequency distribution functions for the same bubble suspension, shown in terms of bubble numbers and bubble volumes (*Data for a typical DAF bubble suspension, taken from Moruzzi and Reali, 2007*).

Macro-Bubbles

A real bubble suspension, whether viewed in the laboratory or at full scale, contains occasional bubbles that are much larger than the rest. These bubbles are so much larger and rise so fast that they are clearly not part of the distributions presented in the literature. They either escape detection by the measurement method used or are deliberately ignored or not reported. This is unfortunate, as macro-bubbles can be seen to break the surface directly above the injection point, often so violently that no float layer can accumulate in these areas.

A laboratory study using nozzles in the form of thin tubes of glass showed that the bubbles smaller than 120 μm accounted for only 52 percent of the total air volume (Ponasse et al., 1998). These values were measured in 16 experiments at a distance of 80 mm from the end of the constriction. The study by Rykaart (1994) provided a more detailed measure of the macro-bubbles. The size of a macro-bubble was arbitrarily classified as all bubbles with diameter larger than 150 μm. More than 16,000 bubbles were measured during 85 experiments, done with numerous injection nozzles at two different saturation pressures. For those experiments conducted at a saturator gauge pressure of 500 kPa, the macro-bubbles (by number) were 3.4 percent of the bubbles counted. For experiments conducted at a lower saturator gauge pressure of 200 kPa, the macro-bubble percentage increased to 7.7 percent. While these macro-bubble percentages may seem low, it should be recognized that one macro-bubble of 200 μm among 100 bubbles of 60 μm represents only 0.5 percent of the numbers, but 27 percent of the total air volume.

After accounting for the presence of macro-bubbles, only a reduced fraction of the precipitated air, which is useful for DAF, remains. If the cut-off size of useful air is taken as 150 μm, the data provided by Rykaart (1994) indicate that useful air varies from as low as 50 percent to a maximum of 90 percent of the total available air. The effective avoidance of macro-bubbles should thus be a high priority when evaluating and choosing injection nozzles.

The direct measurement of bubble sizes in two full-scale DAF reactors confirmed the undesirability of macro-bubbles (Leppinen and Dalziel, 2004). In this case, macro-bubbles were defined as bubbles larger than 1000 μm—much larger than the cut-off size of 150 μm in the previous paragraph. In a tank using needle valves as air injection devices, 148 macro-bubbles were observed in 2000 images, which contained enough air to form 4 million bubbles of 100 μm each.

4-3 BUBBLE FORMATION

A Conceptual Bubble Formation Model

The formation of air bubbles during pressure release is conceptually simplified to three consecutive steps (Rykaart and Haarhoff, 1995). In the first step, numerous nucleation centers are induced by the turbulence in the water as the energy of the saturation pressure is released. During the second step, the available air precipitates evenly onto these nucleation centers until all the available air in the water is depleted. After the second step, there should be an even distribution of similarly sized air bubbles. During the third step, the air bubbles coalesce or break up to eventually yield a bubble suspension with much larger variation in size. The steps are schematically shown in Fig. 4-3 and sequentially discussed in the next sections.

An alternative bubble growth model is the mass precipitation model, that assumes the sudden formation of large gas pockets after pressure release, which then burst into a bubble cloud because of the turbulence in the water (Dupre et al., 1998). This model was formulated after cinematographic observation of air precipitation in thin tubes of glass that were conditioned to be either hydrophilic or hydrophobic. No further development or validation of the mass precipitation theory has since come to the attention of the authors.

Nucleation

In the absence of nucleation sites on solid surfaces, bubbles have to form homogenously on nucleation sites in the liquid phase (Bennett, 1988). These nucleation sites are likely to be the centers of liquid eddies that form as a result of the turbulent depressurization within the nozzle. The pressure is lower in the centers of these eddies than in the surrounding water and is determined by the intensity of the eddies. The efficiency of bubble formation, therefore, depends on the intensity of turbulence created at the point of depressurization (Maddock, 1976). Theoretical and experimental studies confirm that the number of bubble nuclei increases with increasing flow velocity, and decreases with increasing viscosity, nozzle diameter, and surface tension (Schmidt and Morfopoulos, 1983).

Water quality concerns prohibit the manipulation of the surface tension of drinking water by chemical addition, as widely practiced in the mineral beneficiation industry. The only operational variable that remains for controlling the number of nucleation sites, once a specific nozzle had been decided on, is the saturation pressure. Higher pressure pushes the water through the nozzle at higher velocity, leading to more nucleation sites. In terms of the conceptual bubble formation model, more nucleation sites must translate into smaller bubbles. This hypothesis has been confirmed by most studies (e.g., Treille, 1972; Takahashi et al., 1979). The study by Rykaart (1994) found an average bubble size (by number) of 62 μm in a range of experiments conducted at saturator gauge pressure of 500 kPa. When the same experiments were repeated at saturator gauge pressure of 200 kPa, the average bubble size increased to 82 μm.

	t_0	This is the time immediately after pressure release. No air has precipitated yet, but nucleation centres are present upon which air could precipitate.
	t_{100}	This is the time immediately after all the excess air has precipitated out of solution. During the time of air precipitation, the nucleation centers steadily grow and the number of bubbles stay constant. No bubble coalescence has occurred yet.
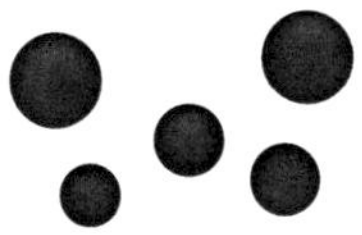	$> t_{100}$	After t_{100}, no more air is precipitated, but bubbles grow in size and diminish in number due to bubble coalescence.

FIGURE 4-3 A conceptual, simplified bubble formation model with subscript referring to the percentage of available air released (*Source*: *Rykaart and Haarhoff, 1995*).

Is there a limit to bubble size manipulation by means of saturation pressure? A tentative answer is provided by the work of Han et al. (2002b), where the saturator gauge pressure was gradually increased from 200 to 600 kPa, while tracking the bubble size with both image analysis and particle counting. From 200 to 350 kPa the bubble size did decrease, but from 350 kPa upward the bubble size stayed the same. This finding suggests 350 kPa as a lower design limit for saturator gauge pressure.

A question remains about the effect of suspended solid particles on bubble formation. If particles are present in the nozzle flow path, they serve as additional nucleation sites and the bubble growth process may be different. The limited experimental evidence is inconclusive (Maddock, 1976; O'Connor et al., 1990). For drinking water applications, however, the question is irrelevant, as the bubble formation process is complete before the nozzle flow is exposed to raw water, as is shown in the next section. Where the presence of particles did affect the behavior of the suspensions in the DAF tank (Haarhoff and Edzwald, 2004), it has to be attributed to reasons other than their effect on bubble formation.

Bubble Growth

How fast does the available air come out of solution to form the primary bubbles at the nucleation centers? No direct measurements have been reported in the DAF literature, but Kitchener and Gochin (1981) observed that bubble growth occurs very rapidly after pressure release and estimated that it happens in less than 1 s. After a series of experiments with different nozzle geometries (Rykaart and Haarhoff, 1995), primary bubble growth time was indirectly inferred to be essentially complete within 1.7 ms of the start of pressure release. The flow velocity through the primary orifice of an injection nozzle, under typical DAF conditions, is of the order of 20 to 30 m/s. In a period of 1.7 ms, at this velocity, the water covers a distance of about 35 to 50 mm, which is less or about equal to the dimensions of a typical nozzle. The primary bubble formation is thus practically complete by the time the water leaves the injection nozzle.

After the bubbles leave the injection nozzle, the bubble size will increase because of the decrease in hydrostatic pressure as the bubbles rise, and the coalescence of the air bubbles. The reduction in pressure as bubbles rise to the surface has a small effect on bubble size. For a typical nozzle depth of 2 m, the pressure at the surface is about 17 percent less than that at the bottom, which translates into a 6 percent increase in bubble diameter.

The air precipitation efficiency is touched on in Sec. 3-5, where the measurement of saturation efficiency is discussed. The incomplete release of the available air is of obvious importance for air injection nozzles as it has an effect on bubble size, as well as the wasteful cost of dissolving air into water, which does not result in useful bubbles for DAF. Incomplete air precipitation was considered in a number of studies, all of which confirmed that 100 percent precipitation could not be achieved (Bratby and Marais, 1974; Maddock, 1976; Takahashi et al., 1979; Lixin and Fan, 1994; De Rijk et al., 1994). The air precipitation efficiency was therefore introduced by Steinbach and Haarhoff (1997) as an additional yardstick for evaluating injection nozzles, as shown in Sec. 4-4 of this chapter.

Bubble Coalescence

The coalescence of bubbles is a complex phenomenon that depends on fluid dynamics and interfacial chemistry. Interfacial chemistry, attraction, and repulsion are covered in Chap. 5. In this section, a brief consideration of the fluid mechanics follows.

Two types of bubble coalescence are distinguished. Heterogeneous bubble growth is characterized by small bubbles colliding and coalescing with large bubbles and is typically found in regions of low turbulence. With more turbulence, homogeneous bubble growth is found when bubbles of approximately equal size collide and coalesce. Bubble suspensions in DAF are believed to be formed by homogeneous bubble growth (Lixin and Fan, 1994). Bubble coalescence is favored either by violent agitation (as found in the narrow confines within a nozzle) or by nonuniform bubble diameters (due to the large bubbles overtaking smaller bubbles as they rise to the surface) (Ramirez, 1979). The differential rise rate may be a contributing mechanism whereby macro-bubbles are formed—once a bubble becomes large enough, its increasing rise rate makes it even more effective in consuming small bubbles until it reaches the tank surface as an unusually large macro-bubble.

Although the conceptual bubble growth model distinguishes between two consecutive modes of bubble formation (growth and coalescence), it is evident that coalescence starts as soon as the first bubbles appear. There is an overlap in time between bubble growth and bubble coalescence. Some coalescence is thus inevitable as the dense suspension of bubbles is exposed to the turbulent conditions in a restricted flow channel. Whereas the bubble suspension may have been fairly homogenous at first, bubble coalescence stretches the suspension to include more large bubbles than before.

Craig et al. (1993) reported that, for bubbles larger than those found in DAF, at low salt concentrations there was bubble coalescence, while at high salt concentrations there was little or no coalescence. For most drinking water applications, salt concentrations are low (ionic strength <0.02 M). The authors are unaware of any such studies on the effect of salt concentration or ionic strength for the small bubble sizes found in DAF. Chapter 13 covers the application of DAF in pretreatment for seawater desalination.

4-4 INJECTION NOZZLES

Number and Spacing

The number of injection nozzles is a compromise between having as many as possible (to ensure rapid, even blending of the saturator flow with the raw water flow in the relatively short time afforded by the contact zone) and as few as possible (to minimize the nozzle cost). Most injection nozzles, once developed and tested, are used unchanged in different installations by only changing the number of nozzles. There is, therefore, no clear or consistent pattern in the number and spacing of the injection nozzles. The number of nozzles relative to the raw water flow seems to be a rational approach to determine the minimum number. A survey of treatment plants that use DAF successfully, excluding those using needle valves (full details in Chap. 11), indicates a range of 2.2 to 3.6 nozzles for every ML/d day (0.6 to 1 for every MGD) of raw water flow. There is no technical constraint on the maximum number, other than their cost. For a small nozzle design, used successfully at several South African treatment plants, up to 12 nozzles per ML/d are common (Ceronio, 2009). In terms of physical spacing, the survey, excluding needle valves, indicates an average spacing of 30 to 110 mm along the distribution manifold, which is substantially less than a minimum spacing of 250 to 500 mm suggested earlier (Stevenson, 1997).

In which direction should the nozzles be pointed? Given that the objective is to blend the recycle water quickly and evenly, it would intuitively seem better if the flow from the nozzle would run directly counter or perhaps at right angles to the raw water flow (analogous to good design practice for chemical dosing). In one of the few studies where this was investigated, such countercurrent blending was indeed found to be superior to cocurrent blending in terms of flotation efficiency (Van Puffelen et al., 1995). This seems to be a minor advantage, as some full-scale DAF plants use cocurrent blending with success.

Nozzle Types

Nozzles with fixed orifices provide a simple, robust option. The nozzles are designed and tested before installation, installed, and then left alone without any further operator attention.

There are no connections to the nozzles other than to the recycle manifold. Numerous variations of fixed-orifice nozzles are reported. One example is shown in Fig. 1-4b; more detailed sketches are shown in Fig. 4-4, and more examples are in Fig. 4-5.

Nozzles with fixed orifices suffer from two potential disadvantages:

- The flow passages are narrow and the orifices are normally small, increasing the potential for clogging. In drinking water applications, where the saturator flow is recycled from the filtered water supply, this is not a concern as nozzle blockage under these conditions is practically unknown. For other applications, where it may be opportune to recycle the saturator flow from the DAF tank outlet or elsewhere, blocking becomes a concern. A solution to the blocking of a fixed-orifice nozzle is offered by a ball valve modification (Takko, 2000). The orifice is positioned such that it is exposed when the valve is closed, as shown in Fig. 4-6a. By manually opening the valve, the blockage is flushed away.
- The second disadvantage of fixed-orifice nozzles is their limited range of flow variation. Water demand variations are inevitable and treatment plants have to be designed to make flow changes with minimal effort and disruption. Due to the square root relationship between flow and pressure, the saturator gauge pressure can be maximally varied from, say, 350 to 650 kPa, then consequently the practical range of flow variation is at most about 35 percent. To overcome this problem, designers specify two or even three manifolds in each DAF tank, which requires the operators to take manifolds in or out of service as the required flow rate goes up or down.

Some of these disadvantages are overcome with nozzles having adjustable orifices. The easiest way to achieve this is by adapting normal valves. The needle valve is used in many successful applications. By opening or closing the valve, the flow can be adjusted while

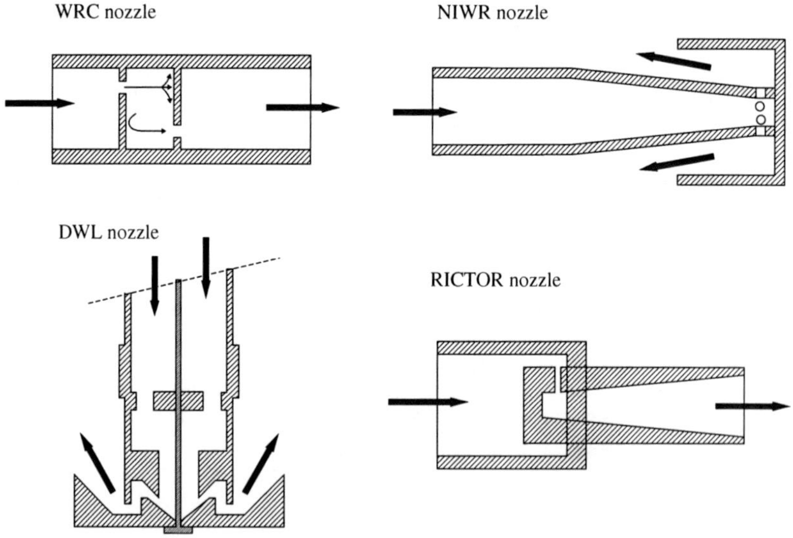

FIGURE 4-4 Cross-sections of typical air injection nozzles; bottom left nozzle is adjustable; the other three are fixed-orifice nozzles (*Source*: *Rykaart and Haarhoff, 1995*).

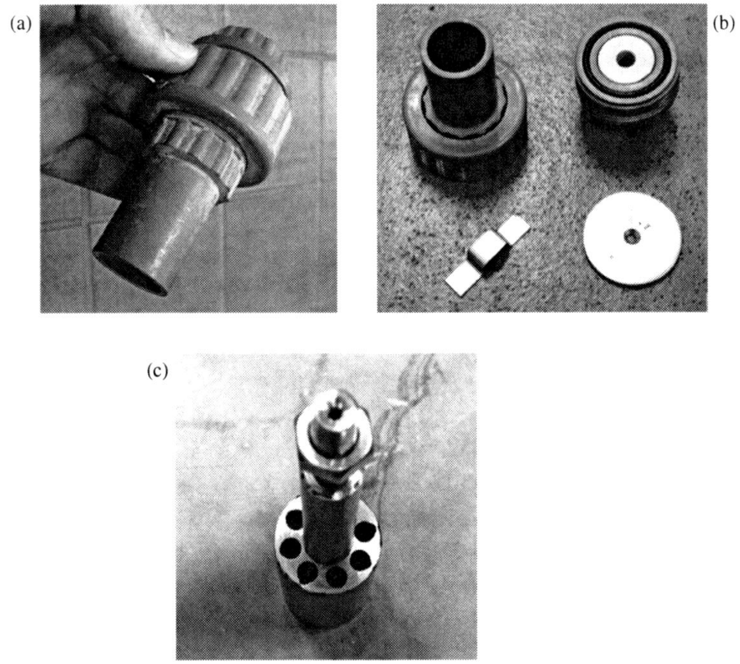

FIGURE 4-5 Examples of full-scale DAF injection nozzles with (a) an assembled nozzle, (b) the same nozzle taken apart to show the inlet piece top right, the machined air orifice bottom right, the impinging surface bottom left, and the outlet shroud top left, and (c) another nozzle viewed from the saturator end.

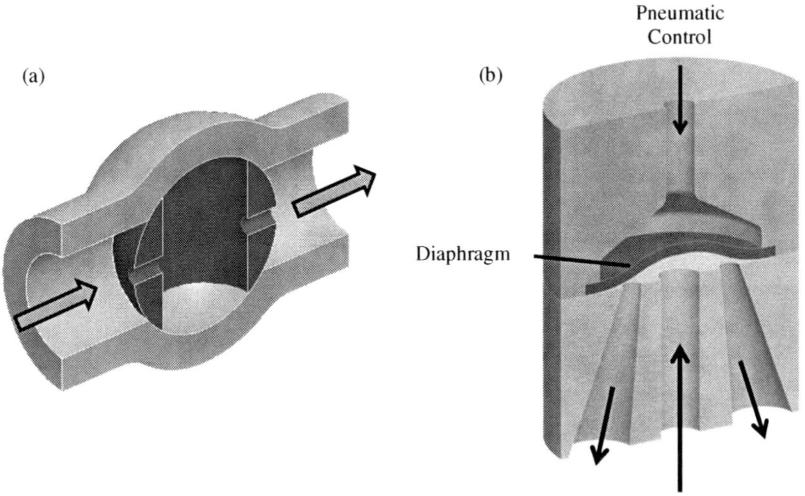

FIGURE 4-6 Cross-sections of nozzles that prevent permanent blockages: (a) a modified ball valve, cleaned by periodic opening of the valve (after Takko, 2000), and (b) a pneumatically controlled nozzle (*based on a sketch from Cromphout and Vanderbroucke, 2000*).

maintaining the desired saturation pressure. When a blockage occurs, the valve is opened for a few seconds and readjusted. Needle valves also have disadvantages:

- The adjustment of individual valves, whenever flow rates change or when the very narrow slit between the valve housing and needle is blocked, increases the burden on operators. After a decade of using both adjustable and nonadjustable nozzles, a large water provider switched to nonadjustable nozzles to avoid the daily adjustment of the nozzles; no difference in efficiency was observed (Kempeneers et al., 2000).
- A second disadvantage is the need to provide a means of adjusting a valve immersed near the bottom of a DAF tank. One option is to have extension spindles protruding above the water level. Another option is to introduce the flocculated water from the side of the DAF reactor, as shown in Fig. 4-7. By putting the distribution manifold and needle valves immediately outside the tank, the valves can be adjusted in the dry. Both options are cumbersome and unworkable at large treatment plants where hundreds of injection points may be required.
- A final disadvantage is the tendency of needle valves to form macro-bubbles when they become partially blocked. In a study where needle valves were compared with fixed-orifice nozzles at full scale, the number of macro-bubbles formed by needle valves was seven times greater (Leppinen and Dalziel, 2004).

Adjustments to some nozzles are made pneumatically (Cromphout and Vanderbroucke, 2000). The orifice is formed by a gap between a solid part of the nozzle body on one side and a flexible rubber diaphragm on the other—a diagram is shown in Fig. 4-6b. By

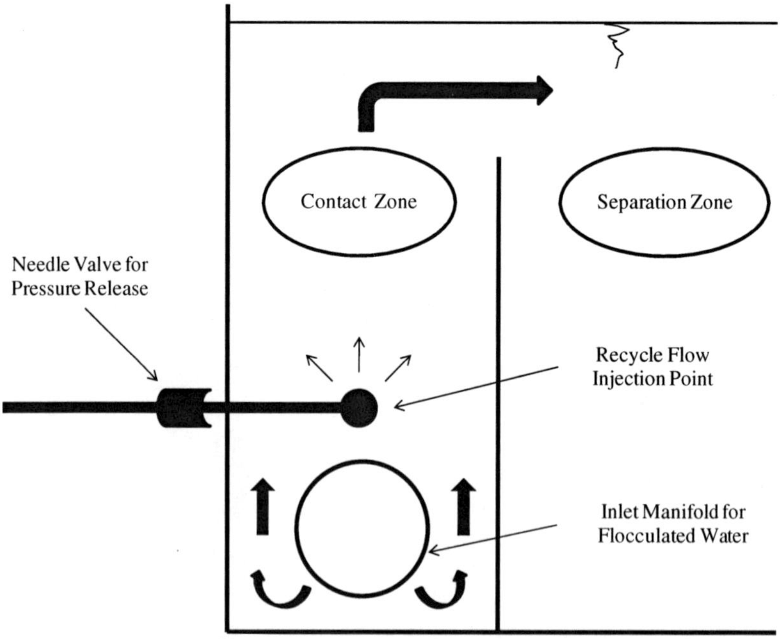

FIGURE 4-7 A practical arrangement for the inlet to the contact zone to allow needle valve operation from outside the DAF reactor.

increasing the pneumatic pressure on the opposite "dry" side of the diaphragm, the diaphragm is forced toward the nozzle body to decrease the orifice opening. The saturation pressure can thus be adjusted by pneumatic control. Also, once the nozzle is set at a specific saturation pressure, the pressure stays the same despite flow rate changes. One setting of the pneumatic pressure, distributed to a number of nozzles on the same manifold, ensures equal distribution among all the nozzles.

A further variation of the pneumatically controlled nozzle, shown in Fig. 4-6b, mainly used for industrial purposes, is to periodically release the pneumatic back-pressure for a few seconds, preset with a timer (Järvenpää, 2000). These sudden, short periods of pressure release allow a sudden large flush through each nozzle, thereby removing any blockages.

Nozzle Design Features and Their Effects

Close inspection of the many available air injection nozzles shows that all nozzle designs share one or more of three objectives, namely to eliminate dead volumes within the nozzle, to gradually reduce the velocity as the flow is guided out of the nozzle, and to provide an impinging surface somewhere in the flow path.

The elimination of dead volumes is important to reduce bubble coalescence. The bubble concentration is extremely large within the nozzle and any unnecessary holdup provides more opportunity for bubbles to collide and coalesce.

It is equally important to reduce the water velocity before it leaves the nozzle. It was pointed out in the earlier section that the velocity through the primary pressure orifice is between 20 and 30 m/s. If this powerful jet is directed onto fragile flocs, floc breakup can be expected (Rees et al., 1980; Williams and Van Vuuren, 1984). Also, a jet directly injected into stagnant water forms small but strong eddies at the interface between the jet and the stagnant water. These eddies are agents that cause coalescence of the bubbles at the interface. It is common therefore to find the flow guided by tapered tubes or bubble shrouds after the pressure has been released. In a relatively new nozzle, two shrouds are used in series—a primary shroud, closely followed by a secondary shroud (Zhang et al., 2009). Ideal nozzle-design guidelines therefore include suggestions on the detention time between pressure release and the end of the nozzle flow path, as well as the maximum exit velocity from the nozzle. These numbers are not readily available, especially for the many proprietary designs in use. A preliminary check of a few treatment plants, where this could be determined, indicated an exit velocity of 0.1 to 0.6 m/s and a detention time of 0.05 to 0.7 s.

There are different ways to incorporate an impinging surface in the nozzle flow path. A surface can be placed in the direct flow path of the jet issuing from the primary pressure reduction orifice. A sudden change in flow direction would serve the same purpose, as would two jets emerging from orifices directly opposite each other (Stevenson, 1997). How far should the impinging surface be away from where the jet leaves the primary pressure reduction orifice? This question was probed experimentally by measuring the bubble sizes as an impinging surface was moved toward a jet issuing from a straight micro-channel (Rykaart and Haarhoff, 1995). At a typical saturation gauge pressure of 500 kPa, the median bubble size by number stayed unchanged from a distance of 10 mm and more. When the distance was decreased from 10 to 5 mm, the median bubble size decreased sharply from about 60 to 40 μm. The most important effect, however, was on the macro-bubble fraction. With no surface, the macro-bubble fraction was 9 percent. With the surface at 20 mm, the fraction was 8 percent. From here it steadily decreased to reach 0 percent at a distance of 5 mm. Based on these observations, the impinging surface should be less than 10 mm and preferably close to 5 mm for the smallest median bubbles and the fewest macro-bubbles.

The conceptual bubble growth model presented earlier may explain the benefit of an impinging surface. The jet smashing into the surface has enough energy to break bubbles apart. The larger bubbles are more prone to breakage due to the reduced effect of surface tension. If the screen is too far away, two things happen. First, there is more opportunity for bubble coalescence (evidenced by the larger median bubble size), and second, there may not be enough energy to effectively break the larger bubbles (evidenced by the larger macro-bubble fraction). By positioning the surface at 5 mm, both these effects seem to be eliminated.

The different nozzle design features were also evaluated in terms of their air precipitation efficiency (Steinbach, 1998). As in the case of the nozzles tested for the air measurement device (Sec. 3-5), the maximum air precipitation efficiency that could be achieved was 95 percent. The most important effect on the air precipitation efficiency was the presence of an impinging surface or bend in the flow path. Without a surface or a bend, the air precipitation efficiency dropped by about 4 percent. The distance of the impinging surface, provided it was 20 mm or less away from where the jet was released, did not matter. Other variables such as the length of the bubble shroud and the length and diameter of the primary pressure release channel did not show an appreciable effect on the air precipitation efficiency.

The air that is not immediately released because of incomplete air precipitation eventually does precipitate in the DAF tank. Two phases of air release are distinguished: the first "quick" release is practically instantaneous in a fraction of a second, and a second "slow" release occurs over tens of minutes (Steinbach, 1998). DAF is driven by the immediate availability of bubbles in the contact zone and the slow release step may therefore be disregarded for practical design. The supersaturated air carried forward to the subsequent filtration step does not lead to noticeable air binding in the media bed. The second author has measured dissolved oxygen concentrations slightly above saturation on numerous occasions at the outlet of DAF reactors, without encountering air problems.

REFERENCES

Bennett, G. F. (1988), The removal of oil from wastewater by air flotation: a review, *CRC Critical Reviews in Environmental Control,* 18 (3), 189–253.

Bratby, J., and Marais, G. v. R. (1974), Dissolved air flotation, *Filtration and Separation,* 11 (6), 614–621.

Ceronio, A. D. (2009), Personal communication to J.Haarhoff, September 10.

Clift, R., Grace, J. R., and Weber, M. E. (1978), *Bubbles, Drops, and Particles*, New York: Academic Press.

Craig, V. S. J., Ninham, B. W., and Pashley, R. M. (1993), Effect of electrolytes on bubble coalesence, *Nature*, 364 (22 July), 317–319.

Cromphout, J., and Vanderbroucke, S. (2000), Design of an automatically adjustable flotation nozzle, *Fourth International Conference on Flotation in Water and Waste Water Treatment*, Helsinki, 11–14 September 2000.

De Rijk, S. E., Van der Graaf, J. H. J. M., and Den Blanken, J. G. (1994), Bubble size in flotation, *Water Research*, 28 (2), 465–473.

Dupre, V., Ponasse, Y., Aurelle, Y., and Secq, A. (1998), Bubble formation by water release in nozzles. I, *Water Research*, 32 (8), 2491–2497.

Edzwald, J. K. (2010), Dissolved air flotation and me, *Water Research*, 44 (7), 2077–2106.

Haarhoff, J., and Edzwald, J. K. (2004), Dissolved air flotation modelling: insights and shortcomings. *Journal of Water Supply: Research and Technology – Aqua,* 53 (3), 127–150.

Han, M. Y., Park, Y. H., and Yu, T. J. (2002a), Development of a new method of measuring bubble size, *Water Science and Technology: Water Supply,* 2 (2), 77–83.

Han, M., Park, Y., Lee, J., and Shim, J. (2002b), Effect of pressure on bubble size in dissolved air flotation, *Water Science and Technology: Water Supply*, 2 (5–6), 41–46.

Järvenpää, V. (2000), Flotation with extremely small micro-bubbles, *Fourth International Conference on Flotation in Water and Waste Water Treatment*, Helsinki, 11–14 September 2000.

Jönsson, L., and Lundh, M. (2000), *Visualization of Particles and Bubbles in a Flotation Basin Using Borescope*. Report 3229, Department of Water Resources Engineering, University of Lund, Sweden.

Kempeneers, S., Van Menxel, F., and Gille, L. (2000), A decade of large scale experience in dissolved air flotation, *Fourth International Conference on Flotation in Water and Waste Water Treatment*, Helsinki, 11–14 September 2000.

Kitchener, J. A., and Gochin, R. J. (1981), The mechanism of dissolved air flotation for potable water: basic analysis and a proposal, *Water Research*, 15 (5), 585–590.

Leppinen, D. M., and Dalziel, S. B. (2004), Bubble size distribution in dissolved air flotation tanks. *Journal of Water Supply: Research and Technology – Aqua*, 53 (8), 531–543.

Lixin, W., and Fan, O. (1994), Hydrodynamic characteristics of the process of depressurization of saturated water, *Chinese Journal of Chemical Engineering*, 2 (4), 211–218.

Maddock, J. E. L. (1976), Research experience in the thickening of activated sludge by dissolved air flotation, *Water Research Centre Conference on Flotation for Water and Wastewater Treatment*, Felixstowe, June 1976.

Moruzzi, R. B., and Reali, M. A. P. (2007), Characterization of micro-bubbles sizes distribution in DAF contact zone by a non-intrusive image analysis system, *Fifth International Conference on Flotation in Water and Wastewater Systems*, Seoul, 11–14 September 2007.

O'Connor, C. T., Randall, E. W., and Goodall, C. M. (1990), Measurement of the effects of physical and chemical variables on bubble size, *International Journal of Mineral Processing*, 28, 139–149.

Ponasse, M., Dupre, V., Aurelle, Y., and Secq, A. (1998), Bubble formation by water release in nozzle. II, *Water Research*, 32 (8), 2498–2506.

Ramirez, E. R. (1979), Comparative physiochemical study of industrial waste water treatment by electrolytic, dispersed and dissolved air flotation technologies, *Proceedings of the 34th Industrial Waste Conference*, Purdue University, 1979.

Rees, A. J., Rodman, D. J., and Zabel, T. F. (1980), Dissolved air flotation for solid/liquid separation, *Journal of Separation Process Technology*, 1 (3), 19–23.

Rykaart, E. M. (1994), *Die Verfyning van Inspuitnossels vir Opgeloste-Lugflottasie (The Optimization of Injection Nozzles for Dissolved Air Flotation)*, M.Ing. dissertation, Rand Afrikaans University Johannesburg, South Africa.

Rykaart, E. M., and Haarhoff, J. (1995), Behaviour of air injection nozzles in dissolved air flotation, *Water Science and Technology*, 31, 25–35.

Schmidt, L., and Morfopoulos, V. (1983), Bubble formation in the dissolved-air flotation process, *Spring National Meeting and Petro Expo '83 of the American Institute of Chemical Engineering*, Houston, 1983.

Stevenson, D. G. (1997), *Water Treatment Unit Processes*, London: Imperial College Press.

Steinbach, S., and Haarhoff, J. (1997), Air precipitation efficiency and its effect on the measurement of saturator efficiency, Dissolved Air Flotation, *Proceedings of an International Conference*, London, April 1997, The Chartered Institution of Water and Environmental Management, 35–49.

Steinbach, S. (1998), *Dissolution and Precipitation of Air in Dissolved Air Flotation*, D.Ing. thesis, Rand Afrikaans University, Johannesburg, South Africa.

Takahashi, T., Miyahara, T., and Mochizuki, H. (1979), Fundamental study of bubble formation in dissolved air pressure flotation, *Journal of Chemical Engineering of Japan*, 12 (4), 275–280.

Takko, P. (2000), Clogging problems of dispersed water nozzles and results with non-clogging nozzle type, *Fourth International Conference on Flotation in Water and Waste Water Treatment*, Helsinki, 11–14 September 2000.

Treille, P. (1972), Pressure flotation of abattoir wastewater using carbon dioxide, *Water Research*, 19 (12), 1479–1482.

Van Puffelen, J., Buijs, P. J., Nuhn, P. N. A. M., and Hijnen, W. A. M. (1995), Dissolved air flotation in potable water treatment: the Dutch experience, *Water Science and Technology*, 31 (3–4), 149–157.

Williams, P. G., and Van Vuuren, L. R. J. (1984), *Windhoek Water Reclamation Plant Process Design for Dissolved Air Flotation*, Pretoria: National Institute for Water Research, 1–20.

Zabel, T. (1984), Flotation in water treatment, in K. J. Ives, ed., *The Scientific Basis of Flotation*, Boston: Martinus Nijhoff.

Zhang, Y., Leppinen, D. M., and Dalziel, S. B. (2009), A new nozzle for dissolved air flotation, *Water Science and Technology: Water Supply*, 9 (6), 611–617.

CHAPTER 5

AIR BUBBLES AND PARTICLES IN WATER

5-1 AIR BUBBLES IN WATER 5-1
Bubble Volume and Number Concentrations 5-2
Bubble-Bubble Interactions and Forces 5-5
5-2 PARTICLES IN WATER 5-6
Sources and Types 5-7
Particle Sizes 5-9
Particle Concentration Measurements 5-10
Particle Stability 5-12
Particle-Particle Interactions and Forces 5-14
5-3 PARTICLE-BUBBLE INTERACTIONS AND FORCES 5-15
DLVO Forces 5-16
Non-DLVO Forces 5-19
Utility of Theories Regarding DLVO and Non-DLVO Interactions 5-23
REFERENCES 5-25

This chapter provides fundamental material and concepts dealing with air bubbles and particles in water. It serves as a foundation chapter for subsequent chapters, and it also supplements fundamental material on air bubbles presented in Chap. 4.

The chapter has several goals. Basic material about bubble concentrations is first presented. The interaction forces between bubbles in water are then considered. These forces affect whether bubbles remain stable or aggregate or coalesce. Summary material is presented on the types and sizes of particles found in water. Particle stability is then considered, which addresses the interaction forces between particles. Finally, the interaction forces between particles and bubbles are presented and discussed.

5-1 AIR BUBBLES IN WATER

Bubble shape is covered in Chap. 4 where it is discussed that the bubbles in DAF processes occur as spheres, and this shape is used: to predict bubble number concentrations (below), to model contact zone performance (Chap. 7), and to predict bubble rise velocities (Chap. 8). Bubble formation, nucleation, growth, coalescence, and the sizes of bubbles found in DAF are subjects also presented in Chap 4. Bubble size is an important parameter for the topics of this chapter. From Chap. 4 we learn that for the contact zone, bubble sizes in the range of 10 to 150 μm can occur with most bubbles at 40 to 80 μm.

Bubble Volume and Number Concentrations

It is common to design for providing mass concentrations of air bubbles (C_b) in the contact zone of 6 to 11 g/m^3 (Gregory and Edzwald, 2010). While many engineers design the saturation system in terms of the mass concentration of air bubbles, we show that the bubble volume concentration is the fundamental variable determining the contact zone efficiency (Chap. 7) and separation zone efficiency (Chap. 8). Guidelines of 7000 to 9000 ppm are presented in Chap. 3 for air requirements as bubble volume concentrations, or also called the volumetric air requirement.

The bubble volume concentration (Φ_b) is calculated from the bubble mass concentration and the moist-air bubble density or simply, bubble density (ρ_b) according to Eq. (5-1). Moist-air densities are used for conditions of 100 percent humidity and with the dew point equal to the water temperature. Bubble densities are tabulated in Table E-4. For example, values for ρ_b at 4 and 20°C are 1.27 and 1.19 kg/m^3, respectively, compared to water densities of 999.98 and 998.21 kg/m^3 for 4 and 20°C, respectively—see Table D-1.

$$\Phi_b = \frac{C_b}{\rho_b} \tag{5-1}$$

The bubble number concentration (n_b) depends on the bubble volume concentration and the volume per bubble, which for spheres is simply $\pi(d_b)^3/6$, where d_b is the bubble diameter.

$$n_b = \frac{\Phi_b}{\left[\pi(d_b)^3/6\right]} \tag{5-2}$$

Figure 5-1 presents bubble volume and number concentrations as a function of mass concentration for two water temperatures. The bubble number concentrations were determined assuming d_b of 60 μm. Bubble sizes in the contact zone are mostly in the range of 40 to 80 μm, as stated above. Thus, an assigned value of 60 μm is reasonable for the purpose of finding approximate bubble number concentrations as a function of mass concentrations. Smaller bubbles yield large numbers while larger bubbles give smaller bubble concentrations. This is a good reason why DAF systems are designed and operated under conditions that produce small bubbles.

What we learn from Fig. 5-1 follows. First, we find slightly greater bubble volume concentrations at higher water temperature due to the lower air density with increasing temperature. Since the bubble volume concentrations increase slightly, the bubble number concentrations also increase slightly with increasing temperature. Second, if we consider bubble mass concentrations in the design range of 6 to 11 g/m^3 (see above), then we find approximate bubble volume concentrations in the range of 4600 to 9200 ppm, which means 4600 to 9200 cubic meters of bubbles per 10^6 m^3 of water being treated. While designers have used mass concentration, in this book in Chaps. 3 and 7 we show that bubble volume concentration is more meaningful. We recommend bubble volume concentrations for design of 7000 to 9000 ppm. These bubble volume concentrations yield bubble number concentrations in the range of 6×10^{10} to 8×10^{10} bubbles per m^3 of water.

It is instructive to compare bubble concentrations to floc particle concentrations in drinking water plants. Since we tend to describe floc particle number concentrations on a mL basis, the bubble number concentrations are converted to mL—6×10^4 to 8×10^4 bubbles per mL. To remove floc particles by collision and attachment to air bubbles in

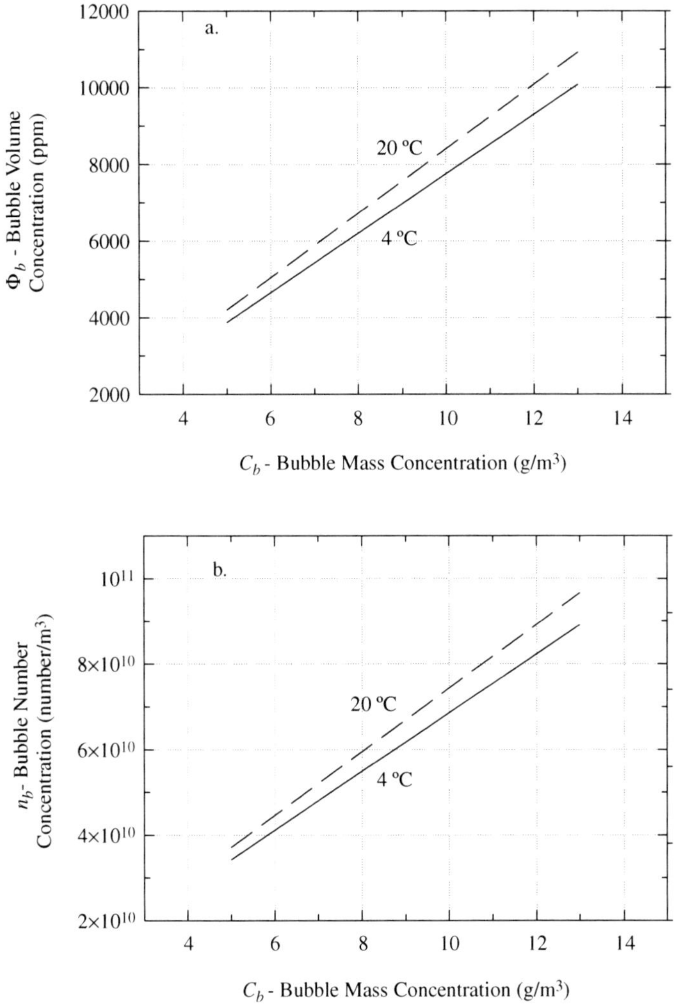

FIGURE 5-1 Bubble volume (a) and number concentrations (b) as a function of bubble mass concentrations for two water temperatures (Condition: bubble diameter of 60 μm).

the contact zone forming floc-bubble aggregates, which rise to the water surface in the separation zone, at least one air bubble must be provided per floc particle. The ratio of bubble to floc numbers is usually much greater than one, thus ensuring high opportunities for collision and attachment of bubbles to particles. The bubble volume concentrations when compared to floc particle volume concentrations yield insights into the ability of air bubbles to reduce the floc-bubble aggregate density below water density so that the aggregates can rise rather than settle. In DAF water plants treating most water types, floc densities are low, nearly the density of water, so a value of 1100 kg/m^3 is assumed. Floc particle mass concentrations are generally less than 100 mg/L. Using the mass concentration of

100 mg/L (high end) and density of 1100 kg/m³ gives a floc particle volume concentration of 91 ppm. Using the above bubble volume concentrations of 7000 to 9000 ppm gives the ratio of bubble volume to floc-particle volume in the range of about 75 to 100. These high ratios ensure adequate bubble volume to reduce the density of the floc-bubble aggregates to less than the water density, and so the aggregates can rise to the surface in the separation zone. Higher ratios of bubble to floc volumes are found for many waters with lower flocculated particle concentrations, as described in the following example.

EXAMPLE 5-1 Comparison of Bubble to Floc Particle Concentrations A DAF plant operating with a mass concentration of air bubbles in the contact zone of 9 g/m³. The flocculated water enters the contact zone of the flotation tank with a floc particle concentration ($n_{p,i}$) of 1.5 × 10³ particles/mL and a floc volume concentration (Φ_p) of 25 ppm. Calculate the bubble volume concentration (Φ_b) and number concentration (n_b) in the contact zone and compare the concentrations of bubbles to floc particles. For these calculations use a water temperature of 20°C.

Solution

Bubble volume concentration Φ_b is calculated from Eq. (5-1). The bubble density (ρ_b) is found in Table E-4 for 20°C and is 1.19 kg/m³.

$$\Phi_b = \frac{C_b}{\rho_b} = \frac{9.0 \text{ mg/L}}{1.19 \text{ kg/m}^3} = 7.56\left(\frac{\text{mg}}{\text{kg}}\right)\left(\frac{\text{m}^3}{\text{L}}\right) = 7.56\left(\frac{\text{mg}}{\text{kg}} \times \frac{\text{kg}}{10^6\text{mg}}\right)\left(\frac{\text{m}^3}{\text{L}} \times \frac{10^3\text{L}}{\text{m}^3}\right)$$

$$\Phi_b = \frac{7560 \text{ m}^3 \text{of air}}{10^6 \text{ m}^3 \text{ of water}} = 7560 \text{ parts per million (ppm)}$$

Bubble number concentration n_b is calculated from Eq. (5-2) using a mean bubble diameter of 60 μm (60 × 10⁻⁶ m)—a typical mean value.

$$n_b = \frac{\Phi_b}{(\pi d_b^3/6)} = \left[\frac{6(7560 \times 10^{-6})}{\pi(60 \times 10^{-6})^3}\right] = 6.7 \times 10^{10}\left(\frac{\text{bubbles}}{\text{m}^3}\right) \text{ or}$$

$$6.7 \times 10^{10}\left(\frac{\text{bubbles}}{\text{m}^3}\right) \times \left(\frac{\text{m}^3}{10^6 \text{ mL}}\right) = 6.7 \times 10^4\left(\frac{\text{bubbles}}{\text{mL}}\right)$$

Ratios of concentrations

$$\frac{n_b}{n_{p,i}} = \frac{6.7 \times 10^4}{1.5 \times 10^4}\left(\frac{\text{bubbles}}{\text{mL}}\right)\left(\frac{\text{mL}}{\text{particles}}\right) = 4.5 \frac{\text{bubbles}}{\text{particle}}$$

$$\frac{\Phi_b}{\Phi_p} = \frac{7560 \text{ ppm}}{25 \text{ ppm}} = 302 \frac{\text{bubble volume}}{\text{particle volume}}$$

This example shows that there are more bubbles than floc particles, which is an essential condition for collisions and attachment of bubbles to particles in the contact zone. Further, the high ratio of bubble- to floc-particle volume ensures that there is sufficient bubble volume to reduce the densities of the floc-bubble aggregates to less than that of water, which is essential for establishing high rise velocities for the separation zone. ▲

Bubble-Bubble Interactions and Forces

Bubble formation is discussed extensively in Chap. 4. After bubbles are formed they rise in water subject to the forces acting on them in which the buoyant force exceeds that of gravity and drag. These forces govern bubble rise velocity, as presented in Chap. 8.

There are also intermolecular and bubble-bubble interaction forces operative over bubble separation distances of 1 to 100 nm that affect bubble stability. Bubble stability is defined as a bubble suspension with little or no tendency for bubbles to aggregate or coalesce and form larger bubbles. As discussed in Chap. 4, there is coalescence of bubbles during their formation downstream of nozzles producing the bubble sizes commonly found in the contact zone of mainly 40 to 80 μm. Bubble-bubble forces then affect bubble stability and restrict further extensive growth and coalescence in the contact zone. The reader should be aware that the forces between bubbles have not been a research subject dealt with by water engineers and scientists interested in flotation, so knowledge of this subject is limited and found mainly in other fields. What is presented next is based on material from a review paper by Edzwald (2010), and it is supplemented with concepts from the important book, *Intermolecular & Surface Forces*, by Israelachvili (1991).

There are four forces or bubble-bubble interactions that affect bubble behavior in terms of coalescence (these forces also affect bubble-particle interaction, which is addressed later). The forces are London-van der Waals (written hereafter simply as van der Waals), electrostatic, hydrophobic, and hydrodynamic. In this section, we describe these forces conceptually to aid our understanding of bubble stability. In Sec. 5-3 particle-bubble forces are discussed, and we develop equations describing these forces.

We begin with van der Waals interaction forces and note that for solid particle-particle interactions, which most people are familiar with, there are a number of intermolecular forces that can cause the van der Waals interaction. The van der Waals force between solid particles may be due to (1) permanent dipole forces, (2) induced dipole forces, and (3) instantaneous induced dipole forces (sometimes called the London dispersion forces). For air bubbles, the gases (mainly N_2 and O_2) are non-polar molecules so the molecular interaction has a London-dispersion origin, which is the weakest intermolecular force. This produces an attractive van der Waals force between air bubbles in water (Israelachvili, 1991; Craig et al., 1993).

Air bubbles in water, without the addition of chemical coagulants, have a negative charge, that is, negative zeta potentials are measured. This causes electrostatic repulsive forces between bubbles. The negative zeta potentials are sometimes attributed to the accumulation of negatively charged surfactants or aquatic humic substances that concentrate at the bubble-water interface. However, even in the absence of surface active agents negative zeta potentials are reported. Since air is non-polar and thus the bubble surface does not contain ionizable functional groups, it is hypothesized that the negative zeta potentials are caused by smaller anions that reside at the bubble-water interface at a greater concentration than larger hydrated cations—for example, deionized water in equilibrium with air contains primarily the ions of HCO_3^- (size of 4 Å or 0.4 nm) and H^+, actually H_3O^+ (size of 0.9 nm). Han and Dockko (1999) and Dockko and Han (2004) have measured zeta potentials for hydrogen bubbles in distilled water and varied the pH. They found that the bubbles had an IEP (isoelectric point, pH of no net charge) at pH < 3, and negative zeta potentials of about −25 mV over the pH range of 6 to 8. Ducker et al. (1994) report a negative zeta potential of −15 mV for large air bubbles (0.5 mm) and state that the zeta potential would be greater (more negative) for smaller bubbles. We can conclude then that air bubbles in water without coagulant addition have electrostatic repulsive forces because of the negative zeta potentials of the bubbles. For negative zeta potentials of −25 mV and more negative, this repulsive force would exceed the van der Waals attractive forces for bubble-bubble separation distances greater than 1 nm and would prevent bubble aggregation.

The charge at bubble surfaces can be changed and made positive. This is common practice in mineral froth flotation, involving addition of cationic surfactants or polyelectrolytes. Malley (1995) demonstrated this for DAF with the direct application of cationic polymers to the saturator recycle water producing positively charged bubbles. More recently, Henderson et al. (2008) added cationic surfactants to the DAF saturator water to produce positively charged bubbles for flotation of algae. Han et al. (2006) have shown that positively charged hydrogen bubbles are produced in the presence of solids precipitated by Al and Mg hydroxide.

A long-range hydrophobic force that is attractive affects bubble-bubble interaction, and has been observed and measured with a range of 150 nm (Israelachvili and Pashley, 1982; Ducker et al., 1994); however, the origin of this force is not understood.

There is a structural force that is repulsive in nature—that is, it inhibits bubble coalescence and bubble-floc attachment. It is the water between bubbles that must be displaced for two bubbles to coalesce or for attachment of a particle to a bubble. In the colloid and filtration fields, it is often referred to as hydrodynamic repulsion or retardation. This force is important at bubble-bubble separation distances of less than 10 nm (Craig et al., 1993). This force depends on bubble sizes, and it ought to be more significant for larger bubbles.

5-2 PARTICLES IN WATER

The purpose of this section is to cover briefly particle types, sizes, concentration measures, and stability of particle suspensions. These subjects are covered in greater detail in various water treatment books: MWH (2005), Gregory (2006), and Edzwald and Tobiason (2010). DAF deals with the removal of particles from water, so it is essential to have a fundamental understanding of particles and particle properties for consideration of pretreatment processes of coagulation and flocculation (see Chap. 6) and for consideration of the DAF process.

Before presenting material about particles in water, it is useful to deal with the terminology of hydrophobic and hydrophilic particles. These terms are often misleading and misunderstood unless defined. The terms are used in two subject areas, and therefore their meaning is related to these subjects. Both subject areas are of interest and relevant to flotation. In the field of colloids, hydrophobic particles are insoluble particles that are dispersed in water. Hydrophilic colloids are defined as water-soluble macromolecules such as synthetic polymers used as coagulants and flocculants (these polymers are discussed in Chap. 6) and natural macromolecules dissolved in water such as some fractions of natural organic matter (NOM), which we are interested in removing in water treatment and we cover in several chapters. The terms hydrophobic and hydrophilic are also used in fields dealing with the contact angle between water and a solid. Solids with large contact angles are hydrophobic (e.g., for Teflon® the contact angle is >90°) while those with low contact angles have a high degree of wettability. (More on contact angles for bubbles in water in contact with solid particles is presented in the Sec. 5-3.)

Small molecules containing a charge are polar and favor the aqueous phase. Larger molecules that are not charged (non-polar) or those that have a structure with both hydrophobic (non-polar groups) and hydrophilic groups (e.g., surfactants, and polymers used as flocculant, flotation, and filter aids) can partition at particle or air bubble interfaces in which the hydrophobic groups orient to the air or particle and the hydrophilic groups orient toward the water.

Sources and Types

We now turn to discussion of the sources and the types of particles in water supplies and plants. Table 5-1 gives a summary that is used in the following presentation. It is important to recognize that particles that enter the flotation process may come not only from the raw water supply, but also from the addition and production of particles within the treatment plant prior to flotation. For our purposes then we will describe three sources: (1) particles from the water supply or raw water source, (2) particles added in DAF pretreatment (e.g., powdered activated carbon [PAC]), and (3) particles produced from precipitation reactions (e.g., Al and Fe hydroxides) in coagulation during DAF pretreatment.

Rivers often carry inorganic mineral particles such as clays and fine sands. These particles are washed into rivers from soil erosion during storm events. The concentrations vary depending on the watershed characteristics (geology, topography, watershed size, amount of urban and agricultural use, and amount of forested and non-forested land), season and

TABLE 5-1 Sources and Types of Particles

Category	Type/Description/Examples	Sources
Inorganic		
Clays	Aluminosilicates (montmorillonite, kaolinite, illite)	Raw water
	Asbestos fibers	
Metal oxides, hydroxides	$Al(OH)_3$ (amorphous) precipitate	Flocs from coagulant addition
	$Fe(OH)_3$ (amorphous) precipitate	Flocs from coagulant addition Fe(II) oxidation
	MnO_x (e.g., MnO_2) precipitate	Mn(II) oxidation
	SiO_2 (mineral)	Raw water
Carbonates	$CaCO_3$ precipitate (calcite)	Raw water
	Ca,$MgCO_3$ precipitate (dolomite)	Raw water
Organic		
Adsorbent	Powdered activated carbon	Added during treatment
Microorganisms	Pathogens (viruses, bacteria, protozoa)	Raw water
	Algae	Raw water
Detritus	Debris from plants and animals	Raw water
NOM	Macromolecular, colloidal NOM	Raw water
	NOM adsorbed to particle surfaces	Raw water, Flocs from coagulant addition

Source: Chapter 3 by Edzwald and Tobiason (2010), reprinted with permission from Water Quality and Treatment: A Handbook on Drinking Water, 6th ed., Copyright 2010, AWWA.

rainfall events, and the flow velocity in the rivers affecting whether the particles remain suspended. Small impoundments of rivers and small lakes can also carry mineral particles but at lower concentrations than rivers. Lakes and reservoirs allow for settling of clay and silt mineral particles so the concentrations of these are much lower. Hard water lakes with moderate algal activity can increase the pH during the day, causing precipitation of calcium carbonate particles.

Cyanobacteria (called blue-green algae in some literature; examples are *Anabaena sp.* and *Microcystis aeruginosa*) and *Actinomycetes* (bacteria) can occur in rivers and lakes, and they are an important cause of tastes and odors, so removal of these particles by DAF and filtration is essential. Algae occur in all lakes; their concentrations vary with season and trophic status. High nutrient or eutrophic supplies contain greater concentrations than low nutrient, oligotrophic supplies. A good summary of the types of algae found in lakes and their effects on water quality is presented in *Water Quality and Treatment: A Handbook on Drinking Water* (see Chap. 3, Edzwald and Tobiason, 2010).

Another class of particulate organic matter is pathogens including viruses, bacteria, and protozoa. Disinfection processes are effective in the inactivation of bacteria, but they are less effective on viruses and protozoa that form cysts and oocysts such as *Giardia* and *Cryptosporidium.* Consequently, particle removal processes together with disinfection are essential in drinking water treatment. The concentrations of pathogens are much lower in raw waters than other particles (addressed in the following text) but because of health effects they must be removed. A problem that is emerging for some water supplies is the possible presence of nanoparticles from manufacturing processes using nanomaterials. These particles may be organic or inorganic in nature.

In addition to particles occurring in the raw water supply, particles may be added during treatment. A good example of this (see Table 5-1) is the addition of PAC, which adsorbs small concentrations of organic compounds; for example, taste and odor compounds or pesticides. This addition of PAC particles is often practiced seasonally so water plant engineers and operators need to account for the effects of these particles on downstream processes such as coagulation and DAF clarification and make operational adjustments.

Water plants produce particles by oxidation of reduced metals and subsequent precipitation, and by the addition of metal coagulants that undergo a phase change and precipitate as metal hydroxide particles (see Table 5-1). For water supplies taken from the bottom of lakes where low or no oxygen conditions occur, reduced Fe and Mn can exist at fairly high soluble concentrations (e.g., 0.1 to 1 mg/L or greater). It is common to oxidize Fe and Mn, which results in precipitation of these metals as $Fe(OH)_3(s)$ and $MnO_x(s)$. The manganese oxide particles can pass through particle separation processes, if coagulation is not adjusted to account for these particles. In other words, knowledge of the size and stability of these particles is important so coagulation can be adjusted to incorporate these particles into other flocculated particles.

The most common way particles are produced in water plants is through the use of metal coagulants. Aluminum and ferric coagulants are commonly used at dosages and pH conditions where they are insoluble, leading to the production of particles in the form of amorphous aluminum hydroxide or ferric hydroxide particles. We call this *sweep floc* coagulation, which is described in more detail in Chap. 6. The flocs produced contain a mixture of particles from the raw water and from precipitated aluminum hydroxide or ferric hydroxide particles. If PAC were added or reduced Fe and Mn oxidized and undergo precipitation of solids, then the flocs would also contain these particles. Almost all water supplies contain dissolved NOM. Coagulation is used to remove NOM by adsorption on the metal hydroxide particles formed by coagulation or by coprecipitation of the NOM in the form of metal-NOM particles. This results in flocs containing inorganic and organic matter. The properties and coagulation of NOM are addressed in Chap 6. Water supplies contain particulate organic matter, especially algae in lake supplies. These organic particles are coagulated and incorporated into flocs.

In summary, the particles that compose flocs reflect those in the water supply, particles added or produced from oxidation-precipitation reactions, the dissolved NOM in the supply that is coagulated, and the metal coagulant used in coagulation.

Particle Sizes

Particle size is an important parameter for several reasons. It affects the light scattering of particle suspensions and thus affects the turbidity measurement—see the following section. It affects the rate of particle aggregation (flocculation), which is discussed in Chap. 6. Particle size also affects collisions and attachment to bubbles in the contact zone (Chap. 7), and particle removal in the separation zone (Chap. 8). It also affects other particle removal processes including sedimentation, granular media filtration, and membrane filtration.

Figure 5-2 shows that particles in water supplies can range in size of over five to six orders of magnitude. We can have the smallest colloidal or nanoparticles at a few nm, to viruses of several to 10s of nm, to bacteria in the order of a μm, to algae from about a μm to 10s of μm. *Cryptosporidium* oocysts are about 4 to 6 μm, and *Giardia* cysts are about 8 to 15 μm. Algae, depending upon type and species, can span from small sizes near a μm to 10s of μm, as shown in Fig. 5-2 (some algae are even larger than 100 μm). Inorganic-type particles occurring in water supplies are not specifically labeled in the figure, but we can state: clay minerals span from roughly 10 nm to several μm; silts are 10s of μm; and sand-type particles are 100 μm and larger. Various metal oxide and carbonate particles can also occur with sizes in the colloidal range.

The size of PAC used in drinking water treatment varies but is mainly <50 μm. It is essential to understand that when a metal precipitates in water treatment (e.g., from the addition of metal coagulant or from the oxidation and precipitation of metals), the metal undergoes a phase change starting with particle nucleation (nm sizes), particle growth by diffusion (submicron to micron sizes), and then flocculation into sizes of tens of μm and larger. In other words, we span a large part of the size range in Fig. 5-2.

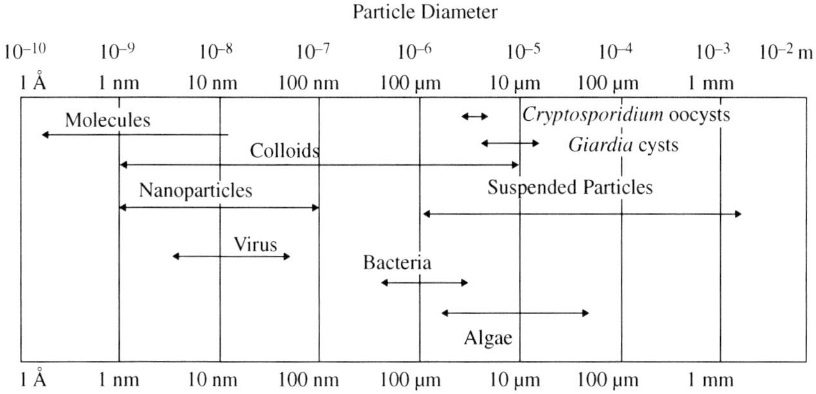

FIGURE 5-2 Particle size ranges for classes of particles in water (*Source: Chapter 3 by Edzwald and Tobiason (2010), reprinted with permission from Water Quality and Treatment: A Handbook on Drinking Water, 6th ed., Copyright 2010, AWWA*).

Particle Concentration Measurements

There are several types of particle measurements used in water treatment. These include mass concentrations or suspended solids (SS), turbidity as a surrogate measurement, and number concentrations. These, as well as particle volume concentration, are presented in the following sections.

Suspended Solids. The mass concentration (dry mass/unit water volume) of particles in water is measured and reported as SS or sometimes as total suspended solids (TSS). It should be noted that SS is an operationally defined measurement since it measures the particles retained on a glass fiber filter with a pore size of about 1 μm. So by definition, as depicted in Fig. 5-2, SS measurements refer to the mass concentrations of particles with sizes greater than about 1 μm, and does not include colloidal particles less than about 1 μm. SS measurements are used in evaluating solids loading for the treatment and handling of residuals (sludges). River water supplies can have SS of 10s to 100s mg/L, while lakes generally have lower SS values, <10 mg/L. Treated waters following DAF and filtration have low SS concentrations (<1 mg/L) that cannot be measured accurately and rapidly, and so it is not used in the routine design and operation of particle clarification and filtration processes.

Turbidity. Turbidity is a common measurement in the design and operation of treatment processes, in monitoring particles across a water plant, and in the finished water. It is a regulated parameter for most countries. It is common to find filtered-water standards <1 NTU with many countries having standards and goals in the range of 0.1 to 0.5 NTU. It is a surrogate parameter because it is not a direct measurement of particle concentration. It can be measured rapidly (online monitors) and inexpensively, and it has a long history of use in the water industry. The reader is referred to Chapter 3 of *Water Quality and Treatment: A Handbook on Drinking Water* (Edzwald and Tobiason, 2010) for a detailed discussion of the turbidity measurement. The material presented in the following text summarizes the main concepts and points from this book.

It is necessary to understand that turbidity is a surrogate measure of the presence of particles in water and what affects the measurement. Turbidity is the amount of visible light (400 to 700 nm wavelength) scattered at a 90-degree angle to the incident light source. Turbidity is commonly reported in nephelometric turbidity units (NTU) because 90-degree scattered light refers to measurements with nephelometers. In the water field we simply call the instrument a turbidimeter. The intensity of scattered light in NTU is based on a standard (but arbitrary) concentration of a primary reference material (formazin), which is used to calibrate a turbidimeter. In some countries turbidity is reported as FTU (formazin turbidity unit) rather than in units of NTU.

The turbidity of a sample does not provide information about fundamental characteristics of particles because the scattered light is the aggregate response for all particles in a suspension, which is a function of the number of particles, as well as size, refractive index, and shape. The particle size, in particular, greatly affects the measured turbidity. Particles with sizes similar to the turbidimeter wavelength of 400 to 700 nm (0.4 to 0.7 μm) scatter more light than smaller and larger particles. Thus, raw waters can have fairly low turbidities if they contain a large number of small particles <0.1 μm, or a large number of algae with sizes of several μm. Changes in particle sizes through a plant affect turbidity. For example, flocculation increases particle sizes causing a decrease in turbidity without any particle removal.

Some discussion follows on raw water turbidity measurements according to raw water sources and for treated waters. Rivers can have highly variable turbidities due to rainfall and runoff events. Rivers can have turbidities as low as about 1 to 5 NTU in the absence of rainfall events, and as high as 100s to 1000s NTU following rain events and during spring runoff, especially with snow melt. Lakes have much lower turbidities with values in the range of 0.5 to 10 NTU. Many large, oligotrophic lakes have turbidities <5 NTU. Turbidity following DAF is usually in the range of 0.3 to 1 NTU. Well-designed and operated granular media filters produce low values of turbidity in the range of 0.01 to 0.3 NTU.

Number Concentration. Particle number concentration (n_p, number of particles per unit water volume) is an important parameter for particle performance models for many treatment processes. The rate of flocculation depends strongly on n_p (Chap. 6), and thus waters with high particle concentrations flocculate rapidly following coagulation. The rate of removal by particles in DAF and in granular media filtration depends on n_p.

While we tend to think of turbidity measurements as our primary measurement of assessing water quality and treatment, various types of particle number concentration measurements are often made. We measure bacteria and viruses using microbiological assays—for example, heterotrophic plant counts, coliform numbers from multiple tube fermentation, and coliphages. We count algae by direct microscopic methods, often by types. Electronic particle counters are sometimes used for online monitoring of water quality, especially filtered water quality. Most electronic counters use a light blockage technique to detect particles. These instruments used in the water field detect particles of 2 μm and larger, although some instruments are available that detect particles somewhat smaller at about 1 μm. These particle counters provide data in terms of the total particle concentration (>2 μm) and the size distributions (i.e., counts in various size ranges). Some information about particle counting is found in Chap. 10, and the reader can find additional material about types of particle counters, particle counting, and handling of particle size data in Edzwald and Tobiason (2010).

In the following summary of particle concentrations, the numbers are for sizes >2 μm corresponding to common use of light blockage–type counters. Particle number concentrations for river supplies are variable, with relatively high numbers in the range of 10^3 to 10^5 particles per mL. Lakes have much lower number concentrations such as 10^2 to 10^3 particles per mL. Following treatment by DAF, particle numbers are often in the 100s per mL, and even much lower after granular media filtration, typically at 10 to 100 particles per mL.

It is worth noting that *Giardia* cysts and *Cryptosporidium* oocysts occur in raw waters at very small number concentrations of, say, <0.1 to 10 per L, which corresponds to 10^{-4} to 0.1 particles per mL. These pathogen numbers are too low relative to the numbers of other particles in water that it is not possible to use electronic particle counters to monitor them.

Particle Volume Concentration. Volume concentration of solids is an important parameter in the design and operation of sludge management systems. In water treatment flocculation, the particle volume concentration (Φ_p) is used in particle flocculation models that describe the aggregation rate when mixing or velocity gradient is the mechanism of providing collisions among particles (see Chap. 6). It can be calculated from mass concentrations or number concentrations from Eqs. (5-3) and (5-4)

$$\Phi_p = \frac{C_p}{\rho_p} \tag{5-3}$$

$$\Phi_p = n_p\left(\frac{\pi d_p^3}{6}\right) \tag{5-4}$$

where C_p is the mass concentration, ρ_p is the density of particles, and d_p is the particle diameter.

Raw water Φ_p values for rivers can be from 10 to 100 ppm (even greater for some rivers after rainfall events). Lakes generally have lower values for Φ_p of about 1 to 10 ppm. A main purpose of coagulation is to increase the floc volume concentration by precipitation of metal hydroxides, thereby increasing the flocculation rate. Depending on whether alum or ferric coagulants are used, and the amount dosed, the floc volume added by the coagulant can be 5 to 50 ppm.

Particle Stability

The stability of particle suspensions is defined with respect to the rate of particle flocculation. A stable suspension is one in which particles flocculate slowly, or not at all for our time scale of interest, and so the particles of the suspension remain as primary particles and do not form flocs. On the other hand, a destabilized suspension is one in which flocculation occurs fairly rapidly (minutes) so that the collisions and attachment of particles cause formation of floc particles, or simply flocs. A primary cause of particle stability is the surface charge that nearly all particles have before coagulation. The origin of particle charge is presented first and how solution chemistry, particularly pH, affects particle charge. Other forces that affect particle stability are discussed in the section following particle charge.

Particle Charge. There are three principal ways by which particles obtain an electrical surface charge: (1) isomorphic substitution, (2) ionization of surface groups, and (3) adsorption of ions.

The best examples of surface charge from isomorphic substitution are the clay minerals. In the formation of clays from chemical weathering of rocks, there can be substitution of Al^{3+} for Si^{4+} in the Si layers composing the clay which produce a charge deficit and a net negative charge. Likewise, there can be substitution of Fe^{2+} or Mg^{2+} for Al^{3+} in the Al layers of the clay producing a negative charge. This charge from isomorphic substitution is not affected by solution chemistry (e.g., pH), but clay particles have edges with hydroxyl groups that are charged depending on pH—covered next under ionization of surface groups. The magnitude of the negative charge of clays in waters depends on clay type and pH. Values of clay negative charge densities (analogous to a cation exchange capacity) for approximately neutral pH conditions are in the range of 0.05 to 0.5 μeq/mg (microequivalent of negative charge per mg of clay).

Particle surfaces usually contain chemical functional groups that ionize and act as acids and bases—that is, they can release or accept protons (H^+). In what follows, the notation ≡XOH represents the surface of the particle with a metal (X) containing the functional group (OH). The change in charge with pH for clays, sand, aluminum hydroxide, and ferric hydroxide particles is illustrated with Eqs. (5-5) and (5-6), where X can be Si, or Al, or Fe. Note that aluminum and ferric hydroxide particles serve as examples of particles formed in water treatment from coagulation. At low pH, (high [H^+]), particles can have a positive charge. As pH increases (high [OH^-]), negatively charged surfaces occur.

$$\equiv XOH + H^+ \rightleftharpoons \equiv XOH_2^+ \tag{5-5}$$

$$\equiv XOH + OH^- \rightleftharpoons \equiv XO^- + H_2O \tag{5-6}$$

Biological organic particles such as bacteria, algae, and protozoan cysts contain functional groups that undergo acid-base reactions so their surface charge is affected by pH.

These functional groups include weak acid carboxyl (COOH) and weak base (amine or NH_2) groups. The charge depends on the number of each functional group and pH.

The pH at which particles have a zero net surface charge is called the pH of zero point of charge (pH_{zpc}) or the isoelectric point (pH_{iep}). The pH of zero point of charge depends on particle material, as well as solution chemistry. Some examples are presented in Table 5-2. Note that clays, sand, and microorganisms carry a negative charge for the pH conditions of most natural waters (pH 6 to 8.5). Amorphous solids of $Al(OH)_3$ and $Fe(OH)_3$ are formed from coagulation. The pH_{zpc} data in Table 5-2 for these solids should be considered as reference values for solids formed in distilled or deionized water. The actual surface charge of floc particles in water plants after coagulation and flocculation is a complex result of the precipitation of metal hydroxides; the precipitation, coprecipitation, and adsorption of NOM; and the adsorption of charged species from coagulant addition. The pH_{zpc} values of aluminum and ferric hydroxide flocs differ from the reference values and are generally lower. Thus, zeta potential or electrophoretic mobility measurements must be made to determine the charge on flocs and the effect of pH.

Specific adsorption of ions onto particle surfaces affects particle charge. The actual reaction can involve surface complexation of ions at surface groups or adsorption of larger molecules such as polymers and macromolecules onto surfaces. In the first case many metal ions such as calcium, aluminum, and iron undergo complexation reactions at the surface affecting the particle surface charge. An example of adsorption of large molecules or macromolecules is the use of cationic polyelectrolytes in water treatment to neutralize the negative charge of particles. Adding additional polymer or overdosing conditions can reverse the particle charge and produce positively charged particles.

Particle surfaces also interact with macromolecules such as dissolved NOM. These macromolecules often adsorb favorably to surfaces and affect surface charge, especially when the NOM is composed of humic and fulvic acids, which carry a negative charge from carboxyl functional groups. Even though humic and fulvic acids carry a negative charge, they are large molecules with some hydrophobic character so some partitioning on negatively charged particles occurs. With adsorption, they increase the negative charge of the particles because their charge density is greater than those of the particles. It is noted,

TABLE 5-2 The pH of the Zero Point of Charge (pH_{zpc}) for Selected Particles

Particle	pH_{zpc}
Mineral Particles	
$Fe(OH)_3$ (amorphous)	~ 8.5
$Al(OH)_3$ (amorphous)	8.0
MnO_2	2.0–4.5
SiO_2 (sand)	2.0–3.5
Kaolinite (clay)	3.3–4.6
Montmorillonite (clay)	2.5
Organic Particles	
Algae	3–5
Bacteria	2–4
Cryptosporidium oocysts	<2 (Kelley, 1996)
	<3.5 (Ongerth and Pecoraro, 1996)

Compiled from MWH (2005) except for *Cryptosporidium* oocysts.

however, that the bulk water contains a far greater concentration of humic and fulvic acids and that they can control coagulant dosing rather than the particles—the role of particles and NOM in coagulation is discussed in Chap. 6.

Other Causes of Particle Stability. There are other phenomena that can cause particle stability. These are hydration effects of particle surface groups and steric effects. Hydration effects are produced by water molecules bound to particle surface groups—sometimes called bound water. As two particles approach each other, their collision and attachment is hindered by hydration repulsion, which is the requirement that the particle surfaces must become dehydrated (bound water must get out of the way) for contact between the particles to occur. Hydration effects can add to the stability of particles, and may be a factor in explaining results that do not follow exactly Derjaguin-Landau-Verwey-Overbeek (DLVO) theory. Repulsive interaction of uncharged silica surfaces is an example of this.

Steric effects sometimes called steric repulsion or steric stabilization are discussed next. The interaction of particles with adsorbed macromolecules that have long tails and loops extending into solution can lead to repulsive steric interactions if the macromolecule segments have segment-water interactions that are more favorable than between segment interactions. In natural waters, polymers excreted by microorganisms can adsorb on particles, affecting their stability. Another example is the adsorption of macromolecules, such as humic acids, onto particle surfaces, causing steric stabilization. In water treatment and the dosing of flocculant aids, if these high–molecular weight polymers are overdosed then they continue to coat the particle, causing steric stabilization—this is covered in Chap. 6.

Particle-Particle Interactions and Forces

The following types of interactions can occur: (1) electrical double layer, (2) van der Waals, (3) hydrodynamic repulsion, and (4) hydrophobic effects.

The first two of the interactions are included in the DLVO theory of particle stability. The other interactions are not taken into account in DLVO theory, and we refer to these as non-DLVO forces. In this section, we describe these forces conceptually to aid our understanding of particle stability. In Sec. 5-3 that deals with particle-bubble interactions, we develop equations describing these forces. Much has been written about particle stability, and the reader is referred to the books by Gregory (2006) and Israelachvili (1991) for details on the forces between particle surfaces.

We start with the charge on particles and the resulting electrical double-layer interaction. The charge on particles, as stated above, is usually the reason particle suspensions are stable. Particles in water with a surface charge have an electrical double layer. The double-layer consists of the primary charge at the surface of the particle and oppositely charged ions (counter-ions) that accumulate near the particle surface in a diffuse double-layer extending away from the surface that neutralizes the primary charge. The extent of the diffuse double-layer is characterized by the double-layer thickness (also called the Debye length) that has values of 10s of nm for low ionic strength waters and less than 1 nm for high ionic strength waters. If particles have the same charge (i.e., negative) as those occurring in water supplies, then when these particles approach each other the diffuse layers begin to overlap and electrostatic repulsion occurs.

The van der Waals force between particles interacting in water is due to intermolecular forces of the particles and include (1) permanent dipole forces, (2) induced dipole forces, and (3) instantaneous induced dipole forces (London dispersion forces). van der Waals forces for similar materials (similar particles) in water are attractive.

The DLVO theory of particle interaction to explain particle stability considers the repulsion from overlapping diffuse double layers as the particles approach each other and the attractive van der Waals force. For fresh water supplies with low to medium ionic strength conditions, there is a repulsive energy barrier causing particle stability and preventing particle aggregation. For high ionic strength systems such as estuarine water and seawater, the diffuse double-layer thickness is reduced so that particles can approach closer allowing van der Waals attractive forces to be dominant. The repulsive energy barrier may be eliminated if the ionic strength is high enough. For most chapters in the book, our interest is in stable particle suspensions for fresh water supplies that occur for lower ionic strength waters. In Chap. 13, we discuss flotation pretreatment in desalination plants; here high ionic strength conditions are applicable.

Some brief comments about the non-DLVO forces follow. We begin with hydrodynamic repulsion. Hydrodynamic repulsion, sometimes called hydrodynamic retardation, refers to the viscous force that exists between two particles in water as they approach each other. This water must be displaced for the particles to collide.

Hydrophobic effects must also be considered. Some of the surface of particles can contain non-polar groups (i.e., no charged functional groups). The water in contact with such a surface differs from water association with charged groups; some would refer to the surface as having some degree of hydrophobic character. As two hydrophobic particles approach, or the parts of the particles with these hydrophobic groups approach, the water between the particles behaves differently from the water in solution, and there is an increased free energy of the water relative to the bulk water resulting in hydrophobic attraction. Surface force measurements indicate that hydrophobic attraction can be important and of longer range than van der Waals attraction (Gregory, 2006). It is a force to consider later in particle-bubble interactions. Another application is in the flotation of mineral particles in mining applications where surfactants are added to adsorb onto particles and increase their hydrophobicity.

5-3 PARTICLE-BUBBLE INTERACTIONS AND FORCES

It is instructive to first discuss briefly the attachment of bubbles to particles for large bubble processes such as dispersed air flotation before presentation of principles regarding the main subject of the book, the smaller bubbles in DAF. Dispersed air flotation is used mainly in mineral separation (see Chaps. 1 and 2). The bubbles are created by mechanical means and have sizes mostly in the range of 0.5 to 2 mm. Bubble attachment to large particles (about 100 μm and larger) is characterized by contact angle measurements, as illustrated in Fig. 5-3. The magnitude of the contact angle is related to particle hydrophobicity and adhesion to the bubble. Particles that are hydrophobic have large contact angles (say 90 degrees and larger) and good separation by flotation occurs—see Gochin and Solari (1983), and Edzwald (1995).

Dissolved air flotation involves smaller bubbles and particles than dispersed air flotation. For DAF, one considers the various particle-bubble forces analogous to the particle-particle forces described above. The important forces are (1) electrostatic repulsion or attraction, (2) van der Waals forces, (3) hydrodynamic effects, and (4) hydrophobic effects. We divide the presentation into consideration of DLVO forces (electrostatic and van der Waals) first, and then move to the discussion of the non-DLVO forces (hydrodynamic and hydrophobic effects). An early reference on the DLVO forces between particles and bubbles for flotation is the classic paper by Derjaguin et al. (1984). While Derjaguin coined the term contactless flotation, it will be shown later that the hydrophobic force is the main attractive force causing particle attachment to bubbles so, while not used in DAF, contact angles are related to hydrophobicity and attachment of particles to bubbles.

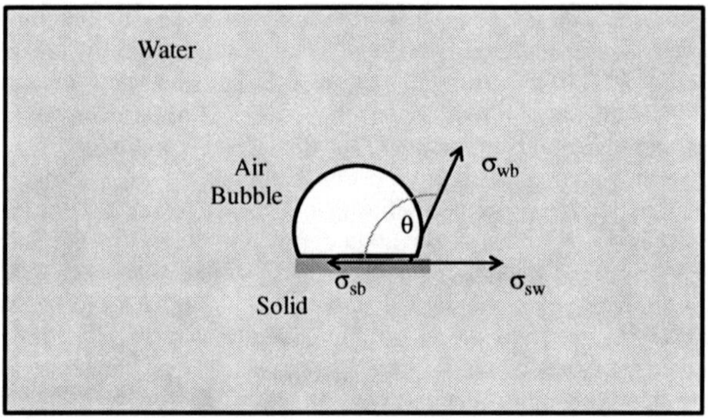

FIGURE 5-3 Contact angle of air bubble on a solid in water: large angle indicates more hydrophobic solid. Notes: θ is the contact angle, σ_{wb} is the surface tension for the water-bubble, σ_{sb} is the surface tension for the solid-bubble, σ_{sw} is the surface tension for the solid-water.

To aid our understanding and to note important variables, we describe the particle-bubble interactions with appropriate mathematical equations. We have utilized the books by Gregory (2006) and Israelachvili (1991) for development of much of the material.

DLVO Forces

Here the DLVO model equations are presented for electrostatic interaction and for van der Waals interaction between bubbles and particles in water. These models were originally developed for colloidal particle-particle interactions. More recently there have been efforts to adapt them for DAF modelling in describing particle-bubble interactions (Han, 2002; Leppinen, 2000). Note that often DLVO interaction energies (V) are considered. In the following section we utilize interaction force (F) expressions; force and energy are related as $F = -dV/dh$, where h is the separation distance between particle and bubble surfaces.

Electrical Double-Layer Interaction Force. The electrostatic forces are repulsive if both particles and bubbles are negatively charged. This would be the case if we attempted to apply DAF without coagulation pretreatment. The electrical double-layer interaction force (F_{edl}) between a bubble and particle for two spheres of unequal size follows:

$$F_{edl} = 2\pi\varepsilon\zeta_b\zeta_p\kappa\left(\frac{d_b d_p}{d_b + d_p}\right)\exp(-\kappa h) \tag{5-7}$$

where ε is the permittivity of water and equals the product of the dielectric constant of water (ε_r—see Table D-1) and the vacuum permittivity (ε_o—see App. C), ζ_b is the bubble zeta potential, ζ_p is the particle zeta potential, d_b and d_p are the bubble and particle diameters, h is the separation distance between the bubble and particle, and κ is the Debye length parameter. By convention for Eq. (5-7), a positive sign is assigned for a repulsive electrical double-layer force. Equation (5-7) applies for systems with surface potentials that are not high depending on the counter-ion valence (Gregory, 2006). For natural waters it is a good approximation for particle and bubble surface potentials less than 25 mV, and hence for zeta potentials of this value and a little greater.

The Debye length is defined by Eq. (5-8)

$$\kappa = \left[\left(\frac{e^2}{\varepsilon k_b T}\right)\left(\sum_i n_i z_i^2\right)\right]^{1/2} \tag{5-8}$$

where e is the electron charge (1.602×10^{-19} C), k_b is Boltzmann's constant, T is the absolute temperature, n_i is the concentration of ion species i in ions per m^3, and z_i is the charge of ion species i. The reciprocal ($1/\kappa$) is a measure of the thickness of the diffuse double layer extending from the particle or bubble surface. Gregory (2006) has a useful approximation at 25°C for κ (units of nm^{-1}) related directly to a water's ionic strength (I).

$$I = \frac{1}{2}\left(\sum_i C_i z_i^2\right) \tag{5-9}$$

$$\kappa = 3.29[I]^{1/2} \tag{5-10}$$

where C_i is the molar concentration of ion species i.

Computing the ionic strength from Eq. (5-9) for raw water supplies or for water within a treatment plant is no easy task as it requires measurements of the concentrations of all ions. There are simple ways, however, to estimate the ionic strength of a water from either conductivity (κ_c in μS/cm) or total dissolved solids (TDS in mg/L) measurements (Snoeyink and Jenkins, 1980).

$$I = 1.6 \times 10^{-5}(\kappa_c) \tag{5-11}$$

$$I = 2.5 \times 10^{-5}(TDS) \tag{5-12}$$

Equation (5-11) applies to waters with specific conductivities of the values that can be found less than about 3000 μS/cm. The TDS correlation equation [Eq. (5-12)] should not be used for TDS greater than 1000 mg/L. Table 5-3 presents double-layer thicknesses ($1/\kappa$) for fresh and salt waters. Most natural fresh waters have ionic strengths between 5×10^{-4} and 10^{-2} M corresponding to conductivities of about 30 to 625 μS/cm or TDS of 20 to 400 mg/L, and diffuse double-layer thicknesses ($1/\kappa$) from slightly over 10 down to 3 nm. As the ionic strength increases, the diffuse double-layer thickness is compressed to the particle or bubble surface. For estuary and seawater ionic strengths, the double-layer thickness is 1 nm or less.

The electrical double-layer interaction force may be repulsive or attractive. Particles in water supplies carry a negative charge such as clays, other inorganic minerals, bacteria, algae, and so on. Likewise bubbles have a negative charge so there is repulsion between the electrical double-layers surrounding the particles and bubbles. Another case of repulsive force interaction can occur under certain coagulation conditions. For coagulants that act by charge neutralization (see Chap. 6), it is possible to overdose, in which excess positively charged coagulant species adsorb on the particles and bubbles, reversing the charge so that both particles and bubbles have a positive charge and hence a repulsive force. Attraction occurs between particles and bubbles of opposite charge (by convention, an attractive F_{edl} is assigned a negative sign in Eq. (5-7)). Malley (1995) and Henderson et al. (2008) demonstrated production of positively charged bubbles by adding cationic polymers to the DAF

TABLE 5-3 Ionic strength (*I*), Conductivity, Total Dissolved Solids (*TDS*), and the Double-Layer Thickness ($1/\kappa$) for Several Water Types

I (M)	Conductivity (μS/cm)	TDS (mg/L)	$1/\kappa$ (nm)
5×10^{-4} low hardness	30	20	13.6
10^{-3} low to moderate hardness	60	40	9.6
5×10^{-3} high hardness	310	200	4.3
10^{-2} brackish water	625	400	3.0
10^{-1} estuarine water[a]	7000	5000	1.0
0.7 Seawater[b]	~ 50,000	~ 36,000	0.4

[a]Extrapolated: exceeds limit of correlation Eqs. (5-11) and (5-12).

[b]Seawater values vary with salinity but are well-known.

saturator water or to the saturator effluent. Han et al. (2006) have also shown that positively charged hydrogen bubbles are produced from Al and Mg hydroxide–precipitated solids.

van der Waals Interaction Force. The van der Waals interaction force equation (F_{vdw}) for interacting spheres is given by Eq. (5-13), where A_{132} is the Hamaker constant and the subscripts 132 refer to the solid particle (subscript 1), water (subscript 3), and air bubble (subscript 2). Equation (5-13) applies for non-retarded conditions. For retarded conditions, Eq. (5-14) applies where there is reduced van der Waals force interaction with particle-bubble separation distance. In this equation, the number 100 has units of nm. By sign convention, these equations are assigned a negative sign for attractive interaction.

$$F_{vdw} = \frac{-A_{132} d_p d_b}{12h^2 \left(d_p + d_b\right)} \tag{5-13}$$

$$F_{vdw} = \frac{-A_{132} d_p d_b}{12\left(d_p + d_b\right)} \left(\frac{1}{1 + (12h/100)}\right) \left(\frac{1}{h^2} + \frac{0.12}{1 + (12h/100)}\right) \tag{5-14}$$

It is a mistake to assume that the van der Waals force between particles and air bubbles are attractive. The first author (Edzwald) in the past has made this mistake. Others have made this error in their models for the DAF contact zone performance dependence on particle-bubble interaction—see Chap. 7. The van der Waals force is attractive for similar solid materials or particles in water, as discussed in an earlier section; however, the Hamaker constant is negative (see the following text) for a solid particle in water interacting with an air bubble. Detailed discussion of the Hamaker constant and why the van der Waals interaction energy is repulsive is found in Lu (1991) and Israelachvili (1991) and is summarized next.

The Hamaker constant (A_{132}) for dissimilar materials (particle-bubble) interacting in water is found from Eq. (5-15). This is applied to a particle-bubble interaction in water where A_{11}, A_{22}, and A_{33} refer to particle (solid), bubble, and water. These are the Hamaker constants for the interactions of each media with itself in a vacuum (Gregory, 2006; Israelachvili, 1991).

$$A_{132} = \left(A_{11}^{1/2} - A_{33}^{1/2}\right)\left(A_{22}^{1/2} - A_{33}^{1/2}\right) \tag{5-15}$$

Because A_{22} (bubbles in a vacuum) is very small compared to particles (A_{33}), A_{22} can be dropped, and the equation simplifies to

$$A_{132} = -A_{33}^{1/2}\left(A_{11}^{1/2} - A_{33}^{1/2}\right) \tag{5-16}$$

The minus sign means that the Hamaker constant is negative so there is a repulsive van der Waals interaction force for solid particles interacting with air bubbles in water; that is, Eqs. (5-13) and (5-14) become positive, indicating repulsive energies. Hamaker constant values (A_{132}) for particle-bubble interaction in water depend on the specific solid particle. For silica particles and bubbles in water, Ducker et al. (1994) report Hamaker constants of -1.0×10^{-20} and -1.16×10^{-20} J. Other values by Lu (1991) and Israelachvili (1991) range from -0.87×10^{-20} to -1.4×10^{-20} J. A more negative value of -3.7×10^{-20} J for Al_2O_3 solids is reported by Nguyen (2007).

Non-DLVO Forces

There are non-DLVO forces to consider that affect the interaction between particles and bubbles. These are the hydrodynamic and hydrophobic forces.

Hydrodynamic Force. In a flotation tank contact zone, as a particle approaches a rising air bubble, the water between them can cause repulsion in the sense that the water between them must move out of the way. This effect is called a hydrodynamic or viscous effect. It adds potentially to the stability of particle-bubble interactions, and its effect has been applied to trajectory analysis of the movement of particles to and around bubbles (Okada et al. 1990; Leppinen, 2000; Han, 2002).

A key assumption in hydrodynamic interaction is that the particles and bubbles are hard impermeable spheres. The hydrodynamic force should not be as large for porous flocs, which would have more permeability and less hydrodynamic resistance. Many people who model particle trajectories ignore this, and so the hydrodynamic force is overestimated for porous flocs. In flotation tanks involving porous flocs and air bubbles, the hydrodynamic force may be a factor affecting particle-bubble interaction, but it is not as important a factor as DLVO forces and the hydrophobic force interaction, and it is not considered in calculations presented in Example 5-2.

Hydrophobic Force. We know that flotation occurs when the zeta potentials of particles and bubbles are small and of the same charge, producing a repulsive electrostatic force. Likewise, as discussed above, the van der Waals force is repulsive for particle-bubble interaction and the non-DLVO hydrodynamic force is repulsive. So, how does flotation occur then if these interaction forces are repulsive? There is an attractive hydrophobic force that can exceed other repulsive forces and bring about flotation.

The exact nature of the hydrophobic force is not clearly understood, but it has been measured (Israelachvili, 1991; Ducker et al., 1994), recently evaluated (Hammer et al., 2010), and used to explain attraction between particles and air bubbles (Israelachvili, 1991; Lu, 1991; Ducker et al., 1994; Nguyen, 2007). Nguyen (2007) presents the following semi-empirical equation for the hydrophobic force ($F_{hydroph}$) between particles and air bubbles

$$F_{hydrop} = \left(\frac{-Kd_p d_b}{2\left(d_p + d_b\right)}\right) \times \left(\exp - \left(\frac{h}{\lambda}\right)\right) \tag{5-17}$$

where K is the hydrophobic force constant and λ is the decay length. In model predictions Nguyen (2007) considered λ with values in the range of 1 to about 10 nm. Nguyen (2007) used values for K of 0.5 to 1.5 mN/m for latex particles and air bubbles. The hydrophobic force decays exponentially with separation distance, but it is long-range, extending to distances exceeding DLVO interactions (Israelachvili, 1991; Nguyen, 2007), as will be shown in Example 5-2.

EXAMPLE 5-2 Particle-Bubble Interaction Forces Model predictions are made for the DLVO and non-DLVO hydrophobic force for two cases: Case A of *No Coagulation* and Case B for *After Coagulation*. Calculations are made in both cases for the following water conditions: 25°C and ionic strength at 2×10^{-3} M (typical of many water supplies). Other conditions and selection of values for variables are presented for each case.

Case A: No Coagulation

PARTICLE AND BUBBLE SIZES For conditions of no coagulation, a particle diameter of 5 μm is assigned. The bubble diameter is set at 60 μm, a mean value for the DAF contact zone.

PARTICLE AND BUBBLE ZETA POTENTIALS For this case, we assign the particle zeta potential (ζ_p) at −25 mV and the bubble zeta potential (ζ_b) at −20 mV. The particle zeta potential is sometimes measured in water treatment practice and the value of −25 mV is based on the authors' experience for raw water particles varying between −15 and −50 mV. The bubble zeta potential is a difficult measurement that is not made in water treatment practice. A value of −20 mV is assigned based on reported values cited earlier in the chapter. Zeta potentials of particles and bubbles depend on solution chemistry and pH. The selected values represent pH conditions of about 6 to 7.

HAMAKER CONSTANT We assign the Hamaker constant (A_{132}) for particle-water-air bubble at -1.0×10^{-20} J based on the range of values reported in the section above on van der Waals forces. Note that the van der Waals force is repulsive, as discussed above. We use the retarded van der Waals force expression, Eq. (5-14).

VARIABLES FOR HYDROPHOBIC FORCE Two variables must be selected: the hydrophobic force constant (K) and the decay length (λ). As presented above, values of 0.5 to 1.5 mN/m for K for latex particles and air bubbles have been used. Of course water supplies do not contain latex particles and by their nature contain various-type inorganic and organic particles. We do not know K for particles in a water supply and must assume a value. We assume a value for K of 0.5 mN/m for Case A and λ at 10 nm.

Case A: Solution The forces are calculated as a function of the separation distance between the particle and bubble and plotted graphically.

1. Electrical Double Layer Interaction Force F_{edl}

To calculate F_{edl}, we must first determine κ, the Debye length from Eq. (5-10).

$$\kappa = 3.29[I]^{1/2} = 3.29[2 \times 10^{-3}]^{1/2} = 0.147 \text{ nm}^{-1}$$

Note that κ is only a function of the ionic strength and so it does not change for Case B. The reciprocal of κ is the diffuse double-layer thickness and is 6.8 nm.

F_{edl} is calculated as a function of the particle-bubble separation distance (h) from Eq. (5-7) for ζ_p at −25 mV, ζ_b at −20 mV, d_p 5 μm, and d_b 60 μm. The permittivity of water

(ε) is the product of the dielectric constant of water (ε_r—see Table D-1) and the vacuum permittivity (ε_o—see App. C). For 25°C, ε equals 6.954×10^{-10} C^2 J^{-1} m^{-1}.

$$F_{edl} = 2\pi\varepsilon\zeta_b\zeta_p\kappa\left(\frac{d_b d_p}{d_b + d_p}\right)\exp(-\kappa h)$$

2. van der Waals Interaction Force, F_{vdw}

F_{vdw} is calculated using the retardation equation [Eq. (5-14)] as a function of h. In the calculations, A_{132} is -1.0×10^{-20} J, d_p and d_b are 5 and 60 μm, respectively.

$$F_{vdw} = \frac{-A_{132} d_p d_b}{12\left(d_p + d_b\right)}\left(\frac{1}{1+(12h/100)}\right)\left(\frac{1}{h^2} + \frac{0.12}{1+(12h/100)}\right)$$

3. Hydrophobic Force, F_{hydrop}

F_{hydrop} is calculated from Eq. (5-17) using the variables of Case A, as explained above—$K = 0.5$ mN/m and $\lambda = 10$ nm.

$$F_{hydrop} = \left(\frac{K d_p d_b}{2\left(d_p + d_b\right)}\right) \times \left(\exp - (h/\lambda)\right)$$

4. Case A: Plots of Forces in Fig. 5-4

F_{edl}, F_{vdw}, and F_{hydrop} are plotted in Fig. 5-4, as well as the sum of these forces, which is the net force (F_{net}). The discussion of these model predictions follows.

- F_{vdw} (van der Waals force) is repulsive for particles and bubbles in water. It increases greatly for separation distances (h) less than 5 nm. It exceeds the repulsive electrical double-layer force (F_{edl}) for h less than 1.75 nm.
- F_{edl} is the greater repulsive force for separation distances (h) greater than 1.75 nm for particles and bubbles with negative zeta potentials of −25 and −20 mV. Obviously, it would be even greater for higher negative zeta potentials. It decays such that it approaches zero ($F_{edl} < 0.02$ nN) for h greater than 30 nm.
- F_{hydrop} is the attractive force that extends over separation distances of about 30 nm. It exceeds both repulsive forces (F_{edl}, F_{vdw}) for separation distances of about 5.5 nm and greater. Thus, there is a small net attractive force (F_{net}) for separation distances greater than about 6 nm, and a net repulsive force (F_{net}) for separation distances less than 6 nm, hindering particle-bubble attachment.

Case B: After Coagulation

PARTICLE AND BUBBLE SIZES After coagulation and flocculation, particle sizes ought to be larger than those in Case A so a particle diameter of 25 μm is assigned—in Chap. 7 it will be shown that this is within the optimum sizes for high DAF contact zone efficiency. The bubble diameter is maintained at 60 μm, the same value as that of Case A.

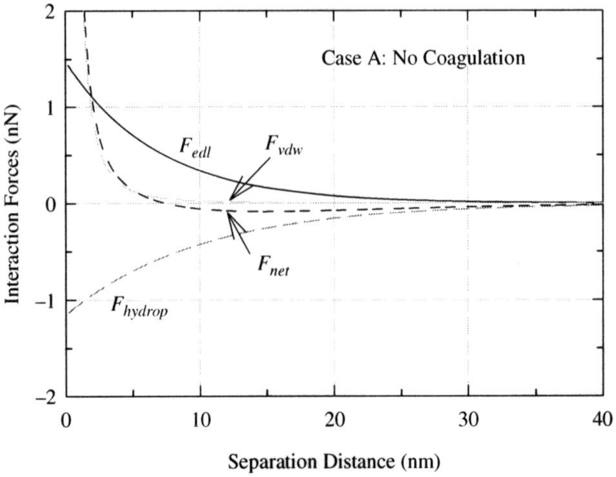

FIGURE 5-4 Electrical double-layer force (F_{edl}), van der Waals force (F_{vdw}), hydrophobic force (F_{hydrop}), and the net force (F_{net}) as a function of the particle-bubble separation distance (h) for Case A of No Coagulation in Example 5-2. (Conditions: particle and bubble diameters of 5 μm and 60 μm, respectively; 25°C; ionic strength of 2×10^{-3} M; particle and bubble zeta potentials of −25 and −20 mV, respectively; Hamaker constant of 1.0×10^{-20} J; and hydrophobic force variables for K of 0.5 mN/m and λ of 10 nm.)

PARTICLE AND BUBBLE ZETA POTENTIALS Coagulation reduces the charge of particles, depending on the type of coagulant, dosage, and pH. We evaluate conditions in which zeta potentials of the particles and bubbles have been reduced greatly but still have small negative zeta potentials: $\zeta_p = \zeta_b = -2$ mV. Experience shows that zeta potentials values of approximately zero, but may be slightly negative or positive, occur for good DAF operation following coagulation—see Chaps. 6 and 10.

HAMAKER CONSTANT We use the same value as that used for Case A: A_{132} of -1.0×10^{-20} J.

VARIABLES FOR HYDROPHOBIC FORCE Here we assume that with coagulation the particles are more hydrophobic than those in Case A, and so we assign a higher K value of 1.5 mN/m than that for Case A. The decay length (λ) is set at 10 nm, same as Case A.

Case B: Solution The forces are calculated as a function of the separation distance (h) between the particle and bubble using the equations presented in Case A. The predictions are plotted in Fig. 5-5. Discussion follows:

- F_{edl} is very small for the low negative zeta potentials of −2 mV for the particles and bubbles. Other calculations (not shown) indicate that various combinations of low zeta potentials yield low values for F_{edl}. This is the case even for particles with a zeta potential of −2 mV and a positive zeta potential of +20 mV for bubbles; F_{edl} is less than −0.12 nN.
- F_{vdw} (van der Waals force) is again repulsive and increases greatly for separation distances (h) less than a few nm.
- F_{hydrop} is an attractive force that extends over separation distances of about 40 nm. It exceeds the F_{vdw} for separation distances of about 1 nm and greater.

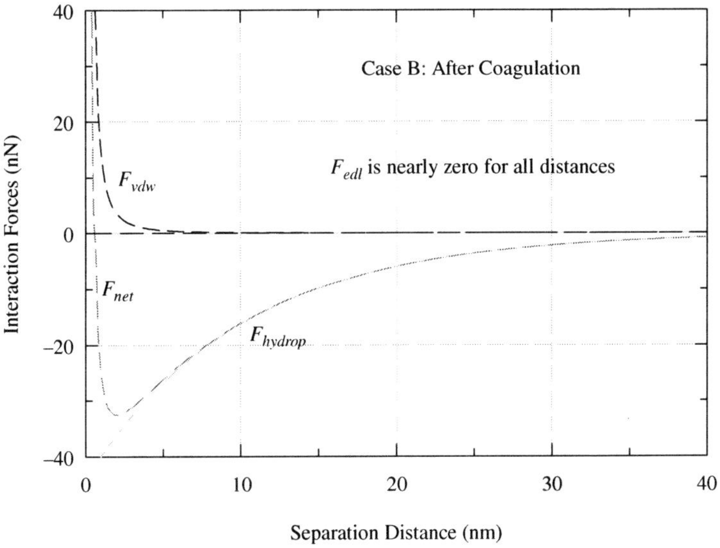

FIGURE 5-5 Electrical double-layer force (F_{edl}), van der Waals force (F_{vdw}), hydrophobic force (F_{hydrop}), and the net force (F_{net}) as a function of the particle–bubble separation distance (h) for Case B of After Coagulation in Example 5-2. (Conditions: particle and bubble diameters of 25 μm and 60 μm, respectively; 25°C; ionic strength of 2×10^{-3} M; particle and bubble zeta potentials of −2 mV; Hamaker constant of 1.0×10^{-20} J; and hydrophobic force variables for K of 1.5 mN/m and λ of 10 nm.)

- There is a net attractive force attraction (F_{net}) for separation distances greater than about 1 nm due to the hydrophobic force. There is maximum negative value for F_{net} at a separation distance of about 3 nm. Thus we conclude that the hydrophobic force accounts for attachment of particles to air bubbles. ▲

Utility of Theories Regarding DLVO and Non-DLVO Interactions

How can we use the theory about DLVO and non-DLVO forces affecting the repulsion and attraction of particles to bubbles? In short, the principles from the theory allow us to gain a fundamental understanding of what forces are involved and what affects these forces. To use as a practical engineering tool to predict flotation performance is another matter, and in short, they cannot yet be used. The equations for the forces have far too many variables that are either unknown or difficult to measure. The usefulness of the theory and the limitations of the theory as a practical engineering tool are well illustrated with Example 5-2. The usefulness and limitations are discussed in the following text.

For particles and bubbles that are of the same charge, the theory shows we have three repulsive forces: both DLVO forces are repulsive (electrical double-layer repulsion (F_{edl}) and van der Waals [F_{vdw}]), and the non-DLVO hydrodynamic force, which is minor compared to the others. F_{edl} [Eq. (5-7)] depends on the particle and bubble sizes and their zeta potentials and the Debye length (κ). Particles and bubbles in water supplies that are not subject to coagulation are negatively charged and so both have negative zeta potentials. The Debye length depends on the ionic strength of the water. F_{vdw} [Eq. (5-14)] depends on the particle and the bubble sizes and Hamaker constant, which depends on the nature of the particles. There is one attractive force without coagulation and that is the hydrophobic force (F_{hydrop}). This force [Eq. (5-17)] depends primarily on the particle and bubble sizes and the hydrophobic force constant (K). The latter depends on the hydrophobicity of the particles. F_{hydrop} extends over large separation distances. Without coagulation, as illustrated in Case A of Example 5-2, the attractive hydrophobic force is smaller than the repulsive forces at short separation distances of about 5 nm and less. This explains why poor particle-bubble attachment occurs without coagulation.

We know from experience that when we coagulate water and produce flocs of low negative and positive charge or no net charge, flotation is effective. Coagulants reduce the charge of particles originally in the water and produce precipitated solids (new particles). This mixture of particles forms flocs, and it is these flocs that enter the flotation tank that should have low (negative or positive) or no charge. The theory, as illustrated in Case B of Example 5-2, shows that particles and bubbles with low zeta potentials have F_{edl} close to zero compared to F_{vdw} and F_{hydrop}. Thus, the attractive hydrophobic force (F_{hydrop}) exceeds the repulsive van der Waals force (F_{vdw}) for separation distances greater than about 1 nm. This accounts for particle-bubble attachment. Note, as discussed in the following text, that we do not know the Hamaker constant for the types of particles and flocs that occur in water treatment practice, so the predicted values for F_{vdw} may be too great, meaning this force is actually lower.

It is noted that some coagulants that act by adsorption of positively charged species, such as cationic polymers and polyaluminum chlorides at acidic pH conditions (see Chap. 6), can reverse the charge of particles and bubbles if overdosed. This should yield positively charged particle and bubble zeta potentials and a repulsive F_{edl}. However, as discussed in Case A of Example 5-2, this repulsive force is small for low zeta potentials. While not used in DAF practice, cationic polymers could be possibly added to the DAF saturator to produce positively charged bubbles. This would yield an attractive F_{edl} because of a positive zeta potential for the bubbles and a negative zeta potential for particles. However, in water treatment practice one does not want to do this. Coagulation is essential to remove dissolved NOM (TOC) and it is also essential for good filtration following DAF—porous media filtration does not work without coagulation. Thus, the practical case for DAF in water treatment is coagulation of the water supply prior to DAF. Coagulation ought to be carried out as stated above (more detail in Chap. 6) to produce flocs with small or no charge.

It is important to examine the limitations of the DLVO and non-DLVO particle-bubble interactions because some researchers use them to model particle-bubble collisions and attachment in the DAF contact zone—their use in DAF contact zone models is addressed in Chap. 7. Equation (5-7) shows the parameters that must be known to calculate the electrical double-layer force (F_{edl}) interaction between a particle and bubble. Some can be assumed, one can be calculated from other measurements, and others can be measured. The permittivity of water (ε) is a physical constant that is calculated from the dielectric constant of water for the temperature of interest. The zeta potential for particles or flocs (ζ_p) can be calculated from electrophoretic mobility measurements (some instruments directly make the calculation and display zeta potential). Not all water plants have zeta potential instrumentation although the instrument is fairly inexpensive so zeta potential is a feasible measurement for particles. Determining zeta potential for bubbles (ζ_b) is entirely another matter. This is a difficult measurement rarely attempted except in research laboratories and requires skillful ways to collect samples with bubbles and then transfer to zeta potential instrumentation.

Bubble size (d_b) is not easily measured, but it can be measured or reasonable assumptions made from the material presented in Chap. 4 and summarized at the beginning of this chapter. Floc or particle size (d_p) can be measured using particle counting and sizing instrumentation. Some water plants have this instrumentation or it can be purchased as the instruments are relatively inexpensive. The Debye length (κ) can be calculated from the ionic strength (I) from Eq. (5-10). The ionic strength requires measurement of the concentrations of all inorganic ions in the water. This is not an easy task. It can be estimated from conductivity or TDS measurements, as presented above—see Eqs. (5-11) and (5-12). The particle-bubble separation distance (h) is then varied to make calculations of F_{edl}.

Equation (5-13) or (5-14) shows the parameters that must be known to calculate the van der Waals interaction force between a particle and bubble. Particle and bubble sizes were addressed above and, of course, the particle–bubble separation distance is varied. This leaves the Hamaker constant. We do not know the Hamaker constant for the flocs formed in water treatment interacting with bubbles in water. The flocs are a mixture of particles (mineral, amorphous solids, organic particles), and the bubbles can have traces of adsorbed organic and inorganic matter affecting the overall Hamaker constant. One is left to assume values, and one can vary them to determine the effect of the constant.

Equation (5-17) describes the variables affecting the hydrophobic force. Particle and bubble sizes were addressed above and, of course, the particle–bubble separation distance is varied. The decay length (λ) is not known exactly but values of about 1 to 10 nm are reported. The most important variable is the hydrophobic force constant (K). Values of 0.5 to 1.5 mN/m have been used for latex particles in water, but we certainly do not know K for the various inorganic and organic particles in water supplies or the flocs formed in water treatment.

Overall, the theory and equations for DLVO and non-DLVO forces are useful in establishing a fundamental understanding of what forces are involved and what affects these forces, but they are not useful as an engineering predictive tool.

REFERENCES

Craig, V. S. J., Ninham, B. W., and Pashley, R. M. (1993), Effect of electrolytes on bubble coalescence, *Nature*, 364 (22 July), 317–319.

Derjaguin, B. V., Dukhin, S. S., and Rulyov, N. (1984), *Surface and Colloid Science* (E. Matijevic, and R. J. Good, eds.), pp. 71–113, New York: Plenum Press.

Dockko, S., and Han, M. (2004), Fundamental characteristics of bubbles and ramifications for the flotation process, *Water Science and Technology: Water Supply*, 50 (12), 207–214.

Ducker, W. A., Xu, Z., and Israelachvili, J. N. (1994), Measurements of hydrophobic and DLVO forces in bubble-surface interactions in aqueous solutions, *Langmuir*, 10, 3279–3289.

Edzwald, J. K. (1995), Principles and applications of dissolved air flotation, *Water Science and Technology*, 31 (3–4), 1–23.

Edzwald, J. K. (2010), Dissolved air flotation and me, *Water Research*, 44 (7), 2077–2106.

Edzwald, J. K., and Tobiason, J. E. (2010), "Chemical principles, source water composition, and watershed protection," in J. K. Edzwald (ed.), *Water Quality and Treatment: A Handbook on Drinking Water*, 6th ed., New York: AWWA and McGraw-Hill.

Gochin, R. J., and Solari, J. (1983), The role of hydrophobicity in dissolved air flotation, *Water Research*, 19 (6), 651–657.

Gregory, J. (2006), *Particles in Water: Properties and Processes*, London: IWA Publishing.

Gregory, R., and Edzwald, J. K. (2010), "Sedimentation and flotation," in J. K. Edzwald (ed.), *Water Quality and Treatment: A Handbook on Drinking Water*, 6th ed., New York: AWWA and McGraw-Hill.

Hammer, M. U., Anderson, T. H., Chaimovich, A., Shell, M. S., and Israelachvili, J. (2010), The search for the hydrophobic force law, *Faraday Discussions*, 146, 299–308.

Han, M. Y. (2002), Modeling of DAF: the effect of particle and bubble characteristics, *Journal of Water Supply: Research and Technology – Aqua*, 51 (1), 27–34.

Han, M., and Dockko, S. (1999), Zeta potential measurement of bubbles in DAF process and its effect on the removal efficiency, *Water Supply*, 17 (3–4), 177–182.

Han, M., Kim, M. K., and Shin, M. S. (2006), Generation of positively charged bubble and its possible mechanism of formation, *Journal of Water Supply: Research and Technology – Aqua*, 55 (7–8), 471–478.

Henderson, R. K., Parsons, S. A., and Jefferson, B. (2008), Surfactants as bubble surface modifiers in the flotation of algae: dissolved air flotation that utilizes a chemically modified bubble surface, *Environmental Science and Technology*, 42 (13), 4883–4888.

Israelachvili, J. (1991), *Intermolecular and Surface Forces*, 2nd ed., New York: Academic Press.

Israelachvili, J. N., and Pashley, R. M. (1982), The hydrophobic interaction is long range, decaying exponentially with distance, *Nature*, 300 (25 November), 341–342.

Kelley, M. B. (1996), *The Removal of Cryptosporidium by Selected Drinking Water Treatment Processes*, PhD Dissertation, Rensselaer Polytechnic Institute, Troy, New York.

Leppinen, D. M. (2000), A kinetic model of dissolved air flotation including the effects of interparticle forces, *Journal of Water Supply: Research and Technology – Aqua*, 49 (5), 259–268.

Lu, S. (1991), Hydrophobic interaction in flocculation and flotation 3. Role of hydrophobic interaction in particle-bubble attachment, *Colloids and Surfaces*, 57, 73–81.

Malley, J. P. (1995), The use of selective and direct DAF for removal of particulate contaminants in drinking water treatment, *Water Science and Technology*, 31 (3–4), 49–57.

MWH (2005), *Water Treatment Design: Principles and Design*, 2nd ed., Crittenden, J. C., Trussell, R. R., Hand, D. W., Howe, D. W., and Tchobanoglous, G., Hoboken, NJ: John Wiley & Sons.

Nguyen, A. V. (2007), One-step analysis of bubble-particle capture interaction in dissolved-air flotation, *International Journal Environment and Pollution*, 30 (2) 227–249.

Okada, K., Akagi, Y., Kogure, M., and Yoshioka, N. (1990), Analysis of particle trajectories of small particles when the particles and bubbles are charged, *The Canadian Journal of Chemical Engineering*, 68 (4), 614–621.

Ongerth, J. E., and Pecoraro, J. P. (1996), Electrophoretic mobility of *Cryptosporidium* oocysts and *Giardia* cysts, *Journal of Environmental Engineering (ASCE)*, 122 (3), 228–231.

Snoeyink, V. L., and Jenkins, D. (1980), *Water Chemistry*, New York: John Wiley & Sons.

CHAPTER 6

PRETREATMENT COAGULATION AND FLOCCULATION

6-1 INTRODUCTION TO COAGULATION 6-2
- Definitions 6-2
- Terminology 6-2
- Optimum Coagulation 6-3
- Framework for Understanding Coagulation 6-4

6-2 CONTAMINANTS 6-4
- Particles 6-4
- Natural Organic Matter 6-5
- Role of Particles versus Natural Organic Matter in Controlling Coagulation 6-8

6-3 BULK WATER CHEMISTRY AND TEMPERATURE 6-9
- pH 6-9
- Alkalinity and Buffer Intensity 6-10
- Hardness 6-12
- Water Temperature 6-12

6-4 RAPID MIXING 6-12

6-5 COAGULATION CHEMISTRY AND MECHANISMS 6-14
- Alum 6-16
- Ferric Coagulants 6-23
- Polyaluminum Coagulants 6-27
- Organic Cationic Polyelectrolytes 6-32

6-6 GUIDANCE ON COAGULANT DOSING FOR DAF 6-33
- Alum Use in DAF 6-34
- Ferric Coagulants and Use in DAF 6-36
- Polyaluminum Chloride Coagulants and Use in DAF 6-36
- Use of Organic Cationic Polymers and Metal Coagulants in DAF 6-38

6-7 INTRODUCTION TO FLOCCULATION 6-41

6-8 FLOCCULATION FUNDAMENTALS 6-41
- Collisions by Brownian Diffusion 6-42
- Collisions by Fluid Shear 6-42
- Discussion 6-43
- Monodispersed Particle Suspension Flocculation 6-45

6-9 FLOCCULATION TANKS 6-46

6-10 FLOCCULANT, FLOTATION, AND FILTER AIDS 6-47

6-11 GUIDANCE ON FLOCCULATION FOR DAF 6-48

REFERENCES 6-49

Coagulation and flocculation are pretreatment processes for particle removal from water. Coagulation is a chemical step involving destabilizing particle suspensions so the particles flocculate readily and involving reactions with natural organic matter (NOM). Flocculation is a physical step whereby the sizes of particles are increased through collisions and attachment of smaller particles leading to the formation of larger particles, called flocs. DAF performance is affected by these pretreatment processes so it is essential to have a good understanding of them.

The fundamentals of coagulation and flocculation are summarized in this chapter, and applications are emphasized with regard to DAF water plants. Guidance on design and operation is also presented. Because coagulation and flocculation are broad subjects, we present material that reflects our extensive research and consulting experience. Several of our key papers and book chapters are referenced throughout the chapter. Three other general sources are recommended and used. They are a chapter ("Coagulation and Flocculation") by Letterman and Yiacoumi (2010) in the book on *Water Quality and Treatment* (Edzwald, 2010), a chapter ("Coagulation, Mixing, and Flocculation") by MWH (2005), and the book *Flocs in Water Treatment* by Bache and Gregory (2007).

6-1 INTRODUCTION TO COAGULATION

Definitions

Coagulation is a chemical pretreatment step. Coagulation is concerned with treating the water so that particles in the raw water supply, particles produced within the treatment plant from precipitation processes (metal hydroxides from coagulation, ferric hydroxide and manganese oxide following oxidation, or precipitated particles from other processes), and particles added to water such as powdered activated carbon (PAC) are destabilized (aggregate or flocculate readily) for downstream flocculation and particle removal processes. Coagulation deals also with removing NOM (measured as DOC and UV_{254}), disinfection by-product (DBP) precursors, and natural color. Coagulation accomplishes this by a phase change converting dissolved organic matter into particles directly by precipitation or by adsorption onto particles produced by the coagulant.

Coagulation is an essential pretreatment process because it affects all downstream particle processes. It affects flocculation (particles do not aggregate well without coagulation), DAF (bubble attachment to flocs and removal of floc-bubble aggregates), and granular media filtration. In sedimentation plants it affects flocculation and particle settling. In membrane plants it affects removal of particles and fouling. In short, water treatment plants do not work well without good coagulation.

Terminology

We deal now with the terminology used to describe various chemicals: *coagulants, coagulant aids, flocculants, flocculant aids, flotation aids*, and *filter aids*. *Coagulants* are used to produce destabilized particles and to react with and remove NOM. Coagulants almost always refer to metals such as aluminum (aluminum sulfate or alum and polyaluminum chlorides [PACls]) and iron (ferric chloride and ferric sulfate). For high-quality water supplies (low turbidity and low TOC), high charge-density cationic polymers can be used as coagulants in direct filtration plants to remove particles, and they can actually remove a small amount of DOC. These cationic polymers are not used as the sole coagulant to treat raw waters in DAF plants; however, they are used with metal coagulation in a dual-coagulation strategy, as described in the following text. There has also been research on their addition to the DAF saturator water or the recycle water exiting the saturator to produce positively charged bubbles in the DAF process. This is a specialized and limited application that is discussed further under cationic polymers.

The term *coagulant aid* is often used loosely to mean the addition of more than one primary coagulant or flocculants or particles such as clays to increase the rate of flocculation. The term coagulant aid is often misleading because it does not describe the function of the chemical or other material added. For example, if two coagulants are used, this practice should be called dual coagulation. A good application of this, which is used in DAF plants, is the use of a metal coagulant (alum or iron) along with a cationic polymer coagulant. Both are added in rapid mixing and both are designed for particle and NOM removal. The cationic polymer coagulant reduces the metal dose thereby reducing the sludge produced. Dual coagulation is discussed in detail in Sec. 6-6. In short, the term coagulant aid is not used in this book. Rather, we use terms describing the function of what the chemical does.

The term *flocculants* refers to use of high molecular weight (MW) polymers with no charge (nonionic) or with low negative charge (anionic polymers). These high MW polymers are not coagulants. They do not convert dissolved NOM into particles and do not work well if applied to raw waters to destabilize negatively charged particles in the water supply. They are used after addition of coagulants to aid particle attachment and removal. The terms we favor align with their function. *Flocculant aids* are high MW polymers added after coagulation (after rapid mixing or within flocculation tanks) to bridge particles, forming larger and stronger flocs. *Flotation aids* are high MW polymers added after rapid mixing to aid attachment of bubbles to flocs or to aid retention of the float layer at the top of the DAF tank. *Filter aids* are high MW polymers added after rapid mixing (can be added for dual role of flocculant aid and filter aid or added solely as filter aid by addition to the filter influent) to improve deposition and retention of particles within the filter.

Optimum Coagulation

We deal now with what we mean by optimum coagulation. Unfortunately, the United States Environmental Protection Agency (USEPA) introduced the term *Enhanced Coagulation* as coagulation conditions designed to improved TOC removal. This terminology has a regulatory basis, but it is not scientifically rational. When metal coagulants are added to water they undergo chemical reactions with both particles and NOM, and we have no choice about it. So one should not describe coagulation in terms of a single objective such as turbidity (particles) removal, and then add an objective regarding TOC, if needed, to meet the USEPA regulation.

We define optimum coagulation conditions in terms of multiple objectives as originally described by Edzwald and Tobiason (1998, 1999). Optimum coagulation is defined as conditions of coagulant dosage and pH that:

- Maximize particle removal (turbidity, particles, pathogens) by clarification (DAF or sedimentation) and filtration.
- Maximize NOM removal (TOC, UV_{254}, and DBP precursors) by clarification and filtration.
- Minimize residual coagulant (in the water after treatment).
- Minimize sludge production.
- Minimize overall operating costs that consider the coagulant, pH-adjusting chemicals, polymers (if used for aiding flocculation, flotation, and filtration), and costs of sludge treatment and disposal.

Framework for Understanding Coagulation

Many people do not understand the coagulation process because they find it complicated or because they consider only one facet. To understand coagulation we must consider the contaminants in the water we wish to remove, the bulk water chemistry and temperature, and the chemistry of the coagulants. A way to think about it is in terms of a chemical equation where on the left-hand-side you have the coagulant and the contaminants undergoing reaction to form destabilized particles and NOM conversion, and you also recognize that these reactions depend on bulk water properties. Coagulation is presented within a framework in which we discuss each of the following separately and integrate the principles as we go along, and finally summarize under Sec. 6-6.

- Contaminants
  - Particles: knowledge about the particles in the raw water or added with treatment, in particular, their concentration and properties
  - NOM: knowledge about the dissolved NOM in the raw water, in particular, the concentration and nature including types and properties
- Bulk Water Properties
  - pH
  - Alkalinity and Buffer intensity
  - Hardness
  - Temperature
- Coagulants
  - Types and chemistry
  - How they accomplish coagulation

6-2 CONTAMINANTS

Particles

In Chap. 5, we discussed at length particle types, sizes, concentration parameters, stability of particle suspensions including interparticle forces between particles, and particle-bubble interaction and forces. Our interest is to prepare particles in the raw water and those added or produced by coagulant addition for effective flocculation and downstream removal by DAF and filtration.

The concentrations of the particles are important as they affect coagulant selection and dosing. For example, a coagulant that accomplishes particle destabilization by charge neutralization could be used to treat a low turbidity or low particle concentration water but the particle flocculation rate may be too low compared to *sweep-floc* coagulation in which additional particles are produced by precipitation of the metal coagulant, increasing the particle concentration and increasing the flocculation rate.

The stability of particles is caused by several mechanisms as described in Chap. 5, but a primary mechanism is the charge on the particle surface. Clays, depending on the type, have a negative charge density in the range of 0.05 to 0.5 μeq/mg (microequivalent of negative charge per mg of clay). Other mineral particles and organic particles such as algae, protozoa, bacteria, and viruses are negatively charged for the pH conditions of water supplies. We must consider the charge density of the particles and the particle concentration to obtain the total suspension negative charge. One can consider that the particle-suspension charge creates coagulant demand that must be satisfied by coagulation. If particles control the coagulant demand and not NOM, then we can see a dependence of coagulant dosing with particle concentration. Usually, however, NOM controls coagulant dosing. We examine the role of particles versus NOM in controlling coagulant dosing in Example 6-1.

Natural Organic Matter

NOM in water can have several effects on water quality and treatment including imparting color to water, causing aesthetic effects; reacting with disinfectants and oxidants, causing formation of DBPs and other by-products that can have health effects; causing coagulant demand; coating particles increasing their negative charge and affecting their stability; causing tastes and odors; causing oxidant demand; causing disinfectant demand; affecting the solubility of metals; decreasing the effectiveness of UV disinfection; causing the fouling of granular activated carbon (GAC) and membranes; affecting corrosion; and serving as a carbon source for biofilm growth in water distribution systems. Extensive coverage of NOM (properties, effects, and removal by various processes) is found in the general references on water treatment by MWH (2005) and Letterman and Yiacoumi (2010). For the many reasons listed here it is important to remove NOM through coagulation followed by particle-separation processes such as DAF and filtration.

Sources and Fractions. *Autochthonous* organic matter is produced within the water body. This organic matter can be particulate (measured as particulate organic carbon, POC) such as algae, bacteria, other microbes, and plant detritus. Algae are the largest contributor of POC, especially for lakes and reservoirs, but also for many rivers. The autochthonous organic matter can be dissolved, measured collectively as DOC. The DOC can be composed of proteins, amino acids, polysaccharides, and other organic matter produced from algal respiration and organic matter decay processes. *Allochthonous* organic matter is that NOM not formed in the water body under study, but formed elsewhere and transported to it. It can come from runoff from the watershed and from washing out of upstream water bodies such as bogs, marshes, swamps, reservoirs, and streams.

There has been considerable work in characterizing NOM into fractions, thereby yielding information on their chemical properties. One widely used fractionation method (Leenheer and Noyes, 1984; Bose and Reckhow, 1998) yields eight fractions: fulvic acid, humic acid, weak hydrophobic acids, hydrophobic bases, hydrophobic neutrals, hydrophilic acids, hydrophilic bases, and hydrophilic neutrals. For convenience in discussing coagulation we simplify the above and discuss two NOM fractions of aquatic humic matter (humic and fulvic acids) and nonhumic matter (the other listed fractions). The decomposition of plant and animal matter on the watershed yields a residual organic matter referred to as humic matter. Soils with a high organic content or areas containing peat impart aquatic humic matter to runoff and water supplies during rainfall events. Likewise bogs and swamps are high in aquatic humic matter that can enter downstream water bodies, especially following rain events. Many nonhumic fractions have properties that do not favor removal through coagulation such as those of lower MW and those with a hydrophilic nature. Some specific organic compounds attributed to nonhumic substances are phenols, carbohydrates, sugars and polysaccharides, proteins, amino acids, and fatty acids. The origin of nonhumic NOM can be allochthonous following rain events but are mostly autochthonous.

Aquatic humic matter is an important NOM fraction. For many water supplies it is the dominant fraction. When the TOC is more than a few mg/L, the presence of the humic matter gives water a yellow-brown color. However, for water supplies in which the yellow-brown color is low, aquatic humic matter can still be an important fraction of TOC.

Humic and fulvic acids are not specific molecules but a mixture with certain properties. Humic acid is more aromatic than fulvic acid. Humic acid is higher in MW (1000s) compared to fulvic acid, which is several 100. Aquatic humic and fulvic acids have similar C, H, and O composition of about 50, 5, and 40 percent, respectively, but humic acid has a greater N content of about 2 percent compared to fulvic acid at about 1 percent. There is a greater amount of aquatic fulvic acid in water supplies than humic acid because it is more soluble in water with its lower MW, while humic acid is retained by soils to a greater degree.

Aquatic humic substances contain acidic carboxyl and phenol-OH groups producing negatively charged macromolecules. The charge density of fulvic acid is about 6 μeq/mg of DOC at pH 5 and increases to 14 μeq/mg of DOC at pH 9 (Van Benschoten, 1988). This negative charge is high compared to the negative charge associated with clay minerals and organic particles such as algae, protozoan cysts, bacteria, and viruses. This high negative charge affects the coagulant demand, as illustrated in Example 6-1. Because of their higher MW compared to nonhumic substances and their hydrophobic properties, coagulation of humic substances is favored and removals of this fraction of NOM can be high.

Total Organic Carbon of Water Supplies. TOC is a collective measurement of organic matter. It is of interest for several reasons including (1) classification of raw and treated water quality, (2) use as surrogate measurement for DBP precursors, and (3) evaluation of water treatment processes for their ability to remove organic matter. TOC, by definition, is the sum of POC and DOC. Except for eutrophic lakes, DOC is 90 to 99 percent of the TOC for surface waters used for drinking water. For groundwaters, POC is nearly zero so the DOC is essentially equal to the TOC.

Table 6-1 summarizes ranges and means for TOC concentrations for various types of water supplies. Groundwaters have the lowest TOC, generally <2 mg/L. Shallow groundwater systems can have higher TOC when under the direct influence of surface waters, and higher TOC can occur for groundwater affected by peat deposits or areas with swamps such as those occurring in Florida. Streams and rivers have TOC of 5 mg/L or less; however, the TOC can be quite high for supplies in watersheds with marshes, bogs, and swamps, especially after rainfall events that wash NOM into the supplies. In some countries, large treated sewage flows can affect the TOC concentration of small rivers. Oligotrophic lakes (low in nutrients and algae biomass) have the lowest TOC concentrations. As lakes and reservoirs progress to mesotrophic and eutrophic states, the TOC increases from the effect of algae. Impounded streams that are downstream of bogs and marshes can have high TOC, especially after high rainfall events.

TABLE 6-1 TOC Concentrations in Various Water Bodies

Water Body	Range (Mean) (mg/L)	Other
Groundwater	0.5–10 (0.7)	Many groundwaters are 2 mg/L or less
Rivers and streams	1.5–20 (ND)*	Typically 5 mg/L or less; higher values downstream of bogs, marshes, and swamps. Some can exceed 20 mg/L
Oligotrophic lakes	1.0–3.0 (2.2)	
Mesotrophic lakes	2.0–4.0 (3.0)	
Eutrophic lakes	3.0–30 (12.0)	

* ND, no data.

Based on authors' experience and from information in MWH (2005) and Edzwald and Tobiason (2010).

***UV_{254}* Absorbance as a Surrogate Parameter for DOC and TOC.** It was shown in the 1980s that UV_{254} can be used as a surrogate parameter for monitoring and predicting DOC, TOC, and THM precursors in raw water supplies and in treated waters by Edzwald et al. (1985). It is now widely used in water treatment as a surrogate parameter for DOC and TOC. Measuring UV_{254} is simple, rapid (grab sampling or on-line instrumentation), and inexpensive, making it a good surrogate parameter. Technically, samples for UV_{254} measurements should be filtered to remove particulate matter that may interfere with the measurement so it ought to be more strongly correlated with DOC, but it has been shown that strong correlations are obtained also with TOC, especially for water supplies low in turbidity and POC and obviously for treated waters where particulate matter has been removed. Our interest is in its utility as a surrogate parameter for DOC and TOC in raw waters, in treated waters following coagulation and particle separation, and in establishing coagulant dosage requirements.

Raw water UV_{254} for water supplies vary from about 0.04 cm^{-1} for water supplies low in TOC and aquatic humic matter to about 0.8 cm^{-1} for water supplies high in TOC and aquatic humic matter. Many water supplies fall often in the range of 0.07 to 0.3 cm^{-1}, which corresponds approximately to TOC concentrations of 2 to 7 mg/L depending on the nature of the NOM in the water supply, as determined by specific UV absorbance (SUVA) which is discussed in the next section. Treated water UV_{254} values depend on the extent of TOC removal and the water treatment process. Edzwald and Kaminski (2009) have shown that full-scale water plants operating with optimum coagulation dosing and pH conditions for both particles (turbidity) and TOC can achieve treated water with UV_{254} of 0.023 to 0.033 cm^{-1} (prior to chlorine or oxidant addition) corresponding to treated water TOC of 1.5 to 2 mg/L.

Specific UV Absorbance. SUVA was developed by Edzwald et al. (1985) to evaluate whether TOC and DBP precursor concentrations can be correlated with UV_{254}, and was then later utilized regarding NOM composition of water and the ability of coagulants to remove TOC (Edzwald and Van Benschoten, 1990; Edzwald, 1993). It provides a simple way to characterize the nature of NOM, and it is calculated from measurements of UV_{254} and DOC. SUVA values are expressed in units of m^{-1} absorbance per mg/L of DOC.

$$\text{SUVA} = \frac{\left(\text{UV}_{254} \text{ in cm}^{-1}\right) \times 100 \dfrac{\text{m}^{-1}}{\text{cm}^{-1}}}{\left(\text{DOC in mg/L}\right)} \tag{6-1}$$

Certain types of organic matter absorb UV_{254} light per unit concentration of DOC to a greater degree than other types. Water supplies contain a mixture of types of NOM, but the SUVA values serve as indicators of what types of organic compounds dominate the DOC or TOC of the water. The composition of NOM for raw waters is presented in Table 6-2 with respect to three ranges of SUVA. The higher the SUVA value, the water's TOC has a greater fraction of aquatic matter and the greater the percentage of TOC removed by coagulation. As SUVA decreases, the TOC composition is greater in nonhumic matter and removal by coagulation is poorer. UV_{254} percentage removals are greater than the TOC removals summarized in Table 6-2. UV_{254} can be correlated to TOC, so TOC removals can be predicted (Edzwald and Kaminski, 2009).

Coagulation preferentially removes aquatic humic matter; consequently, the SUVA of treated waters is much lower than that of the raw waters. SUVA values of 2 or less for treated waters indicate optimum coagulation practice for the removal of NOM.

TABLE 6-2 Characterization of NOM with Respect to Raw Water SUVA and Guidelines on TOC Removals by Coagulation

SUVA (m^{-1} per mg/L of DOC)	Composition	Guidelines on TOC Removals by Coagulation
>4	High fraction of aquatic humic matter	60–80%
	Highly aromatic and hydrophobic character	Higher end for waters with high TOC
	High molecular weight (MW)	
2–4	Mixture of aquatic humic and nonhumic matter	40–60%
	Mixture of aromatic and aliphatic character	Higher end for waters with high TOC
	Mixture of low to high MW	
<2	High fraction of nonhumic matter	<20–40%
	Highly aliphatic and low hydrophobic character	Higher end for waters with high TOC
	Low MW	

Source: Edzwald and Tobiason (2010), reprinted with permission from Water Quality and Treatment: A Handbook on Drinking Water, 6th ed., Copyright 2010, AWWA.

Role of Particles versus Natural Organic Matter in Controlling Coagulation

The traditional view, but invalid one, is that particles (turbidity) control the coagulant dose with the exception of water supplies with moderate to high values of natural color. For most water supplies, whether they contain much natural color or not, NOM controls coagulant dosing. This comes about from recognizing that metal coagulants (Al and Fe) undergo complexation reactions with NOM by which NOM causes a coagulant demand. The good news is that coagulants can remove the dissolved NOM by direct precipitation or by adsorption to metal hydroxide solids. Therefore, we recognize the coagulant demand of NOM and define optimum coagulation conditions as the dose and pH that maximizes removals of both particles and TOC and that minimizes dissolved residual metal coagulant—see section, Optimum Coagulation.

Example 6-1 shows that NOM usually has a far greater coagulant demand than particles based on satisfying the negative charge of the two contaminants. This is true for most water supplies except those low in TOC and with low SUVA values, especially water supplies with high alkalinity and hardness. Generally for water supplies with SUVA of about 2.5 or greater, the TOC (NOM) controls coagulant dose and not particles (turbidity).

EXAMPLE 6-1 Particle Charge versus NOM Charge We compare the negative charge associated with particles to the charge on aquatic fulvic acid. Consider a particle suspension consisting of clays. The negative charge density depends on clay mineral type and pH conditions. For pH conditions of about 6 to 8, we use a typical clay particle charge density of 0.1 μeq/mg of clay. While we are using clays for our model particle suspension, it is reasonable to assume that many other mineral particles and organic particles such as algae, bacteria, and protozoan cysts have negative charges equal to or less than the clays.

For aquatic humic matter, we use aquatic fulvic acid as the primary form. The charge on aquatic fulvic acid depends on pH, as described above. In this example, a negative charge density of 10 μeq/mg of DOC is used in the calculations, which is a reasonable value for near-neutral pH conditions.

Consider then a water supply in which the mass concentration (suspended solids) of particles varies over the year from 5 to 50 mg/L (turbidities, say, of 3 to 60 NTU). Consider also that the TOC varies from 2 to 6 mg/L and corresponding DOC varies from 2 to 5.8 mg/L. The DOC is composed of about 50 percent humic matter in the form of aquatic fulvic acid and the remaining nonhumic NOM is assumed to be uncharged.

Solution

Negative Charge Attributed to the Particles We calculate the particle negative charge for the 5 mg/L suspension case:

$$\text{Particle Negative Charge} = 0.1\ \frac{\mu\text{eq}}{\text{mg}} \times 5\ \frac{\text{mg}}{\text{L}} = 0.5\ \frac{\mu\text{eq}}{\text{L}}$$

A similar calculation is made for the water supply when at a high particle concentration of 50 mg/L, and we obtain a particle suspension negative charge of 5 μeq/L.

Negative Charge Attributed to the NOM For the case of 2 mg/L of TOC or DOC and 50 percent of the DOC composed of aquatic fulvic acid, we obtain:

$$\text{NOM Negative Charge} = 10\ \frac{\mu\text{eq}}{\text{mg TOC}} \times 0.5 \times 2\ \frac{\text{mg TOC}}{\text{L}} = 10\ \frac{\mu\text{eq}}{\text{L}}$$

A similar calculation is made for the high DOC case of 5.8 mg/L and we obtain a NOM negative charge for the water supply of 29 μeq/L. The results for the particles and NOM are summarized in the following table:

Particle Suspension		NOM		
Mass (mg/L)	Negative Charge (μeq/L)	DOC (mg/L)	Negative Charge (μeq/L)	**Ratio of NOM Charge to Particle Charge**
5	0.5	2	10	20
50	5	5.8	29	5.8

The calculations show that the NOM in this water supply has 6 to 20 times greater negative charge than the charge associated with particles. Obviously, the concentrations of particles and NOM can vary among water supplies, but it turns out that for many water supplies the NOM has a far greater negative charge associated with it than particles, and this negative charge must be effectively neutralized in coagulation. In short, for many water supplies coagulant dose is controlled by NOM and not turbidity. ▲

6-3 BULK WATER CHEMISTRY AND TEMPERATURE

pH

The pH of coagulation is an important variable affecting coagulation. When metal coagulants are added in rapid mixing, they act as acids (chemistry is explained in Sec. 6-5) and can decrease the pH depending on the dosage and the water's buffer intensity. The pH of coagulation, whatever the goal, is the pH that results after coagulant addition and should be approximately maintained through flocculation, DAF, and in most cases, granular media filtration.

What effects does pH have on coagulation using alum, ferric coagulants, and PACls?

- pH affects the dissolved coagulant species initially present in rapid mixing, their charge, and their reactions with NOM and turbidity.
- pH affects the solubility of the metal coagulants and therefore affects sweep-floc coagulation and the dissolved residual concentration of the metal coagulant. This is especially true for alum and PACls, less so for ferric coagulants that are insoluble over a wide range of pH conditions.
- pH affects the charge on flocs that are formed through coagulation and flocculation.

What effects does pH have on the contaminants that we are attempting to coagulate? Generally, as addressed earlier in the chapter, particle (mineral and organic) negative charge increases as pH increases; however, more significant is the effect of increasing pH on the negative charge of NOM that increases to a far greater extent than particle charge.

Alkalinity and Buffer Intensity

Alkalinity (Alk) is the acid-neutralizing capacity of water. It is the amount of acid required to titrate a water sample to a pH endpoint of about 4.5. The proper parameter that evaluates buffering is the buffer intensity (Stumm and Morgan, 1996; Benjamin, 2002). While Alk is a capacity measurement, buffer intensity (β) is the resistance to a unit change in pH for a unit addition of strong acid (C_a). Mathematically, it is defined by Eq. (6-2).

$$\beta = \frac{-\Delta C_a}{\Delta pH} = \frac{-\Delta Alk}{\Delta pH} \tag{6-2}$$

It is the inverse of the slope of the alkalinity titration curve. The negative sign is because the slope of this curve is negative. The buffer intensity is more useful in coagulation considerations because it measures resistance to pH change over specific pH intervals.

Figure 6-1 shows the buffer intensity (β) for a moderate-alkalinity water of 115 mg/L $CaCO_3$ and compares it to low-alkalinity water of 30 mg/L $CaCO_3$. Obviously the buffer intensity depends on the alkalinity, but the most buffering or resistance to pH change upon acid addition occurs at pH conditions in the range of pH 5.8 to 6.8, with maximum buffering at pH 6.3. Much less resistance to pH change occurs at higher and lower pH. Thus, if we add a metal coagulant such as alum or an iron salt to the low-alkalinity water initially at pH 7.5, there is little buffer intensity and the pH will decrease rather easily into the low 6s. If the dosage is high, then one can further decrease the pH to low, undesirable conditions. For the moderate alkalinity water initially at pH 8, adding a metal coagulant may decrease the pH to the low to mid 7s. It would be difficult, however, to decrease the pH below 7 because of the high buffer intensity; it would take high coagulant dosages or use of a strong acid.

Buffer intensity is the proper parameter to evaluate the resistance to pH change and one should keep the concept in mind. In practice we tend to use alkalinity as our indicator of buffering. Some guidance is presented on the use of alum and iron coagulants and their effects on consuming alkalinity and pH:

- For low-alkalinity waters of, say, less than 60 mg/L $CaCO_3$:
  - If the metal coagulant dosage is low, then you may end up at the optimum pH condition.
  - If the metal coagulant dosage is higher, then the pH may drop below optimum pH conditions unless you add a base (NaOH or $Ca(OH)_2$) to control pH.

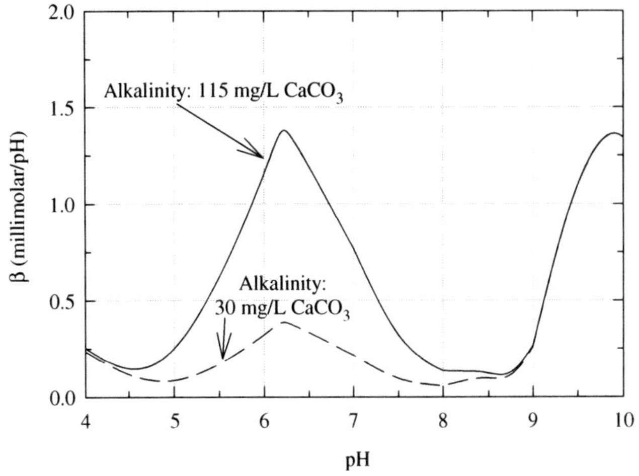

FIGURE 6-1 Buffer intensity (β) at 20°C for a water with a moderate alkalinity of 115 mg/L $CaCO_3$ initially at pH 8 and a water of low alkalinity of 30 mg/L $CaCO_3$ initially at pH 7.5.

- For medium-alkalinity waters of, say, 60 to 120 mg/L $CaCO_3$:
  - Low dosages may give you higher pH conditions than you desire. For these medium-alkalinity waters, you may add additional coagulant and use its acid property to give you the desired pH. This is often the case in practice. Other options are to use carbon dioxide to decrease the pH or a strong acid such as sulfuric acid. The use of carbon dioxide has the additional advantage of increasing the inorganic carbon concentration for corrosion control at the end of the treatment plant.
  - For waters near the lower end of the specified alkalinity range, moderate to high coagulant dosages may yield optimum pH conditions.
  - For waters near the higher end of the alkalinity range, then nonoptimum pH conditions may occur with coagulant dosing unless high doses are used. Addition of a strong acid should be considered to avoid high dosages.
- For high-alkalinity waters of, say, greater than 120 mg/L $CaCO_3$:
  - It is rare to have water supplies with high alkalinity and moderate to high TOC. If this occurs, then acid addition will be required to reach optimum pH conditions and to avoid excessive coagulant dosages to decrease the pH.
  - For water supplies where TOC is low, then one can use coagulants at higher pH conditions without acid addition.

If PACls are used as coagulants, they consume less alkalinity depending on their basicity—see section on PACls. For low- and medium-alkalinity waters, the use of PACls can be a good choice to reach the optimum pH condition without losing pH control since they have less of an effect on decreasing pH than alum and iron coagulants. For some medium- and higher-alkalinity waters, some acid may be needed to reach optimum pH conditions for water supplies with moderate to high TOC. For higher-alkalinity waters, where often TOC is low, the PACls can be effective at higher pH; however, one must also consider the residual coagulant concentration since they are more soluble at higher pH; see section on PACls.

Hardness

There are three things to discuss with respect to hardness. First, the hardness of waters is related to the dissolved solids and alkalinity: the greater the water hardness, the greater the dissolved solids and alkalinity. Since hardness is related to alkalinity, the alkalinity effects apply as spelled out above. Second, waters with high hardness and high dissolved solids have high ionic strength. As the ionic strength of water increases, the stability of particles from electrical double-layer effects is reduced, as explained in Chap. 5. Thus, particles are less stable in high-hardness waters and ought to be easier to coagulate. Third, for water supplies of high hardness one finds generally lower concentrations of aquatic humic matter. This may be due to the effect that calcium is available in the soils and sediments to retain humic matter.

Water Temperature

It is well known that water temperature affects chemical reactions and water properties; however, temperature effects on treatment chemistry are sometimes ignored in the design, operation, and assessment of water plant performance. One must account for water temperature. The following list summarizes temperature effects. In later sections these effects are explained and discussed.

- Water density and viscosity depend on temperature, which have effects on rapid mixing, flocculation, DAF, and filtration.
- Diffusion of small particles is affected by temperature, which can affect flocculation, DAF, and filtration.
- Water dissociation into H^+ and OH^- is affected by temperature.
- The dissolved metal speciation of the coagulant is affected by temperature. Major effects occur in dissolved Al chemistry from use of alum, and there are effects on the dissolved chemistry of iron and PACl coagulants.
- There are major effects on the solubility of aluminum hydroxide solids in sweep-floc coagulation from use of alum and PACls. There is an effect on ferric hydroxide solubility, but since ferric hydroxide is very insoluble over a wide pH range, its effect has small practical importance.

6-4 RAPID MIXING

Some principles and comments are made in a brief form. The reader should consult MWH (2005) for extensive discussion and elaboration of design principles. The purpose of rapid mixing is to disperse quickly the chemicals being added. This can be accomplished in a variety of ways. It can be accomplished by providing turbulent energy from impeller or turbine mixers, pipe injection, in-line static mixers, and open-channel static mixers, as shown in Fig. 6-2. Other ways to disperse chemicals are to inject to the upstream side of pumps, and making use of head loss in water channels, hydraulic jumps, and bends in pipes.

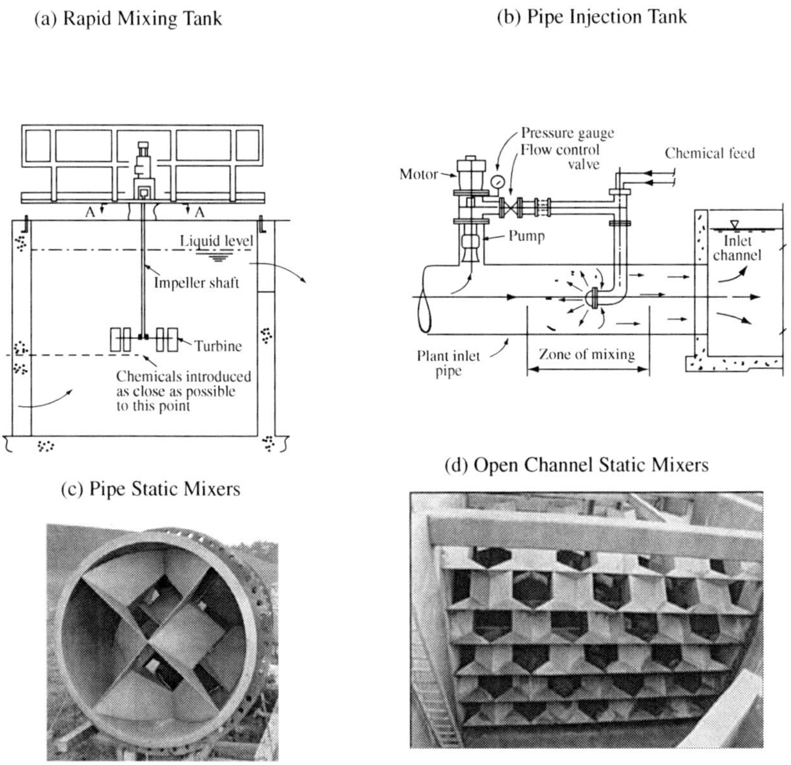

FIGURE 6-2 Types of rapid mixing units. *Credits: (a) and (b) Source: AWWA and ASCE (2005), reprinted from Water Treatment Plant Design, 4th ed., by permission. Copyright © 2005, American Water Works Association*, (c) *Courtesy of Russell Jones, Hazen and Sawyer*; (d) *Courtesy of Statiflo International Ltd.*

The chemicals of interest are acids or bases for pH control and coagulants. To gain an idea of the desired hydraulic detention times to achieve good dispersion of these chemicals, one examines the reaction times of these chemicals. The concept is that you would like to have detention times close to the reaction times so you achieve the desired chemistry. If a coagulant chemical first reacts to form a desired chemistry, but then for longer times undergoes other reactions, you would like to control that by carrying them out in a second reactor or step. Acids and bases are used for pH control. These chemicals involve proton transfer reactions, which are very rapid (much less than a second). Common acids used are strong ones such as sulfuric acid (H_2SO_4) or a weak acid such as carbon dioxide (CO_2). Common bases used are caustic (NaOH) or lime ($Ca(OH)_2$). Mixing times for addition of acids and bases should be as short as possible.

Metal coagulants (Al and Fe) undergo hydrolysis reactions forming dissolved species that are important in charge neutralization reactions of particles and complexation reactions with NOM—these reactions are explained in Sec. 6-5. These hydrolysis reactions occur in reaction times of much less than a second. If one practices sweep-floc coagulation, which is commonly done with alum and ferric coagulants, then a second reaction occurs involving precipitation of metal hydroxides, which takes place over seconds and minutes.

One would like to achieve the dissolved metal hydrolysis chemistry in a uniform way, and so short reaction times are desired. In MWH (2005) they recommend detention times less than a second for alum and ferric coagulants. Depending on the rapid mixing method, this may not be possible, but a short reaction time of seconds is desired.

The degree of mixing intensity is another consideration. Some mixing equipment companies use the power number to characterize mixing from impellers, where the power number depends on the applied power, impeller diameter, and impeller rotational speed. Static-mixer equipment companies sometimes evaluate blending uniformity by the *coefficient of variation* (simply called the variation coefficient), which measures the normalized standard deviation of the desired concentration of chemical in the reactor. Presentation of this parameter is found in MWH (2005). In the water treatment field, Camp's *velocity gradient* (G) is used. It is the root-mean-square velocity gradient and is a global measure of mixing within the system. It is calculated from Eq. (6-3) where P is the power input, μ_w is the water dynamic viscosity, and V is the water volume of the system. Criteria for G vary among design engineers, but it is common to find design values of 400 to 1000 s^{-1}—see MWH (2005) and AWWA and ASCE (2005).

$$G = \left(\frac{P}{\mu_w V} \right)^{1/2} \tag{6-3}$$

The last consideration is the addition sequence of chemicals when adding more than one. For the three cases addressed, water plant operators often do not observe any difference in plant performance on what is added first; nonetheless, the authors' preferences follow.

- *Metal coagulant and acid for pH control:* For waters with moderate to high TOC, add the acid first and then add the coagulant.
- *Metal coagulant and base for pH control:* For waters with moderate to high TOC, add the coagulant first and then add the base.
- *Metal coagulant and high charge-density cationic polymeric coagulants:* Add the cationic polymer first and then add the metal coagulant.

6-5 COAGULATION CHEMISTRY AND MECHANISMS

General information on several coagulants is presented in Tables 6-3 and 6-4. We do not address sodium aluminate and polyferric chlorides and sulfates in this chapter. They are not widely used and the reader can consult other sources such as MWH (2005) and Letterman and Yiacoumi (2010). We consider the following coagulants in this section: alum, ferric salts of ferric chloride and ferric sulfate, PACls, and organic high charge-density cationic polyelectrolytes. Alum is the most widely used coagulant for water treatment so we begin with it. It is a metal coagulant that has some similar chemistry to ferric coagulants so it is covered in greater detail, and then principles of ferric coagulation chemistry are summarized. PACls are specialty coagulant chemicals. They are preneutralized Al products designed to provide polymeric Al when applied under certain conditions. Organic cationic polyelectrolytes are not used as sole coagulants in DAF plants, but they are used as a dual coagulant with metal coagulants. Principles are summarized and some discussion is provided on their application.

TABLE 6-3 Common Inorganic Metal Coagulants and General Properties

Coagulant	Chemical Formula	General Properties of Liquid Products
Aluminum sulfate (alum)	$Al_2(SO_4)_3 \cdot 14H_2O$	Liquid alum: 8.0–8.4% Al_2O_3, s.g.* 1.33, pH 2.0–2.4. Dosing in the U.S. and some countries as dry alum [$Al_2(SO_4)_3 \cdot 14H_2O$ (17.1% Al_2O_3)]; other countries expressed as Al. Note 1 mg/L as Al = 11 mg/L as dry alum.
Sodium aluminate	$NaAlO_2$	20–45% Al_2O_3. It is a base so some use with alum to control pH.
Ferric chloride	$FeCl_3$ (sometimes expressed as $FeCl_3 \cdot 6H_2O$)	12–14.5% Fe^{3+}, s.g. 1.3–1.5, pH < 2 (contains excess HCl). Products contain some Mn.
Ferric sulfate	$Fe_2(SO_4)_3$ (sometimes expressed as $Fe_2(SO_4)_3 \cdot 9H_2O$)	10–13% Fe^{3+}, s.g. 1.4–1.6, pH 1–2 (contains excess H_2SO_4. Products contain some Mn. Available as solid product.
Polyaluminum chloride (PACl)	Various formulations	Important properties are strength (range 9–24% Al_2O_3), basicity, sulfate content, other inorganic and organic constituents).
Polyferric chloride and sulfate	Various formulations	See property data from chemical suppliers.

*Specific gravity.

TABLE 6-4 Types and Properties of Synthetic Organic Polymers Used in Drinking Water Treatment

Type and Charge (μeq/mg)*	Structure	Molecular Weight	Use
Epichlorohydrin dimethylamine (Epi-DMA) Cationic about + 8	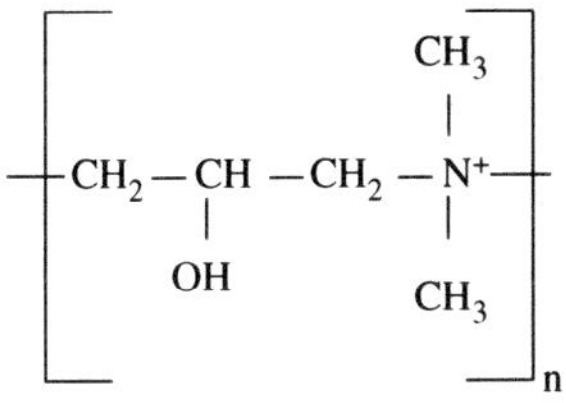	10^4 to 10^5	Primary coagulant
Polydiallyldimethyl ammonium chloride (Poly-DADMAC) Cationic about + 6	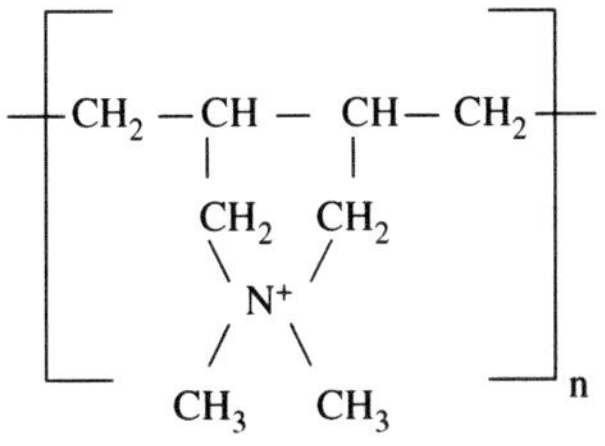	$<10^5$ to 10^6	Primary coagulant

(continued)

TABLE 6-4 Types and Properties of Synthetic Organic Polymers Used in Drinking Water Treatment (*Continued*)

Type and Charge (μeq/mg)*	Structure	Molecular Weight	Use
Polyacrylamide (PAM) No charge	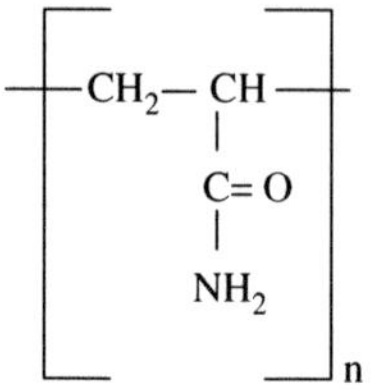	10^6 to 10^7	Flocculant, flotation, and filter aid
Hydrolyzed polyacrylamide (HPAM) Anionic co-polymer (low negative charge)	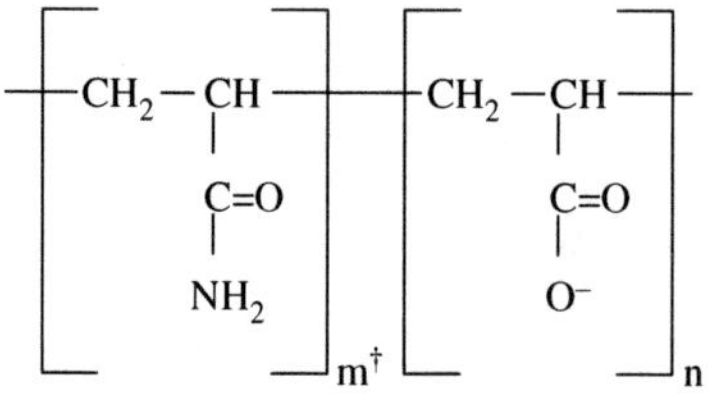	10^6 to 10^7	Flocculant, flotation, and filter aid

* Charge density based on 1 mg of polymer; the cationic polymers are liquid products containing 20 to 50 percent polymer; for example, a 50 percent Epi-DMA polymer product has a positive charge density on a product basis of about + 4 μeq/mg

† m > n

Alum

Figure 6-3 shows a conceptual representation of alum coagulation reactions. This figure follows the Framework for Understanding Coagulation described in Sec. 6-1 and which we use to explain coagulation. Now we present principles about the chemistry of alum and follow it with integration of material on contaminants, bulk water chemistry, and temperature to explain how alum works and to present optimum coagulation conditions for alum.

Alum Chemistry. We find it instructive to present the basics of alum chemistry in a sequence that follows use in a water plant. First, we describe the chemistry of liquid alum of the feed-alum solution added to rapid mixing. Second, we describe the dissolved aluminum chemistry initially present in rapid mixing. Next, we describe the solubility of aluminum hydroxide.

Liquid Alum. Commercial liquid alum is the most common way alum is supplied to water plants. Some water plants may purchase dry alum, and then prepare a liquid alum solution for feeding the coagulant. We describe the chemistry of commercial liquid alum. It is a chemical that is sold with specifications of aluminum content and product specific gravity. The products normally contain 8.0 to 8.4 percent Al_2O_3 and typically contain 8.3 percent Al_2O_3 with a specific gravity of 1.33. The liquid products are typically about 48.5 percent of dry alum on a weight basis. Dry alum is defined as $Al_2(SO_4)_3{\cdot}14H_2O$ and its dissolution is shown in Eq. (6-4).

$$Al_2\left(SO_4\right)_3 \cdot 14H_2O \rightarrow 2Al^{3+} + 3SO_4^{2-} + 14H_2O \qquad (6\text{-}4)$$

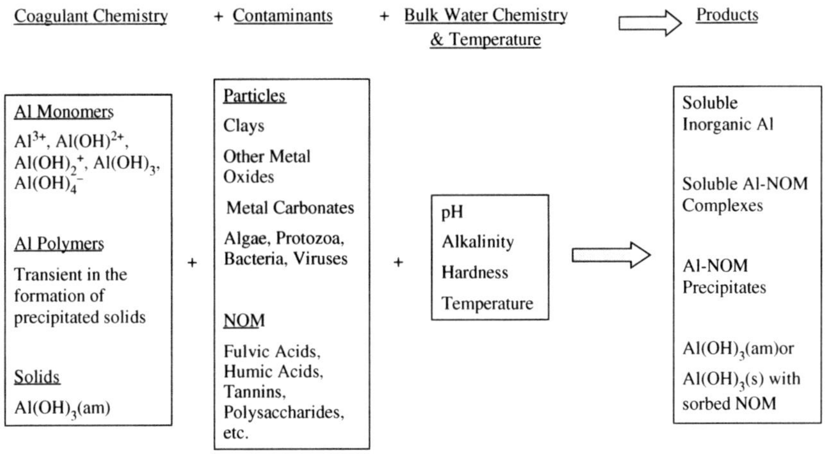

FIGURE 6-3 Conceptual representation of coagulation reactions for alum. *Source: Pernitsky and Edzwald (2006), Journal of Water Supply: Research and Technology – AQUA, 52 (No. 4), 121–141, with permission from the copyright holders, IWA Publishing.*

A few things are worth noting. Alum is made with sulfuric acid so we have dry alum in the form of aluminum sulfate. Alum is an acid because the dissolved metal (Al^{3+}) is an acid (to be shown next) and makes the liquid alum product low in pH, typically about pH 2 to 2.4. The second thing to note is writing dissolved aluminum as Al^{3+} is a shorthand. It does not exist in this form. The positive charge attracts six water molecules to form an aluminum complex with water. So it really exists as $Al(H_2O)_6^{3+}$. So what we feed to the rapid mix process is $Al(H_2O)_6^{3+}$. Sulfate is also added, but at low concentrations. The third thing to note is the reporting of alum coagulant dosages. In the United States and some other countries, dosage is reported as dry alum. In many other countries, the dosage is reported as Al. The authors prefer the latter method of expressing dosing because it relates directly to the metal, which acts as the coagulant and because one can compare dosages on a common basis to other aluminum coagulants such as PACls. Another reason is that it makes monitoring straight-forward of residual Al through the water plant. In the book, we report dosages as Al. Some useful conversions follow:

- 1 mg/L as Al = 11 mg/L as dry alum ($Al_2(SO_4)_3 \cdot 14H_2O$)
- 1 mg/L as Al = 0.037 millimolar (mM) = 3.7×10^{-5} M

Dissolved Aluminum Chemistry. What we describe now is the dissolved aluminum chemistry immediately after addition of alum to rapid mixing. The reactions that occur in microseconds are called hydrolysis reactions and the dissolved aluminum species that are formed are those that can react with contaminants. The first hydrolysis reaction that occurs is shown in Eq. (6-5).

$$Al(H_2O)_6^{3+} + H_2O \rightleftharpoons Al(H_2O)_5(OH)^{2+} + H_3O^+ \qquad (6\text{-}5)$$

This demonstrates that $Al(H_2O)_6^{3+}$ is an acid because it donates a H^+ to the bulk water (forming the hydronium ion, H_3O^+). It also shows that the hydrolysis reaction produces an aluminum-hydroxo complex ($Al(H_2O)_5(OH)^{2+}$) with a smaller charge of +2. It turns out that $Al(H_2O)_6^{3+}$ can lose up to four H^+. Rather than showing each stoichiometric reaction as was done above for the first reaction, we show the entire sequence of reactions and the aluminum-hydroxo complexes formed through Eq. (6-6).

$$\begin{aligned} &Al(HOH)_6^{3+} \xrightarrow{H^+} Al(HOH)_5(OH)^{2+} \xrightarrow{H^+} Al(HOH)_4(OH)_2^{+} \xrightarrow{H^+} \\ &Al(HOH)_3(OH)_3 \xrightarrow{H^+} Al(HOH)_2(OH)_4^{-} \end{aligned} \tag{6-6}$$

The dissolved aluminum-hydroxo complexes formed depend primarily on the pH of coagulation in rapid mixing. Water temperature is also a factor affecting the speciation because it affects the complexation equilibrium and water dissociation. The distribution of dissolved Al species as a function of pH for 20 and 5°C is shown in Fig. 6-4. In this figure for simplification in writing the Al species, the complexed water molecules were excluded; for example, the shorthand Al^{3+} is written rather than $Al(H_2O)_6^{3+}$. The Al species initially

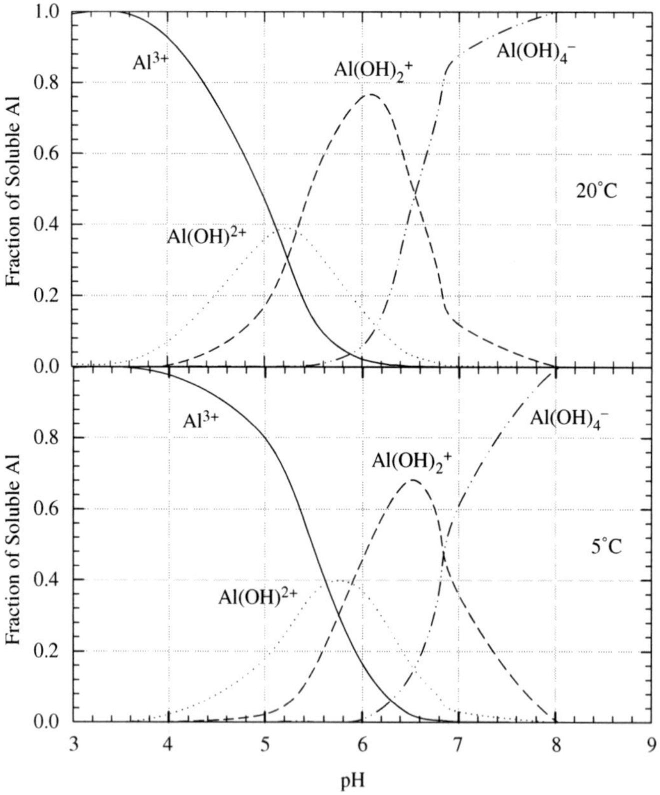

FIGURE 6-4 Dissolved Al speciation for alum as a function of pH for 20 and 5°C. *Source: Pernitsky and Edzwald (2006), Journal of Water Supply: Research and Technology – AQUA, 52 (No. 4), 121–141, with permission from the copyright holders, IWA Publishing.*

available for reactions with contaminants depends on the pH of rapid mixing after we add the alum. Under warm water conditions (say, 20°C) and pH near 6, we see that we have mostly $Al(OH)_2^+$ and then some $Al(OH)^{2+}$ and $Al(OH)_4^-$, with a small fraction of Al^{3+}, which was the primary species in the feed-liquid alum. It is important to note that for cold water coagulation (see distribution of Al for 5°C), there is a shift to the right of the curves, that is, to higher pH. It is discussed later that the optimum pH conditions for alum coagulation depend on temperature. We learn generally from Fig. 6-4 that positively charged Al dominates for pH conditions less than 6.5 to 7 and that we have primarily negatively charged Al for pH greater than 7. This Al speciation chemistry and the effect of pH are important. If we consider that initially we desire positively charged Al to react with negatively charged particles and NOM, then we should practice alum coagulation at acidic pH conditions.

The next chemistry to consider is the solubility of aluminum hydroxide. We will learn that we should use alum at pH 6 to 7 to maximize precipitation of aluminum hydroxide (what we call sweep-floc coagulation) and to minimize dissolved residual Al. To take advantage of the initial presence of positively charged Al, and then shortly thereafter precipitation of aluminum hydroxide, we should carry out coagulation at pH 6 to 6.5 for warm water conditions and at higher pH conditions of 6.5 to 7 for cold waters. Note that the uncharged Al complex is not shown in Fig. 6-4. It was not considered because it is not found in measurements of Al solubility, which is discussed in the next section.

To summarize, three important points are made about dissolved Al chemistry:

- Alum is an acid donating H^+. This can affect the coagulation pH depending on the dosage for water supplies with low and moderate alkalinity, specifically depending on the water buffer intensity.
- Dissolved Al-hydroxo complexes form instantaneously (microseconds) when alum is added, and they are the forms of dissolved Al that initially react with contaminants in rapid mixing.
- The dissolved Al-hydroxo complexes available for coagulation in rapid mixing depend on pH and water temperature.

Finally, the distribution of dissolved Al as discussed here considered only hydrolysis reactions in which Al is complexed with water or OH^-. Other Al complexes can form if certain ligands (anions or molecules that complex with a metal) are present at sufficient concentration. These ligands can be F^-, NOM, or PO_4^{3-}. The one of interest to us for coagulation of raw waters is NOM, and we address it under the section Integration of Alum Chemistry with Reactions with Contaminants.

Solubility of Aluminum Hydroxide. We consider the solubility of amorphous solid hydroxide ($Al(OH)_3$(am)), which is formed when Al is added at a sufficiently high concentration under certain pH conditions. In this presentation, we do not consider the effects of complexing agents such as NOM on Al solubility.

Precipitation occurs in several stages: the dissolved Al undergoes polymerization forming higher-MW Al polymers; this is followed by nucleation in which submicron solid particles are formed, and then aggregation of these submicron particles occurs forming larger particles that we call flocs. For water treatment, heterogeneous nucleation occurs, meaning the particles (turbidity) serve as sites for nucleation to occur and to catalyze the nucleation and precipitation processes. This precipitation of $Al(OH)_3$(am) produces new particles, which can incorporate the raw water particles into the flocs. In water treatment, this is called sweep-floc coagulation.

Figure 6-5 shows solubility diagrams for $Al(OH)_3$(am) for two water temperatures. It is important to note that the solubility diagrams agree with actual Al solubility data for drinking water conditions (Van Benschoten and Edzwald, 1990; Pernitsky and Edzwald, 2003).

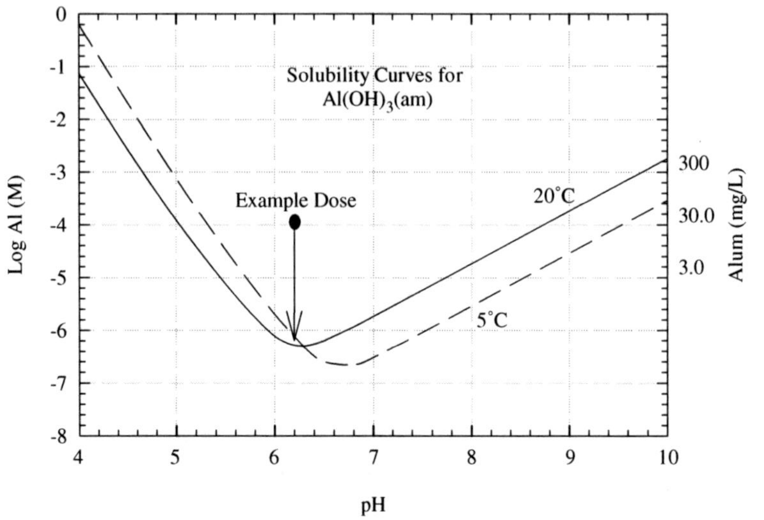

FIGURE 6-5 Solubility of amorphous aluminum hydroxide ($Al(OH)_3$(am)) for alum at 20 and 5°C. Example dose at 10^{-4} M at pH 6.3 and 20°C presented in the text.

If we were to dose alum at 2.7 mg/L (10^{-4} M Al) at pH 6.3 and 20°C, then we would begin coagulation in rapid mixing at the dot (top of arrow in Fig. 6-5) with nearly 85 percent of the dissolved Al species being positively charged with mostly $Al(OH)_2^+$ and some $Al(OH)^{2+}$ (see Fig. 6-4) for reaction with negatively charged particles and NOM. At this concentration of Al and pH, the dissolved Al then decreases with time (seconds and minutes) through precipitation of $Al(OH)_3$(am), as shown by the arrow in Fig. 6-5. Sometimes the solid phase is written showing the appended waters ($Al(OH)_3 \cdot 3H_2O$(am)). As stated above we call this sweep-floc coagulation, and the soluble Al decreases to the solubility line giving a dissolved residual Al of $10^{-6.3}$ M (<20 μg/L).

Some general comments regarding the solubility diagram and its relevance to coagulation concepts follow:

- Note that practicing coagulation at pH 6.3 for 20°C maximizes the concentration of $Al(OH)_3$(am) particles produced and minimizes the dissolved residual Al. This coagulation pH is the pH of minimum solubility.
- Sweep-floc coagulation is best carried out at the pH of minimum solubility, which for 20°C occurs at pH 6 to 6.3 and for 5°C occurs at pH 6.5 to 6.7.

EXAMPLE 6-2 Mass and Volume Concentrations of Solids from Aluminum Hydroxide Precipitation Alum is dosed at 2.7 mg/L as Al. The coagulation pH is 6.3 and the water temperature is 20°C. Calculate the mass and volume concentrations of solids produced.

Solution Figure 6-5 shows that the alum is being added under pH conditions of minimum solubility. The dissolved form of Al is mainly $Al(OH)_2^+$ (see Fig. 6-4). The concentration of dissolved Al is oversaturated with respect to amorphous aluminum hydroxide so the dissolved Al precipitates and the residual dissolved Al after precipitation is about $10^{-6.3}$ M—at the bottom of the solubility curve in Fig. 6-5. The difference between what was added and the residual concentration is the mass of solids produced. The actual dissolved Al species undergoing precipitation are identified above; however, we can express simply in terms of Al^{3+} in the following stoichiometric equation because no matter the form of Al the reactions all involve one atom of Al. In calculating the solids produced, it is usually assumed that aluminum hydroxide has three appended waters.

$$Al^{3+} + 3OH^{-} \rightarrow Al(OH)_3 \cdot 3H_2O(am)$$

Mass Concentration of Solids The amount of Al that will precipitate is the difference between what was added (10^{-4} M or 2.7 mg/L) and the residual dissolved Al concentration of $10^{-6.3}$ M or less than 0.020 mg/L. The difference is 2.68 mg/L. The mass concentration of solids produced is in accordance with the stoichiometric equation in the earlier section.

$$\frac{2.68 \text{ mg Al}}{\text{L}} \times \frac{\text{mmol Al}}{27 \text{ mg}} \times \frac{1 \text{ mmol Al}(OH)_3 \cdot 3H_2O(am)}{\text{mmol Al}}$$
$$\times \frac{132 \text{ mg Al}(OH)_3 \cdot 3H_2O(am)}{\text{mmol Al}(OH)_3 \cdot 3H_2O(am)} = 13.1 \text{ mg/L Al}(OH)_3 \cdot 3H_2O(am)$$

If we calculate on a unit Al concentration basis, then we produce 4.89 mg/L of solids as $Al(OH)_3 \cdot 3H_2O(am)$ at the pH of minimum solubility. On an unit alum basis (recall 11 mg/L alum = 1 mg/L as Al), we get 0.44 mg/L of solids.

Volume Concentration of Solids The density (ρ_s) of $Al(OH)_3 \cdot 3H_2O(am)$ should be low. We assume 1100 kg/m^3. The volume concentration (Φ_s) is calculated as follows for the mass concentration (C_s) of 13.1 mg/L.

$$\Phi_s = \frac{C_s}{\rho_s} = \frac{13.1 \text{ mg/L}}{1100 \text{ kg/m}^3} \times \frac{\text{kg}}{10^6 \text{ mg}} \times \frac{10^3 \text{L}}{\text{m}^3} = 11.9 \times 10^{-6} \frac{\text{m}^3 \text{solids}}{\text{m}^3 \text{ H}_2\text{O}}$$

or 11.9 ppm on a volume basis

Expressing based on a unit concentration of Al, we have 4.89 mg/L of solids produced as indicated above and obtain a volume concentration of 4.44 ppm. On an unit alum basis, we get 0.40 ppm of solids. ▲

Integration of Alum Chemistry with Reactions with Contaminants. In this section, we discuss alum coagulation of particles and NOM and integrate some of the material presented above. Particle suspensions in most drinking water supplies carry a smaller negative charge compared to the negative charge from NOM, as illustrated in Example 6-1. Consequently, particles in the water supply are fairly easy to coagulate because the charge demand for positively charge coagulant species is small. Charge neutralization with alum at low pH could be used to coagulate particles; however, we usually apply alum at higher pH and under sweep floc conditions to coagulate the particles. As stated above, sweep-floc coagulation produces particles so we want these additional particles to be unstable with respect to particle flocculation. If you carry out sweep-floc coagulation at too low a pH, the aluminum hydroxide particles may be positively charged stable particles, and similarly at too high a pH one can produce negatively charged particles. As a rule of thumb, carrying out coagulation near the pH of minimum solubility for aluminum hydroxide (see Fig. 6-5) produces flocs of low or no charge. So if sufficient alum is dosed to satisfy the water's charge demand from particles and NOM and the pH is in the range of 6 to 7, the floc particles usually have little or no charge and flocculate well. This can be verified by making electrophoretic mobility (zeta potential) or streaming-current measurements (see Chap. 10) of the charge of the floc particles following coagulant addition (Letterman and Yiacoumi, 2010).

We now consider the reaction of dissolved Al with NOM. The NOM for most water supplies controls the coagulant demand. The negative charge associated with the NOM must be satisfied from positively charged Al species whose distribution as a function of pH is shown in Fig. 6-4. Dissolved metals such as Al (also true for Fe) are deficient in electrons and thus complex with ligands that contain and share excess electrons. A metal-ligand complex is formed. We have described one of these complexation reactions above, the formation of Al-hydroxo complexes through the hydrolysis reactions. Other possible ligands in water treatment applications are NOM, fluoride (F^-), and orthophosphate. Orthophosphate is low in most water supplies (10^{-6} M and less) so it is not a factor in coagulation reactions. If, however, coagulation is not practiced properly and too high of a dissolved residual Al exists when orthophosphate is added at the end of the treatment plant, then complexation of Al and phosphorus can occur. This can lead to precipitation of aluminum-phosphate solids and deposition on pipes in the distribution system. Fluoride in most raw water supplies is low so complexation reactions with Al are not significant. However, many water plants add fluoride at about 1 mg/L ($<10^{-4}$ M) for control of dental caries. If fluoride at this concentration was added at the front of a water plant where alum coagulation is used, then it could complex Al, requiring additional alum for coagulation, and it would increase the solubility of Al. Thus, fluoride addition is done at the end of the water plant following coagulation, clarification, and filtration. This leaves NOM as the important ligand to consider.

NOM is not a particular organic compound with known structure and MW (see Sec. 6-2 on NOM). While some researchers have used model organic compounds for NOM in an attempt to model complexation with Al and others have determined experimentally complexation constants for model fulvic acids, the work has limited application and even then for low pH conditions that are not applicable to alum coagulation.

Our approach then is to describe conceptually Al complexation for alum coagulation. In Sec. 6-6, we describe an empirical stoichiometry between coagulant dosing and TOC. The conceptual description of alum coagulation involves two possible reaction pathways, as depicted in Fig. 6-6. The top reaction pathway is favored at low pH, probably 5.5 or less, and involves complexation of Al with NOM. When the alum dose is high enough to satisfy the negative charge demand of the NOM, direct precipitation of the NOM with Al occurs. The low pH of 5.5 or less favors complexation and direct precipitation because of the availability of positively charged Al species (Fig. 6-4) and the high solubility of Al (Fig. 6-5). These low pH conditions are not recommended for most water supplies unless the TOC is high and is composed mainly of aquatic humic matter. This is because coagulation at low pH can lead to overdosing and high-dissolved residual Al and positively charged particles, so flocculation, clarification, and filtration are adversely affected.

The bottom reaction pathway of Fig. 6-6 is preferred. Alum is added at pH conditions such that positively charged Al can complex and satisfy the charge demand of NOM and then additional alum results in precipitation of $Al(OH)_3$(am) with removal of NOM either through coprecipitation of Al-NOM(s) or through adsorption of complexed Al-NOM to $Al(OH)_3$(am) particles. The following summary comments are made for sweep-floc coagulation with alum for removals of particles and NOM.

- For water supplies containing NOM with SUVA less than 2.5, TOC does not normally control coagulant dosages but particles (turbidity) control. Dosages should be low when coagulation is carried out at optimum pH conditions.

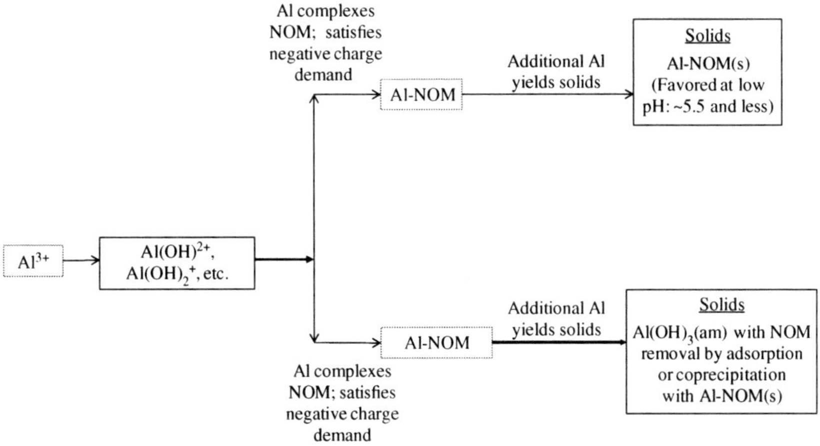

FIGURE 6-6 Conceptual-reaction pathways for Al with NOM: top pathway is favored at lower pH and leads to direct precipitation of Al-NOM solids; bottom pathway is favored for sweep-floc coagulation conditions.

- For waters with SUVA of about 2.5 and greater, TOC controls dosages rather than particles, and there is a stoichiometric relationship between Al and TOC that is discussed in Sec. 6-6.
- If we were to coagulate with alum at pH less than about 6, we would have a greater concentration of positively charged Al species available initially (see Fig. 6-4) for neutralizing negatively charged particles and for reaction with dissolved NOM, but the dissolved residual Al following precipitation of $Al(OH)_3$(am) can be high, as shown by Fig. 6-5.
- If we were to coagulate at pH greater than 7, there would be little positively charged Al species initially available (see Fig. 6-4) for reaction with particles and NOM, and the solubility of Al increases (Fig. 6-5). Thus, high alum dosages would be required to make up for the low fractions of positively charged Al species.
- Optimum pH for alum is 6 to 7 depending on water temperature. This is because positively charged Al is available (Fig. 6-4) for complexation and satisfying the NOM charge demand, maximizing precipitation of aluminum hydroxide (Fig. 6-5), and minimizing dissolved Al.

Ferric Coagulants

The chemistry of ferric iron is similar to that of Al, as described above for alum. Thus, we summarize the key principles and chemistry of ferric coagulants and point out differences. Following the Framework for Understanding Coagulation (presented in Sec. 6-1), we present principles about the chemistry of ferric coagulants and follow it with integration of material on contaminants and bulk water chemistry and temperature to explain how ferric coagulants work and to present optimum coagulation conditions.

Ferric Chemistry. First, we describe the chemistry of liquid feed products; second, the chemistry of dissolved ferric species following hydrolysis and initially present in rapid mixing; and third, the solubility of ferric hydroxide.

Feed Products. Common products are liquid ferric chloride and ferric sulfate. Ferric sulfate is also available as a solid, and then dissolved at the water plant and fed as a liquid. Typical chemical properties of the liquid products are shown in Table 6-3. Because ferric iron is very insoluble, each liquid product contains excess acid to prevent precipitation and provide a stable product. The pH of the liquid products is <2 and with ferric sulfate it can be about pH 1. Unlike liquid alum, in which the soluble form of Al in the liquid coagulant product is Al^{3+}—see above—ferric iron even at these low pH conditions undergoes some hydrolysis, and it is thought to contain some small polymers such as $Fe_2(OH)_2^{4+}$ and $Fe_3(OH)_4^{5+}$. Therefore, we are feeding coagulant products that contains mostly Fe^{3+} with some $Fe(OH)^{2+}$, and some low MW polymeric forms of Fe, as depicted in the box on the left in Fig. 6-7.

A couple of other points about liquid ferric coagulants are important. They contain small quantities of other metals. One important metal is reduced Mn, which can make up 0.01 to 0.1 percent of the product. Consequently, water plants that have Mn in their source waters should seek high-quality ferric products with little Mn to reduce the effect of this source. Dosing of ferric coagulants can be expressed as Fe (preferred way); when using ferric chloride as $FeCl_3$ or $FeCl_3{\cdot}6H_2O$; and when using ferric sulfate as $Fe_2(SO_4)_3$ or $Fe_2(SO_4)_3{\cdot}9H_2O$. Conversions among the mass concentrations follow:

- 1 mg/L as Fe = 2.91 mg/L as $FeCl_3$ = 4.83 mg/L as $FeCl_3{\cdot}6H_2O$
- 1 mg/L as Fe = 3.39 mg/L as $Fe_2(SO_4)_3$ = 5.03 mg/L as $Fe_2(SO_4)_3{\cdot}9H_2O$
- 1 mg/L as Fe = 1.79×10^{-5} M Fe

Dissolved Ferric Chemistry. Dissolved ferric iron undergoes hydrolysis reactions (can consider OH^- as a ligand reacting with dissolved Fe). The liquid ferric coagulant contains mostly Fe^{3+} with some $Fe(OH)^{2+}$ and low MW polymeric Fe, as shown on the left in Fig. 6-7. When the coagulant is added to rapid mixing, it is further diluted and undergoes further hydrolysis instantaneously, as indicated in the second box from the left in Fig. 6-7.

Figure 6-8 shows the fraction of dissolved Fe species as a function of pH. No dissolved polymers are presented because for the concentrations and pH conditions of rapid mixing they would be at low fractions and would not affect the dissolved Fe fractions shown. The dissolved chemistry of Fe has similarities to Al, as presented above. They are similar in that ferric salts are acids and they undergo hydrolysis reactions so that the dissolved form of the coagulant available initially to react with contaminants depends on the coagulation pH, that is, the pH in rapid mixing after coagulant addition. We see from Fig. 6-8 that there is little Fe^{3+} and $Fe(OH)^{2+}$ above pH 4. Above pH 7.5 there is little positively charged Fe and the anion ($Fe(OH)_4^-$) dominates. The availability of positively charged Fe to react with negatively charged contaminants depends on pH and the primary species is $Fe(OH)_2^+$. As the figure shows, the pH must be below about 7.5. To maximize the fraction of positively charged $Fe(OH)_2^+$, the pH should be 5 to 6. The speciation of dissolved Fe is affected by

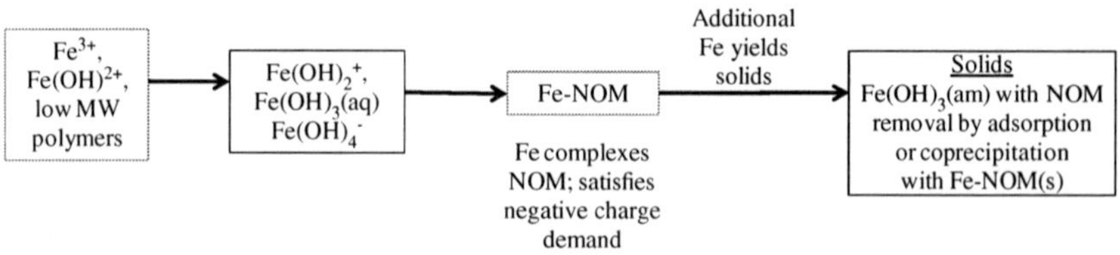

FIGURE 6-7 Conceptual series of reactions for dissolved Fe hydrolysis and complexation with NOM followed by precipitation of ferric hydroxide with NOM removal.

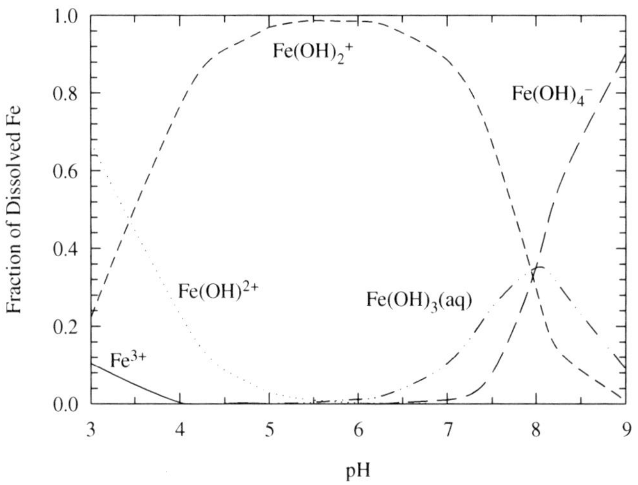

FIGURE 6-8 Dissolved Fe speciation as a function of pH at 25°C.

water temperature; the fractions shown in Fig. 6-8 are shifted to the right for cold waters, so that the range one finds the maximum fraction of $Fe(OH)_2^+$ is pH 5.5 to 6.5.

Solubility of Ferric Hydroxide. We consider the solubility of amorphous ferric hydroxide solid ($Fe(OH)_3$(am)). In this presentation, we do not initially consider the effects of complexing agents such as NOM.

As presented above for Al, precipitation of ferric hydroxide occurs in several stages: involving polymerization, particle nucleation in which submicron solid particles are formed, and then aggregation of these submicron particles occurs forming larger particles that we call flocs. The precipitation of $Fe(OH)_3$(am) produces additional particles, which incorporates the raw water particles into the flocs; that is, sweep-floc coagulation.

Figure 6-9 is the solubility diagram for $Fe(OH)_3$(am). It is presented for a solubility constant (K_{s0} of $10^{-38.7}$) at 25°C. While aluminum hydroxide solubility is affected significantly by temperature, ferric iron is very insoluble and so water temperature has little effect.

Some general comments regarding the solubility diagram and its relevance to coagulation concepts follow:

- While the pH of minimum solubility occurs near pH 8, Fe is very insoluble over a wide pH range. If we consider that ferric coagulation may be carried out for various water supplies in the range of 5 to 8, the dissolved residual soluble Fe would be much less than 1 μg/L at pH 8 and also low at <10 μg/L for pH 5.
- Sweep-floc coagulation can be carried out over a wide range of pH conditions.
- For ferric coagulation of particles and NOM, Fe coagulation above pH 7 is unfavorable because the fraction of positively charged Fe is small (Fig. 6-8). Fe coagulation should be carried out between about pH 5.5 up to the low 6s to maximize the fraction of positively charged Fe with negatively charged NOM under conditions (dosages and pH) that also produce sweep floc.

The concentration of ferric hydroxide solids formed by precipitation in sweep-floc coagulation can be calculated. For every mg/L of Fe added in coagulation, 1.91 mg/L of $Fe(OH)_3$(am) is produced. If the solid contains three appended waters, then 2.88 mg/L of

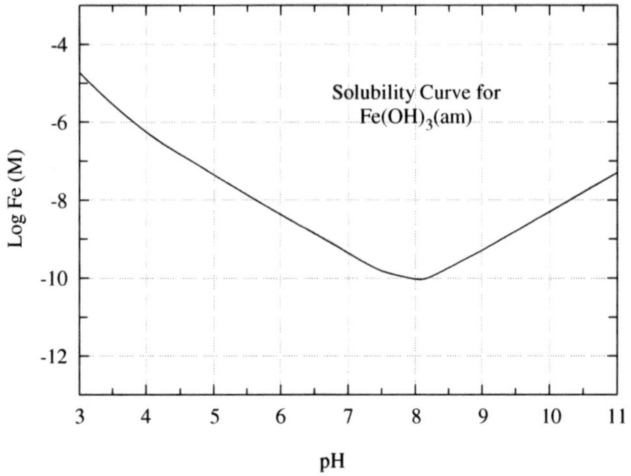

FIGURE 6-9 Solubility of amorphous ferric hydroxide ($Fe(OH)_3(am)$) 25°C.

$Fe(OH)_3{\cdot}3H_2O(am)$ is formed. To this estimate of sweep floc solids, we would have to account for the suspended solids in the raw water and conversion of organic carbon into a solid phase.

Complexation between dissolved Fe hydrolysis species and other ligands such as dissolved NOM, especially aquatic humic and fulvic acids and low MW hydrophilic acids, can occur depending on pH. Because aquatic humic matter in water supplies is a mixture of organic molecules, we cannot model practically the complexation reactions. We show conceptually in Fig. 6-7 that positively charged Fe complexes with negatively charged NOM. In Sec. 6-6, we describe empirical relationships between Fe dosing and TOC.

EXAMPLE 6-3 Mass and Volume Concentrations of Solids from Ferric Hydroxide Precipitation Ferric chloride or sulfate is added at a dosage of 5.6 mg/L as Fe at pH in the range of 5 to 8. Calculate the mass and volume concentrations of solids produced. Note that the Fe dose on a molar basis (10^{-4} M) is the same as that in Example 6-2 for alum as Al. The basis for calculations is similar as in Example 6-2. Here, we consider the stoichiometric equation for $Fe(OH)_3{\cdot}3H_2O(am)$.

$$Fe^{3+} + 3OH^{-} \rightarrow Fe(OH)_3{\cdot}3H_2O(am)$$

Solution

Mass Concentration of Solids The mass concentration of solids produced is calculated:

$$\frac{5.6\text{ mg Fe}}{\text{L}} \times \frac{\text{mmol Fe}}{55.85\text{ mg}} \times \frac{1\text{ mmol }Fe(OH)_3{\cdot}3H_2O(am)}{\text{mmol Fe}}$$

$$\times \frac{160.85\text{ mg }Fe(OH)_3{\cdot}3H_2O(am)}{\text{mmol }Fe(OH)_3{\cdot}3H_2O(am)} = 16.1\text{ mg/L }Fe(OH)_3{\cdot}3H_2O(am)$$

On an equivalent molar dosage basis, we produce more solids with ferric coagulants than alum—see Example 6-2; a 10^{-4} M alum dose as Al produces 13.1 mg/L of solids. If we calculate on a unit Fe basis of mg Fe/L, then we produce 2.88 mg/L of solids as $Fe(OH)_3 \cdot 3H_2O(am)$.

Volume Concentration of Solids Assuming a density (ρ_s) of 1100 kg/m^3 for $Fe(OH)_3 \cdot 3H_2O(am)$, the volume concentration (Φ_s) is calculated as follows for the mass concentration (C_s) of 16.1 mg/L.

$$\Phi_s = \frac{C_s}{\rho_s} = \frac{16.1 \text{ mg/L}}{1100 \text{ kg/m}^3} \times \frac{\text{kg}}{10^6 \text{ mg}} \times \frac{10^3 \text{L}}{\text{m}^3} = 14.6 \times 10^{-6} \frac{\text{m}^3 \text{solids}}{\text{m}^3 \text{ H}_2\text{O}}$$

or 14.6 ppm on a volume basis

This volume concentration is greater than for alum (11.9 ppm, Example 6-2) for an equivalent molar dose. ▲

Integration of Ferric Coagulant Chemistry with Reactions with Contaminants. We discuss ferric coagulation of particles and NOM and integrate some of the material presented earlier. If NOM is present, especially if the NOM contains aquatic humic matter, then the negative charge associated with the NOM must be satisfied from positively charged Fe. The distribution of soluble Fe as a function of pH was presented in Fig. 6-8. To take advantage of the highest fraction of positively charged Fe, present as $Fe(OH)_2^+$, one would want to coagulate at pH 5 to the low 6s if NOM is controlling coagulation with moderate to high values of TOC.

The conceptual description of ferric coagulation for removing particles and NOM, and recognizing that NOM controls dosages, is depicted through the series of reactions in Fig. 6-7. Ferric coagulant should be used at about pH 5.5 to the low 6s such that positively charged Fe complexes and satisfies the charge demand of NOM, and precipitates as $Fe(OH)_3$(am) with removal of NOM either through coprecipitation of Fe-NOM(s) or through adsorption of complexed Fe-NOM to $Fe(OH)_3$(am) particles. The following summary comments are made for sweep-floc iron coagulation for removals of particles and NOM.

- For water supplies containing NOM with SUVA less than 2.5, TOC does not normally control coagulant dosages, but particles (turbidity) control. Dosages should be low and sweep-floc coagulation can occur over a wide pH range of, say, pH 5 to the low 8s.
- For waters with SUVA of about 2.5 and greater, TOC controls dosages and there is a stoichiometric relationship between Fe and TOC that is discussed in Sec. 6-6.
- Optimum pH conditions, when TOC controls, are from about 5.5 to the low 6s. Note that these are lower optimum pH conditions than those for alum.

Polyaluminum Coagulants

There are a variety of polyaluminum coagulants available for drinking water treatment. Most are polyaluminum chlorides (PACls) with some usage of polyaluminum sulfates and polyferric chlorides and sulfates. Coverage is limited to PACls or sometimes called PAXs.

Polyaluminum Chloride Products. The PACl products are proprietary chemicals prepared by carefully controlled reaction of base with aluminum salts, such as $AlCl_3$. They are specialty coagulants in that they can be prepared to produce products with different formulations and chemical properties. The most important characteristic is the degree of neutralization (r) or percent basicity:

$$r = \frac{[OH^-]}{[Al_T]} \quad (6\text{-}7)$$

where $[OH^-]$ is the base added in production to neutralize the total Al ($[Al_T]$) used. Both are expressed in molar concentrations.

$$Basicity = \frac{r}{3} \text{ x } 100\% \quad (6\text{-}8)$$

The basicity affects the acidity of the product because of the preneutralization with base so the products have less effect on consuming alkalinity and decreasing pH than alum and ferric coagulants. The basicity also affects the relative fractions of monomeric and polymeric aluminum such that when used under certain pH conditions, they can have a highly charged polymeric cationic Al species available for coagulation. The polymeric form of Al that is dominant in partially neutralized products is the tridecameric Al_{13} present as $Al_{13}O_4(OH)_{24}(H_2O)_{12}{}^{7+}$ or without the water molecules it is $Al_{13}O_4(OH)_{24}{}^{7+}$ with a MW of 823. This is abbreviated as $Al_{13}{}^{7+}$. The charge of +7 makes it a coagulant species that is strongly attracted to negatively charged contaminants. Most commercial products fall into two groups.

- *Medium-basicity products:* These fall into the range of 40 to 55 percent basicity. The Al species depends on the coagulation pH conditions and would include some monomeric Al and some $Al_{13}{}^{7+}$.
- *High-basicity products:* These are classified with about 60 to 75 percent basicity. Again, the Al species depends on the coagulation pH conditions and would contain some monomeric Al but a higher fraction of the $Al_{13}{}^{7+}$ polymer than medium-basicity products.

A special case is aluminum chlorohydrate, which has a basicity of about 83 percent and a high product strength of 23 to 24 percent Al_2O_3. It is used somewhat as a coagulant in water treatment, but it has a far wider market as an antiperspirant.

PACl products come at various liquid strengths measured as percent Al_2O_3. The range of strengths is about 10 to 24 percent. Most products have strengths of about 10 to 12 percent with specific gravities of 1.2 to 1.4. These contain a higher Al concentration compared to liquid alum, which is 8.3 percent Al_2O_3. PACl doses in practice are usually expressed as the mass concentration of products. PACls come in different strengths so to properly compare PACl products and to compare to alum, doses should be expressed on a common basis as Al. Throughout this chapter, we express doses as Al. Example 6-2 shows how to express product doses in terms of as Al.

Some products may contain other constituents that affect coagulation, flocculation, and clarification performance. In assessing PACls it is important to know whether these constituents are present. Some contain sulfate in addition to the chloride ion. The concentration of sulfate is much lower than that in liquid alum, which is 23.5 percent by weight. In PACls the sulfate is usually about 2 to 3 percent. The presence of sulfate can affect the chemistry of the product. It is thought to aid precipitation of aluminum hydroxide solids and to improve settling (Edzwald et al., 2000; Pernitsky and Edzwald, 2003). Other PACl products are available as blends containing organic cationic polyelectrolytes.

PACl products are fairly stable. They are shipped as liquids and stored onsite. In full-scale water plants they should be fed neat. The chemistry can change with dilution, so if it is necessary to dilute the products in bench-scale jar tests then follow instructions from the supplier.

EXAMPLE 6-4 PACl Product Dosing and as Al Consider a PACl product with the following properties: 10.2 percent strength as Al_2O_3 with a specific gravity of 1.26. Calculate first the product mass concentration and then its mass concentration as Al.

Solution

Product Mass Concentration This follows directly from the specific gravity of 1.26 g/mL or 1.26×10^6 mg/L or 1260 g/L. In the United States, the specific gravity may also be expressed as pounds (lbs) per gallon (gal). It would be 10.5 lbs/gal.

Mass Concentration as Al

$$\frac{10.2 \text{ g } Al_2O_3}{100 \text{ g}} \times \frac{1260 \text{ g}}{\text{L}} = 128.5 \frac{\text{g}}{\text{L}} \text{ as } Al_2O_3$$

Then, we obtain

$$\frac{128.5 \text{ g } Al_2O_3}{\text{L}} \times \frac{54 \text{ g Al}}{102 \text{ g } Al_2O_3} = 68.0 \frac{\text{g}}{\text{L}} \text{ as Al}$$

If we divide the product mass concentration of 1260 g/L by the mass concentration of Al of 68 g/L, we obtain that 18.5 g/L of the product equals 1 g/L as Al. Alternatively, the relationship is 18.5 mg/L of product is 1 mg/L as Al. ▲

Dissolved Al Chemistry for PACls. The chemistry of PACls for water treatment applications has been the subject of considerable study—see Pernitsky and Edzwald (2003) for discussion and listing of these studies. It has been concluded that the primary form of polymeric Al is the Al_{13}^{7+} polymer. The fraction of Al_{13}^{7+} depends on the degree of neutralization (*r*), and generally a greater fraction is found as *r* increases to about 2.1 or a basicity of 70 percent (Bottero et al., 1980a; 1980b). Thus, high-basicity PACls have higher fractions of Al_{13}^{7+} polymer than medium-basicity products. The Al_{13}^{7+} polymer would be available for coagulation when the PACl products are used under certain pH conditions, as described next.

We use Fig. 6-10 to illustrate the speciation of dissolved Al as a function of coagulation pH. The hydrolysis expressions and thermodynamic data used to construct the figure are given in the paper by Pernitsky and Edzwald (2006). It is important to note that because of the presence of polymeric Al, the speciation depends on the total Al (Al_T) concentration. This figure was constructed for Al_T of 1 mg/L, a reasonable value for water treatment practice where dosing is commonly much less than 10 mg/L. For this Al concentration the water is theoretically oversaturated with $Al(OH)_{3(am)}$ for the pH conditions shown in the figure. What we learn from this figure is that the Al_{13}^{7+} polymer dominates at certain coagulation pH conditions. A high fraction of the Al_{13}^{7+} polymer exists for pH 5 to 7.5 for warm waters, while for cold waters there is a small shift to higher pH so the Al_{13}^{7+} polymer dominates from about pH 5.5 to 8. Therefore, if we carry out coagulation between these pH ranges, we have the highly positively charged polymeric Al available for reaction with negatively charged particles and NOM. For the pH regions of oversaturation, the Al_{13}^{7+} polymer is presumably initially available followed by precipitation of aluminum hydroxide.

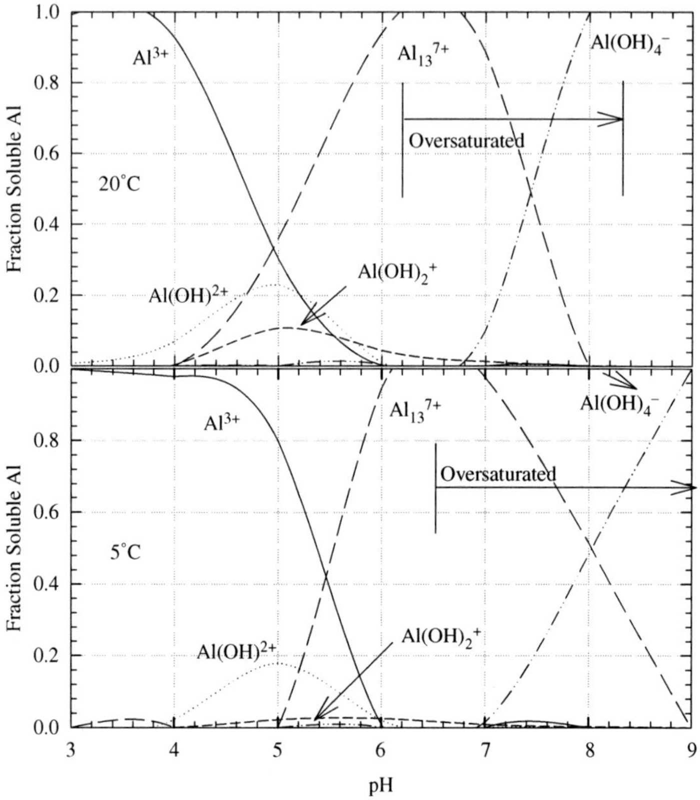

FIGURE 6-10 Dissolved Al speciation for PACl for total Al of 1 mg/L (Note that solutions are oversaturated for $Al(OH)_3$(am) for pH greater than 6.2 at 20°C and pH greater than 6.5 at 5°C. *Source: Pernitsky and Edzwald (2006), Journal of Water Supply: Research and Technology – AQUA, 52 (No. 4), 121–141, with permission from the copyright holders, IWA Publishing.*

Solubility. Figure 6-11 presents theoretical solubility diagrams at 5 and 20°C—see Pernitsky and Edzwald (2006) for the thermodynamic data. These are representative solubility diagrams for high-basicity PACls for the time scale of water treatment, and the solubility curves are in agreement with experimental data presented by Van Benschoten and Edzwald (1990) and Pernitsky and Edzwald (2003). The pH of minimum solubility, as for alum, is pH dependent and it increases with pH with decreasing water temperature. It is in the mid pH 6s for warm waters and increases to the high 6s, near pH 7, for cold waters. For pH conditions greater than the pH of minimum solubility, the curves are identical to alum solubility; see Fig. 6-5. For lower pH conditions; however, Al is much more soluble than alum due to the presence of the Al_{13}^{7+} polymer. As shown from Fig. 6-10, the Al_{13}^{7+} polymer is the dominant soluble Al species between pH 5 for water at 20°C (pH 5.5 for 5°C) and the pH of minimum solubility.

What we learn from the solubility diagrams (Fig. 6-11) and the dissolved speciation diagrams (Fig. 6-10) follows. If we use high-basicity PACls (concepts also apply to

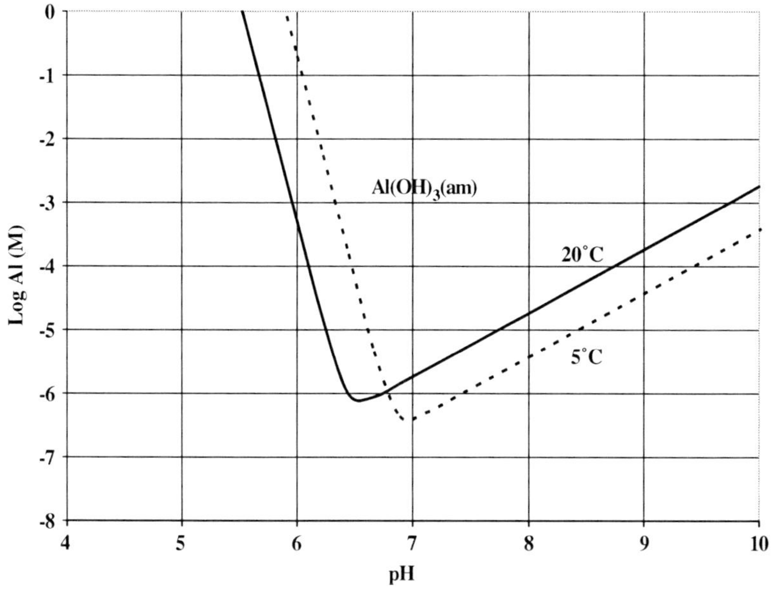

FIGURE 6-11 Solubility diagrams for high-basicity PACls for 5 and 20°C [Considered Al^{3+}, $Al(OH)^{2+}$, $Al(OH)_2^+$, $Al(OH)_4^-$, and $Al_{13}(OH)_{24}^{7+}$ (main polymeric species for PACl) in equilibrium with $Al(OH)_3$(am) as the solid phase]. *Source: Pernitsky and Edzwald (2006), Journal of Water Supply: Research and Technology – AQUA, 52 (No. 4), 121–141, with permission from the copyright holders, IWA Publishing.*

medium-basicity PACls except we would have a lower fraction of the Al_{13}^{7+} polymer) for pH conditions less than the pH of minimum solubility, we would have primarily the Al_{13}^{7+} polymer for coagulation and reaction with negatively charged contaminants. Because the PACl is quite soluble, overdosing and high-residual soluble Al could be problems. Overdosing would be marked by formation of positively charged flocs and their restabilization. If we use PACls at pH conditions much greater than the pH of minimum solubility (say, above 7.5 to 8), then the Al_{13}^{7+} polymer initially present would be small. The best pH conditions for use of medium- and high-basicity PACls are those from the mid 6s to the mid 7s to take advantage of initially having a high amount of the Al_{13}^{7+} polymer (charge neutralization of particles and complexation of NOM satisfying its charge demand) followed by precipitation of aluminum hydroxide solids (sweep-floc coagulation) and control of residual soluble Al.

Integration of PACl Coagulant Chemistry with Reactions with Contaminants. In this section we discuss PACl coagulation of particles and NOM and integrate some of the material presented earlier. The distribution of soluble Al species for PACl as a function of pH is shown in Fig. 6-10, and solubility diagrams are shown in Fig. 6-11.

The following summary comments are made for PACl coagulation for removal of particles and NOM.

- For water supplies in which turbidity controls coagulant dosages meaning TOC is low and the raw water SUVA is less than 2.5, then low dosages of PACl should be effective at pH conditions of 6 to low 7s. The Al_{13}^{7+} polymer is initially present for charge neutralization of negatively charged particles. Sweep-floc coagulation (precipitation) may occur depending on the dose and if the pH of coagulation is near the point of minimum solubility. Coagulation at pH less than 6 is not recommended unless carefully controlled because of the probability of overdosing (stable flocs) causing high-residual soluble Al.

- If NOM is present and the SUVA is about 2.5 or greater, then TOC controls coagulant dosages. Here the negative charge associated with the NOM must be satisfied from positively charged PACl species such as the Al_{13}^{7+} polymer. The concept of the reactions is similar to that described above for alum and ferric coagulants. The negatively charged aquatic humic matter and other negatively charged NOM are organic ligands that complex positively charged Al. Sufficient PACl must be added to satisfy the complexation with NOM and satisfy the charge demand, and if carried out at optimum pH conditions (see next bullet) one can form sweep floc and adsorb the complexed NOM onto the precipitated particles. At lower pH one can precipitate directly the NOM as an Al-NOM solid.
- The optimum pH conditions for use of medium- and high-basicity PACls are those from the mid 6s to the mid 7s to take advantage of initially having a high amount of the Al_{13}^{7+} polymer (charge neutralization of particles and complexation reaction with NOM satisfying its charge demand) followed by precipitation of aluminium hydroxide solids (sweep-floc coagulation) and to control residual soluble Al. Dosages can be low when the TOC or NOM is low. As the TOC increases PACl dosages are greater and there is a stoichiometric relationship between PACl and TOC that is discussed in Sec. 6-6.

Organic Cationic Polyelectrolytes

Coverage is restricted to synthetic cationic polyelectrolytes or, simply cationic polymers. The synthetic polymers are the most widely used. There is some use and interest in natural cationic polymers, especially in developing countries, which are not covered here. Two examples are chitosan and a cationic protein from seeds of the tree, *Moringa oleifera.* Coverage on natural cationic polymers can be found in the review paper on polymers in water treatment by Bolto and Gregory (2007).

Types and Properties. The two main types of cationic polyelectrolytes are polydiallyldimethyl ammonium chloride (Poly-DADMAC) and epichlorohydrin dimethylamine (Epi-DMA). Some properties of these polyelectrolytes are summarized in Table 6-4. These can be prepared with different formulations producing polymers of different MW, hence the ranges shown in the table. A linear structure is illustrated for the Epi-DMA polymer, but in certain formulations a branching agent is added yielding a higher MW polymer. Both cationic polymeric types fall within the general classification of low MW ($<10^5$) to medium MW (10^5 to 10^6). This distinguishes them from the high MW (10^6 and greater) polyacrylamide polymers that have no charge (nonionic) or are slightly negative in charge (anionic). The latter are not coagulants and are used primarily as flocculant, flotation, and filter aids, as indicated in Table 6-4. They are covered in Sec. 6-10.

The cationic polymers carry a positive charge due to quaternary-nitrogen bonding, as shown in Table 6-4. Another important characteristic of the quaternary nitrogen is that the charge is not affected by pH, unlike the metal-hydrolyzing Al and Fe coagulants. The positive charge is about 6 and 8 μeq/mg for the Poly-DADMAC and Epi-DMA polymers for 100 percent polymer products (Edzwald et al. 1987; Bolto and Gregory, 2007). Products are sold as liquids and dosed on a product basis so one must account for the percent of polymer in the product. Polymer strength varies among products and may be in the 20 to 50 percent range. So, for example (see the bottom of Table 6-4), an Epi-DMA product with 50 percent polymer has a cationic charge density of 4 μeq/mg as product.

Reactions with Particles and NOM. It is instructive to describe how cationic polyelectrolytes destabilize particles and how they coagulate dissolved NOM, even though in DAF applications they are not used as sole coagulants but rather as dual coagulants with hydrolyzing Al and Fe coagulants.

Destabilization of negatively charged colloidal particles is accomplished by charge neutralization. The cationic polymer is attracted to and adsorbs onto the particle surface and forms patches of positive charge. The mechanism is called the patch model (Kasper, 1971; Gregory, 1973). Complete neutralization of the negative charge of the particles is not necessary because there is attraction between bare negatively charged areas of the particles and positively charged patches resulting in flocculation. It is observed, however, that the net charge is close to zero. At low cationic polymer doses (underdosing), particle removal is poor because of insufficient cationic polymer to neutralize the colloidal-particle negative charge. As the cationic polymer dose increases, one reaches a point of optimum dose and good coagulation conditions. Greater polymer doses (overdosing) lead to excess polymer adsorption on the particles and formation of positively charged and stable particles that flocculate poorly. This is called particle restabilization.

Cationic polymers can react with dissolved negatively charged NOM leading to formation of an organic solid. Glaser and Edzwald (1979) have described this reaction as a cross-linking of oppositely charged macromolecules involving the negatively charged aquatic humic matter and the positively charged cationic polymer. Cross-linking continues until a large NOM-cationic polymer molecule is formed that precipitates. This reaction is limited to certain MW NOM molecules, probably close to the MW of the cationic polymer molecules. Therefore, removals depend strongly on the SUVA of the water. For waters with SUVA >4 and thus the NOM having a high fraction of aquatic humic matter, TOC removals can reach 40 percent (Edzwald et al., 1987). For water supplies of low SUVA and low fractions of aquatic humic matter, TOC removals are much less.

6-6 GUIDANCE ON COAGULANT DOSING FOR DAF

For drinking water treatment plants, coagulation has two purposes: to destabilize particles present in the raw water and to convert dissolved NOM into particles for downstream removal by DAF and granular media filters. In the contact zone of the flotation tank, collisions occur between bubbles and floc particles. Coagulation is essential to ensure attachment of bubbles to flocs when collisions occur. Optimum coagulation conditions are defined as the required dosages at favorable pH conditions that produce low turbidity, low NOM measured as DOC or UV_{254} as a surrogate of NOM, and low-dissolved residual coagulant (see Optimum Coagulation in Sec. 6-1). Flotation jar testing to evaluate coagulation dosing is discussed in Chap. 10. These optimum coagulation conditions are also those of coagulant dose and pH that produce flocs with charge near zero and with relatively high hydrophobicity. These optimum coagulation conditions cause high bubble attachment efficiency in the contact zone (see Chap. 7).

The coagulants that are effective for flotation are the same that are used in sedimentation plants. Coagulant selection depends not on flotation per se but on raw water quality factors of turbidity, NOM concentration (TOC, DOC, UV_{254}, color) and composition, alkalinity, and water temperature. Coagulant dosing for turbidity and NOM is fundamentally the same for flotation and sedimentation processes. Optimum coagulant doses and pH conditions are generally the same. An exception is that lower coagulant doses for DAF than for sedimentation processes may be found in treating high-quality waters of low turbidity and low to moderate values of TOC. This is because in sedimentation the ability to form large flocs for settling depends on particle concentrations. Achieving rapid flocculation and aggregation of flocs into large sizes often requires adding additional coagulant to increase precipitated solids (additional sweep floc), thereby increasing the flocculation rate (see Sec. 6-8). Large flocs are not required for DAF so additional coagulant for these high-quality water supplies is not necessary. Another important difference is that high MW polymers are not normally used as flocculant aids in DAF plants. An exception is that some plants use small amounts to improve retention of the floated sludge, as discussed further later.

In this section, optimum coagulation conditions are presented for each coagulant and dosage guidelines are presented. Principles of coagulation are reinforced with several figures showing DAF performance.

Alum Use in DAF

The chemistry of alum coagulation is described in Sec. 6-5. From those principles, the following optimum coagulation chemistry for alum is summarized.

- Operate near the pH conditions of minimum solubility for $Al(OH)_3$(am). This maximizes production of sweep floc particles and minimizes soluble Al. Generally these pH conditions when used with coagulant doses, according to the guidelines below, produce floc particles with little or no electrical charge. The pH conditions for minimum solubility depend on water temperature (Fig. 6-5). For warm waters above 10°C, operate at pH 6 to 6.5—say, near 6.5 for 10°C and near 6 for waters as they approach 25°C. For cold waters below 10°C, operate between pH 6.5 and 7—close to 7 for very cold water.
- The optimum alum dose for these pH conditions is that which minimizes turbidity and TOC. For water supplies where TOC controls coagulant dosages, which occurs for raw water with SUVA of about 2.5 and greater, the optimum coagulant dose is proportional to the TOC or UV_{254}. Dosing guidelines with respect to TOC are presented below.
- For water supplies with SUVA less than 2.5, particularly near 2 or less, than the NOM has a low fraction of aquatic humic matter and NOM does not control coagulant doses. The dosages will be controlled by turbidity and the pH and alkalinity of the raw water. If one can dose at favorable pH conditions without using the coagulant to decrease the pH, then dosages should be relatively low. If the pH and alkalinity are high, then additional coagulant, or the addition of a strong acid, may be needed to achieve favorable pH conditions.

Figure 6-12 shows DAF and filter performance for coagulation with alum at pH in the mid 6s according to the principles above (Edzwald et al., 2003). Excellent performance was achieved for DAF turbidity (about 0.4 NTU) and filtration (about 0.05 NTU). UV_{254} removals (not shown) were 70 to 75 percent. The EPM data in Fig. 6-12 show that the raw water particles were negatively charged but after optimum coagulation the flocs had little or no charge.

Guidelines on alum dosages for waters with SUVA of about 2.5 or greater are presented next based on the authors' extensive experience over some 35 years. The reader can refer to some papers on this subject by Edzwald and Van Benschoten (1990), Edzwald (1993), and Pernitsky and Edzwald (2006). These dosages guidelines have been confirmed for full-scale plants by Archer and Singer (2006) and by Edzwald and Kaminski (2009).

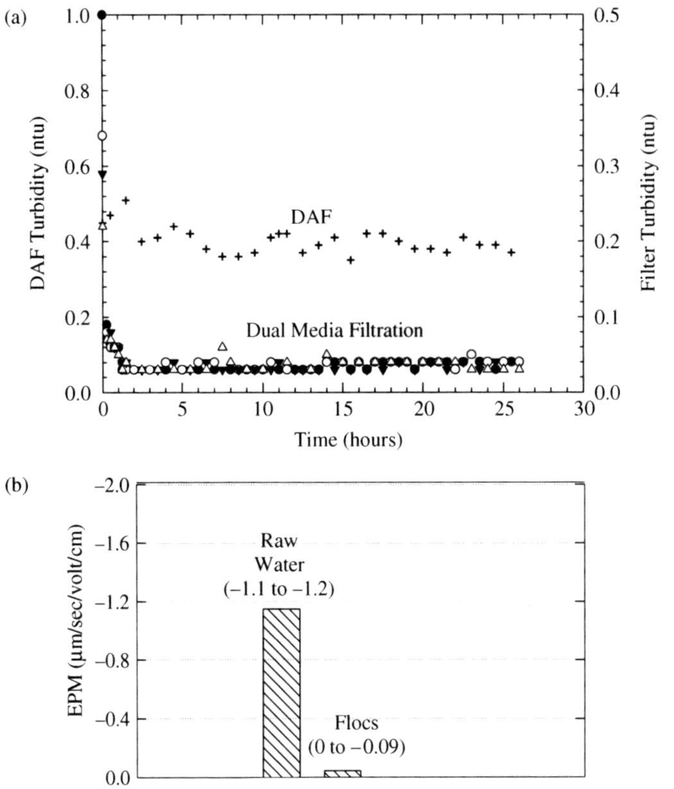

FIGURE 6-12 (a) DAF and filter performance for optimum alum coagulation at pH about 6.5. (b) Particle charge (EPM [electrophoretic mobility] is an indicator) for the raw water and floc charge after coagulation (Edzwald et al., 2003).

- For pH conditions of 6 to 7, alum dosages are about 0.6 to 0.65 mg as Al per mg TOC (in terms of alum, it is 6.6 to 7.2 mg alum per mg TOC). (For waters with high SUVA of 4.5 or greater, the stoichiometry is a little less.)
- For pH conditions of 7 to 7.5, the alum dosage is about 1 mg as Al per mg TOC (11 mg alum per mg TOC).
- For pH conditions of about 5.5, alum dosage is about 0.5 mg Al per mg TOC (5.5 mg alum per mg TOC). These low pH conditions are not recommended, except for waters high in TOC (say about 10 mg/L and greater), because they do not produce maximum sweep floc conditions and can lead to high residual soluble Al.

If a water plant has a correlation between TOC and UV_{254} for its particular water supply, then the alum dosage guidelines can be expressed in terms of alum dose versus UV_{254}. Depending on the raw water SUVA, there are reported stoichiometries of 160 to 290 mg/L alum per UV_{254} in cm^{-1} or 15 to 26 mg/L as Al per UV_{254} in cm^{-1} (Pernitsky and Edzwald, 2006; Edzwald and Kaminski, 2009).

Ferric Coagulants and Use in DAF

As done for alum, we present the optimum coagulation chemistry for ferric coagulants and then present dosage guidelines. The dissolved species chemistry for ferric Fe and solubility of $Fe(OH)_3$(am) were presented Sec. 6-5. From those principles, the following optimum coagulation chemistry for ferric coagulants, whether ferric chloride or ferric sulfate is used, is summarized.

- Positively charged Fe (primarily $Fe(OH)_2^+$ exists mainly between pH 5 and the low 6s. Above pH 7.5 there is little positively charged Fe and the anion ($Fe(OH)_4^-$) dominates.
- Fe is insoluble over a wide pH range, and consequently, water temperature has little effect on the solubility of $Fe(OH)_3$(am). Sweep-floc coagulation can be carried out between pH 5 to the low 8s.

In treating waters in which TOC does not control the coagulant dose, ferric coagulation under sweep-floc coagulation is effective over a wide pH range of about 5 to the low 8s. Guidelines are presented for ferric dosages for water supplies in which TOC controls (say with SUVA of about 2.5 or greater) based on the authors' experience and some other sources (Lefebvre and Legube, 1990; Dennett et al., 1996; Letterman and Yiacoumi, 2010).

- For pH conditions of 5 to 6, ferric dosages are about 1.3 to 1.8 mg Fe per mg TOC.
- For pH 6 to 7, ferric dosages are about 2 to 3 mg Fe per mg TOC.
- For pH 7 to 7.5, ferric dosages are about 3 to 4 mg Fe per mg TOC.

Figure 6-13 is used to illustrate the effect of pH on the coagulation of a water supply in which TOC controls the ferric chloride dose. The reservoir supply has a TOC of 5.1 mg/L and SUVA of 2.7 m^{-1} per mg/L. In one set of experiments, pH was not controlled and the pH decreased with ferric chloride dosage from 7.0 for the raw water to the values shown on the plot. In another set of experiments in which HCl and ferric chloride were added to yield a constant pH of 5.6. For the experiments without pH control, the ferric chloride dosing had to decrease the pH to the low 6s to achieve good coagulation for turbidity and UV_{254} at dosages of 8 mg/L and greater. This is because, as stated above, there is a small fraction above the mid pH 6s of positively charged ferric species (Fig. 6-8) that reacts with the negatively charged NOM. On the other hand for coagulation carried out at a fixed pH of 5.6, we have a high fraction of positively charged Fe (see Fig. 6-8) that initially reacts with the NOM and then proceeds to sweep-floc coagulation with removal of the dissolved TOC by adsorption on $Fe(OH)_3$(am). Good treatment for turbidity and UV_{254} occurs at a lower dose of about 7 mg/L, and the UV_{254} removal is slightly better than ferric chloride at 8 mg/L without pH control.

Polyaluminum Chloride Coagulants and Use in DAF

The chemistry of the dissolved PACl species and conditions of precipitation as a function of pH were set out earlier in the chapter. We present here optimum coagulation chemistry for PACl coagulants and then present dosage guidelines. Optimum coagulation chemistry is summarized as follows:

- For water supplies where TOC is low and the SUVA is less than about 2.5 such that particles control dosages, PACls are effective by charge neutralization from the Al_{13}^{7+} polymer over a wide pH range of about 5 to 8 depending on water temperature. Optimum pH conditions are mid 6s to the mid 7s to take advantage of charge neutralization and precipitation of $Al(OH)_3$(s) solids (sweep floc).

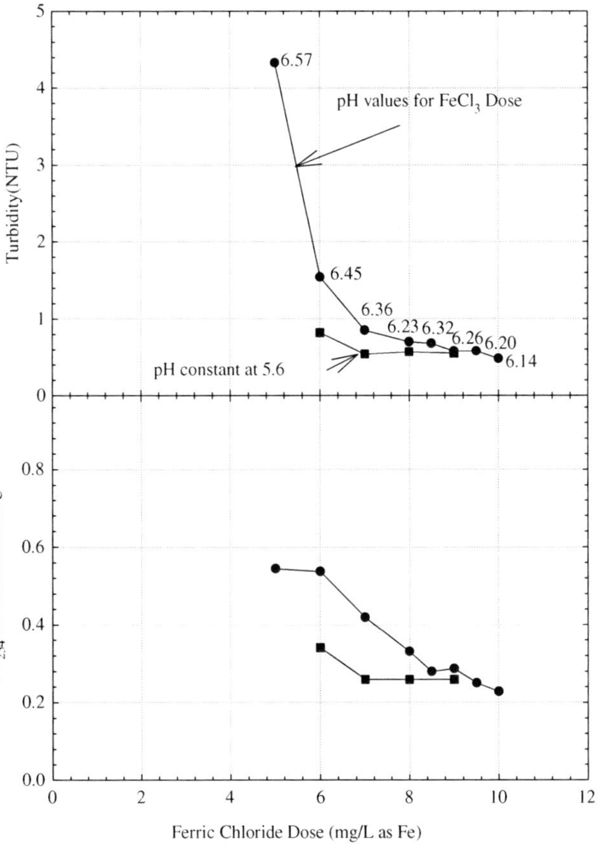

FIGURE 6-13 Coagulation performance for ferric chloride for no pH control and pH fixed at 5.6 (Reservoir supply: turbidity 3.2 NTU, TOC 5.1 mg/L, UV_{254} 0.136 cm^{-1}, SUVA 2.7 m^{-1} per mg/L)

- For supplies that contain turbidity and TOC and the SUVA is about 2.5 or greater so NOM controls coagulant dosages, then the optimum pH conditions are from the mid 6s to the mid 7s. One takes advantage of initially having a high fraction of the Al_{13}^{7+} polymer to react with the negatively charged NOM, and then precipitation of aluminum hydroxide solids (sweep-floc coagulation) and removal of TOC by adsorption of the complexed NOM, and control of residual soluble Al.

Guidelines are presented for PACl dosages for water supplies in which TOC controls (say with SUVA of about 2.5 or greater) based on the authors' experience and references (Edzwald and Van Benschoten, 1990; Edzwald, 1993; Edzwald et al., 1994; and Pernitsky and Edzwald, 2006). These dosing guidelines are for medium- and high-basicity PACls.

- For pH of 6 to 7, PACl dosages are about 0.4 to 0.6 mg as Al per mg TOC.
- For pH 7 to 7.5, PACl dosages are about is 0.7 to 1 mg as Al per mg TOC.
- For pH > 7.5, stoichiometric doses are greater than above.

The PACl stoichiometric doses in terms of Al/TOC ratios is similar to alum but often a little lower. Another feature sometimes observed for PACls, depending on the water supply and the type PACl, is they achieve fairly good removals of NOM at doses less than the optimum. This is demonstrated in Fig. 6-14. While the optimum dose (based on both turbidity and UV_{254}) for alum and the PACl is the same at 6 mg/L as Al, it is shown that the PACl at lower doses achieves better UV_{254} removals than alum.

Use of Organic Cationic Polymers and Metal Coagulants in DAF

As stated above cationic polymers are not used as sole coagulants in DAF applications. Rather, they are used in a dual-coagulant strategy in which the cationic polymer is used along with alum, or PACl, or a ferric coagulant. The ideal and practical way to practice a dual-coagulant strategy is to fix the cationic polymer dosage at a low concentration (typically about a mg/L as product) where overdosing cannot occur because the metal coagulant serves as the primary coagulant. With changes in raw water quality, the metal coagulant dose is adjusted not the cationic polymer. The cationic polymer reduces the metal dosing and sludge production because less metal precipitates. Cationic polymers do not affect pH; therefore, by using less metal coagulant there is less of a decrease in pH, which can be a benefit in treating low-alkalinity waters. By carrying out the dual-coagulant strategy with constant cationic polymer dosing, there is no adverse effect on NOM removal. The reason is that the cationic polymer

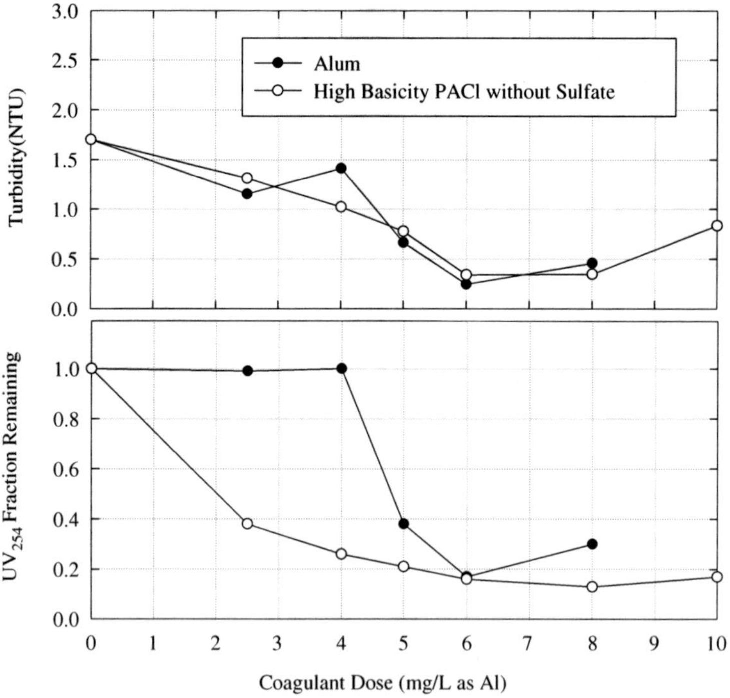

FIGURE 6-14 Comparison of coagulation between alum and high-basicity PACl without sulfate at constant pH of 6.3 (Reservoir supply: turbidity 1.7 NTU, TOC 6.7 mg/L, DOC 6.52 mg/L UV_{254} 0.311 cm^{-1}, SUVA 4.8 m^{-1} per mg/L)

partially satisfies the negative charge demand of the contaminants so less metal coagulant is needed, but overall the effectiveness in removing NOM is not diminished.

The full-scale Warner DAF plant in Fairfield, Connecticut (discussed further in Chap. 11), practices a dual-coagulant strategy of alum and a high charge-density Epi-DMA cationic polymer. The cationic polymer is held constant at 1 mg/L and the alum dose is varied to meet changes in raw water quality. The alum dose varies in response to raw water UV_{254} to achieve a treated water UV_{254} goal between 0.03 and 0.035 cm^{-1}. This dosing strategy achieves excellent DAF and filtered water quality. The use of the cationic polymer at 1 mg/L reduces the alum dosing by 0.6 to 0.9 mg/L as Al, and reduces alkalinity consumption so pH is controlled at optimum conditions in the low to high 6s, depending on water temperature, without addition of base. The use of the cationic polymer also reduces sludge production, Supplemental jar test experiments showed that the dual-coagulant strategy did not diminish removals of NOM (TOC and UV_{254}) compared to alum alone. The Warner plant has coagulation and flocculation followed by DAF over dual media filters.

Alum doses for the Warner DAF plant are presented in Fig. 6-15(a). The data are plant-operating data for an entire year. The alum dose did not respond to raw water turbidity, but responded to changes in the raw water UV_{254}. The alum dose started out relatively high (1.2 to 1.4 mg/L as Al) early in the year when the UV_{254} was high. As the raw water UV_{254} decreased in the spring and summer, the alum dose decreased. Late in the year both UV_{254} and alum doses increased. Fig. 6-15(b) shows good performance for DAF turbidity (average of 0.4 NTU) and excellent filtration (<0.1 NTU). The method of selecting the alum dose to achieve a UV_{254} goal of 0.03 to 0.035 cm^{-1} is met, as it is within this range for most of the year. Average treated water UV_{254} and TOC were 0.033 cm^{-1} and 1.8 mg/L, respectively.

EXAMPLE 6-5 Raw Water SUVA and Estimating the Alum Dosage Consider a water supply with the following characteristics: turbidity 0.7 to 2.6 NTU, TOC 2.9 to 5.3 mg/L, UV_{254} 0.095 to 0.165 cm^{-1}, pH 7.5 to 8.0, and alkalinity of 20 to 40 mg/L as $CaCO_3$. It is a low-turbidity water supply with low alkalinity. Alum coagulation will be practiced at optimum pH conditions between 6 and 7 depending on water temperature, pH of 6 to 6.5 for warm waters above 10°C, and pH of 6.5 to 7 for cold waters. TOC and UV_{254} varies seasonally and so alum doses vary. In this example we estimate the alum dosage for a mean TOC of 4.1 mg/L and UV_{254} of 0.140 cm^{-1}.

Solution

Step 1. Calculate the raw water SUVA We should use the DOC for this calculation, but it was not measured. The DOC is usually 90 to 100 percent of the TOC. Therefore, the mean DOC lies between 0.90 and 1.0 times the mean TOC, or 3.7 and 4.1 mg/L. We calculate from Eq. (6-1) the SUVA for a DOC of 4.1 mg/L.

$$\text{SUVA} = \frac{\left(\text{UV}_{254}\text{ in cm}^{-1}\right)100\ \dfrac{\text{m}^{-1}}{\text{cm}^{-1}}}{\left(\text{DOC in mg/L}\right)}$$

$$\text{SUVA}=\frac{\left(0.14\text{ cm}^{-1}\right)100\ \dfrac{\text{m}^{-1}}{\text{cm}^{-1}}}{\left(4.1\text{ in mg/L}\right)} = 3.41\frac{\text{m}^{-1}}{\text{mg/L}}$$

A similar calculation is made for a DOC of 3.7 mg/L yielding a SUVA of 3.78. We conclude that the SUVA lies between 3.4 and 3.8 m^{-1} per mg/L, and the NOM has a fairly high fraction of aquatic humic matter. We also conclude that the TOC controls coagulant dosages.

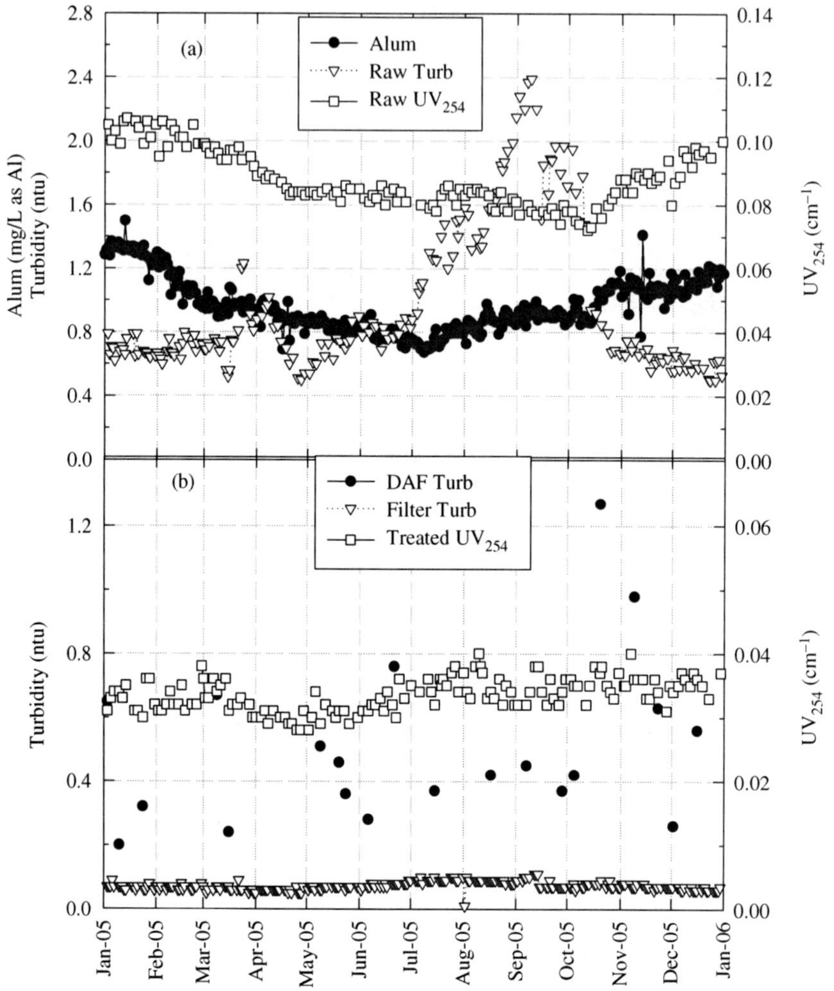

FIGURE 6-15 (a) Alum dose (cationic polymer held constant at 1 mg/L) and raw water quality. (b) DAF and filtered water turbidity and treated water UV_{254} (Fairfield, CT (U.S.) water plant). *Source: Edzwald (2007), reprinted with permission from the Journal of the New England Water Works, Association, Copyright 2009, Holliston, MA.*

Step 2. Estimating the alum dosage from the stoichiometric guidelines Between pH 6 and 7, the stoichiometric guideline is 0.6 to 0.65 mg Al per mg TOC. We will use 0.63 for the example and the mean TOC of 4.1 mg/L.

$$\text{Alum Dosage} = 0.63 \frac{\text{mg AL}}{\text{mg TOC}} \text{ x } 4.1 \text{ mg/L TOC} = 2.6 \text{ mg/L as Al}$$

We can use this estimate in setting up alum dosages for jar tests, pilot-scale studies, or in full-scale plants. Also from Table 6-2, based on SUVA of 3.4 to 3.8, the TOC removal ought to be 50 to 60 percent. ▲

6-7 INTRODUCTION TO FLOCCULATION

The purpose of flocculation is to change the size distribution of particles through collisions and attachment among particles producing on the average, larger particles. It is viewed as a physical step involving particle transport mechanisms producing collisions. The goal of what sizes to produce depends on the downstream particle-separation process. For sedimentation processes, the floc-particle sizes should be 100s of μm. However, it is more efficient to separate smaller flocs in DAF so the goal is to produce floc sizes of 10s of μm. The theory on this concept is presented in Chap. 7, while practical guidance on flocculation for DAF is presented in Sec. 6-11.

Our objectives are limited. They are to first present theory on the mechanisms of particle collisions; second, to describe briefly the types of flocculators used in DAF practice and to discuss the effects of flocculation tank hydraulics on flocculation performance; third, to describe the types of flocculant and flotation aids and their role; and fourth, to provide guidance on flocculation for DAF.

6-8 FLOCCULATION FUNDAMENTALS

Fundamental principles on flocculation kinetics were published in the classic paper of Smoluchowski (1917). These principles have been adapted to water treatment applications and are covered in MWH (2005), Letterman and Yiacoumi (2010), and Bache and Gregory (2007). The governing equation for the rate of floc formation is

$$r_{ij} = \alpha_{\text{floc}} \beta_{ij} n_i n_j \tag{6-9}$$

where α_{floc} is a flocculation attachment efficiency term, β_{ij} is a collision efficiency function, n_i is the number concentration of i-size particles, and n_j is the number concentration of j-size particles. The rate of flocculation is a second-order kinetic expression with respect to particle concentration. Therefore, waters with high particle numbers following coagulation flocculate more rapidly than those with low particle number concentrations.

The rate of flocculation depends on the stability of the particles and is modelled through an empirical attachment efficiency factor term, α_{floc}. The causes of particle stability including the forces between particles as they approach each other are presented in Chap. 5. A major purpose of coagulation is to produce particles that stick or attach on colliding. In principle, α_{floc} has low values approaching zero for raw water particles. In theory, coagulation conditions that produce complete particle destabilization would then have α_{floc} equal to one. Practically, however, good coagulation chemistry conditions yield α_{floc} values of 0.5 to 1.

The collision efficiency function (β_{ij}) depends on the collision mechanism (i.e., how collisions are brought about). Collisions or particle-transport mechanisms for DAF plants occur through Brownian diffusion or motion and fluid shear as depicted in Table 6-5. It is noted that for conventional settling tanks, flocculation by differential settling can occur within the settling tank—this mechanism is not discussed here. In the next section,

TABLE 6-5 Heterogeneous Flocculation Collision Frequency Functions for Brownian Diffusion and Fluid Shear

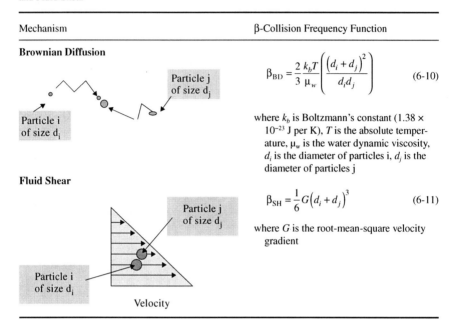

Mechanism	β-Collision Frequency Function
Brownian Diffusion 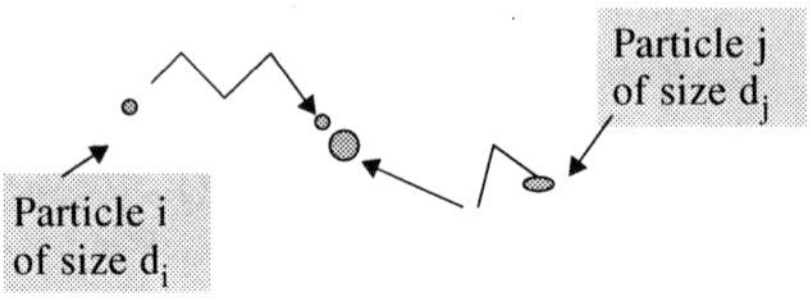	$\beta_{\text{BD}} = \frac{2}{3}\frac{k_bT}{\mu_w}\left(\frac{(d_i + d_j)^2}{d_i d_j}\right)$ (6-10) where k_b is Boltzmann's constant (1.38×10^{-23} J per K), T is the absolute temperature, μ_w is the water dynamic viscosity, d_i is the diameter of particles i, d_j is the diameter of particles j
Fluid Shear 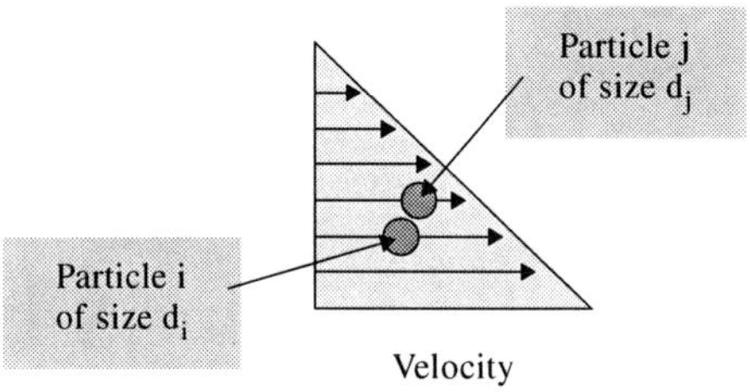	$\beta_{\text{SH}} = \frac{1}{6}G(d_i + d_j)^3$ (6-11) where G is the root-mean-square velocity gradient

expressions for collision efficiency functions are summarized and discussed. Han and Lawler (1992) present an extensive evaluation of the collision efficiency functions and consider particle-particle interaction forces affecting collisions.

Collisions by Brownian Diffusion

Brownian diffusion produces collisions among small particles because of the random movement of water molecules (kinetic energy) striking particle surfaces. If the particles are small enough, order of 1 μm and smaller, then Brownian diffusion occurs. It is important to note that two small particles can collide with both particles subject to Brownian diffusion. On the other hand, small particles undergoing Brownian diffusion can collide with large particles not subject to Brownian diffusion. In some literature, the Brownian diffusion mechanism is called perikinetic flocculation. The collision efficiency function for Brownian diffusion is presented as Eq. (6-10) in Table 6-5.

Collisions by Fluid Shear

The basic equation describing collisions by fluid shear considers a laminar-flow field with different water velocities in space within the system carrying particles, as depicted in Table 6-5. Therefore, particles moving at greater velocities can overtake and collide with particles moving with smaller velocities. The difference in velocities with distance is characterized by a velocity gradient (dv/dy). The water in a flocculation tank is mixed either hydraulically or more commonly mechanically, and the flow conditions are turbulent not laminar as depicted in Table 6-5. Camp and Stein (1943) recognized that turbulent mixing occurs in a flocculation tank, and they decided to characterize the mixing in a flocculation tank through a simple global average called the Camp root-mean-square velocity gradient, G, replacing the dv/dy in the laminar flow concept. Discussion of the turbulent-flow conditions and the use of G are found in the book by MWH (2005). Here, we are interested in concepts so we use the simplified characterization of mixing and fluid shear through G. In spite of its limitations, G is used in water treatment practice to characterize flocculation mixing.

The collision efficiency function for collisions by fluid shear is expressed by Eq. (6-11) in Table 6-5. The Camp velocity-gradient variable, G, can be calculated from Eq. (6-3) presented earlier for rapid mixing in Sec. 6-4. It is widely used in water plant design and operation to characterize mixing. Equation (6-11) shows that the β_{SH} increases with increasing G and thus the flocculation rate increases. There are, however, practical limitations of increasing G to increase the flocculation rate because, at too high G, the intense mixing produces fluid-shear conditions that break flocs apart. From experience G values of 30 to 100 s^{-1} are used for flocculation prior to DAF. This is discussed in Sec. 6-11.

EXAMPLE 6-6 Collision Efficiency Functions and Considerations of Particle Size To evaluate the collision efficiency function (β_{ij}), it is best to hold one particle size constant (d_i, which we call d_1) and vary particle size d_j (will call d_2). Consider the following two cases: (a) d_1 fixed at 0.1 μm, G at 50 s^{-1}, and water temperature at 20°C; (b) d_1 fixed at 10 μm for the same conditions for G and water temperature. Calculate and plot β_{BM} and β_{SH} as well as the sum of these, β_{TOT} (total of the collision efficiency functions).

Solution In Fig. 6-16(a), β_{ij} is plotted as a function of particle size (d_2) holding d_1 constant at 0.1 μm. We consider the conditions for Fig. 6-16(a) representing those immediately after rapid mixing at the beginning of flocculation in which we have mostly small-sized particles. What we learn follows:

- Collisions by Brownian diffusion (β_{BD}) are important for particles less than about 1 μm and mixing (fluid shear) is not important.
- For any particles that may exist above 1 μm, mixing becomes important.

β_{ij} is plotted in Fig. 6-16(b) as a function of particle size (d_2) holding d_1 constant at 10 μm. We consider these conditions an illustration of what happens during flocculation depending on the detention time of each stage. For example, in two-stage flocculation with short detention times this may represent the second stage of flocculation. What we learn follows:

- Brownian diffusion (β_{BD}) is not important for any size particles (small and large) colliding with 10 μm particles.
- Fluid shear produced by the mixing imparted in the flocculation tank is important for all particle sizes colliding with 10 μm particles. ▲

Discussion

The basic rate equation [Eq. (6-9)] describes particle flocculation for heterodispersed suspensions. In other words, there exists a distribution of particles sizes and number concentrations as a function of particle size—simply, a particle-size distribution. What can we learn and use for design and operation of engineered flocculation processes?

To describe the formation of floc particles of some size, say, k, one would have to write differential equations for the change in particle numbers in forming k-size particles from i- and j-size particles. Numerous differential equations would have to be solved

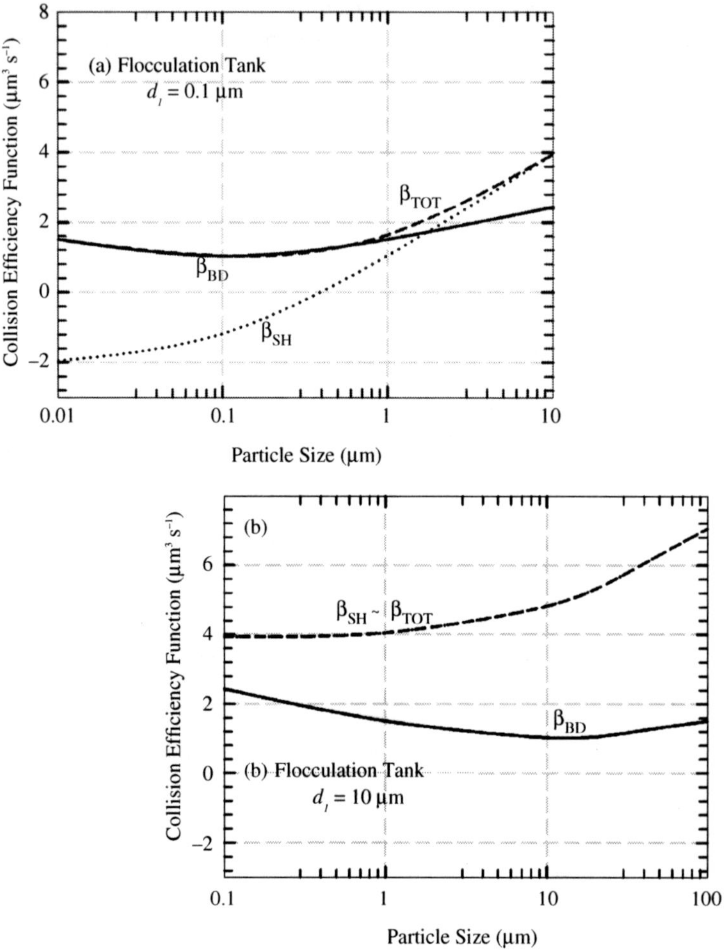

FIGURE 6-16 Collision efficiency functions versus particle size for G of 50 s^{-1} and temperature of 20°C: (a) d_I of 0.1 μm and (b) d_I of 10 μm (see use in Example 6-5).

simultaneously to predict floc formation of various sizes. Thus, it is not practical to develop a practical quantitative heterodispersed-flocculation model. However, the rate equation and collision efficiency function equations [Eqs. (6-10) and (6-11)] for the transport mechanisms are instructive in pointing out what affects the design and operation of engineered flocculation systems. Some points have been discussed above, but they are repeated for completeness.

- Effective coagulation to yield high values of α_{floc} is essential for high rates of flocculation.
- For particle suspensions with similar size distributions, suspensions with higher particle numbers flocculate much more rapidly because flocculation is a second-order rate process with respect to particle numbers, $n_i \times n_j$.

- Particle size affects the flocculation rate process. As shown in Example 6-6, large particles have greater β_{SH} than β_{BD} and thus collisions by fluid shear (mixing) are more important than those by Brownian diffusion.
- The velocity gradient, expressed as the root-mean-square velocity gradient (G), affects flocculation rates. β_{SH} increases with increasing G and thus the flocculation rate increases, but there are practical limits because if G is too high floc break-up occurs. Water temperature affects flocculation as it affects both β_{BD} and β_{SH}. Higher temperatures increase β_{BD} directly with T in the numerator of Eq. (6-10) and by the reduction of viscosity (μ_w) in the denominator. G increases with increasing temperature for a constant power input (P) because of the reduction in viscosity as shown through Eq. (6-3).
- The longer the flocculation time, the more opportunities for collisions and thus larger flocs are formed.

Monodispersed Particle Suspension Flocculation

In raw water supplies and in water plants during flocculation we do not have monodispersed particle suspensions, but it is useful to present the orthokinetic-flocculation equations for this case. Monodispersed particle suspensions are often studied in laboratory flocculation experiments, and they can occur in industrial applications. The equations are instructive and can serve as a rough approximation for flocculation in water plants. The rate equation follows and applies literally when $d_i = d_j$ in Eq. (6-9), but it is a reasonable approximation for d_i/d_j up to 4. The factor ½ appears so that collisions between i and j particles and j and i are not counted twice.

$$\frac{dn}{dt} = -\frac{1}{2}\left(\alpha_{\text{floc}}\beta_{\text{SH}}n^2\right) \tag{6-12}$$

Setting $d_i = d_j$ in Eq. (6-11), β_{SH} becomes

$$\beta_{\text{SH}} = \frac{4}{3}\left(Gd^3\right) \tag{6-13}$$

We substitute for β_{SH} in Eq. (6-12) using Eq. (6-13), and we can also change the second-order rate equation of Eq. (6-12) to a pseudo-first order by assuming particle volume is conserved during flocculation and replacing one of the n's with Φ_p using Eq. (6-14) giving Eq. (6-15).

$$\Phi_p = \frac{\pi d^3}{6} \text{ x } n \tag{6-14}$$

$$\frac{dn}{dt} = \frac{-\,4\alpha_{\text{floc}}G\Phi_p n}{\pi} \tag{6-15}$$

Integrating from t (time) = 0 for an initial particle concentration n_o, to time = t and particle concentration n, we get

$$\frac{n}{n_o} = \exp\left(\frac{-4\alpha_{floc}\Phi_p Gt}{\pi}\right) \tag{6-16}$$

This is the well-known orthokinetic-flocculation equation showing the flocculation dependence on coagulation through α_{floc}, particle volume concentration (Φ_p), G, and detention time (t). These variables are discussed further below.

6-9 FLOCCULATION TANKS

Flocculation tanks are baffled into two or sometimes three stages in DAF plants. Two-stage flocculation is the most common. Mixing is commonly provided in Europe, North America, and elsewhere by mechanical means using paddle or impeller mixers, as shown in Fig. 6-17(a) and (b), but there is also some use of hydraulic flocculation. Hydraulic flocculation is used widely in some countries because of its simplicity and no use of electrical power for mixers. It should be considered for these countries and elsewhere depending on local conditions and preferences. The power term (P) for hydraulic flocculation is calculated from Eq. (6-17) where ρ_w is the water density, g is the gravitational constant, Q is the flow rate, and h is the head loss through the flocculators. From determination of P, the velocity gradient (G) is calculated from Eq. (6-3). A common method is around-the-end channel hydraulic flocculation as depicted in Fig. 6-17(c).

$$P = \rho_w g Q h \tag{6-17}$$

Two good references with regard to DAF plants that consider the type of flocculators, the degree of mixing characterized by G, the number of stages, and mean detention time are those by Valade et al. (1996) and by Gregory and Edzwald (2010). Impeller (propeller) type flocculators are lighter, provide better mixing, and thus have been favored in practice in the last 20 years over paddle (gate) types. The degree of mixing, characterized by G, is in the range of 30 to 150 s^{-1}, with most systems designed and operated at 50 to 100 s^{-1}. There has been some use of tapering G with smaller values for the second stage, but with the short detention times for newer flocculation-tank designs this has become less common. The total mean-hydraulic detention time has decreased over the last 20 years from about 20 min to about 10 min.

The design and optimization of around-the-end hydraulic flocculators are covered in detail in two papers by the second author (Haarhoff, 1998; Haarhoff and Van der Walt, 2001). Hydraulic flocculation of the over-and-under type provides another option, but it is not commonly used.

For conventional sedimentation plants, Camp (1955) suggested the dimensionless product, $G \times t$, be in the range of 2×10^4 to 2×10^5. These criteria were based on good performance using data from full-scale plants. There is a relationship to theory since $G \times t$ appears on the right-hand-side of the orthokinetic-flocculation equation [Eq. (6-16)]. This product represents the engineered aspects of the flocculation tank (mixing intensity and tank size), but neglects the operational aspects of flocculation represented in Eq. (6-16) through α_{floc} (coagulation) and Φ_p (floc-volume concentration). The product, $G \times t$, is not used as a design criterion for flocculation in DAF plants; however, based on typical design values for G and the trend to use about 10 min for flocculation time, $G \times t$ values range from 30,000 to 60,000.

Two-stage flocculation with total mean flocculation detention times of 10 min (i.e., 5 min per stage) have been shown to yield good DAF performance and so these short detention times are now generally accepted in the design of DAF plants for drinking water treatment. Shorter times may be feasible if demonstrated by pilot studies; for example, the Croton water plant for New York City is designed for 5 min based on pilot studies (Crossley et al., 2007).

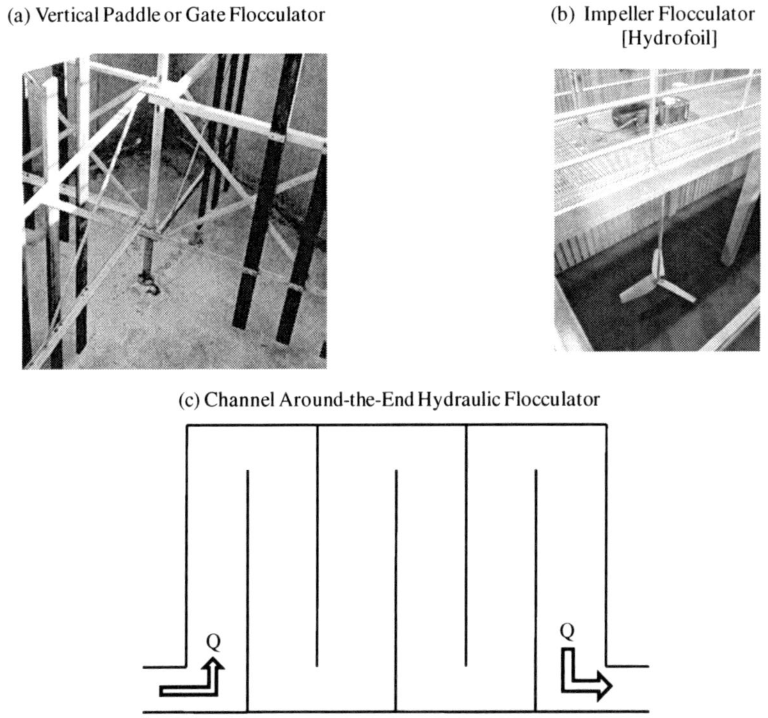

FIGURE 6-17 Types of flocculators: (a) vertical paddle *(Courtesy of Ovivo USA, LLC)*, (b) impeller *(Courtesy of Ian Crossley, Hazen and Sawyer)*, (c) hydraulic flocculator.

The hydraulics of two mixed stages (reactors) produces a distribution in the residence times of water in the flocculation system in which some of the water resides in the flocculation system for shorter and longer hydraulic residence times than the mean time. One would not want to use one-stage flocculation because if it behaves like a CFSTR (continuous-flow stirred-tank reactor) then 63 percent of the water has a residence time less than the mean time, as shown in Fig. 6-18. The figure compares residence times with respect to normalized time [t/τ where t is time and τ is the mean theoretical detention (residence time) calculated from V/Q]. If one had plug flow conditions, then all water molecules have a residence time equal to the mean time, $t/\tau = 1$. With flocculation in stages, it is analogous to CFSTRs in series that dramatically improve the residence time distribution of water. While three-stage flocculation has better hydraulic behavior than two stages (Fig. 6-18), the differences are small and experience has shown that two-stage flocculation yields good DAF performance.

6-10 FLOCCULANT, FLOTATION, AND FILTER AIDS

High MW polymers are used in water treatment as flocculant, flotation, and filter aids. Examples of common polymers used are presented in Table 6-4. They are used not to

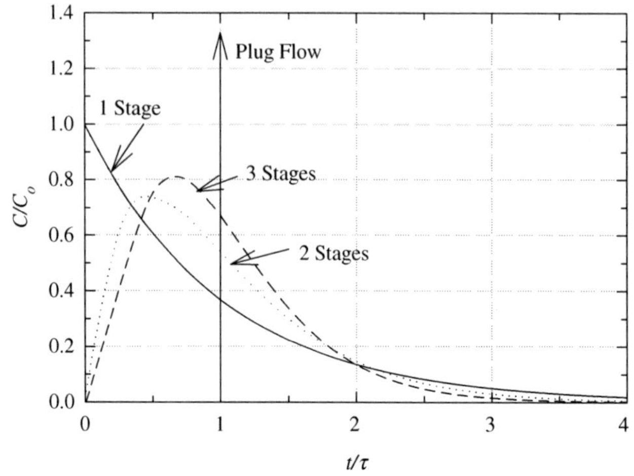

FIGURE 6-18 Comparison of pulse-tracer response times for one, two, and three-stage flocculation; plug flow behavior is shown as a reference (C/C_o is the normalized tracer concentration in which C is the mass tracer concentration in the exit from the system and C_o is the mass of tracer added divided by the reactor volume (for multiple stages it is the mass added to the first stage divided by the volume of the first stage); normalized time (t/τ) is plotted where τ is the mean theoretical hydraulic residence time).

coagulate particles and NOM, but rather as flocculant aids to increase floc size and strength in sedimentation plants. They are also used as filter aids to improve particle attachment within granular media filters, especially with filters operated at higher rates than 10 m/h.

These high MW polymers are not normally used in DAF plants as flocculant aids, but occasionally are used as flotation aids. In the latter case, some water plant operators have found that low concentrations (tenths of mg/L) of high MW polymers improve retention of the floated sludge. This may be especially important for DAF over filtration (DAFF) systems because if floated sludge is not retained the larger particles settle and enter the filter, causing shorter filter runs.

6-11 GUIDANCE ON FLOCCULATION FOR DAF

The purpose of this section is to summarize guidance on flocculation criteria for DAF drinking water treatment plants. The objective of flocculation is to increase the sizes of particles following coagulation. Unlike flocculation pretreatment for sedimentation in which large floc sizes are desired (>100 μm), smaller flocs should be produced for DAF. The objective of flocculation ahead of DAF is to produce flocs of 25 to 50 μm, which have high collision efficiencies with air bubbles (Chap. 7) and the resulting rise velocities for the floc-bubble aggregates are high enough for removal (Chap. 8). The following guidance on flocculation is given: (1) choose two-stage flocculation, (2) choose 10 min for the total flocculation time and split evenly over the two stages, and (3) choose impeller mixers for slow-flocculation mixing with the ability to operate with G of 50 to 100 s^{-1}.

REFERENCES

Archer, A. D., and Singer, P. C. (2006), Effect of SUVA and enhanced coagulation on removal of TOX precursors, *Journal of the American Water Works Association*, 98 (8), 97–107.

AWWA and ASCE (2005), *Water Treatment Plant Design*, 4th ed. (E. E. Baruth, Technical Editor), New York: McGraw-Hill.

Bache, D. H., and Gregory, R. (2007), *Flocs in Water Treatment*, London: IWA Publishing.

Benjamin, M. M. (2002), *Water Chemistry*, New York: McGraw-Hill.

Bolto, B., and Gregory, J. (2007), Organic polyelectrolytes in water treatment, *Water Research*, 41, 2301–2324.

Bose, P., and Reckhow, D.A. (1998), Adsorption of organic matter on preformed aluminum hydroxide flocs, *Journal of Environmental Engineering*, 124 (9), 803–811.

Bottero, J. Y., Cases, J. M., Fiessinger, F., and Poirier, J. E. (1980a), Studies of hydrolyzed aluminum chloride solutions. 1. Nature of aluminum species and composition of aqueous solutions, *Journal of Physical Chemistry*, 84, 2933–2939.

Bottero, J. Y., Poirier, J. E., and Fiessinger, F. (1980b), Study of partially neutralized aqueous aluminium chloride solutions: identification of aluminium species and relationship between the composition of the solutions and their efficiency as a coagulant, *Progress in Water Technology*, 12, 601–612.

Camp, T. R. (1955), Flocculation and flocculation basins, *ASCE Transactions*, 120, 1–16.

Camp, T. R., and Stein, P. C. (1943), Velocity gradients and internal work in fluid motion, *Journal of the Boston Society of Civil Engineers*, 30, 219–237.

Crossley, I. A., Herzner, J., Bishop, S. L., and Smith, P. D. (2007), Going underground–Constructing New York City's first water treatment plant, a 1,100 Ml/d dissolved air flotation, filtration and UV facility, *Proceedings of the 5th International Conference on Flotation in Water and Wastewater Systems*, Seoul National University, Seoul, South Korea.

Dennett, K. E., Amirtharajah, A., Moran, T., and Gould, J. P. (1996), Coagulation: its effect on organic matter, *Journal of the American Water Works Association*, 88 (4), 129–142.

Edzwald, J. K. (1993), Coagulation in drinking water treatment: particles, organics, and coagulants, *Water Science and Technology*, 27 (11), 21–35.

Edzwald, J. K. (2010), *Water Quality and Treatment: A Handbook on Drinking Water*, 6th ed., New York: AWWA and McGraw-Hill.

Edzwald, J. K., Becker, W. C., and Wattier, K. L. (1985), Surrogate parameters for monitoring organic matter and THM precursors, *Journal of the American Water Works Association*, 77 (4), 122–131.

Edzwald, J. K., Becker, W. C., and Tambini, S. J. (1987), Organics, polymers and performance in direct filtration, *Journal of the Environmental Engineering Division, ASCE*, 113 (1), pp. 167–185.

Edzwald, J. K., and Van Benschoten, J. E. (1990), Aluminum coagulation of natural organic matter, in H. H. Hahn and R. Klute, eds., *Chemical Water and Wastewater Treatment*, New York: Springer-Verlag.

Edzwald, J. K., Bunker, D. Q., Jr., Dahlquist, J., Gillberg, L., and Hedberg, T. (1994), Dissolved air flotation: pretreatment and comparisons to sedimentation, in R. Klute and H. H. Hahn, eds., *Chemical Water and Wastewater Treatment III*, New York: Springer-Verlag.

Edzwald, J. K., and Tobiason, J. E. (1998), Enhanced versus optimized multiple objective coagulation, in H. H. Hahn, E. Hoffmann, and H. Ødegaard, eds., *Chemical Water and Wastewater Treatment V*, New York: Springer-Verlag.

Edzwald, J. K., and Tobiason, J. E. (1999), Enhanced coagulation: USA requirements and a broader view, *Water Science and Technology*, 40 (9), 63–70.

Edzwald, J. K., Pernitsky, D. J., and Parmenter, W. (2000), Polyaluminum coagulants for drinking water treatment: chemistry and selection, in H. H. Hahn, E. Hoffmann, and H. Ødegaard, eds., *Chemical Water and Wastewater Treatment VI*, New York: Springer-Verlag.

Edzwald, J. K., Tobiason, J. E., Udden, C., Kaminski, G. S., Dunn, H. J., Galant, P. B., and Kelley, M. B. (2003), Evaluation of the effect of recycle of waste filter backwash water on plant removals of Cryptosporidium, *Journal of Water Supply: Research and Technology – Aqua*, 52 (4), 243–258.

Edzwald, J. K., and Kaminski, G. S. (2009), A practical method for water plants to select coagulant dosing, *Journal of the New England Water Works Association*, 123 (1), 15–31.

Edzwald, J. K., and Tobiason, J. E. (2010), Chemical principles, source water composition, and watershed protection, in J. K. Edzwald, ed., *Water Quality and Treatment: A Handbook on Drinking Water*, 6th ed., New York: AWWA and McGraw-Hill.

Glaser, H. T., and Edzwald, J. K. (1979), Coagulation and direct filtration of humic substances with polyethylenimine, *Environmental Science and Technology*, 13 (3), 299–305.

Gregory, J. (1973), Rates of flocculation of latex particles by cationic polymers, *Journal of Colloid and Interface Science*, 42 (2), 448–456.

Gregory, R., and Edzwald, J. K. (2010), Sedimentation and flotation, in J. K. Edzwald, ed., *Water Quality and Treatment: A Handbook on Drinking Water*, 6th ed., New York: AWWA and McGraw-Hill.

Haarhoff, J. (1998), Design of around-the-end hydraulic flocculators, *Journal of Water Supply: Research and Technology – Aqua*, 47 (3), 142–152.

Haarhoff, J., and Van der Walt, J. J. (2001), Towards optimal design parameters for around-the-end hydraulic flocculators, *Journal of Water Supply: Research and Technology – Aqua*, 50 (3), 149–159.

Han, M. Y., and Lawler, D. F. (1992), The (relative) insignificance of G in flocculation, *Journal of the American Water Works Association*, (84) (10), 79–91.

Kasper, D. R. (1971), *Theoretical and Experimental Investigation of the Flocculation of Charged Particles in Aqueous Solution by Polyelectrolytes of Opposite Charge*, PhD Dissertation, California Institute of Technology, Pasadena, CA.

Leenheer, J. A., and Noyes, T. I. (1984), A filtration method and column adsorption system for on-site concentration and fractionation of organic substances from large volumes of water, *U.S. Geological Survey Water Supply Paper 2230*, USGS, Denver, CO.

Lefebvre, E., and Legube, B. (1990), Iron (III) Coagulation of humic substances extracted from surface waters: Effect of pH and humic substances concentration, *Water Research*, 24 (5), 591–606.

Letterman, R. D., and Yiacoumi, S. (2010), Coagulation and flocculation, in J. K. Edzwald, ed., *Water Quality and Treatment: A Handbook on Drinking Water*, 6th ed., New York: AWWA and McGraw-Hill.

MWH (2005), *Water Treatment Design: Principles and Design*, 2nd ed. (J. C. Crittenden, R. R. Trussell, D. W. Hand, D. W. Howe, and G. Tchobanoglous), Hoboken, NJ: John Wiley & Sons.

Pernitsky, D. J., and Edzwald, J. K. (2003), Solubility of polyaluminum coagulants, *Journal of Water Supply: Research and Technology – Aqua*, 52 (6), 395–406.

Pernitsky, D. J., and Edzwald, J. K. (2006), Selection of alum and polyaluminum coagulants: principles and applications, *Journal of Water Supply: Research and Technology – Aqua*, 55 (2), 121–141.

Smoluchowski, M. (1917), Versuch Einer Mathematischen Theorie der Koagulations-Kinetik Kolloider Losungen, *Zeitschrift für Physikalische Chemie*, 92, 129–168.

Stumm, W., and Morgan, J. J. (1996), *Aquatic Chemistry*, 3rd ed., New York: Wiley-Interscience.

Valade, M. T., Edzwald, J. K., Tobiason, J. E., Dahlquist, J., Hedberg, and Amato, T. (1996), Pretreatment effects on particle removal by flotation and filtration, *Journal of the American Water Works Association*, 88, (12), 35–47.

Van Benschoten, J. E. (1988), *Speciation and Fate of Aluminum in Water Treatment*, PhD Dissertation, University of Massachusetts, Amherst, MA.

Van Benschoten, J. E., and Edzwald, J. K. (1990), Chemical aspects of coagulation using aluminum salts: 1. Hydrolytic reactions of alum and polyaluminum chloride, *Water Research*, 24, 1519–1526.

CHAPTER 7
CONTACT ZONE

The DAF tank is divided into two compartments, as depicted in Fig. 7-1, that have entirely different process functions. The floc particles are carried into the first compartment, called the contact zone, with the influent flow and contacted with air bubbles that are produced by the injection of the pressurized recycle flow. Here, collisions occur between bubbles and floc particles. If the floc particles have the proper surface chemistry, then bubbles colliding with them may attach yielding floc-bubble aggregates. The water carrying the suspension of bubbles, flocs, and aggregates flows through the contact zone into the second compartment of the tank, called the separation zone. Here, bubbles that are not attached to flocs and floc-bubble aggregates can rise to the surface of the tank. It is discussed in Chap. 8 that while clarification is the main function of the separation zone, collisions among bubbles and flocs can continue to occur.

This chapter's goals are to develop a fundamental process model describing the performance of the contact zone and to identify design and operating variables that affect performance. We begin with consideration of the hydraulics of the contact zone, in particular, the residence time distribution (RTD) of fluid (i.e., water) in the contact zone. We then consider attempts to model the contact zone and present features of the white water bubble-blanket model. Next, the white water bubble-blanket model is developed in detail, and we discuss the effects of pretreatment model variables and DAF tank design and operating model variables on contact zone performance. The final section discusses further the practical applications of the contact zone model.

Two things are worth mentioning before we begin. First, the contact zone in some literature prior to the mid 1990s was referred to as the reaction zone; we use the term contact zone. Second, the contact zone performance only addresses the first step, the removal of floc particles from the water phase onto air bubbles through collision and attachment. The

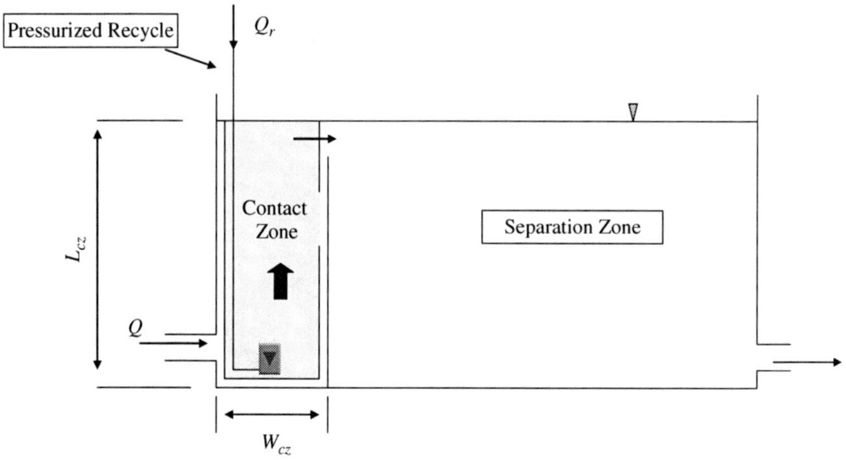

FIGURE 7-1 DAF tank showing the contact and separation zones (the baffle dividing the contact zone from the separation zone is often inclined).

second step occurs in the separation zone, which is the removal of bubbles and floc-bubble aggregates through their rise velocities to the surface of the tank in the form of floated sludge. The overall DAF performance in removing floc particles must consider both steps. Principles pertaining to the separation zone are covered in Chap. 8.

7-1 CONTACT ZONE HYDRAULICS

Fluid Residence Time Distributions

Contact zone performance depends on the reactions taking place and on the reactor (contact zone) hydraulics. The reactions are the collisions between bubbles and flocs, and the attachment of bubbles to flocs upon collision. The hydraulic characteristics deal with the detention time or residence time of the water in the reactor; specifically, we must describe the RTD of water for the reactor. In other words the time spent in the reactor by fractions of the fluid or water; on a microscopic level it would be the time spent in the reactor by each water molecule. We often use a characteristic time called the mean detention time or mean residence time; however, except for the ideal case of plug flow, this is insufficient to describe the RTD of water in the reactor, and thus insufficient to describe reactor performance.

Figure 7-1 shows a simple schematic of a DAF tank. In this schematic the contact zone is depicted in the first compartment with a baffle dividing it from the separation zone. This is an idealization, but one that we use to develop some principles. The baffle is often inclined at a small angle from the vertical, which is not shown in the figure.

We consider the two ideal cases of fluid flow to describe RTDs: the plug-flow reactor (PFR) and the continuous-flow stirred-tank reactor (CFSTR). We also address reactors designed for plug flow but in which some dispersion occurs and discuss how the hydraulic flow behavior of the contact zone can be modelled. What follows are summary presentations for the two ideal cases and the non-ideal case with dispersion. For details about reactor hydraulics and RTDs for ideal and non-ideal flow, the reader can consult the book on water treatment by MWH (2005) and book on wastewater engineering by Metcalf and Eddy (2003), or the book on chemical reaction engineering by Levenspiel (1998).

Ideal Case: Plug-Flow Reactor. A conceptual schematic of a PFR is shown at the top of Fig. 7-2. A PFR is an ideal flow case characterized by no fluid dispersion, no mixing, no short-circuiting of the flow, and no dead spaces in the reactor. All pieces or fractions of fluid or water (e.g., pieces A, B, and C in the top of Fig. 7-2) reside in the reactor for the same time (t), which equals the theoretical mean detention time (τ_{PFR}). For a constant flow rate of Q and reactor volume, V, the ideal mean residence time is given by Eq. (7-1).

$$\tau_{\text{PFR}} = \frac{V}{Q} \tag{7-1}$$

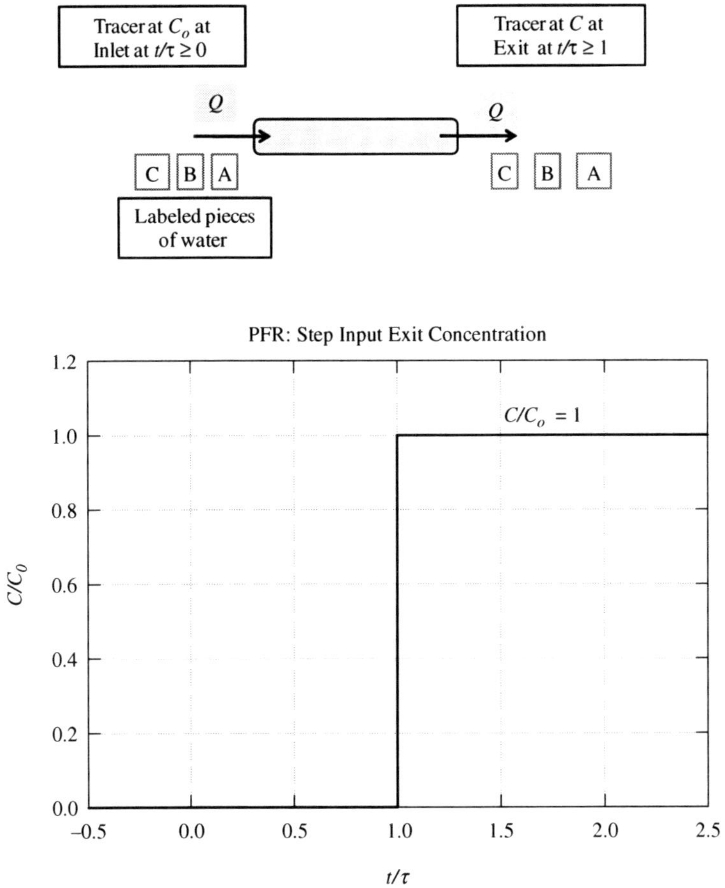

FIGURE 7-2 Plug-flow reactor (PFR): top figure shows schematic of the PFR and the bottom figure shows the normalized exit tracer concentration for a step input feed of tracer.

Tracers are conservative substances (no change in concentration from biological, chemical, or physical reactions) and are used to characterize RTDs. In water treatment, fluoride is often used as a tracer. Other tracers that may be used in pilot plants or in wastewater plants are dyes (e.g., Rhodamine WT) or an inert salt such as NaCl. A continuous feed of tracer at a concentration (C_o) to the inlet of the reactor beginning at time zero ($t = 0$) and continuing throughout the time of the test is called a step input tracer test. The tracer concentration (C) in the reactor exit flow is measured over time. It is convenient to normalize both C and t, and plot an exit tracer curve of C/C_o versus t/τ.

The normalized response curve is presented at the bottom of Fig. 7-2. The response curve shows that none of the tracer leaves the reactor until the time (t), which equals the mean residence time, τ_{PFR}. This means that for the PFR case, all pieces of fluid have the same detention or residence time, τ_{PFR}. The PFR has no distribution of residence times; all water resides in the reactor for the same time, τ_{PFR}. This also means, as stated above, there is no water dispersion, no mixing, no short-circuiting of the flow, and the reactor has no dead spaces.

Ideal Case: Continuous-Flow Stirred-Tank Reactor (CFSTR). The water in a CFSTR is completely mixed so the contents are everywhere the same. An example is a rapid mix tank in which coagulant chemicals are added. A consequence of complete mixing is that the concentrations of everything (chemicals, particles, bubbles, etc.) are everywhere the same in the CFSTR and equal to the exit concentrations. The top of Fig. 7-3 shows a schematic of a CFSTR to which a step feed tracer test is run. The concentration of the tracer in the tank is zero when the test begins. Beginning at time zero, the tracer is fed continuously at a concentration C_o to the inlet of the tank. The flow rate through the CFSTR is constant

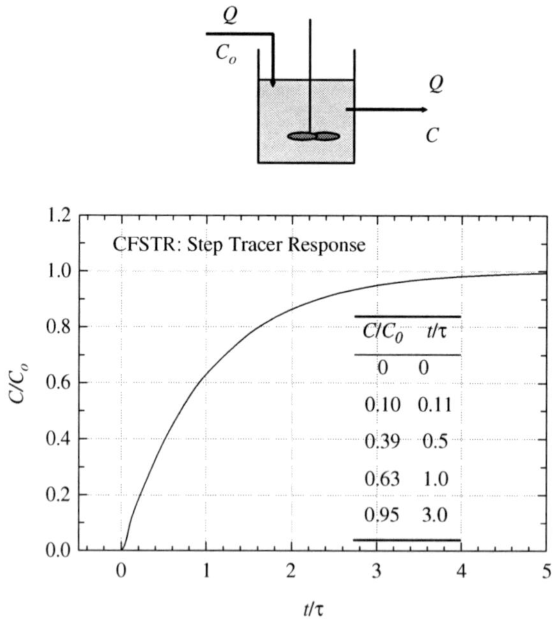

FIGURE 7-3 Continuous-flow stirred-tank reactor (CFSTR): top figure shows schematic of the CFSTR and the bottom figure shows the normalized exit tracer concentration for a step input feed of tracer.

at Q, and the CFSTR has a water volume, V. The tracer concentration (C) in the reactor exit is measured over time. The normalized exit tracer response curve is presented at the bottom of Fig. 7-3. The ideal mean residence time (τ_{CFSTR}) for the CFSTR is given by Eq. (7-2). Equations (7-1) and (7-2) are identical, meaning the mean residence time or mean detention time is identical for a PFR and a CFSTR with the same volume and flow rate. A major difference between the two reactors is that the CFSTR has an RTD; that is, there is a distribution of residence times around the mean time.

$$\tau_{\text{CFSTR}} = \frac{V}{Q} \tag{7-2}$$

To obtain the predictive equation for the RTD of the tracer exiting the CFSTR, a mass balance on the tracer is done around the entire CFSTR volume yielding Eq. (7-3). Equation (7-3) predicts the RTD of the tracer in terms of the fraction of tracer in the CFSTR exit or that remaining in the tank as a function of time with respect to the ideal mean time, τ_{CFSTR}.

$$\frac{C}{C_o} = \left(1 - \exp\frac{-t}{\tau_{\text{CFSTR}}}\right) \tag{7-3}$$

The RTD is shown at the bottom of Fig. 7-3. The tracer curve shows that some tracer or fractions of water reside in the CFSTR for short times and some fractions of water reside for long times. It is instructive to examine the meaning of RTD for a CFSTR using the exact values in the table within Fig. 7-3. A value of C/C_o equal to 0.10 means that 10 percent of the tracer is in the exit so the tracer or water resides in the tank for a time t/τ or less, that is, 11 percent of the ideal mean time, τ. Saying another way, 89 percent of the water has a greater residence time than the mean ideal time, τ. This characteristic time is called the t_{10} time, and it is used in the United States and some other countries to meet disinfection requirements based on the product of a disinfectant residual concentration and residence time (characteristic time being t_{10}). For t equal to $0.5 \times \tau$, we find 39 percent of the tracer in the exit so it resides in this CFSTR for this time or less. We see that for t equal to the ideal mean time τ (or $t/\tau = 1$), the value for C/C_o is 0.63, meaning 63 percent of the tracer is in the exit or 63 percent of the water resides in the tank for the mean detention time or less. For the case of 3 times the mean time ($t/\tau = 3$), 95 percent of the tracer resides in the tank for this time or less. It also means you have not yet reached the time to achieve the steady state concentration for the tracer (i.e., $C/C_o = 1$), but you have reached a time that gives you 95 percent of the steady state concentration. Hence the *rule of thumb* often used in water treatment practice is that it takes at least three detention times after a chemical is added to a process reactor for it to reach approximately a steady state concentration. Or conversely, if you stop dosing a chemical or are flushing a pipe you want at least three detention times to approach washout.

How to Model the Contact Zone Hydraulics

Many water treatment tanks or reactors are designed and assumed to behave as PFRs, but some dispersion (mixing) occurs. If the mixing is considerable, then the RTD is substantially different from a PFR, and the PFR performance is affected adversely. In other situations, the effect of dispersion may be small enough that modelling the reactor with hydraulic behavior approximating plug flow is adequate for reactor design. Most often plug flow is an approximation; an exception is that turbulent flow conditions in pipelines produce plug flow behavior. A couple of examples of treatment processes where plug flow behavior is assumed in design are filters and settling tanks. In the first case, plug flow is a reasonable assumption as the dispersion caused by water passing through the filter media is small. For conventional settling tanks considerable dispersion can take place, so plug flow is not a good approximation. Nevertheless, the Hazen theory of particle removal by settling is based on overflow rates and assumes that the tank hydraulics is plug flow. To account for the non-ideal behavior of settling tanks, conservative overflow rates, based on experience, are used in design.

Now we address the contact zone hydraulics. Contact zone performance depends on the reactions taking place and on reactor (contact zone) hydraulics. The question is can we model the contact zone as an ideal case of plug flow hydraulics? This would make the modelling of the contact zone simpler compared to non-ideal flow behavior. Also for reactors of the same size, a plug flow system achieves higher efficiencies compared to CFSTRs and to reactors that exhibit non-ideal flow behavior. Thus, it is advantageous to design the contact zone as best we can to exhibit plug flow hydraulics.

The contact zone is a baffled compartment within the DAF tank, as shown in Fig. 7-1. In the contact zone the main flow (Q) is vertical and mixes with the recycle flow (Q_r) that enters through nozzles (or other injection devices). Air bubbles are formed with the injection of the recycle flow so some dispersion is expected as the combined flow and bubbles move through the contact zone. One can have some dispersion, however, and still have approximately plug flow behavior. To answer the question, can we model the contact zone as a plug flow process? We examine the contact zone hydraulic flow behavior using three approaches: from tracer tests, from considering the contact zone geometry, and from computational fluid dynamics (CFD).

The authors have conducted numerous DAF pilot-scale studies, and several of these have included tracer (dye) tests. Dye studies showed that the contact zone was approximately plug flow. Other tracer test studies have characterized the hydraulic flow behavior in terms of the dispersion. A measure of the degree of dispersion in a reactor intended to exhibit plug flow hydraulic behavior is the dispersion number (d_{dn}). It is a dimensionless number defined by Eq. (7-4).

$$d_{dn} = \frac{D}{v_x L_{cz}} \tag{7-4}$$

where D is the dispersion coefficient, v_x is the flow velocity in the direction of flow, and L_{cz} is the longitudinal length in the flow direction (for the contact zone, this is the depth—see Fig. 7-1). The dispersion coefficient (D) accounts for all mixing (i.e., dispersion) and includes the effects of molecular diffusion, velocity gradients from advective flow, and turbulent eddies.

The proper way to describe the intensity of axial dispersion (direction of flow) is by use of the dispersion number (d_{dn}). Unfortunately in some books and journals, reference is made to the reciprocal of the dispersion number, which they call the Peclet number. Strictly speaking the Peclet number deals with molecular diffusion, and is not an overall dispersion coefficient.

Equation (7-4) shows clearly that it is not just the magnitude of the dispersion coefficient (D), but it is the ratio of D divided by v_x times L_{cz} that defines the extent of dispersion. That is why turbulent flow conditions in pipes, which is almost always the case for water facilities, exhibit plug flow behavior—the product of v_x times L_{cz} is large, making d_{dn} approximately zero. Plug flow systems are those in which the d_{dn} is small and approaches zero. On the other hand, if we completely mixed the water in the reactor to make it behave like a CFSTR, d_{dn} would have a large number. Fundamentally then, the ideal case of a PFR has a d_{dn} of 0, and for the ideal case of a CFSTR it has a value of infinity. All non-ideal reactors have dispersion numbers between these values. The smaller the value, the more the reactor's hydraulic behavior is approximated by plug flow. As a rough guideline, we can say that a d_{dn} of ≤0.05 indicates a low degree of dispersion and so the reactor has hydraulic behavior close to plug flow. This is a rough guideline because the performance of a reactor depends on both the hydraulics and the kinetics of the reactions. The reader is referred to the books by MWH (2005) and Metcalf and Eddy (2003) on how to determine the dispersion number from tracer tests.

Shawwa and Smith (1998) found a d_{dn} of 0.01 or greater in characterizing the contact zone in a pilot-scale DAF unit. They found smaller values with increasing flow velocity in accordance with principles shown by Eq. (7-4). It is noted that full-scale DAF plants will have smaller d_{dn} than pilot-scale plants because the tanks are deeper (larger L_{cz}). Moruzzi and Reali (2010) also ran tracer tests on a pilot-scale DAF unit and found dispersion numbers between 0.1 and 0.2 depending on the air concentration and the mean detention time. The air concentration reduced the mixing for detention times of 1 min and less.

For full-scale DAF plants with the typical velocities and contact zone ratios for L_{cz}/W_{cz} (see Fig. 7-1), Haarhoff and Edzwald (2004) showed that dispersion numbers (d_{dn}) as high as 0.2 did not have much effect on the contact zone performance, for detention times of about a minute or greater, compared to plug flow conditions.

Lundh et al. (2002) made velocity measurements at the pilot-scale across the width and depth of the contact zone. They found mixing at the bottom of the contact zone where air is introduced. As the water moved away from the bottom toward the top of the contact zone, they found the flow was more like plug flow. With increasing baffle height (in effect, greater L_{cz}), they found better flow behavior more like plug flow.

CFD is a valuable design tool for predicting velocity profiles through process reactors. It has been used to aid the design of DAF tanks (Fawcett, 1997; Ta et al., 2001; Amato and Wicks, 2009). Careful interpretation of the results is required because of the difficulties in modelling three-phase flow (water, air bubbles, and particles). To simplify the modelling, the hydraulics is often done by excluding the particle phase. Various assumptions are made in using CFD, two important ones are bubble size and concentration that affect the velocity predictions. Recognizing that some caution is necessary in evaluating CDF predictions, different baffle arrangements (height, angle, and width of the contact zone) can be evaluated and are valuable in improving the hydraulic design. CFD studies indicate generally that the contact zone exhibits behavior approaching plug flow.

We conclude that assuming plug flow hydraulics for the contact zone is a reasonable approximation for modelling and design. Consequently in the following text, plug flow hydraulics is used for the white water bubble-blanket model.

7-2 CONTACT ZONE MODELLING

Background

Chapter 2 traces the history of DAF technology and its incorporation into drinking water clarification beginning in the 1960s. In the early years the technology suffered from a lack of a firm fundamental understanding, and so water plants were designed and built based on experience and empirical pilot-plant studies. In the 1980s the first attempts began, and continue to the present, to develop fundamental models describing the collisions and attachment of particles and flocs to air bubbles in the contact zone (see Fig. 7-1).

Early work was done by Tambo et al. (1986) that was continued in the 1990s (Fukushi et al., 1995; Matsui et al., 1998). They considered the contact zone to act in an analogous way to a flocculation basin in conventional water treatment. They wrote flocculation rate equations showing dependence on the number concentrations of floc particles and air bubbles, the sizes of floc particles and bubbles, attachment chemistry of flocs to bubbles, and G (root-mean-square velocity gradient)—see Chap. 6 for presentation of concepts and practical material regarding G. The only collision mechanism considered was the mixing brought about by the energy dissipation in the contact zone and characterized by the global mixing parameter, G.

Other flocculation type models were developed by Han (Han et al., 1997; Han, 2002) and Leppinen (1999, 2000). They considered other collision mechanisms and, most important, they considered the effects of particle-bubble interaction forces on particles approaching bubbles and how these forces affected the particle trajectory to the bubble surface. The particle-bubble interaction (Chap. 5) they considered were electrostatic, van der Waals, and hydrodynamic forces.

The details of the contact zone models of Tambo, Han, and Leppenin are found in their papers cited above. Furthermore, Edzwald (2010) has critically reviewed these models. Briefly, the models are instructive but have numerous limitations. They are instructive in that they highlight some important variables affecting DAF performance in the contact zone. Some limitations are listed.

- All of the models were developed for batch conditions and apply only directly to contact zone hydraulics exhibiting plug flow behavior. In the Tambo model, mixing takes place in the contact zone that is characterized by G as a global mixing variable. Thus, one would not have plug flow conditions, and you would have to incorporate their batch kinetic-model equations either into CFSTR (complete mixing in the contact zone) hydraulics or into non-ideal hydraulic flow. Evidence was presented earlier in the chapter that plug flow hydraulics is a reasonable approximation, and so mixing across the contact zone and use of G is incompatible with this view.
- The models of Han and Leppinen with particle-bubble interaction forces contain many parameters that are not known or not easily measured, as covered in Chap. 5. This makes them of limited use as design and operating models.
- Han and Leppinen assumed that the van der Waals force is attractive but, in fact, it is a repulsive force, as discussed in Chap. 5, for bubbles interacting with particles in water. The models, therefore, have no particle-bubble attractive force to bring about attachment since they do not include the hydrophobic force. These models would only have electrostatic attraction for the case of oppositely charged bubbles and flocs, but this is not the normal case in water treatment.
- Flocculation type models predict better performance with larger bubbles, which is not the case—the smaller the bubbles, the better the performance.

Features of the White Water Bubble-Blanket Model

Model Considerations. Models are scientific statements expressed through equations that attempt to describe processes or phenomena. Our interest is the development of a model to describe the performance of the DAF contact zone. Specifically, we wish to describe mathematically the removal of floc particles from water by collisions and attachment to air bubbles. The model must provide a fundamental understanding of the process. A goal is to use the model as a DAF design and operating tool, and so the model must include variables that affect design and operation. The model should avoid adjustable variables. One should be able to measure the variables, and one should verify the model through pilot or full-scale plant testing.

Basics and Analogy to Granular Media Filtration. Edzwald and coworkers (Edzwald et al., 1990; Edzwald, 1995) developed a model for the DAF contact zone using an approach similar to that used in modelling the performance of air filtration (Friedlander, 1977), granular media water filtration (Yao et al., 1971), and other flotation processes (Flint and Howarth, 1971; Reay and Ratcliff, 1973). The DAF modelling approach was examined in a review paper by Haarhoff and Edzwald (2004), and it is also discussed by Gregory and Edzwald (2010). The details of this modelling approach including presentation of fundamental equations are discussed in the next section. Some comparisons are made first between the contact zone and granular media filtration followed by a few descriptive remarks about the modelling approach.

It is instructive to compare features of the DAF contact zone with those of a granular media filter bed. These features are presented in Table 7-1 using typical values for conventional rate filtration and conventional rate DAF systems. First, we note that the DAF contact zone has a far greater hydraulic loading rate than filtration. For it then to collect effectively particles in comparison to filtration, it must have both smaller collectors (bubbles) than the filter grains and more collectors. The information in the table shows that the bubble collectors are about 10 times smaller (40 to 80 μm) than the filter grains (0.5 to 1 mm). A fundamental principle of the model, developed in the next section, is that the smaller the collector, the better the performance. We know this to be true for water filtration. Another feature of the DAF contact zone is that it has a higher collector concentration; bubble numbers are about 10 times greater than the number of filter grains. These features provide insight as to why DAF is an efficient particle-collection process. The modelling approach for the DAF contact zone is that the rising bubble blanket through the contact zone acts analogously to a filter. The separation distance between bubbles is about 200 μm or a little less (see bottom of Table 7-1), which is about the size of pore openings for water to pass through filter grains, making these two processes remarkably similar.

The bubble suspension, being at high concentration, in the contact zone is referred to as the white water blanket. Bubbles are continuously added from the recycle injection and removed (by flow to the separation zone) so that a dynamic steady state results with a constant high bubble concentration in this white water blanket. The bubbles act as collectors of particles (or flocs) as the particles are transported to bubble surfaces and interact with the bubble blanket, much like a filter. The change in floc particle concentration through the white water blanket depends on floc particle and bubble concentrations, floc and bubble

TABLE 7-1 Features of Granular Media Filtration and the DAF Contact Zone

Parameter	Filter	DAF Contact Zone
Hydraulic Loading Rate, m/h	5–15	100–200
Depth, m		
Filter bed	0.5–1	
DAF		2–3.5
Collector size		
Filter, mm	0.5–1	
Bubbles, mm (μm)		0.04–0.08 (40–80)
Number of Collectors		
Filter grains per m^3 of bed	10^9–10^{10}	
Bubbles per m^3 of tank		10^{10}–10^{11}
Pore opening between filter collectors and separation distance between bubble collectors, μm	200–400	160–200

sizes, attachment chemistry, and mechanisms of floc particle transport to bubble surfaces. These particle transport mechanisms can occur by Brownian diffusion, sedimentation, interception, and inertia. Equations for the particle transport mechanisms and model development are presented in the next section.

7-3 WHITE WATER BUBBLE-BLANKET MODEL

General Description

This modelling approach considers a white water blanket of air bubbles within the contact zone with the bubbles acting as collectors of particles. The single collector collision efficiency concept is used to account for the particle transport mechanisms causing collisions with bubbles. The model is developed by considering the blanket of air bubbles to exist in the contact zone at a dynamic steady state at high bubble concentration ($n_b \sim 10^{10}$ to 10^{11} bubbles per m^3, see Chap. 5 and Table 7-1), due to continuous input of air bubbles from injection of recycle from the saturator and continuous output of bubbles and floc-bubble aggregates in the outlet flow to the separation zone. In the development of the following model, the term particle is used in a general sense and refers either to primary particles that enter the flotation tank or to flocculated particles (flocs).

The kinetic rate of particle (n_p) removal by collision and attachment to bubbles follows a second-order rate expression with respect to number concentration of particles and bubbles (n_b)

$$\frac{dn_p}{dt} = -k_c n_p n_b \tag{7-5}$$

where, k_c is the DAF rate coefficient dependent on particle transport to bubble surfaces, the particle-bubble attachment efficiency, and the volume of suspension swept by a single rising bubble per unit time. Eq. (7-6) shows the dependency of k_c.

$$k_c = \left(\alpha_{\text{pb}}\eta_{\text{T}}\right)\left(v_b A_b\right) \tag{7-6}$$

The single collector efficiency (η_{T}) is used to account for particle transport to bubble surfaces or collision mechanisms between particles and bubbles. The particle-bubble attachment efficiency (α_{pb}) accounts for the fraction of collisions between bubbles and particles that result in attachment. The product of the rise velocity of a single bubble (v_b) and the cross-sectional area of a bubble (A_b) yields the volume of suspension swept by a single rising bubble per unit time. A primary feature of the model is the use of the single collector efficiency concept to account for particle transport. This is the subject of the next section.

Single Collector Efficiency

Concept. Particle transport from the bulk water to the bubble surface is modelled using the single collector collision efficiency concept. This concept has been used to model particle deposition onto air filters (Friedlander, 1977), onto water filters (Yao et al., 1971), and onto bubbles in froth flotation (Flint and Howarth, 1971) and in dispersed air flotation (Reay and Ratcliff, 1973) processes. Edzwald and coworkers (Edzwald et al., 1990; Edzwald, 1995) adapted the concept and applied it to DAF. Here we present this concept, then the individual single collector efficiency (η) equations are presented in the next section, and then in the subsequent section the η equations are incorporated into modelling the performance of the contact zone.

We begin with a physical description of the transport of particles to a bubble or collector surface. In the contact zone, water moves upward at a certain velocity with air bubbles rising at a velocity relative to the water motion. This bubble rise velocity (v_b) causes streamlines of flow downward around the bubble, as illustrated in Fig. 7-4. The streamlines are characterized by Stokes flow conditions because the bubbles in the DAF contact zone are generally less than 120 μm and have rise velocities with Reynolds numbers (Re) less than 1. Four particle transport mechanisms can cause particles to move from the bulk water within the separation distances between bubbles. It will be shown that these transport mechanisms are dependent on particle size. Small particles such as depicted by *Particle A* in Fig. 7-4 undergo *Brownian diffusion*. The imparted random motion to small particles can cause collisions with the bubble. The transport of larger particles is influenced by gravity. Particles such as *Particle B* have sufficiently low density relative to water that they move with the streamlines of flow, and can be physically intercepted if in streamlines of flow that move close enough to the bubble surface. This particle collision process with the bubble is called *interception*. Particles like *Particle C* have greater density than water so that they can settle out of the streamlines of flow and collide with the bubble. This is called transport by *settling*. A fourth possible particle transport mechanism is *inertia*; however, it is only significant for larger particles and bubbles than normally found in DAF. It is listed in Fig. 7-4 for the sake of completeness, but not depicted.

Two steps are considered for actual deposition or removal of particles: (1) collision opportunities and (2) particle attachment to the bubbles. The first step involves mecha-

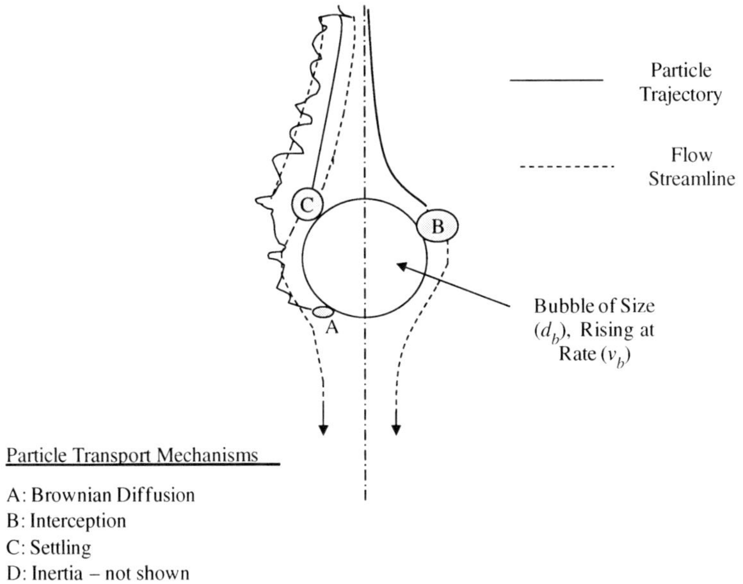

FIGURE 7-4 Single collector efficiency concept depicting particle transport collision mechanisms with the bubble.

nisms of particle transport from the bulk water to the vicinity of the bubble surface and are accounted for by the single collector efficiency (η), which is defined by Eq. (7-7):

$$\eta = \frac{\text{rate at which particles collide with the collector}}{\text{rate at which particles approach the collector}} \tag{7-7}$$

We account for the particle transport mechanisms described above and depicted in Figure 7-4 into mathematical expressions for η. It is noted that the single collector collision efficiency does not account for particle-bubble interaction, only particle transport. The second step of attachment or adhesion of particles is incorporated into an attachment efficiency parameter, α_{pb}. In other words, all possible collisions attributed to particle transport may not result in attachment. Theoretically, particle-bubble interaction forces affect attachment. The interaction forces can prevent or bring about attachment depending on chemistry of the particles, bubbles, and water. These interaction forces are described in Chap. 5 and are discussed later in this chapter. We model the effects of these forces empirically through the use of α_{pb}. Therefore, the removal efficiency of particles by a single collector (E_{sc}) is the product of the collision efficiency and the attachment efficiency, as shown by Eq. (7-8).

$$E_{sc} = (\eta)(\alpha_{pb}) \tag{7-8}$$

Single Collector Efficiency Equations. The individual expressions for the single collector efficiency are derived from the convective-diffusion equation for Brownian diffusion and by particle-trajectory analyses for the other particle transport mechanisms. The reader is referred to other sources for details of these derivations (Reay and Ratcliff, 1973; Yu, 1989).

Equations (7-9) through (7-12) describe the individual collision efficiencies for Brownian diffusion (η_{BD}), interception (η_I), settling (η_S), and inertia (η_{IN}).

$$\eta_{BD} = 6.18\left[\frac{k_b T}{g(\rho_w - \rho_b)}\right]^{2/3}\left[\frac{1}{d_p}\right]^{2/3}\left[\frac{1}{d_b}\right]^{2} \tag{7-9}$$

$$\eta_I = \left(\frac{d_p}{d_b}+1\right)^2 - \frac{3}{2}\left(\frac{d_p}{d_b}+1\right) + \frac{1}{2}\left(\frac{d_p}{d_b}+1\right)^{-1} \tag{7-10}$$

$$\eta_S = \left[\frac{(\rho_p - \rho_w)}{(\rho_w - \rho_b)}\right]\left[\frac{d_p}{d_b}\right]^2 \tag{7-11}$$

$$\eta_{IN} = \frac{g\rho_w\rho_p d_b (d_p)^2}{324(\mu_w)^2} \tag{7-12}$$

where k is Boltzmann's constant (1.3807×10^{-23} J/K), T is absolute temperature (K [273 + °C]), g is gravitational constant (9.806 m/s^2), ρ_w is water density (kg/m^3), ρ_b is air bubble density (kg/m³), ρ_p is particle or floc density (kg/m³), d_p is particle or floc diameter (m), d_b is bubble diameter (m), and μ_w is water dynamic viscosity (N-s/m² or kg/m-s).

The single collector efficiency for inertia (η_{IN}) is small relative to the other η terms for bubbles and flocs less than 100 μm. These are conditions that most often occur for DAF applications in drinking water, and thus η_{IN} is dropped from further consideration. We then obtain the total single collector efficiency, η_T, by summing the individual single collector efficiency mechanisms.

$$\eta_T = \eta_{BD} + \eta_I + \eta_S \tag{7-13}$$

The interception expression [Eq. (7-10)] is more complex than the classical expression of Eq. (7-14) used to model granular media filtration. It can be shown that, if d_p/d_b <<1, then Eq. (7-10) gives the same results as the filtration expression of Eq. (7-14). In other words, Eq. (7-14) is an approximation of Eq. (7-10) that holds for water filtration because the collectors (filter grains) are much larger than the particles being collected, and so this simple equation for interception is used. Equation (7-10) is used for DAF because the approximate expression does not hold for particles (d_p) of tens of microns and bubbles (d_b) of say 60 μm, which apply to DAF.

$$\eta_I = \left(\frac{3}{2}\right)\left(\frac{d_p}{d_b}\right)^2 \tag{7-14}$$

We consider that the total single collector efficiency [Eq. (7-13)] depends on the sum of the individual expressions for particle collisions by Brownian diffusion [Eq. (7-9)], by interception [Eq. (7-10)], and by settling [Eq. (7-11)]. These equations show that η_T depends on particle properties (size and density), bubble size, and temperature (directly and through its effect on water density). It is useful to examine graphically the dependence of η_T on these parameters. We do this for the four cases presented in Table 7-2.

TABLE 7-2 Case Conditions for the Graphical Evaluation (Fig. 7-5) of the Single Collector Efficiency (η_T) as a Function of Particle Size (d_p)

Case	Conditions
A	**Reference Case** d_b = 60 μm ρ_p = 1100 kg/m³ T = 293 K (20°C) ρ_w = 998.2 kg/m³
B	**Examine Bubble Diameter: d_b of 10, 60, and 100 μm** Other conditions: same as Case A
C	**Examine Particle Density: ρ_p of 1100 and 2500 kg/m³** Other conditions: same as Case A
D	**Examine Water Temperature: 4 and 20°C** Corresponding values for ρ_w of 999.98 (say 1000.0) and 998.2 kg/m³ Other conditions: same as Case A

Case A. This is the reference case using conditions of a mean bubble size (d_b) of 60 μm and a typical floc density of 1100 kg/m³ for DAF drinking water applications. Figure 7-5, Case A, plots the total single collector efficiency (η_T) and individual η's as a function of particle or floc size (d_p). Note the meaning of η_T: for example, a value of 0.1 means one particle collides with a single bubble for every 10 particles approaching the bubble.

Some important findings are noted. We find that η_T is at a minimum for particles with a size of a little less than 1 μm. The single collector collision efficiency increases with decreasing particle size <1 μm because Brownian diffusion is more effective. Higher single collector collision efficiencies occur for particles >1 μm due to collisions occurring mainly by interception. The settling mechanism is not important for causing collisions compared to interception for particles or flocs with densities of 1100 kg/m³ or less. More on the effect of higher-density particles is presented under Case C. In short, Brownian diffusion is the main collision mechanism for small particles, interception is the main one for larger particles, and there is a minimum in the total single collector efficiency for particles of about 1 μm.

Case B. Material on bubble sizes in the DAF contact zone is presented in Chap. 4. Most bubbles in the contact zone are in the range of 40 to 80 μm; an average value of about 60 μm is a good choice for model calculations and is used for Cases A, C, and D.

As indicated in Table 7-2 we examine for Case B the dependence of η_T on particle size for bubbles of average size (60 μm, reference of Case A), much smaller (10 μm), and much larger (100 μm). The predictions are plotted in Fig. 7-5, Case B for the total single

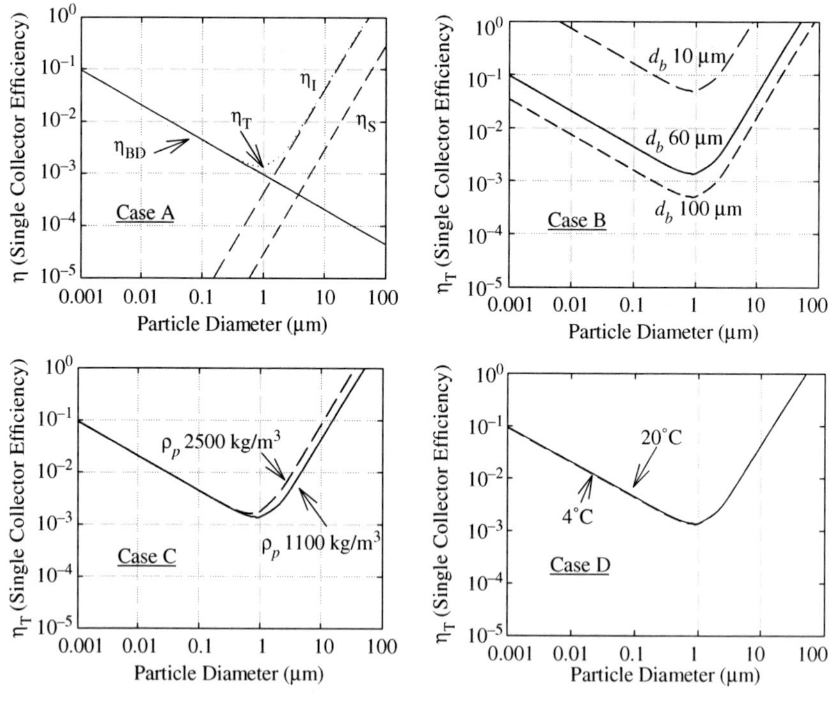

FIGURE 7-5 Single collector efficiency versus particle size for four cases (Case A: reference, Case B: effect of bubble size, Case C: effect of particle density, Case D: effect of temperature—details for the four cases in Table 7-2).

collector efficiency (η_T). The individual single collector efficiencies are not plotted to avoid cluttering the figures with unnecessary plots. We learn from the figure that η_T is strongly dependent on bubble size. In fact, the dependence for all collision mechanisms [Eqs. (7-9) through (7-11)] is

$$\eta \propto d_b^{-2}$$

Greater single collector collision efficiencies occur for smaller bubbles. While smaller bubbles are more efficient, there are practical limitations in designing and operating DAF systems to produce bubbles smaller than the range of 10 to 100 μm—this is addressed in Chap. 4. In summary, smaller bubbles are better. This is an important reason for the use of DAF to treat waters compared to larger-bubble processes like dispersed air flotation, which uses mm (1000 μm) size bubbles. After presenting the contact zone performance model, we return to discussion of the effect of bubble size.

Case C. We examine the effect of particle or floc density in which we compare the reference case density of 1100 kg/m^3 (Case A) against a high-particle density case of 2500 kg/m^3 (see Table 7-2). We would not expect to find particles or flocs with such a high density in treating reservoir or lake water supply sources or supplies with natural color. Particles this dense are found in river sources carrying mineral turbidity; however, even for these sources the flocs formed after coagulation and flocculation and entering the DAF tank would be less than this density of 2500 kg/m^3. We have thus chosen a condition of a high-particle density limit that should not occur for most DAF plants.

Single collector collision efficiencies (η_T) as a function of particle size for the two particle densities are plotted in Fig. 7-5, Case C. There is a small increase in η_T for particles >1 μm for the high-density particles, but the effect is small because the primary collision mechanism for particles >1 μm is interception, not settling. We can conclude that particle density has a small effect on η_T.

Case D. Finally, we use this case to examine the effect of water temperature: 4 versus 20°C. The results are plotted in Fig. 7-5, Case D. In theory temperature affects Brownian diffusion (η_{BD}) especially for small particles <1 μm, but the effect is negligible for 4 versus 20°C. For larger particles (>1 μm), there is no effect because interception (not settling) is the primary collision mechanism and is not dependent on temperature. In summary, temperature has almost no effect on η_T. This only applies to the single collector efficiency. Water temperature, however, does have a small effect on the DAF contact zone performance—this is addressed later in the chapter. Water temperature also affects the rise velocity of bubbles and particle-bubble aggregates with respect to performance in the DAF separation zone. This is addressed in Chap. 8.

Using the analysis of the four cases presented in Fig. 7-5, the following summary statements are made.

- The two most important variables affecting the single collector collision efficiency are bubble size (d_b) and particle or floc size (d_p). Smaller bubbles increase the single collector collision efficiency. Brownian diffusion is the important collision mechanism for particles <1 μm, and interception is the important collision mechanism for particles >1 μm.
- There is a minimum in the total single collector collision efficiency for about 1 μm particles where contributions from Brownian diffusion and interception are low.
- Floc-particle density and water temperature have little effect on the total single collector collision efficiency.

Contact Zone Performance Model

The single collector collision efficiency accounts only for particle collision efficiency of one bubble. As shown by Eqs. (7-5) and (7-6), the kinetics of particle removal by collision and attachment depend on other variables such as the attachment efficiency, the bubble number concentration, the particle number concentration, the bubble rise velocity, and the cross-sectional area of the bubble. To obtain a contact zone performance model, the kinetics must be incorporated into a particle mass balance that considers the hydraulic flow characteristics of the contact zone. Plug flow conditions for the contact zone are assumed so any dispersion is assumed to be small and is ignored, and the flow rate (Q) is constant. The case that plug flow is a good approximation for modelling the contact zone was developed earlier in the chapter. The details of the derivation of the contact zone efficiency [Eq. (7-15)] are presented in the appendix at the end of the chapter.

$$E_{cz} = \left(1 - \frac{n_{p,e}}{n_{p,i}}\right) = \left[1 - \exp\left(-\frac{3/2\left(\alpha_{\mathrm{pb}}\eta_{\mathrm{T}}v_b\Phi_{\mathrm{b}}t_{cz}\right)}{d_b}\right)\right] \tag{7-15}$$

Equation (7-15) describes the removal efficiency (E_{cz}) of particles or flocs onto bubbles within the contact zone, where $n_{p,e}$ is the particle concentration in the effluent, and $n_{p,i}$ is the particle concentration in the influent to the contact zone. The model is instructive in that it identifies important variables affecting contact zone performance. These include the particle-bubble attachment efficiency (α_{pb}), the single collector efficiency (η_{T}), bubble size (d_b), bubble rise velocity (v_b), bubble volume concentration (Φ_{b}), and the contact zone detention time (t_{cz}). Pretreatment processes of coagulation and flocculation affect two of the variables. Flotation tank design and operation affect five of the variables. In the next two sections, we use the model to predict and examine the effects of pretreatment and flotation tank variables on contact zone performance. This yields valuable insights to those design and operating variables that affect DAF performance.

Effects of Pretreatment on Contact Zone Performance

The pretreatment processes of coagulation and flocculation affect DAF contact zone performance—general principles and guidance for DAF are presented in Chap. 6. Pretreatment is accounted for in the contact zone performance model [Eq. (7-15)] through the variables of α_{pb} and η_{T}. Coagulation chemistry affects α_{pb}. Flocculation changes the size distribution of particles or flocs (d_p) affecting η_{T}. These two pretreatment variables are identified in Table 7-3. We refer to this table in the discussion of coagulation and flocculation below.

Coagulation Chemistry and the Conceptual Consideration of $\boldsymbol{\alpha_{\mathrm{pb}}}$**.** α_{pb} is the particle-bubble attachment efficiency. In other words, not all of the possible collisions brought about by the particle transport mechanisms yield attachment and removal of particles by bubbles. Thus, the single collision efficiency (η_{T}) is multiplied by the attachment efficiency to account for the overall removal efficiency by a single collector, as presented above in Eq. (7-8).

Conceptually, α_{pb} is the fraction of successful collisions. α_{pb} can have values between 0 (no collisions lead to attachment) and 1 (all collisions result in attachment). Fundamentally, α_{pb} depends on particle-bubble forces when particles approach bubbles, as addressed in Chap. 5 and listed in Table 7-3.

TABLE 7-3 Contact Zone Model Variables Affected by Pretreatment

Variable	Dependence	Comments
α_{pb} (particle–bubble attachment efficiency)	Particle–bubble interaction affected by electrical double layer, van der Waals, hydrodynamic, and hydrophobic forces	Coagulation chemistry including dose and pH are critical in maximizing particle attachment to bubbles; α_{pb} approaches 1 with optimum coagulation
η_T (total single collector collision efficiency)	Brownian diffusion: η_{BD} depends on $d_p^{-2/3}$ Settling and interception: η_S and η_I depend on d_p^2	Flocculation increases d_p; desire flocs of 10s of microns

The effect of α_{pb} on contact zone performance [Eq. (7-15)] is addressed. We use typical DAF design conditions for bubble size (d_b of 60 μm), bubble volume concentration (Φ_b of 8500 ppm), and detention time (t_{cz} of 2 min). A floc density of 1100 kg/m^3 is assigned. Figure 7-6 shows the contact zone efficiency as a function of particle or floc size for two cases of α_{pb}: 0.5 for good coagulation chemistry and particle-bubble attachment and 0.05 for poor coagulation. The predictions displayed in Fig. 7-6 show that for poor coagulation chemistry, the contact zone efficiency is small (<20 percent) for particles or flocs <10 μm and that high efficiencies are not obtained unless flocs approach 100 μm in size. For good coagulation chemistry (favorable particle-bubble attachment, α_{pb} of 0.5), there is a minimum in the contact zone efficiency for particles or flocs of about 1 μm. This is due to particle transport collision mechanisms and was discussed earlier under the single collector collision efficiency material (see Fig. 7-5). In a water plant operating under good coagulation chemistry conditions, particles flocculate. The idea then is to provide sufficient flocculation time to allow particles to aggregate into flocs with sizes of tens of μm. Figure 7-6 shows that for good coagulation chemistry (α_{pb} of 0.5), contact zone efficiencies are high for particles or flocs of tens of μm, and there is no benefit in forming large flocs of say about 100 μm and greater. More on the effects of flocculation and floc size in the following section on flocculation.

Our modelling approach separates the step of particle transport to bubbles from particle-bubble forces and interactions that were discussed in Chap. 5. Hence, we have separate variables for the particle transport collision efficiency (η) and for attachment efficiency (α_{pb}), as shown in Eqs. (7-8) and (7-15). The attachment of particles to bubbles is modelled by the empirical variable, α_{pb}, that accounts for particle-bubble interactions. This is done for practical reasons: α_{pb} can be measured as discussed later in the chapter. In some flocculation-based models, attempts are made to write equations to predict the effects of electrical double layer, van der Waals, and hydrodynamic interaction on what the authors (Leppinen, 1999, 2000; Han et al., 1997; Han, 2002) call the collision efficiency, but actually they are evaluating the theoretical attachment efficiency.

The particle-bubble interaction forces are affected mostly by coagulation, and therefore, for practical reasons we use the empirical variable, α_{pb}, to account for them. It is instructive to discuss these interparticle forces and account for coagulation effects. As particles are transported from the bulk water to close distances near the bubble surface, forces between the bubble and particle affect attachment. These forces include the DLVO electrostatic forces from overlapping of electrical double layers and the van der Waals forces, and the non-DLVO hydrodynamic and hydrophobic forces. These particle-bubble interaction forces are discussed in Chap. 5. A brief summary is presented here.

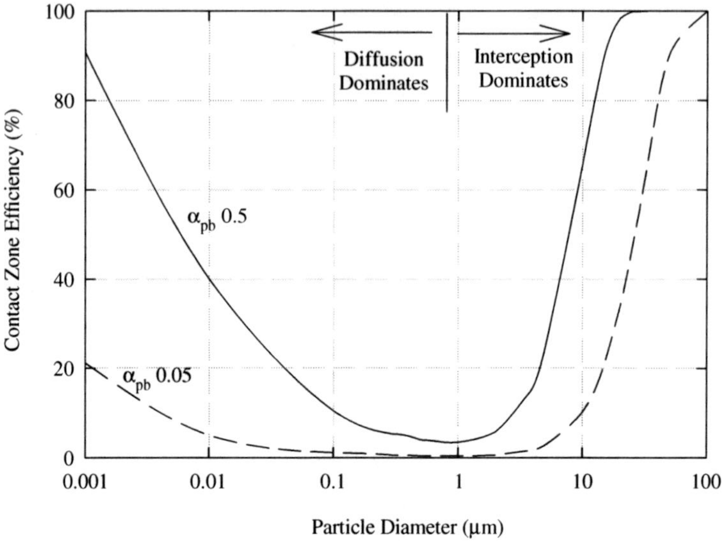

FIGURE 7-6 Contact zone efficiency as a function of particle or floc size for good coagulation (α_{pb} = 0.5) and poor coagulation chemistry (α_{pb} = 0.05) for conditions of t_{cz} = 2 min, ρ_p = 1100 kg/m^3, temperature = 20°C, d_b = 60 µm, and Φ_b = 8500 ppm.

We begin with the DLVO forces. Without coagulation, both air bubbles and particles carry negative zeta potentials. When particles approach air bubbles the electrical double layers surrounding the particles and bubbles overlap causing a repulsive force. This is the case for no coagulant addition or insufficient coagulant dose to neutralize the negative charges of particles. It is possible to add cationic polymers or surfactants to the saturator effluent and produce bubbles of opposite charge to the particles (see Chap. 5), which would produce electrostatic attraction, but this is not practiced in drinking water treatment. Rather for most good coagulation chemistry conditions, as practiced for drinking water applications of DAF, it is expected that the flocs have little or no electrical charge. This is illustrated later in the chapter. The van der Waals force between dissimilar materials (solid particle and an air bubble) in water is repulsive (see Chap. 5). Thus, unlike particle-particle coagulation processes where we rely on attractive van der Waals forces to bring about attachment of particles, for particle-bubble attachment we must consider the hydrophobic interaction between particles and bubbles.

The hydrophobic force has to do with bubbles favoring hydrophobic surfaces relative to the water—that is, the bubbles do not favor being in water and seek a hydrophobic particle surface phase—see Chap. 5. Its importance does depend on the electrostatic force because a reduction in particle charge leads to less electrostatic repulsion, giving the particle a more hydrophobic character. Bubble attachment is favored for hydrophobic particles with no charge; however, even particles carrying a small charge can have hydrophobic spots on the surface that can provide opportunity for bubble attachment.

Hydrodynamic interaction may cause another repulsive force. As a particle approaches a bubble, the water between them must be displaced for collisions to occur. As pointed out in Chap. 5, it has less importance in DAF applications involving bubbles and porous flocs than the DLVO forces and the hydrophobic force.

Flocculation Flocculation pretreatment affects both the particle size distribution of particles and the particle number concentration (see Chap. 6). The mean particle or floc size (d_p) increases and the floc-particle concentration (n_p) decreases through flocculation. The contact zone model [Eq. (7-15)] does not show an explicit relationship between efficiency and particle or floc size; however, there is an important effect of particle size on the single collector efficiency as shown, by Eqs. (7-9) through (7-11). The dependence of η on d_p for each collision mechanism is shown below and summarized in Table 7-3.

$$\eta_{\mathrm{BD}} \propto d_p^{-2/3}$$

$$\eta_{\mathrm{I}} \propto d_p^2$$

$$\eta_{\mathrm{S}} \propto d_p^2$$

Example 7-1 is presented to show the effects of particles with two different sizes on the contact zone performance efficiency.

EXAMPLE 7-1 Pretreatment Effect of Floc Size on Contact Zone Performance We examine the effect of mean particle size on the contact zone efficiency. Two particles (or flocs) sizes are considered: 1 versus 30 μm. The contact zone efficiency is calculated for the following conditions:

Good coagulation:	$\alpha_{\mathrm{pb}} = 0.5$
Temperature:	20°C
Water density:	$\rho_{\mathrm{w}} = 998.21$ kg/m^3 (Table D-1)
Air bubble density:	$\rho_{\mathrm{b}} = 1.19$ kg/m^3 (Table E-4)
Water viscosity:	$\mu_{\mathrm{w}} = 1.002 \times 10^{-3}$ kg/m-s (Table D-1)
Particle or floc density:	$\rho_{\mathrm{p}} = 1100$ kg/m^3
Bubble size:	$d_b = 60$ μm
Bubble volume concentration:	8400 ppm
Contact zone detention time:	2 min

Comment

First, a comment is made about setting the bubble volume concentration rather than the mass concentration of air bubbles in the contact zone. Some design engineers consider the mass concentration as a primary design variable; however, as shown with Eq. (7-15), the bubble volume concentration (Φ_{b}) is the fundamental variable relating air concentration to the contact zone efficiency. Additional discussion of this important concept is found later in the chapter. From Eq. (5-1), the mass concentration can be calculated, and it is 10 g/m^3.

Solution Before applying Eq. (7-15), two variables must be determined: they are the bubble rise velocity (v_b), and the total single collector efficiency (η_{T}).

Step 1: Determine the bubble rise velocity (v_b).

This is calculated from the classical Stokes expression that is presented in Chap. 8.

$$v_b = \frac{g\left(\rho_{\mathrm{w}} - \rho_{\mathrm{b}}\right) d_b^2}{18\mu_{\mathrm{w}}}$$

$$v_b = \frac{9.806(998.21 - 1.19)(60 \times 10^{-6})^2}{18(1.002 \times 10^{-3})} \left(\frac{\text{m}}{\text{s}^2}\right)\left(\frac{\text{kg}}{\text{m}^3}\right)\left(\text{m}^2\right)\left(\frac{\text{m s}}{\text{kg}}\right)$$

$$v_b = 1.952 \times 10^{-3}\,\frac{\text{m}}{\text{s}} \times 3600\,\frac{\text{s}}{\text{h}} = 7.03$$

Before proceeding, the Reynolds number is calculated for the rise velocity of the bubbles.

$$\text{Re} = (v_b d_b)\left(\frac{\rho_w}{\mu_w}\right)$$

$$\text{Re} = (7.03)(60 \times 10^{-6})\left(\frac{998.2}{1.002 \times 10^{-3}}\right)\left(\frac{\text{m}}{\text{h}}\,\frac{\text{h}}{3600\ \text{s}}\right)\left(\text{m}\right)\left(\frac{\text{kg}}{\text{m}^3}\right)\left(\frac{\text{m s}}{\text{kg}}\right)$$

$$\text{Re} = 0.12$$

The Reynolds Number is <1 indicating that the water streamlines flowing downward around the rising bubble are laminar in nature. This was assumed in the trajectory analysis for the single collector efficiency concept (see also Fig. 7-4).

Step 2: Determine the single collector collision efficiency (η_T) for particle sizes of 1 and 30 μm. We calculate the individual single collector collision efficiencies from Eqs. (7-9) through (7-11), and then sum them to obtain η_T [Eq. (7-13)]. We go through the calculations for the particle size of 1 μm, and then list the results for the 30 μm particle size case. Recalling Eq. (7-9) we have.

$$\eta_{BD} = 6.18\left[\frac{k_b T}{g(\rho_w - \rho_b)}\right]^{2/3}\left[\frac{1}{d_p}\right]^{2/3}\left[\frac{1}{d_b}\right]^{2}$$

η_T is dimensionless so paying strict attention to units for the parameters for the η equations is important, and it is especially true when working with η_{BD}. Note that the Boltzmann constant is 1.38×10^{-23} J/K, where a joule (J) is 1 kg m²/s².

$$\eta_{BD} = 6.18\left[\frac{(1.38 \times 10^{-23})(293)}{9.806(998.23 - 1.194)}\right]^{2/3}\left[\frac{1}{10^{-6}}\right]^{2/3}\left[\frac{1}{60 \times 10^{-6}}\right]^{2} = 9.55 \times 10^{-4}$$

Let's check the units.

$$\left(\frac{\text{kg m}^2}{\text{s}^2}\,\frac{\text{K}}{\text{K}}\,\frac{\text{s}^2}{\text{m}}\,\frac{\text{m}^3}{\text{kg}}\right)^{2/3}\left(\frac{1}{\text{m}}\right)^{2/3}\left(\frac{1}{\text{m}}\right)^{2} \text{ or}$$

$$\left(\text{m}^4\right)^{2/3}\left(\frac{1}{\text{m}}\right)^{2/3}\left(\frac{1}{\text{m}}\right)^{2} \text{ or } \left(\text{m}^{8/3}\right)\left(\text{m}^{-2/3}\right)\left(\text{m}^{-2}\right) \text{ or } \left(\text{m}^{8/3}\right)\left(\text{m}^{-8/3}\right)$$

Next, calculate η_I from Eq. (7-10).

$$\eta_I = \left(\frac{d_p}{d_b}+1\right)^2 - \frac{3}{2}\left(\frac{d_p}{d_b}+1\right)+\frac{1}{2}\left(\frac{d_p}{d_b}+1\right)^{-1}$$

Here we can use units of μm and because we have ratios of sizes, the units cancel.

$$\eta_I = \left(\frac{1}{60}+1\right)^2 - \frac{3}{2}\left(\frac{1}{60}+1\right)+\frac{1}{2}\left(\frac{1}{60}+1\right)^{-1} = 4.14\times10^{-4}$$

Next, calculate η_S from Eq. (7-11).

$$\eta_S = \left[\frac{(\rho_p-\rho_w)}{(\rho_w-\rho_b)}\right]\left[\frac{d_p}{d_b}\right]^2$$

$$\eta_S = \left[\frac{(1100-998.2)}{(998.2-1.19)}\right]\left[\frac{1}{60}\right]^2\left(\frac{\text{kg}}{\text{m}^3}\frac{\text{m}^3}{\text{kg}}\right)\left(\frac{\mu\text{m}}{\mu\text{m}}\right)^2 = 2.84\times10^{-5}$$

From Eq. (7-13), η_T is the sum of the individual values.

$$\eta_T = \eta_D + \eta_I + \eta_S$$

$$\eta_T = 9.55\times10^{-4} + 4.14\times10^{-4} + 2.84\times10^{-5} = 1.40\times10^{-3}$$

The following table summarizes the results for the two particle sizes of 1 and 30 μm.

d_p	η_{BD}	η_I	η_S	η_T
1 μm	9.55×10^{-4}	4.14×10^{-4}	2.84×10^{-5}	0.0014
30 μm	9.88×10^{-5}	3.33×10^{-1}	2.55×10^{-2}	0.36

Step 3: Finally, the efficiency of the contact zone is determined from Eq. (7-15).

$$E_{cz} = \left(1-\frac{n_{p,e}}{n_{p,i}}\right) = \left[1-\exp\left(\frac{-3/2\left(\alpha_{pb}\eta_T v_b \Phi_b t_{cz}\right)}{d_b}\right)\right]$$

Recall those parameters that apply to both particle size cases. The values are $\alpha_{pb} = 0.5$, $d_b = 60$ μm, $t_{cz} = 2$ minutes, $\Phi_b = 8400$ ppm, and $v_b = 7.03$ m/h. We calculate the contact zone efficiency for the particle size of 1 μm. Specific to this particle size is the total single collector efficiency, η_T. From the table at the end of Step 2, its value is 0.0014. Before solving Eq. (7-15), the units for v_b are changed to m/min to be consistent with minutes used for t_{cz}. v_b becomes 0.1172 m/min. Then, for a particle size of 1 μm we get

$$E_{cz} = \left(1 - \frac{n_{p,e}}{n_{p,i}}\right) = \left[1 - \exp\left(\frac{-3/2\left[(0.5)(0.0014)(0.1172)\left(\frac{8400}{10^6}\right)(2)\right]}{60 \times 10^{-6}}\right)\right] = 0.034$$

The fraction of particles removed from the bulk water in the contact zone by collision and attachment to air bubbles is 0.034 or 3.4 percent. Thus, 96.6 percent of the particles entering the contact zone leave in the effluent not collected by air bubbles. Solving Eq. (7-15) for 30 μm particles yields a fractional efficiency of 0.9998. By providing a small flocculation time and producing flocs with a size of 30 μm, increases the contact zone efficiency to 99.98 percent. Only 0.02 percent of the flocs leave the contact zone not collected by air bubbles. ▲

Example 7-1 shows that floc size has a major effect on the contact zone efficiency. It also demonstrates how complex the calculations are when carried out by hand. Calculations by computer using a spreadsheet to solve the various equations are recommended to make full use of the contact zone performance model.

We can learn much from a graphical display of examining the effect of particle size. From Fig. 7-6 for good coagulation chemistry ($\alpha_{pb} = 0.5$), we see there is a minimum in the contact zone efficiency for particles of about 1 μm (actually at 0.85 μm). Particle transport by Brownian diffusion is the controlling mechanism for particles <1 μm, while particle transport by interception controls for particle sizes >1 μm. Interception, and not settling, is the important particle transport mechanism even for flocs with much greater density than the 1100 kg/m^3 used here. This was shown earlier with Fig. 7-5, Case C.

For good coagulation chemistry conditions, the sub-micron particles flocculate quickly into sizes approaching 1 μm in rapid mixing or early on in the flocculation tank. This is the size at which we have a minimum in the contact zone efficiency, and thus undesirable. We wish to carry flocculation further to produce flocs of sizes where the contact zone efficiency is high. Figure 7-6 shows that the larger the floc size, the greater the efficiency of the contact zone. In fact, even for poor coagulation the efficiency approaches 100 percent when floc sizes exceed 100 μm. To be shown in Chap. 8, the rise velocity of floc-bubble aggregates decreases as floc size increases unless there is multiple bubble attachment to each floc. So larger flocs introduced into the contact zone are not necessarily better. If we optimize floc size and produce relatively small flocs, what floc sizes should be produced to obtain high contact zone efficiency? Figure 7-6 shows high contact zone efficiencies, for good coagulation chemistry (α_{pb} of 0.5), for flocs of tens of μm. More specifically, model calculations show efficiencies of 98.4 percent for 20 μm particles or nearly 100 percent for 35 μm flocs (see Example 7-1).

In summary, an important finding from the contact zone model predictions is that the desired and optimum sizes of flocs are about 25 to 50 μm. Floc particles of tens of microns should be prepared by the pretreatment flocculation process for the influent to DAF tanks. This model finding provides valuable insights into the design, operation, and performance of the DAF contact zone, and how pretreatment affects the performance.

Effects of DAF Tank Variables on Contact Zone Performance

Important design and operating variables associated with the flotation tank and recycle system are summarized in Table 7-4. Each of these variables and their effects on DAF contact zone design and operation are presented here.

Rise Velocity and Water Temperature. The rise velocity, v_b, depends on d_b (d_b is an important flotation tank variable in itself that can be changed by design and operation and is discussed separately). For a fixed bubble size (d_b), water temperature affects the rise velocity, and it also affects the single collector collision efficiency, as discussed earlier with Fig. 7-5 (Case D), where it was shown that water temperature has a minor effect on η_T.

In addition to the bubble rise velocity (v_b), temperature affects Brownian diffusion. Temperature effects, overall, on the contact zone efficiency [Eq. (7-15)] are examined with Fig. 7-7. This figure shows the effect of water temperature with other contact zone variables fixed at typical design conditions. The contact zone efficiency decreases for the winter season (4°C) compared to the summer season (20°C). For submicron particles, lower water temperature has two effects: Brownian diffusion decreases (small effect on η_D, see Fig. 7-5 [Case D]) and there is a decrease in the rise velocity—directly affecting the efficiency, as shown explicitly in Eq. (7-15). For larger particles, for example, the contact zone efficiency for removing 10 μm floc particles is about 50 percent in the winter compared to about 70 percent in the summer. Therefore, if everything else is equal regarding the operation of the DAF contact zone, the performance is less in the winter. Why is this? Interception is the primary collision mechanism and temperature has no effect. However, temperature decreases v_b, affecting the sweeping rate of bubbles in colliding with particles.

TABLE 7-4 Contact Zone Model Variables Affected by DAF Tank Design and Operation

Variable	Dependence	Comments
v_b (bubble rise velocity)	Affected mainly by the size of bubbles (d_b^2); effect of water temperature: decreases with viscosity for colder waters	Describes motion of bubbles relative to the water flow in the contact zone
η_T (total single collector collision efficiency)	Affected primarily by floc and bubble sizes Floc size as in Table 7-3: η_{BD} depends on $d_p^{-2/3}$, η_S and η_I depend on d_p^2 Bubble size: η_T depends on d_b^{-2}	Minimum η_T for floc particles with size of ≈ 1 μm Flocculation should produce floc particles with sizes of 10s of μm
d_b (mean bubble diameter)	Controlled mainly by pressure difference across the injection device; also an effect from the nozzle type or injection device	Desire microbubbles, smaller bubbles better the performance. Most 40 to 80 μm, mean of 60 μm
Φ_b (bubble volume concentration)	Affected by recycle rate and saturator pressure	Controlled mainly by the recycle rate. Increasing Φ_b, increases collision opportunities (more bubbles or collectors)
t_{cz} (contact zone detention time)	Determined by the water velocity and depth of the contact zone.	In practice, detention times are in range of 1 to 2.5 min; hydraulic behavior approximately plug flow

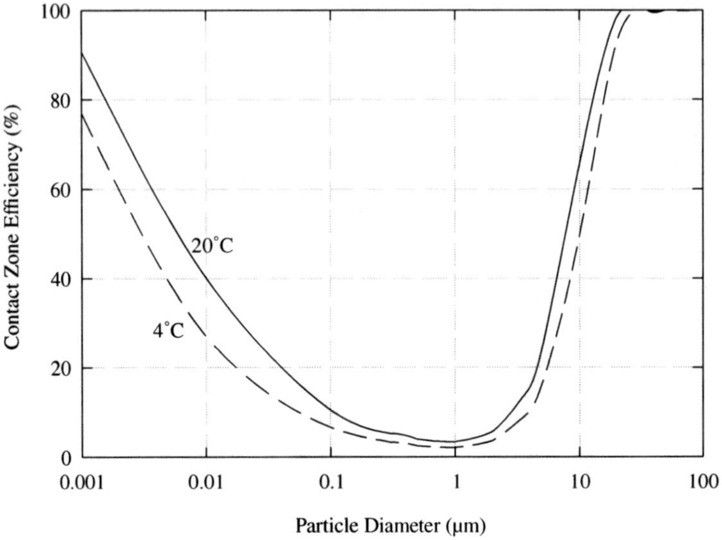

FIGURE 7-7 Effect of water temperature on the contact zone efficiency as a function of particle or floc size (Conditions: $t_{cz} = 2$ min, $\alpha_{pb} = 0.5$, $\rho_p = 1100$ kg/m^3, $d_b = 60$ μm, and $\Phi_b = 8500$ ppm).

For colder water conditions, the water viscosity increases, slowing down this sweeping rate and decreasing collisions.

Now let's examine the other design and operating variables associated with the flotation tank and recycle system that affect the performance of the contact zone that are identified in Table 7-4.

Total Single Collector Efficiency (η_T). Particle or floc size affects the total single collector efficiency (η_T), and was addressed under pretreatment flocculation above. The other major variable affecting η_T is the bubble size, which is addressed next.

Bubble Diameter. We previously examined the effect of bubble diameter on η_T in Fig. 7-5, Case B. Here we extend the effect of bubble size with respect to the contact zone model, Eq. (7-15). The RHS of Eq. (7-15) shows that the contact zone efficiency depends directly on the inverse of the bubble size, d_b^{-1}. Bubble size also affects η_T, as stated above, and the rise velocity (v_b). Thus the overall dependence of the efficiency with respect to bubble size is

$$\text{Efficiency} \propto \frac{\eta_T v_b}{d_b}$$

Substituting into the RHS, the dependence of η_T and v_b on bubble size, we obtain

$$\text{Efficiency} \propto \frac{\left(1/d_b^2\right)\left(d_b\right)^2}{d_b} \quad \text{or Efficiency} \propto \frac{1}{d_b}$$

The net outcome is that the contact zone efficiency depends on d_b^{-1}, so smaller bubbles are better. Bubble sizes and control of bubble size are discussed in Chap. 4. Briefly, the average bubble size is controlled principally by the pressure difference across the recycle injection device and also by the injection device itself (nozzle type or other control valve). DAF systems produce a suspension of microbubbles with average bubble diameters of 40 to 80 μm.

For our standard DAF conditions, we have been using a mean bubble diameter of 60 μm. In Fig. 7-8 we examine contact zone model predictions for this standard bubble size, as well as smaller (10 μm) and larger bubbles (100 μm). The contact zone efficiency improves with decreasing bubble size for all particle or floc sizes. Considering the efficiency of 60 μm bubbles compared to 100 μm bubbles for removing flocs of 10 μm, one can see from Fig. 7-8 that the efficiency increases from about 50 percent for 100 μm bubbles to 70 percent for 60 μm bubbles and approaches 100 percent efficiency for 10 μm bubbles. This principle that smaller bubbles are better for the contact zone agrees with observations that DAF is more efficient than dispersed air flotation. Bubble sizes of about 40 to 80 μm are common for the contact zone of DAF systems and these small sizes contrast to the large bubbles of about 1 mm for dispersed air flotation.

Bubble Concentration. There are three measures of air bubble concentration: mass, number, and volume. These measures and relationships among them are covered in Chaps. 4 and 5. We summarize briefly the importance and use of the three bubble concentration measures. Then we address how the bubble concentration affects contact zone performance.

The mass concentration (C_b) is important in several subjects of flotation—some of these topics were covered in Chaps. 3 and 4. First, it is used to describe the equilibrium solubility of air in water. This applies to evaluating whether the incoming water to the plant is saturated with air. Second, in a saturator air is dissolved into the water under pressure where the maximum concentration that can be dissolved is the solubility limit or saturation

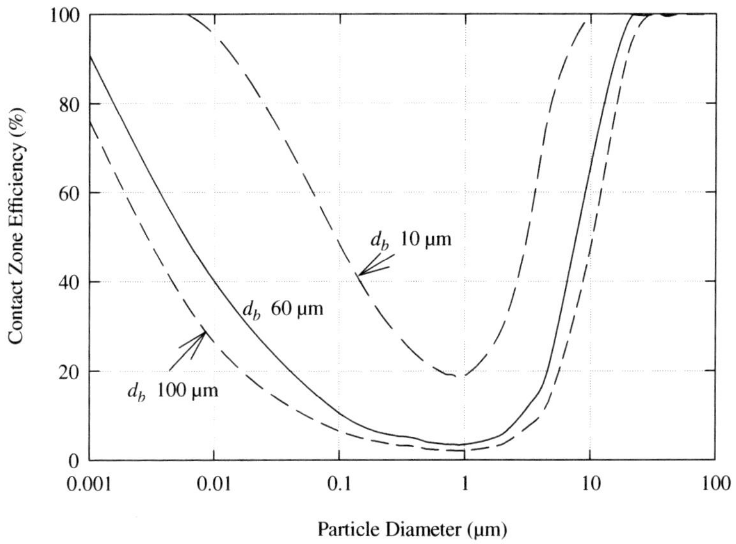

FIGURE 7-8 Effect of bubble diameter on contact zone efficiency as a function of particle or floc size (Conditions: t_{cz} = 2 min, temperature = 20°C, α_{pb} = 0.5, ρ_p = 1100 kg/m^3, and Φ_b = 8500 ppm).

TABLE 7-5 Contact Zone Bubble Concentrations Examined for Range of Design and Operating Conditions (bubble diameter of 60 μm, 20°C)

Φ_b (ppm)	C_b (g/m^3)	n_b (bubbles per m^3)	Comments
4250	5	3.8×10^{10}	Poor design and operation
8500	10	7.5×10^{10}	Good design and operation
10,000	12	8.8×10^{10}	Higher than standard design

concentration. There are kinetic limitations in dissolving the air, so the rate of mass transfer of air must be considered in saturator design. Third, the mass concentration of air in the contact zone is determined from a mass balance of air released from the recycle flow mixing with the influent water from flocculation.

The bubble number concentration (n_b) in the contact zone is an important fundamental measure. We saw earlier that the kinetic rate of collisions and attachment of bubbles to particles depends on n_b—see Eq. (7-5). Further, it is important in that the bubble number concentration must at least equal, but it is better to exceed, the particle number concentration for bubbles to collide with and attach.

In the development of the white water bubble-blanket contact zone model, n_b was replaced with the bubble volume concentration (Φ_b). Consequently, it is the bubble volume concentration that is our fundamental variable affecting contact zone performance. We examine the effect of varying it in the following text. It is worth mentioning that the bubble volume is also important in evaluating the rise velocity of floc-bubble aggregates in the DAF tank separation zone (Chap. 8). The floc-bubble aggregates cannot rise unless sufficient bubble volume is attached to flocs to reduce the aggregate density to less than the water density.

It has been common to design saturators and recycle injection systems to provide a minimum mass concentration of air in the contact zone; however, the bubble volume concentration should be used. As said above, the bubble volume concentration is a fundamental variable affecting the contact zone efficiency, and in Chap. 8 it is shown that it is a fundamental variable affecting the separation zone efficiency. Guidelines are given in Chap. 3 for bubble volume concentration requirements of 7000 to 9000 ppm (or 7 to 9 mL/L) for drinking water applications. In daily plant operation, the operator can vary the air delivered by changing the recycle flow or the saturator pressure. We consider bubble volume concentrations for three conditions: 4250 ppm (poor design and operation), 8500 ppm (good design and operation), and 10,000 ppm (higher than standard design). These concentrations and corresponding bubble number and air mass concentrations are presented in Table 7-5. The number concentration (n_b) is a fundamental measure of air concentration. You can see from the values in the table that n_b is high and should exceed floc-particle concentrations, especially for the good design case and the higher than standard design case. The bubble volume concentration (Φ_b) is the fundamental variable in the white water contact zone model accounting for the effect of air concentration on performance. In Example 7-2, we examine the contact zone efficiency for removing 15 μm size flocs for two of the bubble volume concentrations listed in Table 7-5.

EXAMPLE 7-2 Air Concentration Effect on Contact Zone Performance We examine the effect of the air bubble volume concentration on the efficiency of the contact zone. Two bubble volume concentrations are used: 4250 ppm (poor design and operation) and 8500 ppm (good design and operation)—see Table 7-5. We determine the contact zone efficiencies for a particle or floc size of 15 μm for the following conditions that also applied to Example 7-1.

Good coagulation:	$\alpha_{pb} = 0.5$
Temperature:	20°C
Water density:	$\rho_w = 998.21$ kg/m^3
Air bubble density:	$\rho_b = 1.19$ kg/m^3
Water dynamic viscosity:	$\mu_w = 1.002 \times 10^{-3}$ kg/m-s
Particle or floc density:	$\rho_p = 1100$ kg/m^3
Bubble size:	$d_b = 60$ μm
Contact zone detention time:	2 min

Solution Before applying Eq. (7-15), the bubble rise velocity (v_b) and the total single collector efficiency (η_T) must be determined.

Step 1: The bubble rise velocity (v_b) for 60 μm bubbles was found in Example 7-1. It is 7.03 m/h or 0.1172 m/min.

Step 2: Determine the single collector collision efficiency (η_T) for a particle or floc size of 15 μm. Example 7-1 showed how to calculate the individual single collector collision efficiencies from Eqs. (7-9) through (7-11), and then sum according to Eq. (7-13) to obtain η_T. Here, the results are summarized in the following table.

d_p	η_{BD}	η_I	η_S	η_T
15 μm	1.57×10^{-4}	8.75×10^{-2}	6.38×10^{-3}	9.45×10^{-2}

Step 3: Finally, the efficiency of the contact zone is determined from Eq. (7-15) for the two cases of bubble concentrations. These type calculations were done for Example 7-1 so only the results are given.

$$E_{cz} = \left(1 - \frac{n_{p,e}}{n_{p,i}}\right) = \left[1 - \exp\left(-\frac{3/2\left(\alpha_{pb}\eta_T v_b \Phi_b t_{cz}\right)}{d_b}\right)\right]$$

For Φ_b of 4250 ppm (poor design case) the contact zone efficiency is only about 70 percent, but increases to about 90 percent for Φ_b of 8500 ppm (good design). These calculations are for 15 μm size flocs and a 2-min detention time. The efficiencies are greater for flocs of 25 to 50 μm and lower for smaller flocs. In the next section, we consider both the bubble volume concentration and the detention time in evaluating the contact zone performance for various size particles or flocs. ▲

Contact Zone Detention Time. The contact zone is designed for mean detention times between 1 and 2.5 min (Table 7-4). The range in values is because of the ranges of velocities and depths chosen in the design of the baffled contact zone. The hydraulic behavior of the contact zone is approximated by plug flow; this was addressed earlier in the chapter. For the ideal case of plug flow, all fluid elements (water parcels) have the same residence time or detention time in the reactor (here, the contact zone) and is the contact zone detention time.

In this section we examine the effect of detention time on performance. We use the model [Eq. (7-15)] to make predictions for bubble volume concentrations: 4250 ppm (poor design and operation) and 8500 ppm (good design and operation)—refer to Table 7-5. We do this for four cases of particle or floc sizes: **Case A**: 1 μm, **Case B**: 10 μm, **Case C:** 25 μm, and **Case D**: 50 μm.

Model predictions presented in Fig. 7-9 show the dependence of contact zone efficiency as a function of detention time. Poor efficiencies occur for particles of 1 μm (**Case A**) for both bubble volume concentrations for typical design detention times (1 to 2.5 min) and much greater times. Clearly, we wish to avoid poor flocculation conditions that would introduce particles or small flocs of this size into the DAF contact zone. For 10 μm particles or flocs (**Case B**), the efficiency of the contact zone improves for both bubble volume concentrations. Still, the contact zone efficiency is not good for the low bubble volume concentration of 4500 ppm except for detention times greater than used in practice. Even for the higher bubble concentration of 8500 ppm, the efficiencies are not good unless detention times exceed 4 min.

Of particular interest are the model predictions for 25 μm (**Case C**) and 50 μm (**Case D**) flocs shown in Fig. 7-9. For flocs of 25 μm (**Case C**), we see the efficiency is greater than 95 percent for a 1-min detention time and approaches 100 percent for 2 min when the bubble volume concentration is 8500 ppm (good design and operation). The performance decreases for the lower bubble volume concentration, but is reasonably good at 80 to 95 percent for detention times of 1 to 2 min. **Case D** for 50 μm flocs shows excellent performance for both volume concentrations as long as the detention time is about 1 min or greater. Poor performance occurs for detention times <30 s.

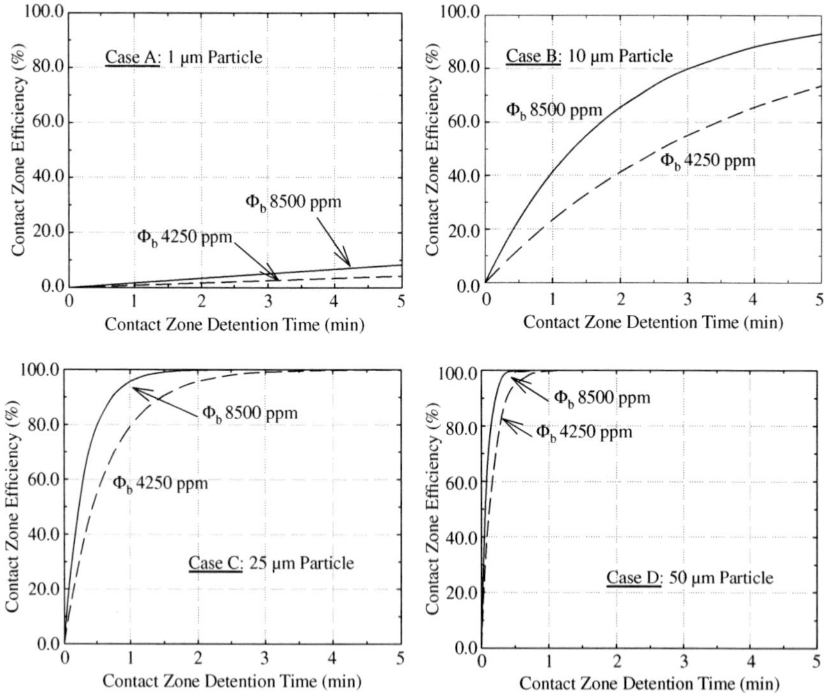

FIGURE 7-9 Effect of contact zone detention time on the contact zone efficiency for two bubble volume concentrations (Φ_b): 8500 ppm (good design and operation) and 4250 ppm (poor design and operation) for four cases of floc size: Case A: 1 μm, Case B: 10 μm, Case C: 25 μm, Case D: 50 μm (Conditions: temperature = 20°C, $\alpha_{pb} = 0.5$, $\rho_p = 1100$ kg/m^3, and $d_b = 60$ μm).

In summary, for contact zone detention times of 1 min or greater and bubble volume concentration of 8500 ppm, excellent performance is predicted for floc sizes of 25 to 50 μm applied to the DAF contact zone. Furthermore, contact zone efficiency is high and insensitive to detention times exceeding 1 min. Calculations are not shown here but the contact zone efficiencies are 99 percent or greater for bubble volume concentrations of 7000 to 9000 ppm for flocs of 25 μm and larger for detention times of 2 min.

7-4 PRACTICAL APPLICATIONS OF THE CONTACT ZONE MODEL

In this section, the contact zone model is discussed in terms of its use as a design and operating tool. A model must include the fundamental variables affecting the process—here the DAF contact zone. Other criteria follow in judging the utility of process models. The model should avoid adjustable parameters. The model variables should be either well known or easily measured. Finally, one should verify the model through pilot-scale or full-scale plant testing.

We first present previously published data verifying the white water bubble-blanket model. We discuss these data and present a practical discussion of material on the particle-bubble attachment efficiency (α_{pb}) and optimum coagulation chemistry. Next, we present and discuss the model variables affecting design and operation. This discussion includes what can be controlled by the designer and operator, what can be varied, and desired values.

Model Verification

Before discussing the agreement of experimental data with the model, some information is given about the source water and pilot plant used in these studies. The pilot studies were conducted at the William S. Warner 190 ML/d DAF plant of Aquarion Water Company of Connecticut in Fairfield, Connecticut (United States). The water supply is Hemlocks Reservoir, which is a protected low-alkalinity (20 to 30 mg/L $CaCO_3$) supply of high quality with low turbidity and low TOC (2.5 to 3 mg/L). It was demonstrated that the pilot plant yields the same results as the full-scale plant in terms of DAF effluent quality and filter performance. Descriptions of the full-scale plant and its performance are presented in Chap. 11.

The pilot plant had a flow rate of 190 L/min and consisted of static mixers for dispersion of coagulants, two-stage flocculation (total detention time of 11.5 min), and a DAF nominal hydraulic loading of 15 m/h. The DAF recycle rate was 8 percent with a saturator pressure of 550 kPa. The raw water quality for the experiments was temperature 5.5°C, turbidity 0.7 NTU, and particle counts 6000 particles/mL (1 to 200 μm; most particles < 20 μm). A dual-coagulant strategy of alum and a high charge-density cationic polymer was used with alum at 1.5 mg/L as Al and cationic polymer dose of 1 mg/L (liquid product). This coagulant strategy is used by the full-scale plant, whereby the cationic polymer reduces the alum dosage by 0.7 mg/L as Al (compared to using only alum) without affecting plant performance, and this practice reduces sludge production. The pH after coagulation for these cold water conditions was controlled at 6.7 to 6.8. The dual-coagulant dosages and pH conditions produced flocs with a charge of approximately zero, considered optimum coagulation conditions for floc-bubble attachment—discussed under the section Coagulation Chemistry and the Conceptual Consideration of α_{pb}, and discussed again in the following text.

Figure 7-10 compares model predictions of the contact zone efficiency using Eq. (7-15) to experimental data as a function of particle size. To compare model predictions to experimental data some variables had to be assigned values (uniform bubble size of 60 μm, floc density

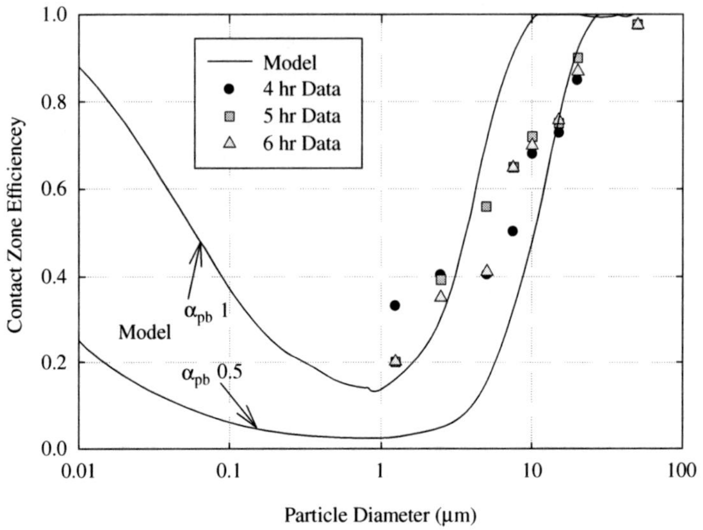

FIGURE 7-10 Contact zone model predictions versus experimental data (model assumptions: d_b = 60 μm, ρ_p = 1100 kg/m^3, and α_{pb} of 0.5 and 1; experimental conditions: DAF loading rate 15 m/h, t_{cz} = 1.9 min, temperature 5.5°C), Φ_b= 7840 ppm) (*Source: Haarhoff and Edzwald (2004), Journal of Water Supply: Research and Technology – AQUA, 53 (3), 127–150, with permission from the copyright holders, IWA Publishing*).

of 1100 kg/m^3), and the bubble volume concentration (Φ_b of 7840 ppm) was calculated from the saturator pressure and recycle rate. The particle-bubble attachment efficiency (α_{pb}) was unknown, but for optimum coagulation it was assumed to be high. Two α_{pb} values of 0.5 and 1 were tested for the model predictions. The experimental data are in general agreement with the model, and it appears that α_{pb} between 0.5 and 1 is a good assumption for optimum coagulation. The data show an increase in efficiency with increasing particle size greater than about 1 μm, as predicted by the model. No particle data are presented for sizes below about 1 μm because of the limitation of the electronic particle counter to detect smaller particle sizes. The agreement between theory and data is good considering that the flocs density and a uniform bubble size were assumed—that is, the frequency distribution of bubble size was ignored. Furthermore, plug flow was assumed, and while this may be a reasonable approximation, as discussed earlier in the chapter, there is some dispersion.

Recall that most water treatment process models make similar assumptions regarding plug flow, when in fact some dispersion occurs. Here, a uniform bubble size was assumed similar to assuming a uniform media size in water filtration models. A model need not be exact in all the details, but it should account for the fundamental variables and make predictions in agreement with observations; this model does that. Next, its usefulness as a design and operating tool is addressed.

Design and Operating Tool

Design and operating variables that affect the contact zone performance have been taken from the model and are summarized in Table 7-6. Each is discussed in practical terms in the following sections.

TABLE 7-6 Design and Operating Variables Affecting the Contact Zone Performance

Variable	Model	Design and Operation
α_{pb} (particle–bubble attachment efficiency)	Primary variable	Controlled by coagulation chemistry of coagulant type, dosage, and pH Desire floc particles with little or no charge yielding high α_{pb} values of 0.5 to 1
d_p (particle or floc size)	Affects η_T	Controlled by flocculation tank detention time, number of stages, and mixing Desire flocs with sizes of 25 to 50 μm
d_b (mean bubble diameter)	Primary variable and affects η_T	Controlled mainly by pressure difference across the injection device; also an effect from the nozzle type or injection device Desire small bubbles but for saturator pressures >400 kPa designer and operator have little control on bubble size; most bubbles are 40 to 80 μm
Φ_b (bubble volume concentration)	Primary variable	Controlled by the recycle rate and saturator pressure Desire for design purposes saturator pressures of 400 to 600 kPa; operator can vary Φ_b, by changing the recycle rate; recommend 7000 to 9000 ppm
t_{cz} (contact zone detention time)	Primary variable	Controlled by the flow rate and volume of the contact zone Desire approximately plug flow and t_{cz} >1 min

Particle-Bubble Attachment Efficiency **(α_{pb}).** Coagulation is an essential chemical pretreatment process affecting α_{pb}. The role of coagulation is to reduce repulsive charge interactions between particles or flocs and bubbles and thus produce favorable attachment conditions (high α_{pb} values). Flocs with zero or low charge or zeta potentials should be produced by coagulation. Under these conditions, attractive hydrophobic forces can prevail leading to particle attachment to bubbles as shown theoretically in Chap. 5.

Figure 7-11 illustrates the effect of coagulant dose on floc charge and flotation efficiency for turbidity and for the removals of DOC and UV (254 nm) absorbance (surrogate for dissolved natural organic matter). Electrophoretic mobility (EPM) data are presented as an indicator of floc charge—as an approximation the zeta potential is 13 × EPM. The data show clearly that underdosing of coagulant is associated with flocs carrying a negative charge causing poor particle removal, as shown by the small difference in turbidities before and after flotation. Minimum turbidity after flotation is associated with a floc charge of about zero and one obtains good removal of natural organic matter, both DOC and UV_{254}.

Good coagulation chemistry is essential to obtain favorable particle attachment to bubbles. Coagulation chemistry is the most important operating control variable affecting flotation performance. The operator must ensure that coagulation is effective for good flotation. High values of α_{pb} are desired, and while the operator does not measure this variable, there is considerable empirical evidence that with good coagulation chemistry α_{pb} is about 0.5 to 1, as discussed next.

Values for α_{pb} have been obtained from DAF studies with pilot-scale and full-scale plants. For the pilot-plant data presented above for model verification (Fig. 7-10), it was found that α_{pb} had values between 0.5 and 1 for optimum coagulation chemistry (dose and pH). Shawwa and Smith (2000) found α_{pb} values of 0.35 to 0.55 in pilot-scale studies for good coagulation conditions, defined as providing good DAF treatment. Schers and Van Dijk (1992) found α_{pb} values of 0.2 to 1 from data for six full-scale DAF plants in the Netherlands.

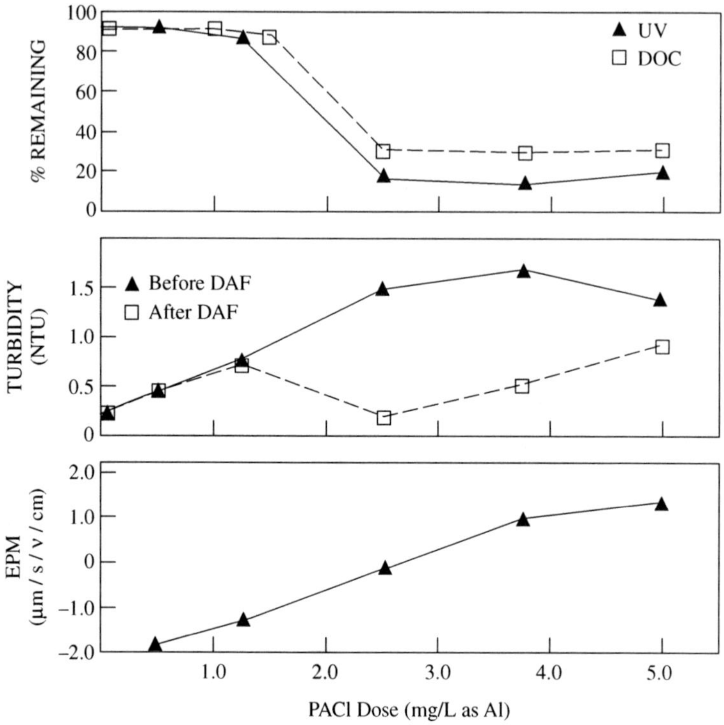

FIGURE 7-11 Effect of polyaluminum chloride dose on the electrophoretic mobility (EPM: indicator of charge) of flocs and flotation performance for turbidity and natural organic matter—DOC and UV (254 nm) absorbance (Conditions: high basicity PACl at constant coagulation pH of ~5.5, 4°C, 5 mg/L TOC of fulvic acid, and low initial turbidity) (*Source: Edzwald and Malley, 1990*).

In practice, measurements of floc-particle charge by plant operators are rarely made to determine optimum coagulation conditions. They are not necessary. Coagulant dosing may be evaluated using standard sedimentation or flotation jar test apparatus in which the optimum dose is that which produces low turbidity and low NOM measured as DOC or UV_{254}—see Chap. 6 for discussion of coagulation and guidelines on dosing and pH conditions for various coagulants and Chap. 10 for discussion of flotation jar tests and the use of zeta potential measurements.

Floc-particle Size (d_p). From the material presented above in the discussion on flocculation, it was concluded that flocs of 25 to 50 μm are desirable for good contact zone performance. The experimental data presented in Fig. 7-10 for the model verification support this guideline on desired floc sizes. Floc size is not shown explicitly as a variable in the model [Eq. (7-15)], but it has a great effect on the collision mechanism (i.e., the single collector efficiency, η_T—see Tables 7-3 and 7-6). Floc size depends on design and operation of the flocculation tank. The design engineer sets the number of flocculation stages, detention time for design flow conditions, and the mixing intensity. The detention time through flocculation depends on flow demands (seasonal effect) and the number of treatment trains in service. A detention time of 10 min is recommended for design based on the theory above of producing small flocs and experience. The operator, when not faced with operating the plant at design flows, can use additional parallel flocculation trains to increase the detention time or take trains out of service to decrease the detention time. The operator can also vary the mixing intensity to control floc size. Floc size measurements are rarely made so practical guidance is to produce *pinpoint* size flocs.

Bubble Size (d_b). In the discussion on the contact zone performance as a function of bubble size, it was concluded that the efficiency increases with decreasing bubble size. An important characteristic of DAF is that small bubbles are used compared to, say, dispersed air flotation. The bubble size is controlled mainly in design. Table 7-6 contains some summary comments about this important variable. The sizes of bubbles depend mainly on the pressure difference across the nozzle or the recycle injection device. The nozzle design also affects the sizes of bubbles and the distribution of sizes as discussed in Chap. 4. Most well-designed DAF systems produce bubbles in the size range of 40 to 80 μm—see Chap. 4. The operator can change the pressure difference across the nozzle by changing the saturator pressure and, in principle, this can affect bubble size. However, a considerable body of information presented in Chap. 4 indicates that DAF systems should be operated above a minimum pressure of 350 kPa and that at pressures above this there is a small effect on decreasing the bubble size. In practice then, while operators can vary the saturator pressure and there is some variation in normal operation, it is not varied to control bubble size. Normal operating pressures are 400 to 600 kPa.

Bubble Volume Concentration (Φ_b). The bubble volume concentration in the contact zone depends on the saturator pressure and the recycle rate. The contact zone model uses the bubble volume concentration as a fundamental variable (see Table 7-6). The designer provides for some operational flexibility in controlling the saturator pressure and the recycle rate so the operator can vary the air delivered or air bubble concentration. While design engineers typically specify the delivered air as a mass concentration (C_b), and not the bubble volume concentration, in this chapter we recommend the bubble volume concentration (referred to as the volumetric air concentration) as the primary way to describe air concentration requirements. (Note also in Chap. 8 that it is the bubble volume concentration that affects separation zone efficiency.) We recommend bubble volume concentrations of 7000 to 9000 ppm, as mentioned earlier in the chapter and in Chap. 3.

The operator can vary the bubble volume concentration by changing either the saturator pressure and dissolving more air or the recycle rate. As mentioned earlier, the saturator pressure can be varied within a certain range but it is not the better means to vary the air delivered. The better way is to vary the recycle rate. Plant operators do not normally measure the bubble volume concentration, and so they use the recycle rate as a surrogate operational parameter. One can show that the mass concentration of air (and the bubble volume concentration) depends on the recycle rate and a number of factors (see Chap. 4) including the saturator pressure, the saturator efficiency, the water temperature, the loss in hydraulic head between the saturator and the recycle injection in the DAF contact zone, and any air deficit in the water leaving flocculation and entering flotation so the calculation is not normally done. Rough guidelines, however, are the mass concentrations in g/m^3 are approximately equal to the recycle rates. Consequently, recycle rates of 8 to 11 percent deliver approximately 8 to 11 g/m^3 or 7000 to 9000 ppm as bubble volume concentrations.

Contact Zone Detention Time (t_{cz}) This is affected by the total flow rate (flow from the flocculation tank plus the recycle flow) and the contact zone volume, as indicated in Table 7-6. The designer sets the detention time for design flow rates. A minimum design value of 1 min was identified in the theory section above. The operator has some control over this variable in that plant operation is usually not for design flow rates. As water demands increase, it is important for the operator to bring on additional DAF trains to maintain detention times in excess of the minimum and to maintain acceptable hydraulic loadings for the separation zone (Chap. 8). The operator can increase the contact zone performance when treating difficult water quality, such as may occur in the winter or spring with cold waters or difficult to remove algae by utilizing additional DAF trains to increase the contact zone detention time. This would also decrease the separation zone loading rate and improve separation performance.

REFERENCES

Amato, T., and Wicks, J. (2009), The practical application of computational fluid dynamics to dissolved air flotation, water treatment plant operation, design and development, *Journal of Water Supply: Research and Technology – Aqua,* 58 (1), 65–73.

Edzwald, J. K. (1995), Principles and applications of dissolved air flotation, *Water Science and Technology*, 31 (3–4), 1–23.

Edzwald, J. K. (2010), Dissolved air flotation and me, *Water Research*, 44 (7), 2077–2106.

Edzwald, J. K., and Malley, Jr., J. P. (1990), Removal of humic substances and algae by dissolved air flotation, *EPA/600/2-89-032*, U.S. Environmental Protection Agency, Cincinnati.

Edzwald, J. K., Malley, Jr., J. P., and Yu, C. (1990), A conceptual model for dissolved air flotation in water treatment, *Water Supply*, 8, 141–150.

Fawcett, N. S. J. (1997), The hydraulics of flotation tanks: computational modeling, in *Dissolved Air Flotation*, London: Chartered Institution of Water and Environmental Management.

Flint, L. R., and Howarth, W. J. (1971), The collision efficiency of small particles with spherical air bubbles, *Chemical Engineering Science*, 26, 1155–1168.

Friedlander, S. K. (1977), *Smoke, Dust, and Haze*, New York: John Wiley & Sons, Inc.

Fukushi, K., Tambo, N., and Matsui, Y. (1995), A kinetic model for dissolved air flotation in water and wastewater treatment, *Water Science and Technology*, 31 (3–4), 37–47.

Gregory, R., and Edzwald, J. K. (2010), Sedimentation and flotation, in J. K. Edzwald, ed., *Water Quality and Treatment: A Handbook on Drinking Water*, 6th ed., New York: AWWA and McGraw Hill.

Haarhoff, J. and Edzwald, J. K. (2004), Dissolved air flotation modelling: insights and shortcomings, *Journal of Water Supply: Research and Technology – Aqua,* 53 (3), 127–150.

Han, M. (2002), Modelling of DAF: the effect of particle and bubble characteristics, *Journal of Water Supply: Research and Technology – Aqua*, 51 (1), 27–34.

Han, M., Dockko, S., and Park, C. (1997), Collision efficiency factor of bubble and particle in DAF, *in Dissolved Air Flotation, London*: Chartered Institution of Water and Environmental Management.

Leppinen, D. M. (1999), Trajectory analysis and collision efficiency during microbubble flotation, *Journal of Colloid and Interface Science*, 212, 431–442.

Leppinen, D. M. (2000), A kinetic model of dissolved air flotation including the effects of interparticle forces, *Journal of Water Supply: Research and Technology – Aqua*, 49 (5), 259–268.

Levenspiel, O. (1998), *Chemical Reaction Engineering*, 3rd ed., New York: John Wiley & Sons, Inc.

Lundh, M., Jönsson, L., and Dahlquist, J. (2002), The influence of the contact zone configuration on the flow structure in a dissolved air flotation pilot plant, *Water Research*, 36, 1585–1595.

Matsui, Y., Fukushi, K., and Tambo, N. (1998), Modeling, simulation and operational parameters of dissolved air flotation, *Journal of Water Supply: Research and Technology – Aqua*, 47 (1), 9–20.

Metcalf & Eddy, Inc. (2003), *Wastewater Engineering Treatment and Reuse*, 4th ed., New York: McGraw-Hill.

Moruzzi, R. B., and Reali, M. A. P. (2010), Characterization of micro-bubble size distribution and low configuration in DAF contact zone by a non-intrusive image analysis system and tracer tests, *Water Science and Technology*, 61(1), 253–262.

MWH (2005), *Water Treatment: Principles and Design*, 2nd ed., Hoboken, NJ: John Wiley & Sons, Inc.

Reay, D., and Ratcliff, G. A. (1973), Removal of fine particles from water by dispersed air flotation, *Canadian Journal of Chemical Engineering*, 51 (2), 178–185.

Schers, G. J., and Van Dijk, J. C. (1992), Dissolved-air flotation: theory and practice, in R. Klute and H. H. Hahn, eds., *Chemical Water and Wastewater Treatment II*, New York: Springer Verlag, 223–246.

Shawwa, A. R., and Smith, D. W. (1998), Hydrodynamic characterization in dissolved air flotation (DAF) contact zone, *Water Science and Technology*, 38 (6), 245–252.

Shawwa, A. R., and Smith, D. W. (2000), Dissolved air flotation model for drinking water treatment, *Canadian Journal of Civil Engineering*, 27, 373–382.

Ta, C. T., Beckley, J., and Eades, A. (2001), A multiphase CFD model of DAF process, *Water Science and Technology*, 43 (8), 153–157.

Tambo, N., Matsui, Y., and Fukushi, K. (1986), A kinetic study of dissolved air flotation, *World Congress of Chemical Engineering* (Tokyo), 200–203.

Yao, K. M., Habibian, M. T., and O'Melia, C. R. (1971), Water and wastewater filtration: concepts and Applications, *Environmental Science and Technology*, 5 (11), 1105–1112.

Yu, T. (1989), *Modeling of Dissolved Air Flotation for Drinking Water Treatment*, M.S. Thesis, University of Massachusetts, Amherst, MA, USA.

APPENDIX: DERIVATION OF THE WHITE WATER BUBBLE-BLANKET CONTACT ZONE MODEL

To obtain the performance equation describing the efficiency of the contact zone, one must perform a mass balance, actually a particle number balance. This particle balance is done around a control volume as depicted in Fig. A7-1. Plug flow for the contact zone is assumed so any dispersion is assumed to be small and is ignored. The case that plug flow is a good approximation for modelling the contact zone was developed earlier in the chapter. The flow rate (here generically Q, actually $Q + Q_r$) is constant.

In words, the particle balance statement follows:

Rate of Change in Particle Number through the Control Volume	=	Input Particle Rate to the Control Volume	Minus	Output Particle Rate from the Control Volume	Minus	Particle Rate of Removal by Collision and Attachment to Bubbles in the Control Volume

The control volume is $A \cdot \Delta L$. The removal rate term (last box) comes from Eq. (7-5) and is $k_c \cdot n_p \cdot n_b$. Writing now the differential equation for the particle balance, we obtain

$$\frac{\partial n_p}{\partial t}\left(A\Delta L\right) = Qn_p - Q(n_p + \Delta n_p) - k_c n_p n_b \left(A\Delta L\right) \tag{A7-1}$$

Combining the first two terms on the right-hand-side (RHS) of the equation yields

$$\frac{\partial n_p}{\partial t}\left(A\Delta L\right) = -Q\Delta n_p - k_c n_p n_b \left(A\Delta L\right) \tag{A7-2}$$

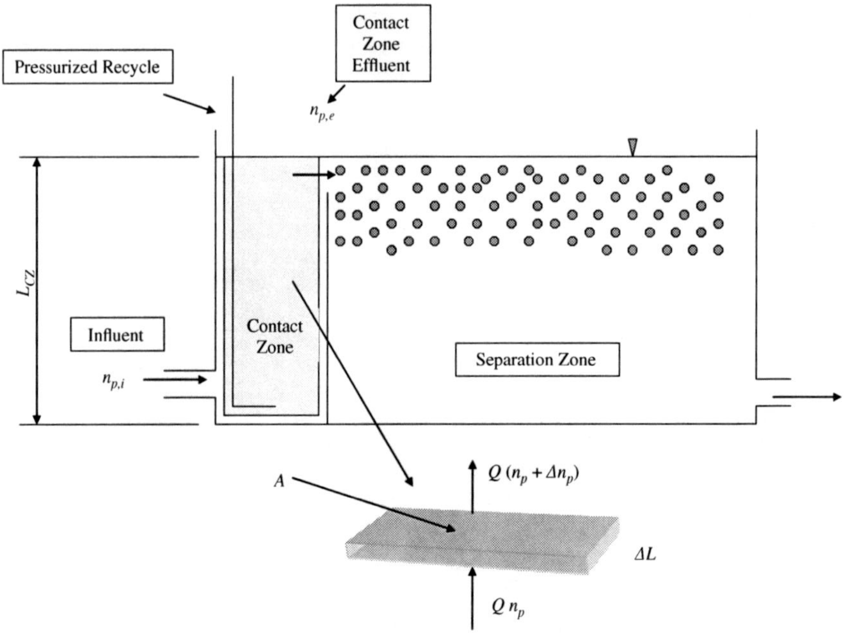

FIGURE A7-1 DAF tank schematic highlighting the contact zone and notation for the model (cutaway figure shows the control volume for the mass balance).

This equation is solved for steady state conditions, which means that the left-hand-side (LHS) is set to zero: $(\partial n_p / \partial t) = 0$. So the above equation becomes Eq. (A7-3)

$$0 = -Q\Delta n_p - k_c n_p n_b (A\Delta L) \tag{A7-3}$$

Dividing through by the control volume yields

$$0 = -\frac{Q\Delta n_p}{A\Delta L} - k_c n_p n_b \frac{(A\Delta L)}{A\Delta L} \tag{A7-4}$$

We replace Q/A with the flow through velocity, v, and express Δn_p and ΔL in differential form—that is, as dn_p and dL

$$0 = -v\frac{dn_p}{dL} - k_c n_p n_b \tag{A7-5}$$

Moving the first term of the RHS to the LHS of the equation, we obtain

$$v\frac{dn_p}{dL} = -k_c n_p n_b \tag{A7-6}$$

Now, we replace v and dL with the time of water travel through the contact zone by recognizing that

$$v = \frac{dL}{dt} \quad \text{or} \quad \frac{v}{dL} = \frac{1}{dt} \tag{A7-7}$$

Substitution of Eq. (A7-7) into Eq. (A7-6) yields the final differential expression

$$\frac{dn_p}{dt} = -k_c n_p n_b \tag{A7-8}$$

Take note of two things. First, the time in the differential equation is the time of water travel or the detention time through the contact zone. For a tank with a depth of L_{cz}, then the detention time of the contact zone (t_{cz}) is simply

$$t_{cz} = \frac{L_{cz}}{v} \tag{A7-9}$$

The differential equation [Eq. (A7-8)] expressed in terms of the time of travel does not violate the steady state condition invoked earlier. Often there is confusion on this point. Invoking steady state means that at a particular location of the contact zone, there is no time dependence of performance. For example, if we set the location at the exit of the contact zone, then we mean that the contact zone has a fixed detention time and its performance is time independent—that is, the performance today will be the same tomorrow, and so on. The second point is that the final differential equation [Eq. (A7-8)] is identical to Eq. (7-5) in the text that describes the kinetics of collisions and attachment to bubbles in a batch reactor. This means that the performance of a batch reactor is identical to a PFR, such as our continuous flow contact zone DAF process. It is well known from reactor theory that a batch reactor and an ideal PFR yield the same performance (Levenspiel, 1998; Metcalf and Eddy, 2003; MWH, 2005).

We take the differential equation [Eq. (A7-8)] and now integrate it for the condition of $t = 0$ at the entrance to the contact zone, the influent particle concentration is $n_{p,i}$ (see Fig. A7-1). At the exit of the contact zone, we have $t = t_{cz}$ and the effluent particle concentration of $n_{p,e}$.

$$\int_{n_{p,i}}^{n_{p,e}} \frac{dn_p}{n_p} = -\int_{t=0}^{t_{cz}} k_c n_b dt \tag{A7-10}$$

$$\frac{n_{p,e}}{n_{p,i}} = \exp\left(-k_c n_b t_{cz}\right) \tag{A7-11}$$

Recalling Eq. (7-6) in the text for the rate coefficient k_c, and multiplying it by n_b, we get

$$k_c n_b = \left(\alpha_{\text{pb}} \eta_{\text{T}}\right)\left(v_b A_b\right) n_b \tag{A7-12}$$

We then substitute for the cross-sectional area of the bubble (A_b) and express the bubble number concentration (n_b) in terms of the bubble volume equation (Φ_b), this last equation then becomes

$$k_c n_b = (\alpha_{pb}\eta_T)v_b\left(\frac{\pi d_b^2}{4}\right)\left(\frac{\Phi_b}{\pi d_b^3/6}\right) \quad \text{(A7-13)}$$

Finally substituting the RHS of Eq. (A7-13) for $k_c n_b$ into Eq. (A7-11), we obtain the fraction of particles remaining in the exit of the contact zone

$$\frac{n_{p,e}}{n_{p,i}} = \exp\left(\frac{-3/2(\alpha_{pb}\eta_T v_b \Phi_b t_{cz})}{d_b}\right) \quad \text{(A7-14)}$$

We re-write the equation in terms of the fraction of particles removed or the contact zone efficiency (E_{cz}) and obtain Eq. (A7-15) which is presented in the text as Eq. (7-15) for the contact zone efficiency or performance equation.

$$E_{cz} = \left(1 - \frac{n_{p,e}}{n_{p,i}}\right) = \left[1 - \exp\left(\frac{-3/2(\alpha_{pb}\eta_T v_b \Phi_b t_{cz})}{d_b}\right)\right] \quad \text{(A7-15)}$$

CHAPTER 8
SEPARATION ZONE

8-1 THE BOUNDARY BETWEEN CONTACT AND SEPARATION 8-1
8-2 THE RISE RATE OF AIR BUBBLES 8-2
8-3 THE NATURE OF FLOC-BUBBLE AGGREGATES 8-5
8-4 THE RISE RATE OF FLOC-BUBBLE AGGREGATES 8-6
Behavior of Flocs 8-6
Behavior of Floc-Bubble Aggregates 8-6
8-5 AN OVER-SIMPLIFIED VIEW OF DAF SEPARATION 8-9
8-6 SEPARATION-ZONE MEASUREMENTS AND MODELLING 8-10
Tracer Studies 8-10
Velocity Measurements 8-11
Computational Flow Dynamics (CFD) 8-11
Position of the Bubble Bed 8-11
8-7 INLET AND OUTLET CONSIDERATIONS 8-12
Outlet Considerations 8-12
The Effects of Air 8-12
Inlet Considerations 8-13
8-8 SEPARATION-ZONE FLOW PATTERNS 8-14
Flow Pattern at Low Hydraulic Loading 8-14
Stratified Flow at High Hydraulic Loading 8-14
The Bubble Bed 8-15
REFERENCES 8-17

8-1 THE BOUNDARY BETWEEN CONTACT AND SEPARATION

The DAF process is usually conceptualized as two distinctly different steps taking place in separate parts of the reactor, one following the other. In the *contact* zone, preformed flocs come into contact with a fine air bubble suspension, forming floc-bubble aggregates. This process is the theme of Chap. 7. The suspension then moves to the second part of the reactor, called the *separation* zone, where the aggregates separate from the water and become concentrated in a float layer at the top of the tank, while collecting clear subnatant at the bottom of the tank. Although there is no, or only a partial, physical barrier between these zones, a virtual *cross-flow* boundary is imagined.

In reality, the contact and separation steps are not exclusive. After the suspension moves into the separation zone, the interaction between flocs, bubbles, and aggregates continues. Flocs break up and reflocculate, building new floc-bubble aggregates. Bubbles may detach, re-attach, or coalesce to form larger bubbles. The traditional separation zone (the part of the DAF reactor downstream of the baffle—see Fig. 1-3) therefore hosts both contact and separation processes. So where is the boundary between contact and separation? Contact takes place wherever *white water* (water containing a large concentration of small air bubbles) is found, which extends over the largest part of the DAF reactor. Separation, in turn, takes place wherever the float layer is found, which also covers almost the entire DAF reactor. There is thus extensive overlap between contact and separation.

The CoCoDAFF technology (see Chap. 11) allows a counter-current flow pattern (Eades et al., 1997). The flocculated water is introduced at multiple points near the top of the tank, and the recycle stream at multiple points at a lower depth, both evenly spread over the entire footprint of the reactor. The general flow pattern of the flocculated water is thus vertical toward the filter bed below, while the recycle stream with its high air content rises to the surface. In this case, it is clear that the contact and separation zones occupy the same space—an extreme case of the overlap discussed above.

The direct observation of contact and separation processes in full-scale DAF reactors is not possible. The objective of this chapter is first to use simplified rise-rate models to complement our understanding of how separation is possible, and secondly to consider the effect of the complex hydraulic flow pattern in the separation zone on the overall efficiency of DAF.

8-2 THE RISE RATE OF AIR BUBBLES

The distribution of bubble sizes and typical mean sizes found in DAF tanks are discussed in Chap. 4. Typical bubbles in the separation zone are a little larger than the bubbles in the contact zone. This is partly due to their decompression as they rise from a hydrostatic head 2 to 3 m to the surface, as well as some bubble coalescence. For the purpose of modelling separation zone rise rates and performance, we made a case of selecting a mean bubble diameter of 100 μm, used in this chapter. Based on photographic evidence, the air bubbles are assumed to be perfectly spherical (Jönsson and Ljunggren, 2003).

The rise rate of an air bubble is a response to two opposing forces. First, the differential densities of air and water generate a net upward buoyant force. Second, the bubble encounters a drag force resisting the upward movement. For a constant rise rate (no acceleration), these forces are in balance:

$$F_B = F_D$$

$$(\rho_w - \rho_b) g V_b = \frac{C_D A_b \rho_w v_b^2}{2} \tag{8-1}$$

F_B and F_D are the forces due to buoyancy and drag, respectively; ρ_w and ρ_b are the water and air bubble densities, respectively; V_b is the bubble volume; A_b is the projected area of the bubble in the direction of movement; g is the Earth's gravity acceleration (9.806 m/s^2); v_b is the uniform rise velocity of the bubble; and C_D is the drag coefficient of the rising bubble.

Knowing that DAF bubbles are spheres, V_b is replaced with $(\pi d_b^3)/6$ and A_b with $(\pi d_b^2)/4$. The Reynolds number (Re) for the rising bubble is defined as:

$$\mathrm{Re} = \frac{\rho_w v_b d_b}{\mu_w} \tag{8-2}$$

The dynamic viscosity of water is μ_w. Laminar flow is indicated if Re ≤ 1. For laminar flow, the drag coefficient C_D is estimated as a function of Re:

$$C_D = \frac{K}{\mathrm{Re}} \tag{8-3}$$

Substitution of Eqs. (8-2) and (8-3) into Eq. (8-1) yields Stokes' law:

$$v_b = \frac{4g(\rho_w - \rho_b)d_b^2}{3K\mu_w} \tag{8-4}$$

For solid spheres, $K = 24$:

$$v_b = \frac{g(\rho_w - \rho_b)d_b^2}{18\rho_w} \tag{8-5}$$

Equations (8-2) and (8-5) indicate that the Reynolds number = 1 for bubbles of 125 μm at 20°C and 150 μm at 4°C. For the bubble sizes encountered in DAF and for typical temperatures for water treatment, the assumption of Re < 1 thus holds.

Air bubbles moving through water, even though they act as rigid spheres, have less drag than solid particles. Air bubbles rising in water, especially in clean water, have internal circulation of the gases causing rotation of the bubbles as they rise thereby reducing the drag (Clift et al., 1978; Matsui et al., 1998). This is known as a slipping condition at the bubble surface with $K = 16$ for Re ≤ 1. Stokes' law now takes the form of Eq. (8-6). Rise velocity predictions using Eq. (8-6) are 50 percent higher than those using Eq. (8-5). Ljunggren et al. (2004) found agreement between Eq. (8-6) and experimental rise rate measurements for bubbles with sizes of 85 μm and smaller. They had only one measurement for larger bubbles of 115 μm, and they found that the measured rise rate was less than predicted by Eq. (8-6) but greater than the prediction of Eq. (8-5).

$$v_b = \frac{g(\rho_w - \rho_b)d_b^2}{12\mu_w} \tag{8-6}$$

Rise velocities are plotted in Fig. 8-1 for no slipping conditions [$K = 24$, Eq. (8-5)] and for slipping conditions [$K = 16$, Eq. (8-6)] as a function of bubble diameter for cold (4°C) and warm (20°C) waters. The following points are made:

- As stated above, rise velocities are 50 percent higher for slipping conditions compared to no slipping conditions because of less drag on the rising bubbles.
- Bubbles rise at a slower rate in colder water because of the increase in water viscosity and drag resistance. Other parameters being equal, bubble rise velocities at 4°C (winter) are 64 percent of those at 20°C (summer).

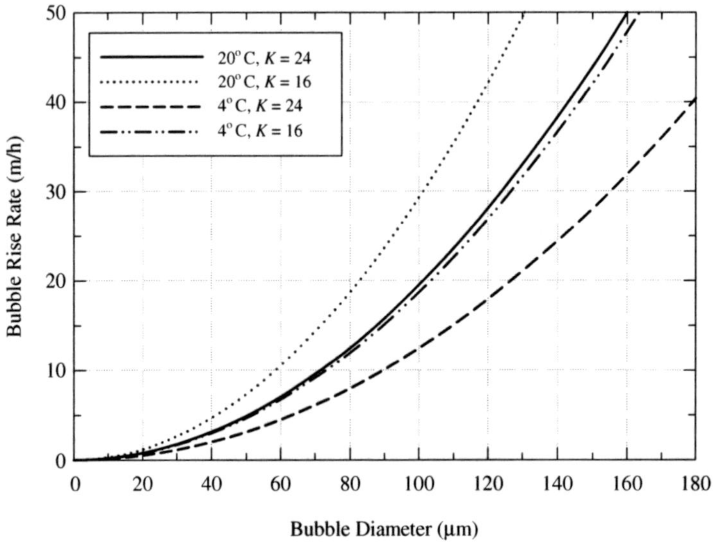

FIGURE 8-1 Bubble rise rates calculated for different temperatures (4 and 20°C) and for different slip assumptions ($K = 24$ for no slip; $K = 16$ for slip).

- Chapter 7 deals with the contact zone in which bubbles are typically 40 to 80 μm and a mean size of 60 μm is used for the contact zone model. For 20°C and bubbles of 60 μm, the rise velocities are 7.0 to 10.5 m/h depending on no slipping versus slipping conditions.
- In this chapter, our interest is with the bubbles in the separation zone. As discussed earlier and in Chap. 4, these bubbles are mostly about 100 μm. For 20°C, the rise rates are about 20 and 30 m/h for no slipping and slipping conditions, respectively. To be conservative, we conclude that 100 μm bubbles in the separation zone rise at about 20 m/h.

EXAMPLE 8-1 Bubble Rise Velocity Calculate (a) the rise velocity of a free bubble (no attached floc) for a 60-μm-diameter bubble, typical size for the DAF tank contact zone and (b) the rise velocity of a 100-μm-diameter bubble, a larger size that is more typical of the DAF tank separation zone. Make the calculations for 20°C with a conservative $K = 24$.

(a) Rise velocity of a 60-μm bubble: To solve the problem requires bubble and water densities, and the viscosity of water, all at 20°C. The water properties are obtained from Table D-1: ρ_w is 998.21 kg/m^3 and μ_w is 1.002×10^{-3} N-s/m^2. The bubble density (ρ_b) from Table E-4 is 1.19 kg/m^3. Two notes about units: (1) a 60-μm bubble in meters is 60×10^{-6} m, and (2) the viscosity unit of N-s/m^2 is equivalent to kg/m-s.

$$v_b = \frac{g(\rho_w - \rho_b)d_b^2}{18\mu_w} = \frac{9.806(998.21 - 1.19)\left(60 \times 10^{-6}\right)^2}{18\left(1.002 \times 10^{-3}\right)} \frac{\text{m}}{\text{s}} \times \frac{3600\ \text{s}}{\text{h}} = 7.03\ \frac{\text{m}}{\text{h}}$$

(b) Rise velocity of a 100-μm bubble: Repeating the above calculation except using a bubble diameter of 100×10^{-6} m yields

$$v_b = \frac{g(\rho_w - \rho_b)d_b^2}{18\mu_w} = \frac{9.806(998.21 - 1.19)\left(100 \times 10^{-6}\right)^2}{18\left(1.002 \times 10^{-3}\right)} \frac{\text{m}}{\text{s}} \times \frac{3600\ \text{s}}{\text{h}} = 19.5\frac{\text{m}}{\text{h}}$$

8-3 THE NATURE OF FLOC-BUBBLE AGGREGATES

The exact nature of the floc-bubble aggregates in DAF is difficult to determine. If preformed bubbles are brought into contact with preformed flocs, the bubbles either attach to the flocs (if conditions are favorable for attachment) or get enmeshed within the flocs (if conditions are less favorable for attachment, or when the flocs are large). For small flocs (of the same magnitude as the bubbles or smaller), attachment is the only possible mechanism as the flocs do not provide enough room for mechanical enmeshment. At least one bubble can be attached to a floc, even if a floc is much smaller than the bubble; an observation supported by photographic evidence (Ives, 1995; Jönsson and Ljunggren, 2003). Larger flocs offer opportunities for the enmeshment of bubbles—in this case either attachment, depending on the floc-bubble surface forces (Chap. 5), or enmeshment could take place.

Bubbles can be incorporated within the flocs in two additional ways. One possibility is that the bubbles precipitate on nucleation sites within the floc. This is unlikely in conventional DAF, as all the air had already been precipitated when they come into contact with the flocs in the contact zone. The other possibility is when flocs continue to grow in the presence of bubbles—the bubbles are then incorporated into the structure of the larger flocs.

These mechanisms are not exclusive. Photomicrographs of wastewater clarification by DAF, where large and small flocs are present, show that both attachment and enmeshment mechanisms are at work in the same suspension (Jönsson and Ljunggren, 2003). For drinking water treatment, either mechanism could be dominant due to the variability of the primary particles in the raw water, the choice and dose of coagulant, the flocculation conditions, the floc strength, and the relative concentrations of flocs and bubbles.

The geometry of a single bubble attached to a small floc is simple and predictable—such an aggregate can be modelled with reasonable certainty, as shown in the next section. For larger flocs, with numerous bubbles attached, the geometry of the floc-bubble aggregates is complex, uncertain, and variable. Examples of these large aggregates were photographed by Ljunggren et al. (2004), and Leppinen and Dalziel (2004). The latter authors introduced the term bubble clusters to describe the large floc-bubble aggregates. They argued that the success of DAF relies predominantly on the large floc-bubble aggregates which act as "nets" as they rise to sweep the smaller aggregates toward the surface.

The modelling of the rise rate of these large aggregates necessarily rely on simplistic assumptions which may be questioned. The value of the next modelling section is therefore not found in the absolute value of the results, but in the identification of the variables affecting the rise rate of the floc-bubble aggregates.

8-4 THE RISE RATE OF FLOC-BUBBLE AGGREGATES

To estimate the rise rate of floc-bubble aggregates, a conventional approach is followed (Haarhoff and Edzwald, 2001). The rise rate of bubbles was calculated in Sec. 8-2. Next, the movement of flocs in water is considered, which draws on the many studies done on the settling of flocs. With a better understanding of the behavior of bubbles and flocs on their own, the combined behavior of bubbles and flocs as floc-bubble aggregates will be estimated.

Behavior of Flocs

The settling rate of a floc can be modelled starting with the Stokes' equation, used earlier for bubbles (see Sec. 8-2). As Stokes' equation is derived for solid spheres, its use for flocs has been questioned, as it does not allow for two important considerations:

- The floc is not impermeable, allowing some water to flow through the floc. The assumption of a solid floc, therefore, underestimates the settling velocity.
- The drag coefficient of the floc is dependent on the shape of the floc as well as the internal structure of the floc. Flocs have a fractal structure and a wide range of fractal dimensions have been reported. Loose, dendritic flocs allow more flow through than compact flocs. A drag coefficient that ignores the fractal dimension is therefore incomplete.

These objections become important for large, heavy flocs with loose dendritic structures. For DAF, the best flocs, as described in Chaps. 6 and 7, are produced at G values of 50 to 100 s^{-1} and are in the size range of 25 to 50 μm. These small, dense flocs are not likely to allow much internal flow and have lower drag due to their compact shape. The use of Stokes's equation should therefore not introduce significant error in the derivation that follows.

For perfect solid spheres, it was noted earlier that $K = 24$. For floc particles with a sphericity of 0.8, Tambo and Watanabe (1979) suggested a value of $K = 45$. The drag force on a floc is therefore about twice as much as for a solid sphere. Using this value in Eq. (8-4), the settling rate of flocs is about half that of perfect spheres:

$$v_f = \frac{4g(\rho_f - \rho_w)d_f^2}{135\mu_w} \tag{8-7}$$

The predictions made by Eq. (8-7) depend greatly on the floc density. Floc density is very variable according to floc size, due to the fractal nature of flocs. For the purposes of predicting the rise rate of the floc-bubble aggregates, the choice of floc density is fortunately not critical. Average floc densities in the range of just above 1000 to 1100 kg/m^3 are commonly reported in the literature and are used in the next section.

Behavior of Floc-Bubble Aggregates

A deciding rise rate parameter is N, the number of bubbles that can be attached to each floc. There is a lower as well as an upper limit to N. The lower limit $N_{available}$ is determined by the number of available bubbles, which becomes a constraint when there are more flocs than bubbles (typically when the solids are distributed among numerous small flocs, such as at the onset of flocculation). For typical DAF applications, this is never a constraint. It is shown in Chap. 5 that there are typically 50 to 100 times more bubbles than flocs.

The upper limit N_{max} is determined by the respective sizes of the bubbles and flocs. As pointed out earlier, at least one bubble can be attached to a floc, even if a floc is much smaller than the bubble. It is also known that bubbles will not attach to flocs in layers (Tambo et al., 1986). For larger flocs, therefore, the surface area of the floc limits the number of bubbles that can be attached. The maximum number of bubbles that can attach to a floc is approximated by assuming that each bubble occupies a square of $d_b \times d_b$ on the floc surface (Tambo et al., 1986). This provides a theoretical upper limit to N, but it is unrealistic to assume such dense bubble packing. Photomicrography of floc-bubble aggregates with multiple bubbles showed that only about 10 percent of the floc surface is covered with bubbles (Jönsson and Ljunggren, 2003). A coverage of 50 percent was proposed as an upper limit for calculation purposes, which is adopted for the derivation below (Gregory and Edzwald, 2010):

$$N_{\max} = \frac{\pi}{2}\left(\frac{d_f}{d_b}\right)^2 \tag{8-8}$$

The equivalent diameter d_{fb} of the floc-bubble aggregate produced from N bubbles of size d_b adhering to a single floc particle with size d_f is estimated by assuming that the bubble and floc volumes are additive:

$$d_{fb} = \left(d_f^3 + N d_b^3\right)^{1/3} \tag{8-9}$$

The aggregate density ρ_{fb} is estimated from the weighted average of the air and floc densities:

$$\rho_{fb} = \frac{\rho_f d_f^3 + N \rho_b d_b^3}{d_f^3 + N d_b^3} \tag{8-10}$$

Once the aggregate size and density are found, a rise rate equation can be developed analogous to Eq. (8-4):

$$v_{fb} = \frac{4g(\rho_w - \rho_{fb}) d_{fb}^2}{3K\mu_w} \tag{8-11}$$

In the case of individual bubbles and flocs discussed earlier, the flow regime remains laminar for typical DAF applications. In the case of floc-bubble aggregates, the rise rates could be substantially higher. As long as the flow regime remains laminar, which is characterized by Re ≤ 1, Eq. (8-11) is valid. At higher rates, in the transition zone where 1 ≤ Re ≤ 50, the drag coefficient for particles with sphericity of 0.8 is estimated with (Brown, 1963):

$$C_D = \frac{K}{\mathrm{Re}^{0.75}} \tag{8-12}$$

The rise rate for the transition zone between laminar and turbulent flow is obtained by substitution of Eq. (8-12) into Eq. (8-1):

$$v_{fb} = \left(\frac{4}{3K}\right)^{0.8}\left(\frac{g^{0.8}(\rho_w - \rho_{fb})^{0.8} d_{fb}^{1.4}}{\rho_w^{0.2}\mu_w^{0.6}}\right) \tag{8-13}$$

The choice of an appropriate value for K, which determines the drag coefficient, is a challenge. This is due to the uncertainty and the variability of, the shape and configuration of the aggregates. In the case of a small floc of say 25 μm attached to a bubble of 50 μm, the bubble forms the leading edge of the aggregate with the floc completely shielded in the wake of the bubble. In this case, the drag coefficient of the aggregate is almost identical to that of the bubble alone; K is between 16 and 24 as discussed earlier. At the other extreme, a number of bubbles of 50 μm could be enmeshed in a large floc of say 200 μm. In this case, the drag coefficient of the aggregate is almost identical to that of the floc alone; K is 45. For modelling purposes, it is assumed that K varies smoothly from $K = 20$ at a floc size at or below 50 μm to $K = 45$ at a floc size at or above 150 μm.

In DAF, the rise rate of the floc-bubble aggregates is much higher than the settling rate of the flocs. The buoyancy force on a 75-μm bubble is an order of magnitude larger than the suspended weight of a 100-μm floc (Ljunggren et al., 2004). This means that the density of the floc has a minor effect on the rise rate. Typical values between 1000 and 1100 kg/m^3 are considered to be adequate for the calculation of the floc-bubble rise rate.

Figure 8-2 shows the rise rate of the floc-bubble aggregate for a bubble size of 100 μm and floc density of 1020 kg/m^3, taking into account all the assumptions and adjustments

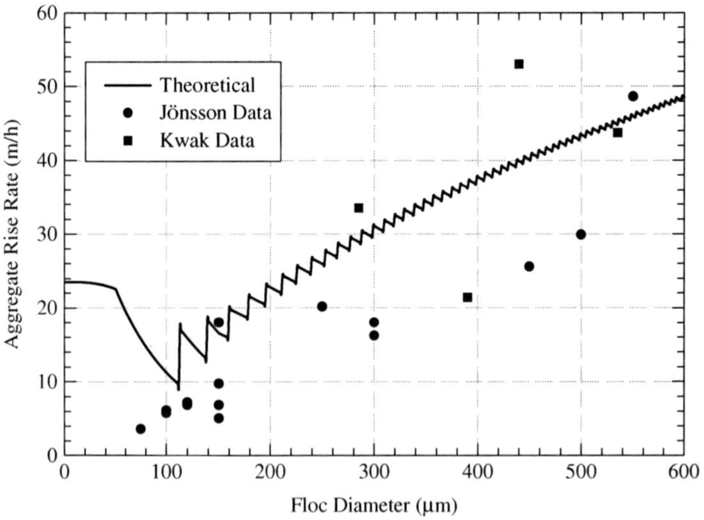

FIGURE 8-2 Theoretical floc-bubble aggregate rise rates compared to actual rise rates as a function of floc diameter (theoretical rates based on bubble diameter of 100 μm, floc density of 1020 kg/m^3, 50% bubble coverage on large flocs, and changing drag coefficient between floc size of 50 and 150 μm; *Source: the data points are taken from Jönsson and Ljunggren (2003) and Kwak et al. (2007).*

above—the changing drag coefficient, Reynolds number, aggregate size and limitations of the number of attachable bubbles. The data points on the graph, taken from two independent experimental studies (Jönsson and Ljunggren, 2003; Kwak et al., 2007), indicate that real suspensions contain aggregates with large variations in their rise rates. Broadly speaking, the data support the plausibility of the rise rate model presented.

EXAMPLE 8-2 Floc-Bubble Aggregate Rise Velocity Calculate the rise rate of (a) a 30-μm floc and (b) a 130-μm floc after contact with a suspension of air bubbles, all 100 μm in diameter. Make the calculations for 20°C, a floc density of 1020 kg/m³ and bubble coverage of 50 percent.

(a) Floc diameter 30 μm: With the bubble diameter larger than the floc, only one bubble is attached to the floc. The solution requires bubble and water densities, and the water dynamic viscosity, all for 20°C. The water properties are obtained from Table D-1: ρ_w is 998.21 kg/m³ and μ_w is 1.002×10^{-3} N-s/m². The bubble density (ρ_b) from Table E-4 is 1.19 kg/m³.

Eq. (8-8) (rounded up):	$N = 0.141 \rightarrow 1$
Eq. (8-9):	$d_{fb} = 100.9$ μm
Eq. (8-10):	$\rho_{fb} = 28.0$ kg/m³
Floc size below 50 μm, therefore	$K = 20$
Eq. (8.11):	$v_{fb} = 0.00645$ m/s = 23.2 m/h
Eq. (8.2):	Re = 0.65 ≡ laminar flow, therefore Eq. (8-11) is applicable.

(b) Floc diameter 130 μm: Using the same properties in the first part:

Eq. (8-8) (rounded up):	$N = 2.65 \rightarrow 3$
Eq. (8-9):	$d_{fb} = 173.2$ μm
Eq. (8-10):	$\rho_{fb} = 431.9$ kg/m³
Floc size below 150 μm, interpolate	$K = 40.8$
Eq. (8-11):	$v_{fb} = 0.00544$ m/s = 19.6 m/h
Eq. (8-2):	Re = 0.94 ≡ laminar flow, therefore Eq. (8-11) is applicable.

8-5 AN OVER-SIMPLIFIED VIEW OF DAF SEPARATION

The traditional analysis of the DAF separation zone, as presented in most books and papers, assumes perfect plug flow as the water flows from the inlet to the outlet. This allows an analysis analogous to Hazen's sedimentation theory, which concludes that successful separation can only follow if the vertical particle rise rate is less than the surface loading (flow rate divided by the plan area). Figure 8-3 illustrates this principle.

It is pointed out in Chap. 1 that hydraulic loadings for DAF are expressed in either nominal or net terms. The nominal hydraulic loading [raw water flow rate divided by entire area of the DAF reactor, as indicated by Eq. (1-2)] is useful for the general classification of DAF plants. For the more detailed analysis of the separation zone, the net hydraulic loading [raw water plus recycle flow rates divided by area A_{sz} of the separation zone in Fig. 8-3,

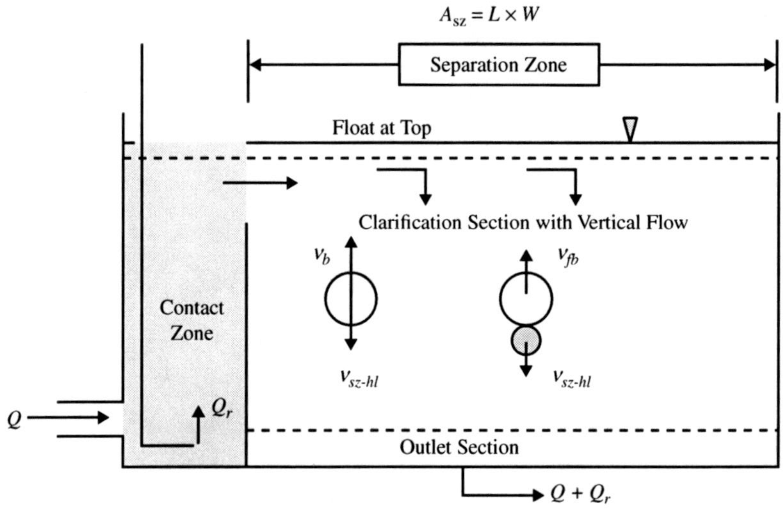

FIGURE 8-3 Idealized DAF tank showing separation zone divided into three sections: float or sludge layer, vertical flow section for clarification, and the outlet (*Source: Gregory and Edzwald (2010), reprinted with permission from Water Quality and Treatment: A Handbook on Drinking Water, 6th ed., Copyright 2010, AWWA*).

mathematically expressed by Eq. (1-3)] is more appropriate. The remainder of the chapter refers to the net hydraulic loading.

It is noted that the direct application of the above over-simplified approach is not sufficient. The flow patterns in the separation zone are complex, and a great deal of effort was invested during the past twenty years to gain a better understanding of the effects of flow patterns on performance. Flow patterns are discussed in Sec. 8-8.

8-6 SEPARATION-ZONE MEASUREMENTS AND MODELLING

It is difficult to make observations and measurements in the separation zone. The great number of small air bubbles severely limits visual or light penetration into the white water zone. Visual observations have to be made through transparent sides into small pilot tanks, which raise concerns due to the reduced scale. Despite these difficulties, researchers have used an assortment of methods to get a better understanding of the complex separation-zone processes, briefly reviewed in this section.

Tracer Studies

Chapter 7 presents flow patterns for the two ideal cases of a plug-flow reactor (PFR) and a continuous-flow stirred-tank reactor (CFSTR). The use of tracers offers a well-known method to characterize the flow pattern in a reactor. Good DAF separation requires adequate time for the floc-bubble aggregates to rise to the surface, best achieved by plug-flow conditions. Longitudinal dispersion, or deviations from plug flow, is rapidly and easily characterized by tracer studies. The characterization of dispersion, using the dispersion number (d_{dn}), is discussed in Chap. 7.

Velocity Measurements

By knowing the magnitude and direction of the flow at many points within the separation zone, the flow pattern can be clearly discerned. Two methods have been used to measure these point velocities in three dimensions, namely acoustic Doppler velocimetry (ADV) and laser Doppler velocimetry (LDV). In the case of ADV, short acoustic pulses are directed into the water to form a sample volume of 250 mm^3 about 50 mm away from the sensor (Lundh and Jönsson, 2002), while LDV shines laser beams into the water to form a much smaller sample volume of about 0.1 mm^3 about 45 mm away from the sensor (Hague et al., 2001). ADV measures the water velocity directly (after correction for the presence of air bubbles), while LDV measures the velocity of the air bubbles. As the rise rate of air bubbles is small compared to the water velocity in the white water zone, LDV measurements can be interpreted as being representative of the water velocity (Hague et al., 2001).

Both ADV and LDV were reported to provide credible results in two different studies. The ADV experiments were done in a pilot-scale DAF reactor, with a separation zone 700 mm wide, 1350 mm deep, and about 1200 mm long (Lundh and Jönsson, 2002). The LDV study was done in an even smaller tank of 300 mm wide, 350 mm deep, and 500 mm long (Hague et al., 2001). In the case of ADV, an earlier study showed it to be effective in DAF white water (Adlan et al., 1997), while the LDV measurements were found to agree closely with results from mathematical flow modelling (Hague et al., 2001).

Computational Flow Dynamics (CFD)

With the advent of easily accessible high-speed computing, CFD became a feasible tool for predicting fluid flow patterns by modelling the reactor volume as thousands of small fluid elements. CFD analysis of single-phase flow in water treatment reactors has become commonplace and finds useful application, for example, in simulating tracer studies. CFD studies have also been performed for DAF reactors by using two-phase modelling—water and air (e.g., Fawcett, 1997; Crossley et al., 2001; Hague et al., 2001; Guimet et al., 2007). A powerful aspect of CFD is that it can depict different patterns in the DAF reactor: the magnitude and direction of velocity, the air concentration at any point, tracer response, etc. The principal weakness of CFD is that it is limited to water and bubbles only, without consideration of the flocs and particles. Flocs and particles are known to affect the flow pattern in the separation zone, as shown later in this chapter.

Position of the Bubble Bed

There is a clearly discernible interface between the bubble bed and the clear water below, as observed in small-scale pilot tanks with transparent sides. Until recently, it was not possible to observe the depth of the bubble bed in full-scale tanks. A new technique developed in South Korea makes it possible to locate the underside of the bubble bed in full-scale tanks. A sampling probe is inserted into the separation zone from the top and slowly traverses the separation zone in a vertical direction. Water is continuously withdrawn from the tip of the probe and directed to a particle counter, which measures the bubble concentration. The bottom of the bubble bed is detected, within a few centimeters, at the point where a sudden change in bubble concentration is measured (Han et al., 2007a). This technique is used to measure the effects of different design and operating conditions on the depth of the bubble bed.

8-7 INLET AND OUTLET CONSIDERATIONS

Outlet Considerations

It is assumed throughout this discussion that the clear water from the separation zone is evenly withdrawn from the bottom of the tank. During the early days of DAF development, it was found that a single outlet, at the bottom of the far end opposite to the inlet, could lead to a large circulation pattern in the separation zone, upsetting the separation process (Haarhoff and Van Vuuren, 1993). Even-flow withdrawal is usually accomplished with a number of pipe manifolds evenly distributed over the floor of the separation zone, unless DAFF is used, where the flow distribution is perfectly even due to the uniform resistance of the underlying filter media layer. For the later high-rate DAF systems, the hydraulic loading is too high for filtration, which necessitates a stand-alone DAF reactor. In this case, even-flow distribution is achieved with a thin perforated plate suspended above the floor of the separation zone.

The Effects of Air

Without air bubbles, the flow in a DAF tank, despite even-flow withdrawal, has a tendency to assume a rotary pattern. The momentum of the water entering at the top produces a horizontal flow toward the far boundary at the top of the tank. At the far wall, the flow deflects toward the bottom, and deflects again at the floor toward the inlet. This rotary pattern was confirmed by modelling (e.g., Fawcett, 1997; Guimet et al., 2007) as well as by experiment (Lundh et al., 2000).

With air bubbles, the flow patterns in the DAF separation zone are much different. This is primarily due to the large density gradients induced by the differences in air concentration in different parts of the separation zone. It is well known that the flow patterns in sedimentation tanks are easily upset by density currents, caused by sudden changes in turbidity, salinity, or temperature. A density change of 0.1 percent is caused by any of the following: a temperature change of 5°C, a TDS change of 1200 mg/L, or a SS change of 1000 mg/L. This should be compared with the density change of about 0.8 percent (eight times more than the extreme examples above) caused by the typical air concentrations used for DAF.

Water from the contact zone should enter the separation zone near the top to minimize the distance between the floc-bubble aggregates and the surface of the tank—refer to Fig. 8-3. This is normally achieved by a baffle in the tank with its top edge a short distance from the surface. The water rises in the contact zone, changes direction at the top of the baffle and enters the separation zone at the top in a horizontal direction. The incoming layer of water is rich in air, has low density, and therefore stays near the surface. This layer spreads evenly over the top part of the separation zone as it flows toward the far end of the separation zone. As it is rich in bubbles, this layer is observed as white water. For the typical air concentrations in DAF, the Richardson number (a stability parameter derived from fluid mechanics) indicates a large degree of stability of this top layer (Jönsson et al., 1997).

The effect of air on the separation-zone flow pattern is strikingly demonstrated by tracer studies (Lundh and Jönsson, 2005). Figure 8-4 shows three tracer responses. (For comparison, tracer responses for ideal plug-flow conditions and CFSTRs in series are shown in

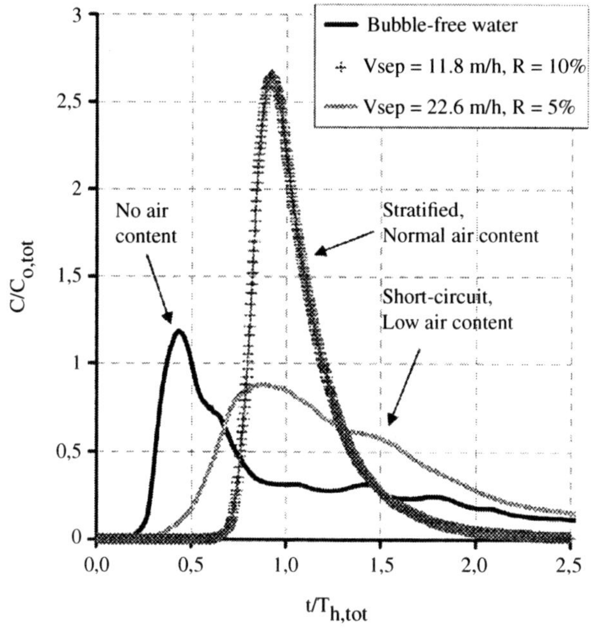

FIGURE 8-4 Tracer tests performed on a DAF separation zone for three cases: no air, stratified flow, and flow under breakthrough conditions. *(Source: Lundh, M., and Jönsson, L. (2005) Residence time distribution characterization of the flow structure in dissolved air flotation, Journal of Environmental Engineering, 131 (1), 93–101, with permission from ASCE).*

Fig. 6-18.) Without bubbles, there is substantial short-circuiting with the peak response reached at half the theoretical retention time, followed by a tail to wash out the remaining tracer. With normal air content at a low loading rate, the tracer pulses sharply at about the theoretical retention time—the desired condition for efficient separation. However, with less air and higher hydraulic loading, the plug-flow conditions break down and short-circuiting becomes evident once more, with a long washout tail.

Inlet Considerations

The inlet conditions to the separation zone are determined by the baffle separating the contact and separation zones, as depicted in Fig 8-3. The opening between the top edge of the baffle and the surface controls the horizontal velocity and momentum of the water entering the separation zone. This cross-flow velocity was therefore deemed to be an important design parameter with initial suggested limits of 20 to 100 m/h (Haarhoff and Van Vuuren, 1993). Later experimental work confirmed the importance of the cross-flow velocity. For high hydraulic loading, the cross-flow velocity should be great enough to ensure plug-flow conditions. It is therefore important to have a firm and even horizontal velocity into the separation zone. To accomplish this, a minimum cross-flow velocity of 40 m/h is suggested (Lundh et al., 2002a).

If all operating conditions are kept constant, the depth of the bubble bed is determined by the height of the baffle between the contact and separation zones. The bubble bed depth is directly controlled by the baffle height. If the top of the baffle would be moved up by 50 mm, the bubble bed depth would decrease by 50 mm. The inclination of the baffle has no effect on the bubble bed depth, confirmed by both modelling (Amato and Wicks, 2009) and experiment (Han et al., 2007a; Lundh et al., 2002a).

8-8 SEPARATION-ZONE FLOW PATTERNS

Flow Pattern at Low Hydraulic Loading

At low hydraulic loading, say, <10 m/h, the flow pattern in the separation zone is fairly simple. The white water from the contact zone enters the separation zone with a reasonable cross-flow velocity toward the far wall. Due to the density difference between the incoming water and the water already in the tank, the white water will stay in a narrow, stable horizontal layer as it flows toward the far wall. In the time that it takes to cover the distance between the inlet and the far wall, the floc-bubble aggregates rise to the surface and collect on the surface. As the floc-bubble aggregates rise out of the white water into the float layer, the top layer slows down as it "loses" water which flows vertically down toward the outlets. When the top layer reaches the far wall, the velocity is low and the water is deflected downward to the outlet. This pattern is confirmed by LDV measurements, as well as CFD modelling (Hague et al., 2001).

Stratified Flow at High Hydraulic Loading

At high hydraulic loading, the water from the inlet flows in a surface layer toward the far wall, the same as described in the previous paragraph. Due to the shorter residence time (because of higher hydraulic loading), there is some air remaining in the water, with density less than that of clear water, when the water reaches the far wall. At the far wall, the water is deflected downward. Due to its lower density, the water cannot easily penetrate the denser water below and is forced back in a horizontal direction immediately below the top layer, but in the opposite direction—called the return flow. One can thus visualize the separation zone as three horizontal layers—the top layer flows toward the outlet end, a return flow immediately below flow back toward the inlet, and a relatively air-free layer of clear water at the bottom. This three-layer structure is known as the *stratified flow structure* and it is schematically depicted in Fig. 8-5. Our understanding of the stratified flow structure is largely due to the systematic experimental investigations at the Lund University in Sweden (Haarhoff and Edzwald, 2004).

Stratified flow does not automatically follow at high hydraulic loading. It is possible to disrupt the stratification which allows the water from the inlet to follow a shorter path to the outlet, rather than following the to-and-fro flow path at the surface. This is termed *breakthrough*, which is obviously detrimental to the efficiency of DAF as some air, together with floc-bubble aggregates, could be drawn into the outlet. The work at Lund University uncovered a most important condition for stratified flow, summarized in Fig. 8-6. Higher hydraulic loading can only be sustained by an increase in the air concentration. At a hydraulic loading of about 12 m/h (the left-hand column of the figure) the flow structure changes from stratified at the top (15 percent recycle) to breakthrough at the bottom (5 percent recycle). At the intermediate hydraulic loading of about 18 m/h (middle column) the breakthrough at 5 percent recycle (bottom) is more severe. At the highest hydraulic loading of 24 m/h, breakthrough is evident in all cases, even at 15 percent recycle (top). Although the tests were performed in a pilot tank only 1.2 m deep and separation zone length of

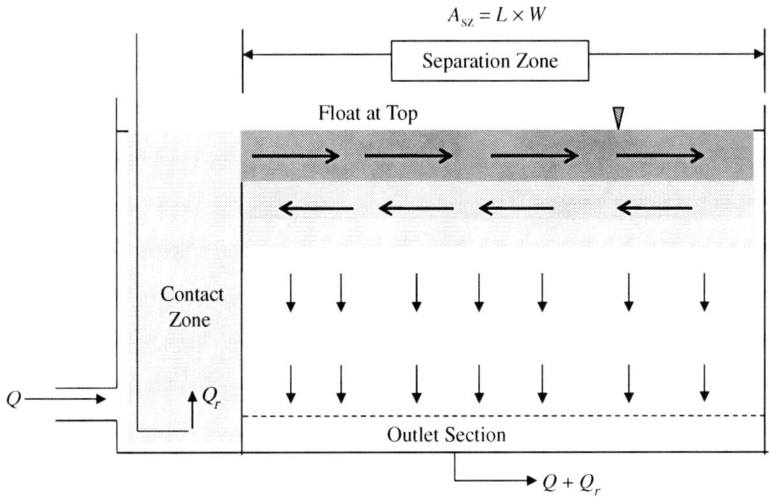

FIGURE 8-5 Conceptual horizontal stratified flow pattern near the top of the separation zone and vertical plug flow below the bubble blanket.

1.5 m, the general conclusions and their design implications are clear—higher loading rates should go hand in hand with greater air concentrations.

It should be noted that the return flow path was not detected by LDV measurements, made at a hydraulic loading of 11 m/h (Hague et al., 2001). The LDV study used a recycle percentage of only 4.5 percent at a saturation pressure of 400 kPa, which represents a low air concentration. The measurements were therefore probably made at breakthrough rather than stratified flow. CFD simulation also failed to detect a return flow, which may be due to the CFD assumption of two-phase flow of discrete bubbles and water, without taking flocs and particles into consideration.

The presence of floc-bubble aggregates, rather than air bubbles only, has an effect on the flow structure, but has not yet been illuminated by systematic investigation. The first author of this book observed a clearly discernible effect of coagulant addition in a pilot plant treating water from a reservoir. As soon as coagulant was added, the white water layer stabilized at a higher position in the separation zone. Without coagulant addition, the blanket was drawn deeper into the tank and closer to the outlet. Whether this was a surface charge, or a particle-induced phenomenon, is not clear. This observation is supported by the experimental work of Lundh et al. (2002b) in treating wastewater. A few tests were conducted with SS concentrations varying from 0 mg/L (tap water) to as high as 102 mg/L. It was found that particle removal was better at high SS concentration, all other conditions being the same. A higher SS concentration seems to improve the flow structure from short-circuit flow toward stratified flow, but no rational explanation has been offered yet.

The Bubble Bed

The top two layers of the stratified flow structure, shown in Fig. 8-5, are collectively called the bubble bed. Below the bubble bed, the water remains free of bubbles. There are two aspects of the bubble bed depth to consider. The first is the bubble bed depth, as enough

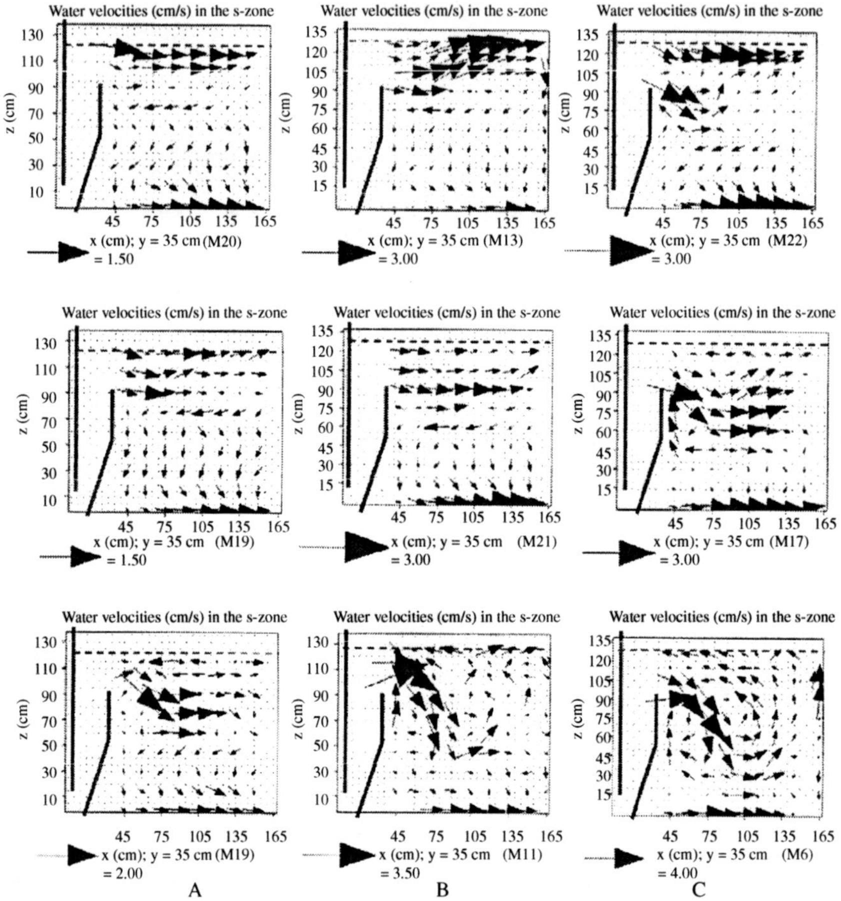

FIGURE 8-6 Experimentally measured flow patterns in the separation zone. The recycle fraction is 15% (top row), 10% (middle row) and 5% (bottom row). The net hydraulic loading is 11.3 to 12.4 m/h (left column), 16.9 to 18.5 m/h (middle column) and 23.2 to 24.7 m/h (right column). *Source: Taken from Lundh, 2000.*

depth in the separation zone should be provided to ensure that the bubble bed is not drawn into the outlet. The second is the bubble concentration in the bubble bed. The bubble concentration has a bearing on the separation efficiency.

At a constant air concentration, a higher hydraulic loading rate drives the bubble bed deeper into the separation zone (Kiuru, 2001). In a small-scale laboratory tank of 100 mm wide, 500 mm deep and 550 mm long, the bubble bed depth increased from 100 to 300 mm as the hydraulic loading was doubled from 5 to 10 m/h, all other conditions being constant (Han et al., 2007a). As the number of bubbles remained constant, the higher hydraulic loading led to a lower bubble concentration as the bubble bed was stretched in a vertical direction.

For a fixed hydraulic loading, the bubble supply can be manipulated by changing either the saturation pressure or the recycle ratio. An increase in saturation pressure increases the bubble concentration, without increasing the hydraulic loading of the separation zone. An increase in the recycle ratio increases the bubble concentration and, to a smaller extent, the hydraulic loading of the separation zone. The effects of these parameters were systematically considered at pilot scale (Han et al., 2007b). At a constant hydraulic loading rate of 7 m/h, a doubling of the saturation pressure from 3 to 6 atmospheres (thus a doubling of the bubble supply) increased the bubble bed depth by about 40 percent and the bubble concentration by 43 percent. A tripling of the recycle ratio from 10 to 30 percent (thus a tripling of the bubble supply) increased the bubble bed depth by about 50 percent and the bubble concentration by 100 percent. Whether the air supply is increased by either a higher saturation pressure or a higher recycle ratio, the effects are the same—the bubble bed depth increases, as well as the bubble concentration in the bubble bed.

The bubble concentration in the bubble bed has an effect on the removal efficiency of the separation zone. At a low hydraulic loading of 5 m/h, there was a small difference in removal efficiency at different bubble bed depths. The removal efficiency dropped from 99 to 92 percent as the bubble bed decreased from 320 to 160 mm. The effect was more pronounced at a higher hydraulic loading of 10 m/h, where the removal efficiency dropped from 92 to 69 percent for a similar change in bubble depth (Han et al., 2007b).

The bubble bed in the separation zone therefore acts like a filter collector, analogous to the contact zone model developed in Chap. 7. It is most efficient when the bed is deep and the bubble concentration is high. When the hydraulic loading is increased, the bubble bed expands—this should be countered by more air.

REFERENCES

Adlan, M. N., Elliott, D. J., Noone, G., and Martin, E. B. (1997), Investigation of velocity distribution in dissolved air flotation tank, in *Dissolved Air Flotation*, London: CIWEM.

Amato, T., and Wicks, J. (2009), The practical application of computational fluid dynamics to dissolved air flotation, water treatment plant operation, design and development, *Journal of Water Supply: Research and Technology – Aqua*, 58 (1), 65–73.

Brown, G. G. (1963), *Unit Operations*, New York: John Wiley & Sons.

Clift, R., Grace, J., and Weber, M. E. (1978), *Bubbles, Drops, and Particles*, New York: Academic Press.

Crossley, I. A., Valade, M. T, and Shawcross, J. (2001), Using lessons learned and advanced methods to design a 1,500 ML/Day DAF water treatment plant, *Water Science & Technology*, 43 (8), 35–41.

Eades, A., Jordan, D., and Scheidler, S. (1997), Counter-current dissolved air flotation filtration, in *Dissolved Air Flotation*, London: CIWEM.

Fawcett, N. S. J. (1997), The hydraulics of flotation tanks: computational modelling, in *Dissolved Air Flotation*, London: CIWEM.

Gregory, R., and Edzwald, J. K. (2010), Sedimentation and flotation, in J. K. Edzwald, ed., *Water Quality and Treatment: A Handbook on Drinking Water*, 6th ed., New York: McGraw-Hill.

Guimet, V., Broutin, C., Vion, P., and Glucina, K. (2007), CFD modeling of high-rate dissolved air flotation. *Proceedings of the 5th International Conference of Flotation in Water and Wastewater Systems*, September 11–14, Seoul, South Korea, 113–119.

Haarhoff, J., and Van Vuuren, L. R. J. (1993), *A South African Design Guide for Dissolved Air Flotation*, Water Research Commission Report No. 322, Pretoria, South Africa.

Haarhoff, J., and Edzwald, J. (2001), Modelling of floc-bubble aggregate rise rates in dissolved air flotation, *Water Science and Technology*, 43 (8), 175–184.

Haarhoff, J., and Edzwald, J. (2004), Dissolved air flotation modelling: insights and shortcomings, *Journal of Water Supply: Research and Technology – Aqua*, 53 (3), 127–150.

Hague, J., Ta, C. T., Biggs, M. J., and Sattary, J. A. (2001), Small scale model for CFD validation in DAF application, *Water Science & Technology*, 43 (8), 167–173.

Han, M., Kwak, D., Kim, T., Park, S., and Kim, H. (2007a), Measurement of bubble bed depth in DAF using a particle counter, *Proceedings of the 5th International Conference of Flotation in Water and Wastewater Systems*, September 11–14, Seoul, South Korea, 97–104.

Han, M., Kwak, D., Kim, T., Park, S., and Kim, H. (2007b), A crucial importance of bubble bed in high rate DAF, *Proceedings of the 5th International Conference of Flotation in Water and Wastewater Systems*, September 11–14, Seoul, South Korea, 309–315.

Ives, K. J. (1995), Inside story of water treatment processes, *Journal of Environmental Engineering ASCE*, 121 (5), 846–849.

Jönsson, L., Gunnarsson, M., and Rångeby, M. (1997), Some hydraulic aspects on flotation tanks, in *Dissolved Air Flotation*, London: CIWEM.

Jönsson, L., and Ljunggren, M. (2003), *Studies of Rise Velocities of DAF Bubble/floc Aggregates*, Report 3244, Department of Water Resources Engineering, Lund Institute of Technology, Lund University, Sweden.

Kwak, D. H., Jung, H. J., Lee, J. W., Kwon, S. B., Kim, S. J., Yoo, S. J., and Won, C. H. (2007), Rise velocity verification of bubble–floc agglomerates using population balance in DAF process, *Proceedings of the 5th International Conference of Flotation in Water and Wastewater Systems*, September 11–14, Seoul, South Korea, 121–134.

Kiuru, H. J. (2001), Development of dissolved air flotation technology from the 1st generation to the newest or 3rd one (very thick micro-bubble bed) with high flow rates (DAF in turbulent flow conditions), *Water Science & Technology*, 43 (8), 1–7.

Leppinen, D. M., and Dalziel, S. B. (2004), Bubble size distribution in dissolved air flotation tanks, *Journal of Water Supply: Research and Technology – Aqua*, 53 (8), 531–543.

Ljunggren, M., Jönsson, L., and La Cour Jansen, J. (2004), Particle visualisation—a tool for determination of rise velocities, *Water Science & Technology*, 50 (12), 229–236.

Lundh, M. (2000), *Flow Structures in Dissolved Air Flotation – Experimental Study in a Pilot Plant*, Licentiate thesis, Department of Water and Environmental Engineering, Lund University, Sweden.

Lundh, M., and Jönsson, L. (2002), Performance evaluation of the acoustical Doppler velocimeter in water with high contents of micro-bubbles, Paper B in a doctoral dissertation entitled *Effects of Flow Structure on Particle Separation in Dissolved Air Flotation*, Lund University, Sweden, 2002.

Lundh, M., and Jönsson, L. (2005), Residence time distribution characterization of the flow structure in dissolved air flotation, *Journal of Environmental Engineering*, 131 (1), 93–101.

Lundh, M., Jönsson, L., and Dahlquist, J. (2000), Experimental studies of the fluid dynamics in the separation zone in dissolved air flotation, *Water Research*, 34 (1), 21–30.

Lundh, M., Jönsson, L., and Dahlquist, J. (2002a), The influence of contact zone configuration on the flow structure in a dissolved air flotation pilot plant, *Water Research*, 36 (6), 1585–1595.

Lundh, M., Jönsson, L., and Dahlquist, J. (2002b), Relating the separation of biological floc from the kaldnes moving bio bed process to the flow structure in a dissolved air flotation pilot tank, Paper G in a doctoral dissertation entitled *Effects of Flow Structure on Particle Separation in Dissolved Air Flotation*, Lund University, Sweden, 2002.

Matsui, Y., Fukushi, K., and Tambo, N. (1998), Modeling, simulation and operational parameters of dissolved air flotation, *Journal of Water Supply: Research and Technology – Aqua*, 47 (1), 9–20.

Tambo, N. and Watanabe, Y. (1979), Physical characteristics of flocs – I. The floc density of aluminum floc, *Water Research*, 13 (5), 409–419.

Tambo, N., Matsui, Y., and Fukushi, K. (1986), A kinetic study of dissolved air flotation, *World Congress of Chemical Engineering*, Tokyo, 200–203.

CHAPTER 9
FLOAT LAYER REMOVAL

Successful DAF separation leaves the bulk of the contaminants floating on the surface of the separation zone, with the clarified water leaving as underflow to the next process. The float layer cannot be left to accumulate indefinitely—it has to be intermittently or continuously removed in a careful and controlled manner. The design of the float removal system does not have a theoretical basis and is guided by empirical experience gathered over years of close observation and careful experimentation by many. The most important primary references regarding float layer removal date to the 1970s and 1980s, with relatively few reports since. The objective of this chapter is to collate and interpret the available published information for the benefit of DAF designers and operators.

9-1 THE NATURE OF THE FLOAT LAYER

The float layer is a complex matrix of raw water particles, precipitates formed by chemical dosing, trapped air, and water. The mechanical properties of the float layer are therefore highly variable as it depends on both the raw water source and the chemical dosing strategy. Furthermore, other than the solids content, the water industry does not have commonly accepted quantitative descriptors of the float layer rheology, which plays an obvious role in how the float layer is best handled.

The air content of the float layer allows it to float partially above the water level. The solids content at the top of the float layer increases as water drains away and air escapes to the atmosphere. The drainage of water from a thick float layer is a highly effective way to thicken the float layer to a considerable degree, and exploited for the thickening of waste activated sludge (e.g., Bratby and Jones, 2007). For a float layer formed during drinking water treatment, which was allowed to reach an unusually large thickness of 250 mm, the dry solids at the top of the layer reached 5.3 percent, while it was 3.9 and 1.4 percent at its midpoint and bottom, respectively (De Groot and Van Breemen, 1987). The solids content of the float layer cannot be changed by adding more air (Gregory and Edzwald, 2010).

Little is known about the distribution of air in the float layer. Attempts were made at Lund University in Sweden to observe the float layer at close range from below with a borescope (Jönsson and Lundh, 2000). The observations were made at a wastewater treatment plant where DAF was used to polish the treated wastewater before discharge to the ocean. The first set of observations was made closely to the inlet from the contact zone. Here there was evidence of many small bubbles (roughly 50 μm in diameter) on the underside of the float layer, providing the buoyant forces for pushing the float layer upward. The second set of observations was made 3 m downstream of the inlet and a large change was observed. Large circular shapes of about 500 μm in diameter were observed, which were probably air accumulations. These air pockets were firmly embedded in the underside of the float layer. These observations suggest that, once the air reaches the underside of the float layer, there is rapid coalescence of the air bubbles into much larger air pockets on the underside of the float layer to keep it afloat. This is in line with an observation made by the second author on many occasions—when a glass rod or small sampling bottle is pushed into a stable float layer, there is a gentle "burp" of a fairly large volume of air, which supports the presence of air pockets much larger than the microbubbles associated with DAF.

A stable float layer, without any mechanical disturbance, is subject to desiccation from above, and hydraulic erosion from below. How long can the layer remain afloat before partial or complete disintegration? Without a continuous supply of air from the saturation system, the float layer rapidly becomes unstable. For a float layer rich in algae, which is normally very stable, the float layer started to break up only 2 h after the air supply was stopped (Williams et al., 1985). This means that the float layer on all the DAF tanks should be completely removed at each shutdown. Alternatively, the recycle pumps have to be kept running during interruptions in raw water flow. For uninterrupted operation, the time before float layer breakup varies greatly, depending mostly on the nature of the raw water treated. The shortest period was reported for a low-turbidity, colored raw water where breakup of the float layer was observed after only 30 min, with a consequent increase in the head loss development rate of the downstream filters (Rees et al., 1979). By contrast, much longer times were reported for float layers rich in algae. Float layer buildup could be allowed for up to five days without problems for water from a South African eutrophic impoundment (Van Vuuren and Van der Merwe, 1988). Where float layers show such stability, it is exploited in DAFF applications, where the float layer is commonly left undisturbed on the surface and washed out during the filter backwash cycle, with as much as 40 h between filter backwash cycles for a reported case treating water from a Finnish lake (Rosen and Morse, 1976). Figure 9-1 shows two DAF plants with float layers which are ready for hydraulic removal.

The stability of the float layer could not be improved by the addition of flocculant aid, an observation made during studies in England and the Netherlands (De Groot and Van Breemen, 1987).

The above paragraphs apply to stable float layers without any mechanical interference. As soon as the float layer is pushed around by mechanical scrapers, or flushed toward the sludge outlet, the potential for float-layer breakup and knockdown into the water below escalates. This is a matter taken up in the following sections.

9-2 HYDRAULIC OR MECHANICAL FLOAT LAYER REMOVAL?

The float layer can be removed either hydraulically by flushing the layer into a sludge channel or mechanically by scraping the float layer over a beach plate into a sludge channel. How do designers choose between these options? In short, hydraulic removal offers the simplest solution with minimal mechanical equipment and with little or no impact on the treated water quality, but comes at the expense of larger water wastage and the production of a large volume of thin, watery sludge. This is not a problem if the sludge is discharged directly to a sewer, but could be if the sludge has to be thickened on site before disposal. Mechanical removal produces a denser sludge with less water wastage, but will impair the treated water quality while requiring an expensive sludge removal mechanism. Both

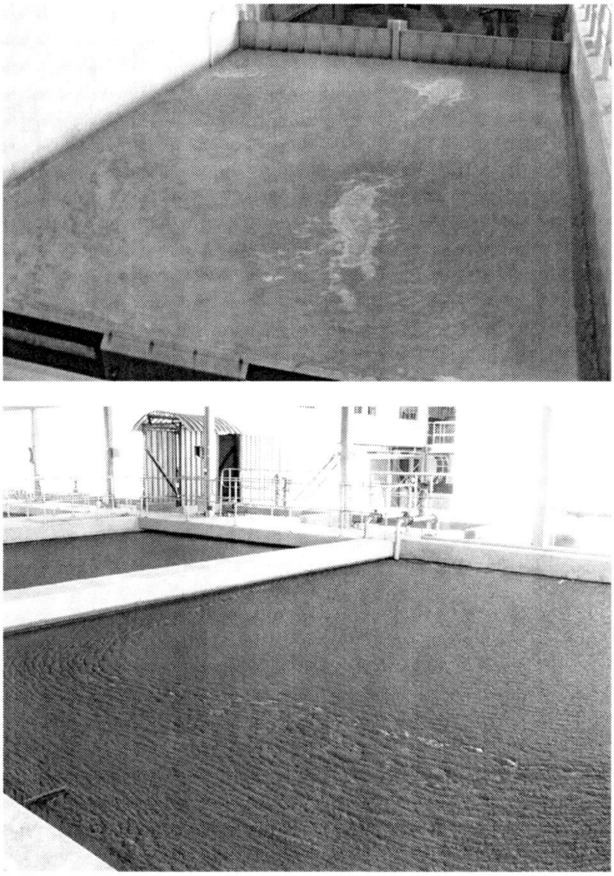

FIGURE 9-1 Two typical float layers prior to hydraulic removal. (The top shows a smaller DAF reactor at the Flemish Water Supply plant near Kluizen in Belgium, and the bottom, a larger DAF cell at the Goreangab plant in Windhoek, Namibia.)

hydraulic and mechanical methods, however, will waste less water than from a settling tank (Stevenson, 1997).

The dry solids content of the floated sludge is an important consideration. Besides the floated sludge, there are other waste streams from the treatment plant to consider, along with their final form and point of disposal. The overall strategy of how to deal with the treatment plant waste streams will determine the importance of the concentration of the floated sludge. The dry solids content of hydraulically removed sludge can be as high as 0.8 percent (Stevenson, 1997), but is more commonly reported as 0.5 percent (AWWA and ASCE, 2005) or even 0.3 percent (MWH, 2005). The experience in the Netherlands indicated a range between 0.15 and 0.5 percent (Schers, 1991). The dry solids content of mechanically removed sludge can be substantially higher. Values as large as 10 percent are reported, but these concentrated sludges are not practical for handling by gravity after removal. The sludge, as it comes off the beach plate, may be so thick that it requires the addition of flush water to make it flow under gravity. A maximum practical limit of 5 percent is reported (Stevenson, 1997), while typical operational values are in the range of 1 to 3 percent (Gregory and Edzwald, 2010; MWH, 2005). The difference between hydraulic and mechanical removal is large in terms of sludge volume and water wastage. A typical dry solids content of 0.3 percent for hydraulically removed sludge means that the sludge volume will be ten times more in comparison with the same float layer removed mechanically at say 3 percent dry solids. The larger sludge volume is not necessarily a problem. For sludge lagoons, with recovery of the decanted water, the sludge volume and water wastage do not really matter—the size and cleaning frequency of the lagoons are dictated by the mass of dry solids. For gravity thickeners, it is pointed out that the thickness of the sludge does not necessarily affect the size of the sludge thickener if it is limited by the underflow flux (Stevenson, 1997).

There are some common design guidelines to heed, whether the float layer is removed hydraulically or mechanically. The gelatinous nature of the float layer causes it to stick to the sides of the separation zone. When the float layer moves relative to the tank, it has to break loose from the sides, with inevitable knockdown as a result. By spraying a thin sheet of water onto the edge of the float layer, or trickling some water down the sides of the tank, the float layer is cut away from the sides, which not only prevents unnecessary knockdown, but also clears the float layer quicker if the float layer is hydraulically removed. It is also suggested that the channels used to lead the sludge away are designed with a shaped bottom profile to ensure fast flow of the sludge to minimize sludge buildup, with overflow outlets rather than bottom outlets to prevent the accumulation of foam. DAF reactors should always be covered with roofs or light-weight plastic domes to prevent float layer disruption by wind or rain, or freezing in cold climates (Stevenson, 1997).

9-3 HYDRAULIC FLOAT LAYER REMOVAL

There is a remarkable lack of published guidelines on the design of hydraulic removal systems, despite the fact that it was the most common method of float layer removal in Scandinavia and Finland during the early years of DAF (Rosen and Morse, 1976) and remains a popular method all over the world.

Hydraulic removal is effected by one of two methods. The first and simplest method is to close the DAF effluent outlet valve. The water level in the DAF reactor rises at the nominal hydraulic loading until the sludge overflow weir is reached. The second method requires the DAF reactor to be operated at a level slightly higher than the sludge weir, with outlet control on the sludge channel. During normal operation, the sludge-channel outlet valve is closed and the sludge channel is submerged. When float layer removal is required, the sludge-channel outlet valve is opened, which starts the hydraulic removal process. At the same time, the DAF effluent valve is closed to sustain the flow toward the sludge channel. In both cases, the water level stabilizes at the point where the flow rate into the DAF reactor equals the flow rate out. For the first method, the flow rate over the weir therefore starts out slow and reaches a maximum when the equilibrium level is reached. For the second method, the initial flow rate over the weir may be more or less than the sum of the raw water and recycle flow rates, but eventually reaches equilibrium as described above.

There is a concern about the flow variation on the other DAF reactors when one is being flushed. This is the same concern when one filter from a bank of filters is taken out of service for backwash. It suggests that a minimum of three or four DAF reactors should be used when hydraulic float layer removal is used. More DAF reactors reduce the flow variation.

DAF performance is less sensitive to flow rate changes than filtration, so this is not a major concern. If it is, the washing of a filter or the flushing rate of a DAF reactor can be accompanied with a reduction of the raw water flow rate.

Should the float layer flushing be continuous or intermittent? An important consideration is to provide adequate overflow depth. The float layer, as it reaches the sludge channel, breaks up in clumps which may get stuck on the edge of the weir if the overflow depth is too shallow. A practical suggestion is to provide a minimum overflow depth of at least 20 mm (Stevenson, 1997). Continuous flushing at an overflow depth of 20 mm will lead to an unacceptably large water loss of roughly 20 percent for an average-sized flotation cell. Hydraulic flushing is therefore done intermittently.

The frequency of flushing is determined by the raw water quality since this governs the chemical treatment and consequently the rate of sludge solids production. A stable float layer can be left for a long period without flushing, with only little deterioration in water quality. An example is provided in Fig. 9-2. An early report found that the float layer could be satisfactorily flushed at intervals of 12 h for a particular type of water (Zabel, 1978). This correlates with the current practice at South African treatment plants dealing with algal-rich raw waters, where the average flushing frequency is between 8 and 24 h for water abstracted from a river (Kruger, 2011) and between 4 and 10 h for water abstracted from a reservoir (Strydom, 2011). The main reason for not waiting longer between flushes has nothing to do with the stability of the float layers, but is to prevent the float layer from

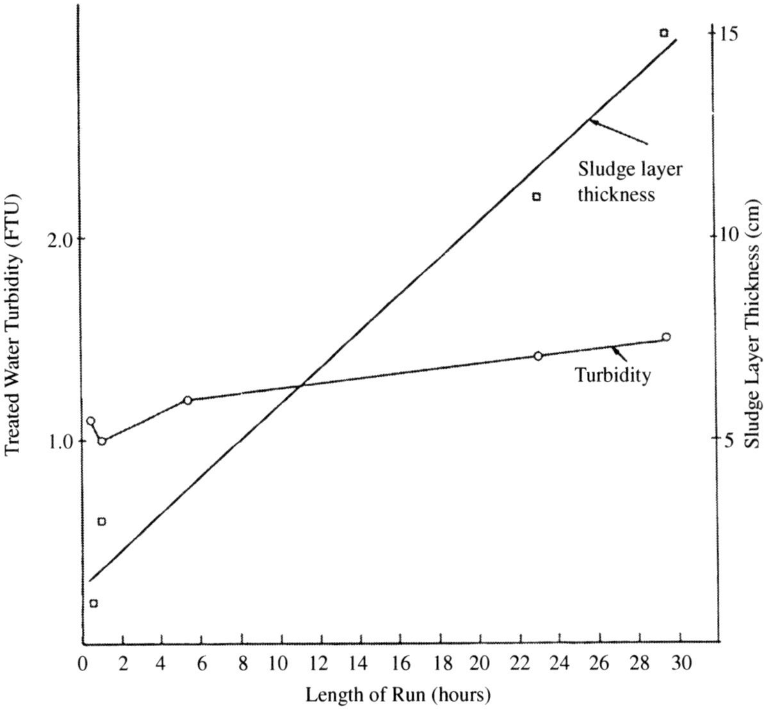

FIGURE 9-2 The effect of float layer retention on float layer thickness and DAF effluent turbidity (*Source: from Zabel and Rees, 1976*).

going anaerobic in warm weather. A secondary reason is that longer periods between flushing leaves a partly desiccated-float layer which does not break up evenly, and therefore does not move as easily across the tank surface.

Although strictly not related to float layer removal, a comment is appropriate on the bottom desludging of DAF reactors. It is good design practice to allow a means of draining out the bottom sludge, which inevitably builds up over time to a smaller or larger extent. A DAF plant treating river water in South Africa observes an inverse correlation between the need to withdraw bottom sludge and the sludge from the float layer. During times of high mineral turbidity, the draining of the bottom sludge becomes more important; during high-algal concentration with low-mineral turbidity, there is little or no bottom sludge (Kruger, 2011).

The water loss is proportional to the total time of flushing. The aim should therefore be to wait as long as possible between flushes, and to make each flush as short as possible. Designers can take some precautions to minimize the water loss due to flushing.

- The frequency of flushing should be an operational variable to allow operators to compensate for raw water quality changes by manual adjustment of a timer.
- The duration of flushing, likewise, should be under operator control, as the time required to clear the surface is variable. The typical time to clear an algal-rich float layer off the surface ranges from 2 to 10 min (Strydom, 2011; Zabel, 1978). An interesting observation is that the float layer clears faster on cool, humid days. On hot, dry days, the likelihood is larger for a thin dry crust to develop on the surface, which hinders the breakup and removal of the float layer (Kruger, 2011).

An unusual case was reported for a small DAF plant which was operated at a nominal hydraulic loading of 7.8 m/h, having no float-layer removal whatsoever. Every 72 h, when the float layer accumulated to about 150 mm, each tank was completely drained to get rid of the float layer, and the cycle started from scratch (Williams et al., 1985). With a DAF reactor depth of 3 m, this system has a low calculated water loss of 0.5 percent. For the more usual case of intermittent flushing, the upper limit of water losses should be about when there are, say, three flushing events of 8 min each over a period of 24 h. Batch removal of the float layer in this fashion may therefore be a practical option for small plants with one or two DAF reactors. The calculated water loss associated in this case is $(3 \times 8)/(24 \times 60) = 1.7$ percent. This agrees well with the practical experience in the Netherlands after some years of operation. Hydraulic removal of the float layer was found to have no negative effects on the DAF effluent quality, and therefore recommended in preference to scraping. The different sites which used hydraulic removal of the float layer showed similar results, namely water losses of 1.0 to 1.6 percent (Schers, 1991).

9-4 MECHANICAL FLOAT LAYER REMOVAL

There are three types of mechanical scrapers used in DAF as illustrated in Fig. 9-3. The first is a chain-and-flight scraper, with scrapers fixed at regular intervals to a continuous chain mounted above the DAF reactor. The second is a reciprocating scraper, with a number of scrapers fixed to a rigid frame which oscillates above the reactor—it pushes the float layer toward the outlet for some distance, pulls out, reverses, plunges into the float layer for the next cycle, etc. Both chain-and-flight and oscillating scrapers may cover the entire DAF reactor surface, or only the last part of it. The third type is a beach scraper, which is a simpler, revolving unit mounted directly above the beach plate which pushes the edge of the float layer into the sludge channel (MWH, 2005; AWWA and ASCE, 2005). For float

Chain-and-Flight Scraper

Reciprocating Scraper

Beach Scraper

FIGURE 9-3 Pictures of the three types of mechanical scrapers (chain-and-flight scraper on a pilot unit, others for full-scale plants).

layers with high solids content of 1 to 3 percent, the sludge may be too thick to flow on its own toward the beach plate—a beach scraper is not appropriate in this case (Gregory and Edzwald, 2010).

For chain-and-flight and reciprocating scrapers, the depth of penetration of the blades is important. If the blades do not penetrate far enough, the delicate matrix of the float layer is disturbed with solids settling back into the separation zone and into the outlet. If the blades are set too deep, they could create secondary-flow patterns directly below the float layer, detrimental to floc separation and effluent quality (AWWA and ASCE, 2005). It is a tricky design problem to keep the depth of blade penetration constant in relation to the depth of the float layer. The water level in the DAF reactor varies with the flow rate, because the effluent withdrawal is done through orifices in manifolds or suspended floor plates. Some degree of flow variation through a treatment plant is inevitable, which means that the water level in the DAF reactors is not constant. Some means must be provided to maintain the same blade penetration during flow variations. This requires an adjustable outlet weir—an additional mechanical complication which needs careful adjustment.

The scraper speed should closely match the forward velocity of the water in the separation zone, to minimize scraper-induced flow disturbances. It was suggested in Sec. 8-7 that the cross-flow velocity at the entrance to the separation zone should be between 40 and 100 m/h. For stratified flow in the separation zone, it was also shown that the incoming water stays in the top layer, moving at approximately the same speed until it reaches the outlet end of the tank. The cross-flow velocity is therefore a good indicator of what the scraper speed should be. Practical scraper speeds are indeed within this range. For algal water, a scraper speed of 60 m/h was recommended (De Wet, 1980)—a recommendation echoed for scraping in general (Stevenson, 1997). For a float layer from colored, low-alkalinity water, the optimum scraper speed was 30 m/h (Gregory and Edzwald, 2010). An example of the effect of scraper speed on the effluent quality is shown in Fig. 9-4.

Scrapers allow a choice between continuous and intermittent scraping. An important consideration, as with hydraulic sludge removal, is the stability of the float layer. For stable float layers from algal waters, hours may elapse between scraping intervals. For more fragile float layers, the scraping intervals have to be reduced. There is a correlation between the solids content of the sludge and the scraping interval. If the scraping intervals allow long periods without scraping, the float layer is allowed to thicken on the surface of the tank. With short intervals or continuous scraping, the float layer does not have time to consolidate and thinner sludge ends up in the sludge channel.

Figure 9-5 shows the surge in turbidity that follows a period of scraping. For this example, the DAF effluent turbidity returned to normal after about 15 min. This suggests that the intervals between scraping should be made as long as possible. A recent study,

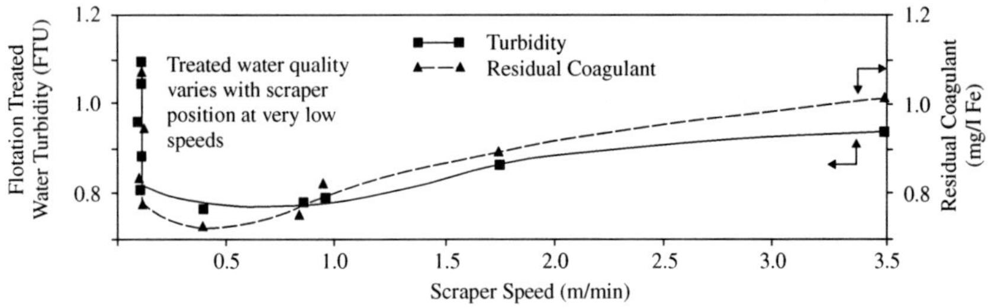

FIGURE 9-4 The effect of scraper speed on the DAF effluent quality (*Source: from Rees et al., 1979*).

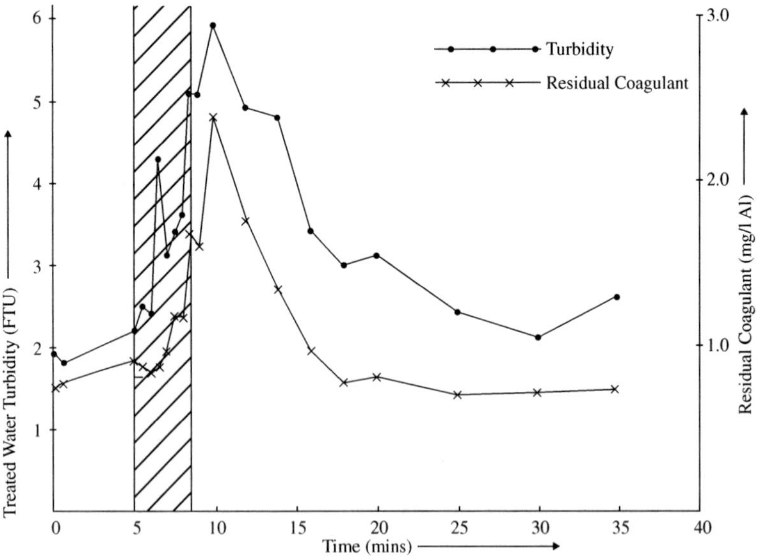

FIGURE 9-5 The deterioration in water quality following a period of float layer scraping during the period between 5 and 8 min (*Source: from Zabel and Rees, 1976*).

which used particle counting (5 to 10 μm size range) in the water below the float layer, studied this question in more detail for a full-scale plant with a reciprocating scraper (Kim et al., 2007). It was found that longer resting periods cause a smaller increase in particle counts. This is attributed to the fact that the longer resting time caused a denser, stronger float layer which can better withstand the shocks induced by the scraper movement. With a constant resting period fixed at 30 min, two successive movements cause less disruption than four or six successive movements. A final finding is that more particles are measured in the vicinity of the beach plate, which suggests that equipment designers should pay special attention to the transfer of the floated material onto the beach plate.

The float layer can be stabilized with a grid, constructed of galvanized sheet metal. The grid provides square cells of 150 mm × 150 mm when seen in plan, and is 150 mm deep (Van Vuuren and Van der Merwe, 1988). The grid serves two purposes. First, the float layer is squeezed upward through the cells which results in greater solids content than that would have been found without the grid. Second, the scraper can skim the top of the float layer protruding above the grid without the usual "rolling" of the float layer, with less effect on the quality of the DAF effluent. The grid has some disadvantages. The upward force on the grid by the buoyant float layer is large, with obvious complications for structural design and costs. Moreover, the long-term maintenance of thin metal strips in a corrosive environment is costly, and the use of the grids had been discontinued at the treatment plants in South Africa where they had been installed.

REFERENCES

AWWA and ASCE (2005), in E. E. Baruth, ed., *Water Treatment Plant Design*, 4th ed., New York: McGraw-Hill.

Bratby, J. R., and Jones, G. (2007), How far can you take flotation thickeners? *Proceedings of the 5th International Conference of Flotation in Water and Wastewater Systems*, September 11–14, Seoul, South Korea, 231–238.

De Groot, C. P. M., and Van Breemen, A. N. (1987), *Ontspanningsflotatie en de Bereiding van Drinkwater (DAF and the Treatment of Drinking Water)*, Report 11, Faculty of Civil Engineering, Delft University of Technology, July 1987.

De Wet, F. J. (1980), *Flotation of algal waters*, Paper presented at a meeting of the Institute of Water Pollution Control, Pretoria, IWPC, 1–28.

Gregory, R., and Edzwald, J. K. (2010), Sedimentation and flotation, in J. K. Edzwald, ed., *Water Quality and Treatment: A Handbook on Drinking Water*, 6th ed., New York: AWWA and McGraw-Hill.

Jönsson, L., and Lundh, M. (2000), *Visualization of Particles and Bubbles in a Flotation Basin Using Borescope*. Report 3229, Department of Water Resources Engineering, Lund University, Sweden.

Kim, H., Kim, T., and Kim, H. (2007), Effects of scraper's operation on efficiency in the dissolved-air-flotation process, *Proceedings of the 5th International Conference of Flotation in Water and Wastewater Systems*, September 11–14, Seoul, South Korea, 371–374.

Kruger, M. (2011), Personal communication, Midvaal Water Company, Stilfontein, South Africa.

MWH (2005), in J. C. Crittenden, R. R. Trussell, D. W. Hand, D. W. Howe, and G. Tchobanoglous, eds., *Water Treatment Design: Principles and Design*, 2nd ed., Hoboken, NJ: John Wiley & Sons.

Rees, A. J., Rodman, D. J., and Zabel, T. F. (1979), *Water Clarification by Flotation – 5, Technical Report TR114*, Water Research Centre, Medmenham, April 1979.

Rosen, B., and Morse, J. J. (1976), Practical experience with dissolved air flotation on various waters in Sweden and Finland, *Flotation for Water and Wastewater Treatment, Proceedings of a Water Research Conference*, Felixstowe, 1977, 338–338.

Schers, G. J. (1991), *Flotatie: de theorie en de praktijk (Flotation: Theory and Practice)*, Thesis submitted to the Delft University of Technology, August 1991.

Stevenson, D. G. (1997), *Water Treatment Unit Processes*, London: Imperial College Press.

Strydom, B. (2011), Personal communication, Magalies Water, Vaalkop, South Africa.

Van Vuuren, L. R. J., and Van der Merwe, P. J. (1988), Thickening of sludges by dissolved air flotation, *Water Science and Technology*, 21, 1771–1774.

Williams, P. G., Van Vuuren, L. R. J., and Van der Merwe, P. J. (1985), Dissolved air flotation upgrades a conventional plant treating eutrophic water, *Proceedings of the 11th International Convention of the Australian Water and Wastewater Association*, Melbourne, 1985, 189–196.

Zabel, T. F., and Rees, A. J. (1976), The WRC flotation working group's 2300 m^3/d pilot plants, *Proceedings of the Water Research Conference, Flotation for Water and Wastewater Treatment*, Felixstowe, 1977, 245–293.

Zabel, T. (1978), Flotation. *Proceedings of the 12th Congress of the International Water Supply Association*, 1978, London, PF1-PF10.

CHAPTER 10
PROCESS SELECTION AND DESIGN

10-1 DAF SELECTION BASED ON RAW WATER QUALITY CONSIDERATIONS 10-1
Why Consider DAF? 10-2
Treatment Selection Guidelines 10-2
10-2 BENCH-SCALE TESTING 10-5
Performance Parameters, Pretreatment and Flotation Variables 10-6
Benefits 10-7
Conditions for DAF Jar Tests 10-7
10-3 PILOT-SCALE TESTING 10-9
The Need for Pilot Testing 10-9
Design Variables Appropriate for Pilot Testing 10-10
Examples of Pilot Testing 10-11
10-4 SUPPLEMENTAL MEASUREMENTS 10-12
Zeta Potential 10-13
Streaming Current 10-14
Flocculation Index 10-15
Particle Size Distributions 10-15
Volume of Air Delivered 10-16
REFERENCES 10-16

We begin this chapter with general reasons to consider DAF for drinking water treatment and guidelines for selecting it or sedimentation for clarification. Next, we lay out what DAF operating variables can be evaluated with bench-scale tests and how to conduct the tests. Then we examine pilot-scale testing, and what operating and design variables can be evaluated and how to conduct the pilot tests. Finally, we describe the basis for and the information gained for some particle and floc measurements and for measurement of the air delivered in DAF.

The chapter serves those who are considering DAF for new water plants or in retrofitting existing plants. The chapter leads them through the first part, which is essentially a paper study of the feasibility of DAF, through bench-scale studies in which they can gain information rapidly and inexpensively to direct what will be studied during the pilot-scale phase. The chapter also serves those with existing DAF plants who wish to optimize and improve their performance through bench-scale or pilot-scale testing or both. The measurements described in the last section provide useful information for both new and existing DAF facilities.

10-1 DAF SELECTION BASED ON RAW WATER QUALITY CONSIDERATIONS

For water utilities and consulting engineering companies involved in selecting water clarification processes for new water plants or in the expansion of existing plants, it is helpful to have some guidance on process selection with respect to raw water quality. This is addressed below, but first we present general background material on the reasons why DAF should be considered as a clarification process.

Why Consider DAF?

There are several reasons for considering DAF for clarification of waters rather than sedimentation. They are listed and then discussed: (1) low DAF effluent turbidities and particle counts, (2) low particle loading to the filters, (3) high removal of algae, (4) removal of some taste and odor compounds, (5) high removal of pathogens, (6) rapid start-up and adjustment to flow changes, (7) relatively high sludge solids, (8) small flocculation tanks, and (9) small plant footprint.

DAF is more efficient in removing low-density floc than sedimentation processes. For well-operated DAF plants, it is common to find DAF effluent turbidities of 0.2 to 0.5 NTU. Because of the lower turbidity and lower number of particles being applied to filters, one can design granular media filters at higher filtration rates yielding smaller filter areas. If plants choose to use lower filtration rates, then one can obtain long filter runs with less frequent filter backwashing.

DAF is particularly effective in removing algae from water supplies. This prevents problems of algae clogging filters and reducing filter run times that can occur in other types of plants. General discussion and case studies are presented in Chap. 11 on the removal of algae. The air bubbles in the DAF tank can also strip some taste and odor compounds from the water, which is a major secondary benefit of the process.

DAF is more effective than sedimentation in removing pathogens such as *Giardia* cysts and *Cryptosporidium* oocysts. DAF is the main barrier in the removal of these type of pathogens, and filtration serves as a polishing process (Edzwald et al., 2000a; 2000b; 2003). The ability of DAF to remove these pathogen cysts is covered in Chap. 11.

Conventional DAF plants have a smaller footprint compared to sedimentation plants. There are several reasons. The flocculation tanks are smaller. The DAF tanks are smaller, especially compared to conventional low-rate sedimentation, and also compared to high-rate tube and plate sedimentation. In some applications, DAF is placed over filtration reducing the plant footprint further.

DAF can produce floated sludges with a higher percent solids compared to sedimentation process sludges thus possibly reducing sludge treatment and disposal. Additional information on the floated sludge is found in Chap. 9.

Although there are several reasons for considering DAF as listed above, one must consider the energy required to dissolve air in water by pumping the recycle water to the saturator under pressure, the static head between the recycle pump and the saturator, frictional head loss, and the energy used for the air compressor supplying the air. For reference, the energy required is about 1.3 kW per ML/d (5 kW per MGD of plant flow) (Crossley, 2010). The energy is for saturator pressure conditions used for design and would be less for lower operating pressure conditions. It is common to find that the overall energy requirement for a DAF plant is about the same or less than a sedimentation plant. This is because there are energy savings for DAF plants through smaller motors to run the flocculators with the smaller flocculation tanks, and a large savings in energy because of longer filter runs and less filter backwashing of filters. The energy required for the entire plant must be evaluated, but this provides some guidance.

Treatment Selection Guidelines

There are several unit processes, or combinations of processes, that can be used to treat a given raw water supply and achieve specified treated water quality goals. It is important for water utility managers and design engineers to identify the most promising configurations at the earliest stage of planning. This strategy saves time and ultimately treatment costs by directing the design work to the most promising treatment processes. The strategy identifies what bench-scale and pilot-scale studies should be undertaken. Here, treatment selection guidelines are presented for direct filtration plants and filtration plants with sedimentation or DAF as pretreatment. This material is developed from the published literature in which both the authors had major roles. Before presenting the guidelines, some brief comments are made about the literature.

Haarhoff et al. (1992) presented a process selection diagram based on South African experience of the treatment of waters with algae versus those with turbidity. Thus, the guidelines were presented in terms of treatment plant type as functions of raw water chlorophyll *a* and turbidity. Treatment plant types included: slow sand filtration, direct filtration, two-stage filtration, DAF followed by filtration, and sedimentation followed by filtration. The diagram was qualitative showing preferred treatment plant types with increasing chlorophyll *a* and turbidity. For example, slow sand filtration and direct filtration were suggested for treating waters with low values of chlorophyll *a* and turbidity. For higher chlorophyll *a* and for turbidity from algae or low-density particles, DAF was preferred. For higher-turbidity waters, especially those containing silts, sedimentation was preferred.

Janssens and Buekens (1993) used the above concept and developed a treatment process selection diagram with boundaries between preferred processes for specified values of chlorophyll *a* and turbidity. They chose the boundaries based on European experience and their judgment, and like Haarhoff et al. (1992) they considered treatment of waters with algae versus turbidity. For example, sedimentation followed by filtration was recommended for waters with turbidities of >100 NTU regardless of the concentration of chlorophyll *a*. For raw waters with turbidity between 20 and 100 NTU, the boundaries for DAF versus sedimentation depended on the chlorophyll *a* and turbidity values, while DAF was preferred for raw waters with an algae problem and turbidities <20 NTU.

Building upon the above work, Valade et al. (2009) prepared two treatment selection diagrams. These are presented and recommended. Rather than examining chlorophyll *a* (a parameter that many water utilities do not measure), the diagrams are based on routinely measured parameters of turbidity and organic matter content, the latter expressed as TOC and true color. UV_{254} absorbance as a surrogate of dissolved NOM in water was also considered although not shown on the diagrams. This parameter is explained in Chap. 6. The diagrams were developed based on raw water quality data and plant types for 400 plants in the United States and Canada. The database was supplemented with data from equipment manufacturers and pilot-plant studies for water utilities that recently integrated new treatment schemes such as DAF and ballasted-sand sedimentation.

The two process selection diagrams are for average raw water quality (Fig. 10-1) and maximum raw water quality conditions (Fig. 10-2). Direct filtration (with or without flocculation [contact filtration]) is recommended only for the highest quality water supplies with low average levels (Fig. 10-1) of turbidity (<5 NTU) and low levels of organic matter (true color <20 CU and TOC <3 mg/L). Valade et al. (2009) relate the TOC of 3 mg/L to UV_{254} <0.07 cm^{-1} as an operating surrogate parameter for TOC. Maximum raw water quality values in Fig. 10-2 are defined as those occurring for a short-period event. Maximum turbidities for selection of direct filtration should not exceed 30 NTU and maximum color and TOC should not exceed 35 CU and 5 mg/L.

Although direct filtration is suitable for treating high quality type supplies, it has the disadvantage of having only one physical barrier to prevent the passage of pathogens. For this reason, its selection is less frequent than in the past. One option is to consider it along with additional disinfection barriers such as ultraviolet light disinfection.

DAF is recommended for the following average raw water quality conditions based on Fig. 10-1: supplies with mineral turbidity at levels <10 NTU but up to 100 NTU for

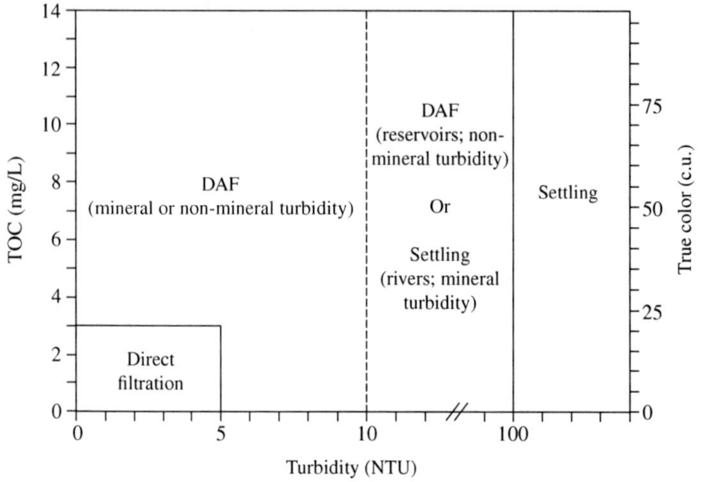

FIGURE 10-1 Process selection diagram based on average raw water quality conditions (*Source: reprinted from Valade et al. (2009), Journal of Water Supply: Research and Technology AQUA, 58 (6), 424–432, with permission from the copyright holder, IWA Publishing).*

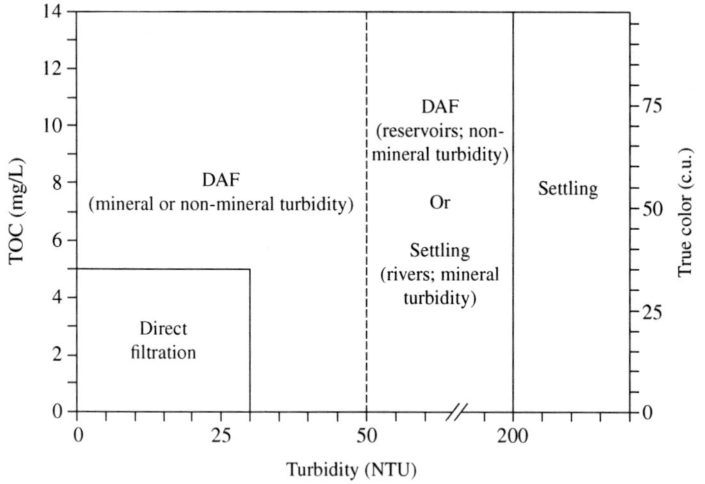

FIGURE 10-2 Process selection diagram based on maximum raw water quality conditions (*Source: reprinted from Valade et al. (2009), Journal of Water Supply: Research and Technology – AQUA, 58 (6), 424–432, with permission from the copyright holder, IWA Publishing*).

supplies with non-mineral turbidity caused by organic matter such as algae. There is no upper limit on true color and TOC for DAF selection. For maximum turbidities, DAF is recommended based on Fig. 10-2: supplies with mineral turbidity up to 50 NTU and up to 200 NTU for non-mineral turbidity. In short, DAF plants are robust and can treat high quality water supplies, can treat supplies of low to moderately high turbidity depending on its nature, and can treat supplies of low to high TOC.

As Figs. 10-1 and 10-2 show, conventional sedimentation processes (including high-rate plate and tube settling) are recommended for water supplies with average mineral turbidities >10 NTU or supplies with mineral and non-mineral turbidities >100 NTU. For maximum turbidities, sedimentation is recommended for supplies with mineral turbidities >50 NTU or supplies with mineral and non-mineral turbidities >200 NTU. Solids-contact sedimentation processes such as ballasted-sand, floc-blanket clarification, and upflow-filtration through plastic media (also referred to as adsorption clarification) were also considered and discussed by Valade et al. (2009). They are not identified on the process selection diagrams. They perform well over a wide turbidity range, but they are not as good as DAF in treating high color and TOC supplies.

In summary, DAF is a good clarification process for the treatment of reservoir water supplies. DAF can be considered a *Best Available Technology* for clarification of water supplies with algae. It is also a good process for the treatment of rivers with low mineral turbidity and rivers with natural color.

10-2 BENCH-SCALE TESTING

A bench-scale DAF jar tester is recommended for water utilities that have existing or proposed DAF plants, consultants involved in DAF design, and DAF equipment companies. Figure 10-3 shows a picture of a DAF jar tester used by the first author for many years. This particular one was made in England, but it is no longer available commercially. EC Engineering (Edmonton, Canada) is one company (there may be others in other countries) that has a 6-jar batch unit that comes with a pressure-vessel saturator. The following topics on bench-scale testing are presented below: (1) what performance parameters are evaluated on an absolute basis versus a relative basis, and what pretreatment and flotation variables can be evaluated, (2) the benefits of bench-scale testing, and (3) how to conduct DAF jar tests.

FIGURE 10-3 DAF bench-scale jar test unit.

Performance Parameters, Pretreatment and Flotation Variables

Bench-scale testing (or DAF jar tests) is done in a batch mode of operation—no flow of water through the process. The scale of the process or flow does not affect chemistry for reactions that are completed within the time period of the tests (i.e., kinetics of reactions are not important). So chemical reactions that take place in seconds or minutes can be studied in batch bench-scale tests and applied to flow-through systems such as pilot plants and full-scale plants. Therefore, the coagulation conditions of dosing and pH selected from bench-scale tests apply to pilot and full-scale plants (refer to Sec. 6-6, for optimum chemistry conditions for the various coagulants). It also means that performance parameters that depend only on chemistry and not on flocculation or flotation can be evaluated on an absolute basis. These parameters are those that involve dissolved constituents. The main one and the one that should be tested on a routine basis is UV_{254} (surrogate for NOM). The removal of dissolved organic matter depends entirely on coagulation chemistry. With coagulation, the dissolved organic matter is converted to particles which are filtered by a membrane as part of the sample treatment for measurement of UV_{254}. Therefore, the treated water UV_{254} found in bench-scale tests applies directly to pilot-scale and full-scale plants for the same coagulation conditions. Other performance parameters dependent entirely on coagulation chemistry are DOC and dissolved residual coagulant.

Turbidity is a performance parameter that depends on coagulation chemistry and the physical processes of flocculation and flotation. These physical processes are affected by the hydraulics of flow-through tanks (ideal versus non-ideal flow, extent of mixing), flocculation detention time, and DAF hydraulic loading. The floated water turbidity from batch-scale testing cannot be applied directly to what would be achieved by flow-through systems. Therefore, turbidity is evaluated in bench-scale testing on a relative basis, not on an absolute basis.

In principle, can flocculation mixing intensity and detention time be studied in bench-scale tests? Common velocity gradient (G) values of 50 to 100 s^{-1} are used in practice (see Sec. 6-9). While some study of mixing intensity can be evaluated in bench-scale tests, it is recommended that it be examined in pilot-scale, and more attention devoted in bench-scale for testing the effect of flocculation time. Flocculation time in the range of 5 to 20 min is recommended for bench-scale study (see Chap. 6).

Flotation tank variables that may be considered for bench-scale testing are saturator pressure, recycle ratio, and flotation detention time. Hydraulic loading, and not detention time, is the basis for sizing DAF tanks but it cannot be studied in bench-scale testing. The flotation time is the complementary variable for bench-scale testing. A time of 10 min is recommended as a standard value. Some variation of the time can be incorporated into bench-scale studies. The amount of dissolved air delivered to the flotation jars can be varied through the saturator pressure and the recycle ratio. It is recommended, however, that saturator pressure be held constant at about 500 kPa (saturator pressures vary typically in practice between 400 and 600 kPa [see prior chapters]) because of differences in efficiencies of saturators and nozzles for bench-scale versus full-scale plants. Recycle ratio can be varied in bench-scale testing. Recycle ratios in the range of 6 to 12 percent are recommended for study. Higher values should be considered for water supplies with high turbidities or high particle counts, and to evaluate the use of DAF with powdered activated carbon (PAC).

As stated above, an important flotation tank variable that cannot be studied in bench-scale tests is the hydraulic loading on the DAF tank; it must be evaluated in pilot-plant evaluations. Likewise, the effect of flotation on filter performance (filter effluent quality and filter run time) cannot be studied at bench-scale.

Benefits

Bench-scale testing can be done more rapidly and inexpensively than pilot-scale studies. The main variables that can be studied are (1) coagulation, (2) pretreatment flocculation mixing and especially time, and (3) recycle ratio. It is best to study coagulation at the bench-scale in which various coagulants can be evaluated and optimum dosages and pH conditions can be determined.

The data from the bench-scale can be used to focus the pilot-plant work by eliminating some variables and selecting those variables for evaluation at the pilot-scale including values or ranges of values. Optimum coagulation conditions for one or more types of coagulants can be confirmed at the pilot-scale. Pilot-scale studies should focus on the effect of hydraulic loading on DAF performance and filter performance (filter effluent quality and filter run time). Pilot-scale studies are often required to demonstrate effective treatment in response to seasonal changes in raw water quality such as turbidity events, algae problems, TOC, and water temperature. Effects of seasonal changes in water quality on coagulation dosages and pH conditions are best determined at bench-scale and selected for pilot-scale where it can be fine-tuned, if necessary. This would allow seasonal pilot-scale studies to focus on other variables such as flocculation time or hydraulic loading.

It is recommended that existing full-scale DAF plants make use of bench-scale DAF testing. The bench-scale unit can have many of the same benefits described above in assessing various coagulants and determining optimum coagulation dose and pH. It can also provide valuable data on the effect of recycle ratio on flotation performance, for example, with changes in raw water quality. In addition, one can vary the recycle ratio and determine any improvement in floated water turbidity with increasing recycle ratio. The recycle ratio at which there is no further improvement indicates an optimal condition at which there is essentially complete coverage of the flocs with bubbles.

Conditions for DAF Jar Tests

The following conditions are recommended for DAF jar tests.

Rapid Mixing: Add the coagulant using intense mixing (200 to 400 rpm depending on jar tester) for one minute. If pH adjustment is being studied, add the base or acid in the sequence with the coagulant following the recommendations in Sec. 6-4. Try to add the chemicals as quickly as possible, if it takes more than a minute to add the coagulant and pH adjusting chemicals to the several jars, then continue the rapid mixing period until all chemicals are added.

Flocculation: When evaluating full-scale plants use the same mixing intensity and flocculation time. If unknown, then use a paddle speed of 50 rpm (commercial bench-scale units usually provide relationships between mixing paddle speed and *G*; for the EC Engineering DAF jar tester a paddle speed of 50 rpm corresponds to a *G* of about 65 s^{-1} at 20°C). As stated above, it is better to study mixing intensity effects in pilot-scale. If flocculation time is being evaluated, then vary the time as desired—see above. If flocculation time is not being studied, a time of 10 min is recommended.

Flotation: As noted above the effect of saturator pressure should be studied at the pilot-scale and not in bench-scale studies. Saturator pressure should be held constant at the typical pressure used for an evaluation of an existing full-scale plant or at a saturator pressure of 500 kPa for studies for new plants. Recycle ratio can be evaluated in the range of 6 to 12 percent or higher as noted above. If not being studied, a recycle ratio of 8 percent is usually sufficient and is recommended. A flotation time of 10 min is recommended as a standard value. Some variation of the time can be incorporated, if desired, into bench-scale studies.

Raw Water Measurements: Measure the water temperature, pH, turbidity, and UV_{254} (surrogate for dissolved NOM or DOC). These parameters should be measured in all bench-scale tests. For specific studies, one may be interested in one or more of the following parameters: true color, TOC, total and dissolved Al or Fe depending on the coagulant under investigation. Data on the raw water alkalinity and knowledge of the buffer intensity (see Sec. 6-3) are also useful in assessing coagulant dosages and achieving the desired coagulation pH.

Measurements after Flotation: In routine jar tests the primary performance variables of turbidity and UV_{254} are measured. Other measurements such as true color, TOC, DOC, and residual coagulant may be made for specific studies. For turbidity and TOC sampling, samples should be taken from tubes near the bottom of the jars—commercial DAF jar testers come with sampling tubes for this purpose. Samples for UV_{254}, DOC, and true color may be taken from the bulk water of the jars since these samples are filtered before measurement—see Example 10-1. For soluble residual coagulant measurements, these samples can be handled the same way as true color and UV_{254}.

EXAMPLE 10-1 Determination of Optimum Alum Dosage from DAF Jar Tests DAF jar tests were run at room temperature (say, 20°C) to determine the optimum alum dosage for summer water quality and temperature conditions. The jar tests were carried out with pH control (addition of base [NaOH] immediately after alum) at pH 6.3 ± 0.2—within optimum pH conditions for alum coagulation, see Sec. 6-6.

Raw Water Quality: temperature 20°C, pH 6.83, turbidity 1.7 NTU, and UV_{254} 0.31 cm^{-1}.

Jar Test Conditions: rapid mixing at 400 rpm for 30 s to 1 min; flocculation mixing at 50 rpm for 10 min, saturator pressure at 500 kPa, recycle ratio at 8 percent, and flotation time of 10 min.

Measurements after Flotation:

1. Samples were taken for turbidity from the sampling tubes near the bottom of the jars.
2. Samples were taken for UV_{254} from the bulk water in the jars. These samples were filtered through a glass fiber membrane and UV_{254} measured according to Standard Methods (APHA et al., 2005).
3. Temperature was measured directly in one of the jars, and pH was measured directly from each of the jars.

Data Presentation and Discussion:
Turbidity and UV_{254} fraction remaining are plotted as a function of alum dosage in Fig. 10-4. The optimum dose is selected for conditions where the turbidity and UV_{254} reach minimum values such that increasing the alum dosage there is little change in performance. The optimum dosage is about 5 mg/L as Al. ▲

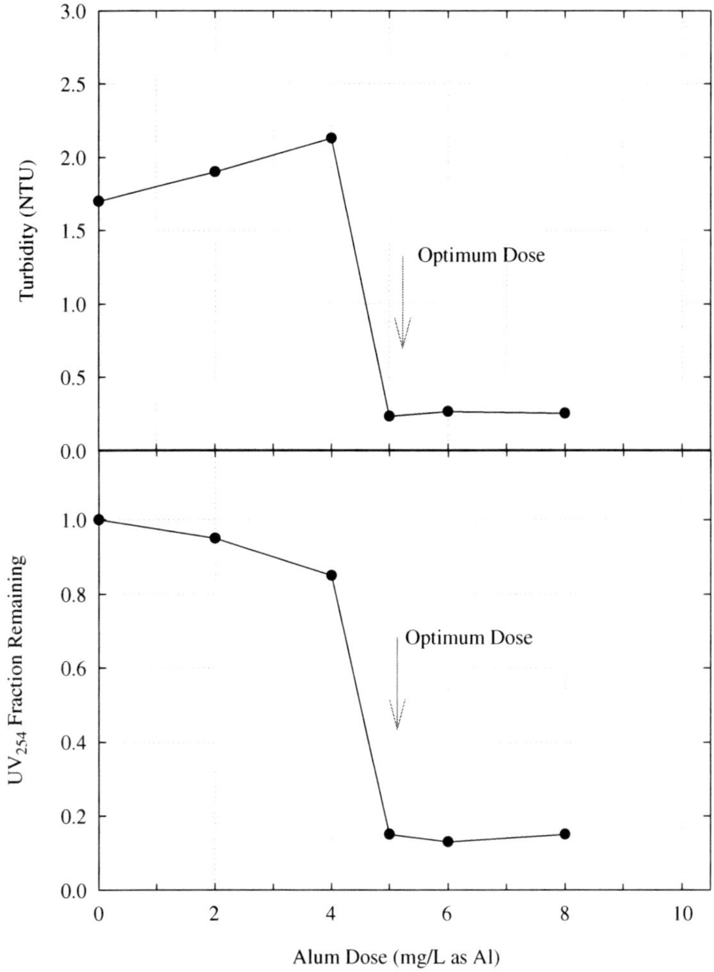

FIGURE 10-4 DAF jar test data showing turbidity and UV_{254} fraction remaining as a function of alum dose at pH 6.3; raw water: high TOC reservoir supply, turbidity 1.7 NTU, UV_{254} 0.31 cm^{-1}.

10-3 PILOT-SCALE TESTING

The Need for Pilot Testing

DAF as a technology has matured so much that the first question is whether pilot testing should be done at all? Pilot testing is expensive and could make up a large part of the project cost if the treatment facility is small. The extent and complexity of a typical DAF pilot plant is shown in Fig. 10-5. Moreover, the cost of pilot testing is roughly proportional to the period over which testing is conducted. Raw waters from a stable, oligotrophic

FIGURE 10-5 Layout of a modern pilot plant, showing the extensive facilities for proper chemical pretreatment and flocculation, with the DAF reactors in the far background (*Courtesy: Simon Breese, AECOM*).

impoundment may show little variation in quality and a pilot project of a few weeks will probably suffice. Raw waters from eutrophic impoundments or those subject to flood periods with high turbidity peaks, can have large and sudden quality changes and it is necessary to continue pilot projects for months to capture the swings in water quality. Even then, there is no guarantee that the worst scenario is captured.

When the water quality is well-characterized and DAF design parameters are used which fall within well-established, conservative ranges, there is generally no need for pilot testing, unless required by the regulatory agency. Pilot testing is recommended if any or more of the following is true (Pernitsky, 2010):

- The water supply is close to the limits for DAF shown in Figs. 10-1 and 10-2;
- The design parameters are pushing the envelope toward higher DAF hydraulic loading rates and/or lower flocculation times to reduce costs; and
- Unconventional process combinations are being considered, such as DAF combinations with adsorbents, oxidants, or membranes.

Design Variables Appropriate for Pilot Testing

It is customary to test DAF in conjunction with the subsequent filtration step. With DAF alone, its performance is measured at its outlet in terms of turbidity or particle counts. When piloted in conjunction with filtration, it is advisable to use filtered-water turbidity and unit filter run volume (UFRV) as the primary evaluation criteria, rather than DAF effluent turbidity. An UFRV target is set (say 400 m^3/m^2 [10,000 gal/ft^2] or greater), and the operating variables are systematically adjusted to find the point where the pilot unit fails to meet the UFRV target (Pernitsky, 2010).

With the coagulant selection and dose determined from bench-scale testing, three important design parameters remain to be optimized, namely flocculation time, hydraulic loading, and air dosing rate. It is difficult to vary the flocculation time independent of hydraulic loading. Directing the full flow through both the flocculation tank and the DAF reactor, the flocculation time is reduced when the hydraulic loading is increased. It is prudent to size the floc tank for a total detention time of 10 min (well-established as an adequate flocculation time for most plants) for the target maximum hydraulic loading rate and realize that lower hydraulic loadings could be studied by maintaining the same flow through flocculation and then wasting flow as it enters the DAF reactor. Due to the interrelationship between air dosing and hydraulic loading, explained in Secs. 8-7 and 8-8, the air dosing rate is fixed and the limiting hydraulic loading is found, followed by another setting of the air dosing rate, etc. With the best hydraulic loading and air dosing rate thus determined, the flocculation time is minimized, also in terms of the UFRV.

Most modern plants have a large degree of operational flexibility, for example, flocculation energy input, recycle flow rate, saturator pressure, etc. These variables are best optimized during commissioning of the full-scale plant and need not be necessarily fine-tuned during pilot testing. Some questions are best answered with testing in a laboratory, for example, the evaluation of different air injection nozzles, geometrical changes to the contact zone, etc. Visual observation of the DAF reactor during piloting remains important—much of our current understanding of DAF was pieced together from visual observation of the complex flow patterns in the DAF contact and separation zones, such as shown in Fig. 10-6.

Examples of Pilot Testing

A number of published papers provide good examples on the planning and execution of pilot studies, and how their results in most cases were used for the design of full-scale facilities. Table 10-1 provides a summary of these studies. The studies were performed with either pilot plants constructed by the utilities themselves or using mobile pilot units provided by equipment manufacturers or consultants. The pilot studies for drinking water

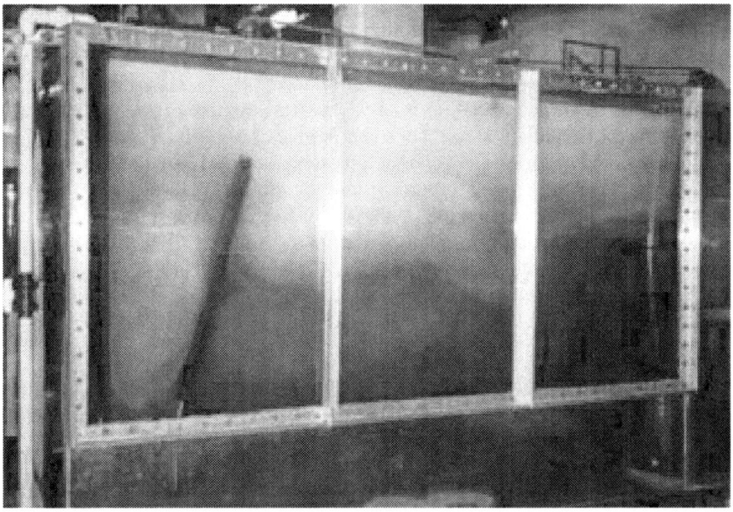

FIGURE 10-6 The contact and separation zones of a DAF pilot unit, viewed through the transparent side (*Courtesy: David J. Pernitsky, CH2M Hill Inc.*).

TABLE 10-1 DAF Pilot Studies Reported at International Conferences

Location	Objective	Reference
Winnipeg, Canada	Comparing DAF to direct filtration	Wobma et al. (1997)
	Optimize design parameters for high-rate DAF	Pernitsky et al. (2007)
West Chester, Pennsylvania	Verify the suitability of DAF	O'Connell et al. (1997)
Wachusett Reservoir, Massachusetts	Maximize hydraulic loading	Shawcross et al. (1997)
	Optimize design parameters	Crossley et al. (2001)
Cambridge, Massachusetts	Comparing DAF to sedimentation	Prendiville et al. (1997)
Croton, New York	Select treatment train, optimize design parameters	Nickols et al. (2000)
Lower Lliw Reservoir, Wales	Select treatment train, optimize design parameters	Miles et al. (2000)
Hemlocks Reservoir, Connecticut	Assess DAF for *Cryptosporidium* removal	Edzwald et al. (2000a)
Five U.S. sites	Optimize DAF design parameters for treatment of spent filter backwash water	Eades et al. (2000)
Rhode Island	Assess DAF for membrane pretreatment	Farmerie (2007)
San Luis Obispo, California		

treatment were run intermittently for the various seasons often taking a year or more to complete. For special problems like the treatment of filter backwash water, the testing time was shorter. Almost all the drinking water studies were piloted in tandem with the next process step—either GAC or granular media filters, or membranes. In the cases with GAC filters, intermediate ozonation was also included.

A variety of performance indicators were used in the studies listed. For DAF-granular filter combinations, water was sampled after both processes and tested for turbidity and particles. When DAF was compared to direct filtration, the UFRV and the filter head loss development rate were used as performance indicators. To quantify the benefits of DAF as a membrane-pretreatment process, the flux rate, backwashing time, and the chemical-enhanced backwash frequency of the membranes were considered.

The studies quoted in Table 10-1 reported favorably on DAF. Where DAF was compared to sedimentation, DAF was found to perform more consistently throughout the year, notably during times of high algal activity. When compared to direct filtration, DAF brought an improvement in the filter performance in terms of higher UFRV and lower head loss development rate. During summer, high DAF hydraulic loading rates could be sustained, but often had to be reduced during winter conditions.

10-4 SUPPLEMENTAL MEASUREMENTS

There are several supplemental measurements that are helpful in assessing DAF treatment and improving performance. The following are discussed: (1) zeta potential, (2) streaming current, (3) particle size distributions, (4) flocculation index, and (5) the amount of air delivered by the saturation system.

Zeta Potential

Zeta potential is calculated (some instruments display the calculated value) from electrophoretic mobility (EPM) measurements. A small water sample containing the particles or flocs is inserted into an electrophoresis cell. The cell contains two electrodes and when an electric field is applied, the particles or flocs move with a certain velocity (mobility) to the electrode of opposite charge of the particles. This is measured as the EPM that accounts for the applied voltage and is reported in units of (μm/s)/(V/cm). The EPM is proportional to the zeta potential of charged particles, and both are indicators of the magnitude of the particle or floc charge. The zeta potential is calculated from the Smoluchowski equation, which applies to water treatment applications (Gregory, 2006), where ζ_p is the zeta potential, μ_w and ε are the water's dynamic viscosity and permittivity.

$$\zeta_p = \frac{EPM \times \mu_w}{\varepsilon} \tag{10-1}$$

Inserting values for μ_w and ε at 20°C, the ζ_p in mV is simply about 14 times the EPM. Several commercial instruments are available such as the Zeta-Meter and Zetasizer. Figure 10-7 shows a picture of a Zeta-Meter instrument.

(a) (b)

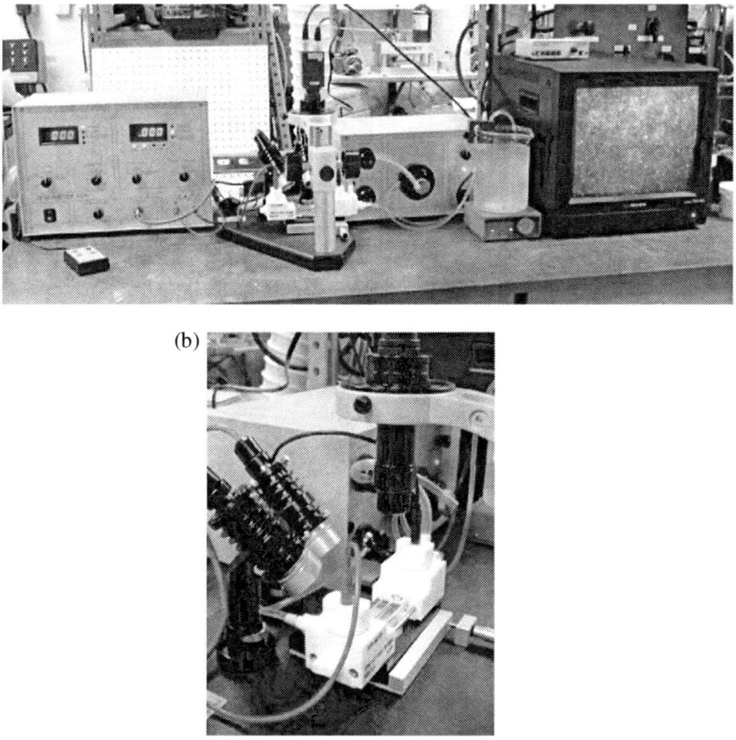

FIGURE 10-7 (a) Zeta potential equipment including power supply, microscope and electrophoresis cell, and video monitor for tracking the electrophoretic mobility of colloids, (b) Close-up of the microscope and electrophoresis cell (*Courtesy: Ashley Wille, CH2M Hill Inc.*).

There is no need to use the DAF jar tester when making the EPM measurements. One can simply use several small jars or beakers with a magnetic stir bar and vary the coagulant dose and control pH as required. Use of zeta potential instruments in the field for drinking water treatment plants has shown that only a very short rapid mix period (<15 s) is required to form discrete floc particles that are easily visible under the microscope, which is part of the instrumentation. Longer slow-mix flocculation can form large, amorphous floc particles that are difficult to distinguish and track.

Optimum coagulation conditions for DAF are those of coagulant dose and pH that produce flocs with charge near zero (see Chap. 6). These conditions produce flocs that have sufficient hydrophobic characteristics so the hydrophobic force can cause attachment of flocs to bubbles (see Chap. 5). The guidance summarized in Sec. 6-6 regarding dosages and pH conditions is useful, but can be supplemented with EPM measurements to ensure optimum coagulation conditions for DAF and downstream granular media filtration efficiency. EPM measurements or zeta potentials can be useful in optimizing coagulation in bench-scale and pilot-scale studies and for existing full-scale plants.

An example of the use of EPM measurements is shown with the data in Fig. 10-8. These data show EPM as a function of coagulant dose for two coagulants (alum and a high-basicity PACl) at pH 6.2. The PACl dose to achieve flocs with no net charge is 4.2 mg/L as Al, and for alum the dose is about 4.7 mg/L. A good practical example of the application of EPM measurements for DAF is shown in Fig. 6-12.

Streaming Current

Streaming current measurements are sometimes employed by water plants in an on-line continuous mode. A good description of the principles behind the measurement is found in Letterman and Yiacoumi (2010). The AWWA Manual M37 (AWWA, 2011) has a good description of both principles and applications. In summary, a water sample flows through the cell chamber where a reciprocating piston causes water to flow inside the annular space between the piston and cylinder chamber. As the water sample moves back and forth, the electroneutrality of the suspension is affected by the presence of the electrical double layer

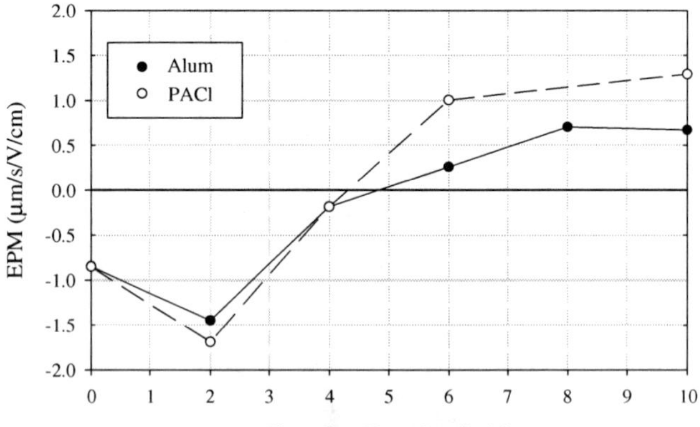

FIGURE 10-8 EPM as a function of coagulant dose for alum and a high-basicity PACl at pH 6.2; raw water: high TOC reservoir supply, turbidity 0.7 NTU, TOC 6.1 mg/L (*Source: Data taken from Pernitsky and Edzwald, 2006*).

of the particles and the attraction of counter-ions to the electrodes in the cell. The movement of the counter-ions with the pressure-driven flow produces a streaming current that is related to the zeta potential of the particles.

These measurements are made downstream of coagulant addition. They can accomplish two things: (1) provide indicator measurements of floc charge and therefore control coagulant dose and (2) provide an alarm when the coagulant feed is interrupted. For use in controlling coagulant dose, the plant operator must independently determine optimum coagulant dose and pH conditions for removing turbidity and TOC in the plant. These conditions are used to establish the streaming current set-point. The plant operator can then adjust the instrument to a streaming current of zero for these optimum coagulation conditions. Subsequently, a negative streaming current indicates coagulant underdosing and positive values indicate overdosing.

Flocculation Index

John Gregory conducted research and developed instrumentation that provides a sensitive measure of particle aggregation (Gregory and Nelson, 1984). The authors are not aware of the use of the instrument in DAF research studies, pilot-scale studies, or in full-scale plants. However, the instrument is commercially available (Photometric Dispersion Analyzer) and may be beneficial to evaluate floc breakage and the probability of regrowth of floc that may occur in flocculation tanks. Another problem where it may be useful is in evaluating retention of floated sludge. For some water supplies, a flotation aid polymer is used to retain the floated sludge and this instrument may be beneficial in solving this problem.

Particle Size Distributions

Particle counters provide a measure of the number of particles in a water sample, grouped in different size fractions. Particle sizes range from about 1 to 200 μm, and users have the option to set between 8 and 32 different size channels within this range. For water treatment applications, the most widely used are laser-based counters which measure the light blockage as the particle moves through an illuminated counting chamber.

The great volumes of data generated by particle counters have both advantages and disadvantages. From a research perspective, particle counters provide detailed insights of the particle size distributions which are useful to track complex processes such as flocculation, floc breakup and differential particle removal according to their sizes. For routine use during water treatment operation, the data coming continuously from multiple particle counters may be overwhelming. For such applications, the particle concentration in one or two specific channels are selected (say all particles below 10 μm) for continuous observation. With experience at a specific plant, operators can use this information to detect non-optimal coagulant dosage conditions, filter breakthrough events, etc.

A recent example shows how particle counting is used to optimize design and operation (Kim et al., 2007). In this case, the number of particles in the size range 5 to 10 μm was used to determine the best mode of operation of the float scraper. The particle counts in the outlet were first used to find the best timing of the scraper movement, and second to determine where in the separation zone the float layer was most likely to be knocked down (incidentally, in the area closest to the beach plate).

A novel application of particle counting, of great relevance to DAF, is the measurement of bubble size distributions. This technique, earlier described in Sec. 4-2, was first reported by researchers from South Korea (Han et al., 2002). Due to the speed at which bubble size distributions can be determined with this method, it becomes a feasible method to use at pilot scale. Two parameters may be assessed:

- The bubble size distribution produced by the nozzle(s). When more than one nozzle type is considered, the nozzle best suited may be selected after relatively brief testing at pilot scale. Nozzles in full-scale installations are usually small and may be installed in the pilot plant without a need to scale them down.
- Once the nozzle best suited is selected, the effect of saturation pressure on bubble size can be investigated. There are indications in the literature that bubble sizes are relatively constant for the saturation pressure above 350 kPa (discussed in Sec. 4-3), but this needs to be verified at more locations with different nozzles.

Nozzle assessment tests with particle counting are done at pilot scale. For the duration of the nozzle tests, the DAF reactor should be run with clear water without coagulants, to eliminate the possibility of measuring the size of flocs along with bubbles.

Volume of Air Delivered

The volume of air made available by the saturation system can be volumetrically measured with a simple method. The equipment needed, the supporting theory, and the data analysis procedure are presented in Sec. 3-5.

Why measure the available air volume? Although most of the parameters of the saturation system can be easily measured, such as saturation pressure and recycle flow rate, there are some that are not. For open-end saturator pumps, the air flow rate into the pump is difficult to determine. For dead-end saturators, although the transfer efficiency of the saturator can be approximated with the mass transfer model of Chap. 3, the end effects remain unknown. There are good reasons to measure the available air volume:

- During pilot testing, the air dosing has to be exactly known to allow scale-up to the full-scale plant;
- During plant commissioning and acceptance of the full-scale plant, the transfer efficiency (dead-end) or saturator pump capacity (open-end) has to be precisely determined for enforcing contractual specifications and for future operational needs; and
- During routine operation, periodic measurement will indicate reduced recycle flow or partial blocking of the packing material, to allow timely corrective action.

REFERENCES

APHA, AWWA, and WEF (2005), *Standard Methods for the Examination of Water & Wastewater*, 21st ed., Washington, DC: APHA.

AWWA (2011), *Operational Control of Coagulation and Filtration Processes*, AWWA Manual M37, 3rd ed., Denver: AWWA.

Crossley, I. (2010), Personal communication, Hazen and Sawyer, PC, New York.

Crossley, I. A., Valade, M. T., and Shawcross, J. (2001), Using lessons learned and advanced methods to design a 1500 ML/day DAF water treatment plant, *Water Science & Technology*, 43(8), 35–41.

Eades, A., Bates, B. J., and MacPhee, M. J. (2000), Treatment of spent filter backwash water using dissolved air flotation, *Proceedings of the 4th International Conference: Flotation in Water and Waste Water Treatment*, September 11–14, Helsinki, Finland.

Edzwald, J. K., Tobiason, J. E., Dunn, H., Kaminski, G., and Galant, P. (2000a), Removal and fate of *Cryptosporidium* in dissolved air drinking water treatment plants, *Proceedings of the 4th International Conference: Flotation in Water and Waste Water Treatment*, September 11–14, Helsinki, Finland.

Edzwald, J. K., Tobiason, J. E., Parento, L. M., Kelley, M. B., Kaminski, G. S., Dunn, H. J., and Galant, P. B. (2000b), *Giardia* and *Cryptosporidium* removals by clarification and filtration under challenge conditions, *Journal of the American Water Works Association*, 92 (12), 70–84.

Edzwald, J. K., Tobiason, J. E., Udden, C., Kaminski, G. S., Dunn, H. J., Galant, P. B., and Kelley, M. B. (2003), Evaluation of the effect of recycle of waste filter backwash water on plant removals of *Cryptosporidium*, *Journal of Water Supply: Research and Technology – Aqua*, 52 (4), 243–258.

Farmerie, J. E. (2007), Dissolved air flotation for membrane pretreatment, *Proceedings of the 5th International Conference of Flotation in Water and Wastewater Systems*, September 11–14, Seoul, South Korea, 199–202.

Gregory, J. (2006), *Particles in Water: Properties and Processes*, London: IWA Publishing.

Gregory, J., and Nelson, D. W. (1984), A new optical method for flocculation monitoring, in J. Gregory, ed., *Solid-Liquid Separation*, Chichester: Ellis Horwood.

Haarhoff, J., Langenegger, O., and Van der Merwe, P. J. (1992), Practical aspects of water treatment plant design for a hypertrophic impoundment, *Water SA*, 18 (1), 27–36.

Han, M. Y., Park, Y. H., and Yu, T. J. (2002), Development of a new method of measuring bubble size, *Water Science and Technology: Water Supply*, 2 (2), 77–83.

Janssens, J., and Buekens, A. (1993), Assessment of process selection for particle removal in surface water treatment, *Journal of Water Supply: Research and Technology – Aqua*, 42 (5), 279–288.

Kim, H., Kim, T., and Kim, H. (2007), Effects of scraper's operation on efficiency in the dissolved-air-flotation process, *Proceedings of the 5th International Conference of Flotation in Water and Wastewater Systems*, September 11–14, Seoul, South Korea, 371–374.

Letterman, R. D., and Yiacoumi, S. (2010), Coagulation and flocculation, in J. K. Edzwald, ed., *Water Quality and Treatment: A Handbook on Drinking Water*, 6th ed., New York: AWWA and McGraw-Hill.

Miles, J. P., Davies, P. H., and Murphy, I. (2000), Improving the security and performance of the largest water treatment plant in Wales: upgrading of Felindre WTW with the addition of first stage CO-CO-DAFF clarification-filtration, *Proceedings of the 4th International Conference: Flotation in Water and Waste Water Treatment*, September 11–14, Helsinki, Finland.

Nickols, D., Schneider, O., and Leggiero, S. (2000), Pilot-testing of high-rate DAF for New York City, *Proceedings of the 4th International Conference: Flotation in Water and Waste Water Treatment*, September 11–14, Helsinki, Finland.

O'Connell, J. K., Phillips, N. R., and Lutz, C. A. (1997), Pilot testing and implementation of full-scale dissolved air flotation, intermediate ozonation, and high rate filtration for public water supply in the United States of America—case study, in *Dissolved Air Flotation*, London: Chartered Institution of Water and Environmental Management.

Pernitsky, D. J., and Edzwald, J. K. (2006), Selection of alum and polyaluminum coagulants: principles and applications, *Journal of Water Supply: Research and Technology – Aqua*, 55 (2), 121–141.

Pernitsky, D. J., Breese, S., Wobma, P., Griffin, D., Kjartanson, K., and Sorokowski, R. (2007), From pilot tests to design on Canada's largest DAF treatment plant, *Proceedings of the 5th International Conference of Flotation in Water and Wastewater Systems*, September 11–14, Seoul, South Korea, 203–210.

Pernitsky, D. J. (2010), Personal communication, CH2M Hill, Calgary, Canada.

Prendiville, P. W., Lo, S. H., Stoops, R. A., and Nicoloro, M. A. (1997), High-rate dissolved air flotation treatment of cold waters of low alkalinity in the northeastern USA, in *Dissolved Air Flotation*, London: Chartered Institution of Water and Environmental Management.

Shawcross, J., Tran, T., Nickols, D., and Ashe, C. R. (1997), Pushing the envelope: dissolved air flotation at ultra-high rate, in *Dissolved Air Flotation*, London: Chartered Institution of Water and Environmental Management.

Valade, M. T., Becker, W. C., and Edzwald, J. K. (2009), Treatment selection guidelines for particle and NOM removal, *Journal of Water Supply: Research and Technology – Aqua*, 58 (6), 424–432.

Wobma, P., Bellamy, B., Pernitsky, D., Kjartanson, K., Adkins, M., and Sears, K. (1997), Effects of dissolved air flotation on water quality and filter loading rates, in *Dissolved Air Flotation*, London: Chartered Institution of Water and Environmental Management.

CHAPTER 11

CONVENTIONAL APPLICATIONS FOR DRINKING WATER TREATMENT

11-1 DAF CONFIGURATIONS 11-1
Reactor Shapes and Sizes 11-1
High-Rate DAF 11-2
Integration of DAF with Filtration 11-3
Counter-Current DAF 11-4
Combinations of Sedimentation and DAF 11-4
11-2 DAF TREATMENT OF VARIOUS WATER QUALITY TYPES 11-6
Treatment of Low Turbidity, Low Color, Low TOC Water Supplies 11-7
Treatment of Low Turbidity, High Color, High TOC Water Supplies 11-11
Treatment of Water Supplies with Algae 11-13
Tastes and Odors 11-17
Giardia and Cryptosporidium Removals 11-18
REFERENCES 11-20

The chapter objectives are (1) to show and describe various DAF plant configurations and (2) to present and evaluate DAF applications in the treatment of various types of water supplies with respect to raw water quality characteristics. The chapter begins with descriptions of the placement of DAF in a horizontal layout separated from granular media filtration, DAF over filtration (DAFF), a special case of the latter called Counter-Current DAFF (CoCoDAFF), and others. For the horizontal-layout DAF plant, we comment on conventional and high-rate DAF. The major features of all configurations are described, and design and operating parameters are compared. The next section considers the applicability of DAF in treating water supplies according to general raw water quality characteristics. General comments are made on the feasibility of treating and DAF performance followed by one or more case studies. Some design and operating criteria are presented for the case studies, and treatment performance is summarized. This section also includes discussion of DAF applications dealing with tastes and odors and the removals of the protozoa pathogens, *Giardia* and *Cryptosporidium*.

11-1 DAF CONFIGURATIONS

Reactor Shapes and Sizes

The flow patterns in the DAF separation zone are driven primarily by large-density gradients induced by the air suspension. When compared to sedimentation, there is thus a reduced dependence on the inlet conditions and reduced vulnerability to density gradients caused by changes in temperature, salinity, or suspended solids (as pointed out in Sec. 8-7). Furthermore, the depth required for good flotation is less than that for good sedimentation, and no extra depth is required for sludge storage, as the DAF float layer is continuously or intermittently removed. These factors make the DAF process adaptable to many geometrical shapes and sizes, and thus a good candidate for retrofitting and plant upgrades within the constraints posed by existing tanks and site limitations.

Circular tanks are often used when converting sedimentation tanks that had a central flocculation chamber. The flocculation chamber is retained, and the DAF contact zone is provided in an annular ring around the flocculation compartment. From the contact zone, the water flows radially outward into the separation zone. Size is not a limitation for the use of circular DAF tanks, as they are found at treatment plants large and small. Circular tanks are well-suited when large volumes of floated sludge have to be removed in sludge thickening applications, as the design and operation of circular scrapers of a central pivot is mechanically more robust. As a general guideline, circular tanks are not suggested for new DAF plants, as the controlled introduction of flocculated and recycle water at the center of a large tank is inherently more difficult than the usual side entry used for rectangular tanks.

Rectangular tanks are the preferred option for drinking water treatment. There is no minimum limit to the size of a single DAF reactor, as DAF pilot plants with a footprint of a few square meters perform just as efficiently as large units. There is no commonly accepted maximum size for a DAF unit, although it is recognized that uniform float removal and withdrawal of product water are more challenging with increasing size. For DAFF units, the maximum footprint is dictated by filtration and backwash considerations. The DAFF unit footprints for the water plants, in Lysekil (Sweden), Fairfield (Connecticut United States), and New York City (Croton plant, United States) are 42, 68 and 85 m^2. The Fairfield and New York City plants are very large. The first two plants are case studies presented later in this chapter, and the third was reported by Crossley et al. (2007). For stand-alone DAF units, the footprints are larger. For St Johns (Canada) and Winnipeg (Canada) case studies presented later in this chapter, the footprints are 121 and 132 m^2, respectively. The second author has observed a number of large DAF units with footprints between 100 and 200 m^2, while a case is known where the DAF footprint is 225 m^2 for a single unit (Amato, 2011).

The aspect ratio of DAF units is controlled by the hydraulic loading rate and the design of the float layer removal system. Where the float layer is scraped, the structural design of the moving scraper bridge dictates fairly narrow and therefore long DAF units. For the hydraulic removal of the float layer, the units have aspect ratios closer to one, with the minimum distance usually in the flow direction of the floated sludge. At conventional hydraulic loading rates, the aspect ratio is not a critical factor, as DAF works well, independent of aspect ratio. The exception is with high-rate DAF, where some designers require an aspect ratio close to one (Kiuru, 2000). As more practical experience with high-rate systems are reported, the limiting aspect ratio may be relaxed for future applications.

High-Rate DAF

Chapter 2 traces the history and evolution of DAF. From its early use in the 1960s for drinking water clarification until the late 1990s, the hydraulic loading increased from about 5 to 15 m/h. We define these as conventional hydraulic loading rates. Edzwald et al. (1999) showed through pilot studies that the hydraulic loading can be increased to rates of 30 to 40 m/h (8 to 16 gpm/ft^2) depending on water temperature and still achieve good DAF effluent and filtered water quality. About the same time a company in Finland (Rictor Oy) developed a high-rate DAF tank featuring a porous plate floor for collection of the clarified water. The Rictor technology has been licensed to Infilco Degremont, and it is available under the trade name of AquaDAF®. The case study for Haworth, New Jersey, described later in the chapter, uses high-rate AquaDAF®. Other DAF equipment companies have also developed high-rate DAF systems. These include Clari-DAF™ from ITT WWW (Leopold, USA) and Enflo-vite® from Enpure (UK). Earlier Purac Ltd. (now Enpure) and Purac AB (now Purac Sweden) developed a high-rate DAF system called DAFRapide® directed to the industrial market that has either tubes or plates near the bottom to aid in the separation of floc-bubble aggregates and bubbles.

While there is no clear demarcation in classifying conventional rate and high-rate DAF, we define conventional rate generally in the range of 5 to 15 m/h (maybe, little greater) and high rate at nominal loading rates of 15 to 30 m/h (maybe a little greater), and with separation zone loading rates of 20 to 40 m/h. Hydraulic loading sets the size of the DAF tank so there are advantages in reduced plant footprint with high loading rates.

Beside the loading rate, what are the differences in DAF plants with conventional rate versus high rate? Gregory and Edzwald (2010) compared design and operating parameters for these systems. There are no differences in (1) pretreatment (coagulation and flocculation), (2) recycle rates, nozzles, and saturators although each equipment company has their preferences, and (3) floated sludge withdrawal although again there are company preferences. The only difference between conventional and high-rate systems is that deeper tanks are preferred in high-rate systems. Another difference for high-rate systems with hydraulic withdrawal of the floated sludge is the aspect ratio (length to width) is close to one.

Chapter 8 presents principles for the rise rates of floc-bubble aggregates and free bubbles in the separation zone. Based on a simplified view of rise rates and the hydraulic flow pattern through the separation zone, one would not expect for high-rate DAF that bubbles and aggregates would be removed since their rise rates are less than the hydraulic loading. However, the flow patterns through DAF systems with high hydraulic loadings develop a stratified flow pattern allowing for removal of bubbles and aggregates—see Chap. 8, Gregory and Edzwald (2010), and Edzwald (2010). Another consideration for high-rate systems is that they are generally a little deeper, which is a factor in retaining the bubble blanket within the DAF tank. The AquaDAF® system has a porous plate floor so the flow leaving the tank is uniformly distributed. This, undoubtedly, aids the hydraulic flow pattern at the bottom of the tank.

For high-rate DAF systems some bubbles can exit the bottom of the DAF tank, but this should not be a problem. As water flows from the DAF tank to the filters through channels and over weirs, there are plenty of opportunities for the bubbles to escape. Edzwald et al. (1999) showed that some bubbles entering the top of filters mostly escaped from the water column above the filter media and the few that entered the filter media had no adverse impact on filtered water quality or head loss.

Integration of DAF with Filtration

The hydraulic loading rates for conventional-rate DAF and filtration are in the same range, making the integration of DAF and filtration an early and obvious option, as described in Sec. 2-6. With DAF separation taking place in the headspace above the filter media, there are savings in space and capital, together with the stabilization of the hydraulic flow pattern induced by the perfectly uniform flow into the filter media. This DAFF option (also Flofilter™ or flotation over filtration) is commonly used and illustrated in Fig. 11-1. Between 1992 and 2000, nine treatment plants in South America were retrofitted with DAF, of which eight used the DAFF process combination (Richter and Gross, 2000). In Australia, 20 DAFF plants were constructed between 1980 and 1996 (Finlayson and Huijbregsen, 1997). At larger, modern treatment plants, DAFF continues to play a role, such as at the

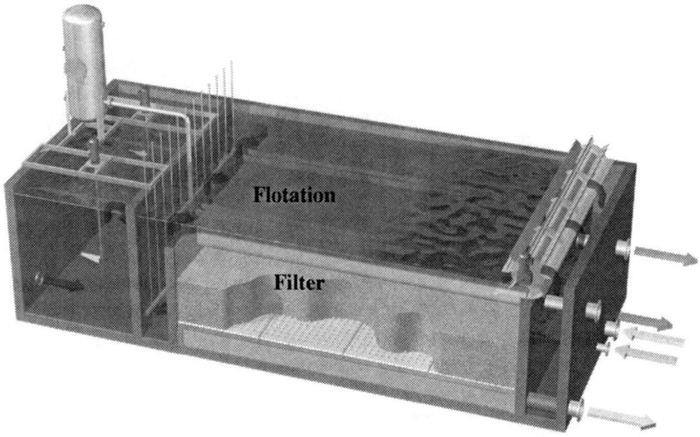

FIGURE 11-1 Schematic of the Flofilter™ (*Courtesy: Purac, Sweden*).

Fairfield plant in the United States (a case study later in the chapter), and the Croton treatment plant in New York (Crossley et al., 2007).

Counter-Current DAF

Most DAF systems rely on co-current flow where both flocs and bubbles are introduced at the bottom of the contact zone and form agglomerates as they ascend toward the float layer. In the middle of the 1980s, a package-plant supplier in South Africa installed some plants with counter-current flow, where the flocculated raw water was introduced at the top of the separation zone and the saturated recycle at the bottom, thus obviating the need for a separate contact zone. This principle was applied to both small DAF and later DAFF reactors (Offringa, 1995). During the 1990s the independent development of counter-current DAFF, under the wings of a large international supplier, took a huge step forward when it was comprehensively tested at pilot-scale for three years and re-engineered for large-scale municipal water treatment, under the label CoCoDAFF. The process was first used at full scale in 1995 and has since been applied at some large treatment plants in South Africa and the United Kingdom (Eades et al., 1997; Officer et al., 2001). The most recent application is at the 175 ML/d (46 MGD) Glencorse treatment plant in Scotland (Journal AWWA, 2010). A schematic of the process is shown in Fig. 11-2 with a photograph of a full-scale installation in Fig. 11-3.

Combinations of Sedimentation and DAF

The principal weakness of DAF as a phase separation process is that it cannot readily deal with high amounts of inorganic (mineral) turbidity–the flocs become too heavy and unstable to be buoyed up for long periods. This problem is discussed in Sec. 10-1. So what to do if the raw water has intermittent spikes of inorganic turbidity, which is too high for DAF? Such raw waters are commonly found with river supplies and with impounding reservoirs subject to seasonal rains. During the rainy season for the impounding reservoirs, occasional flooding causes severe turbidity peaks. During the dry season, the turbidity settles in the reservoirs and the opposite problem of algal growth prevails.

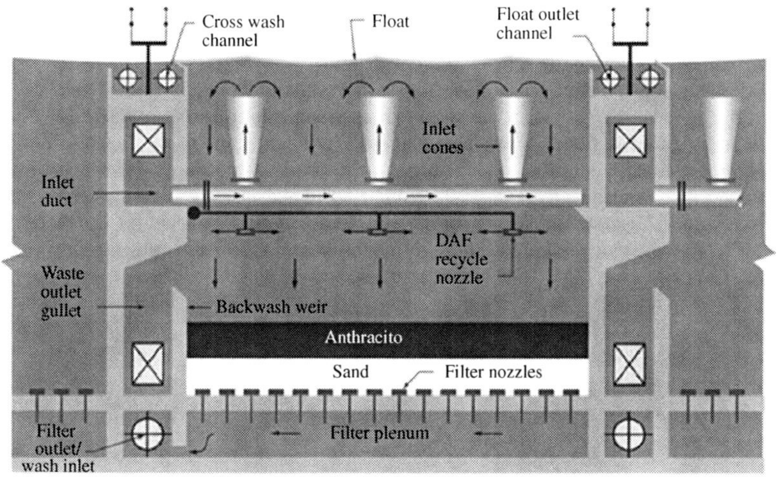

FIGURE 11-2 CoCoDAF (*Source: from Crossley and Valade (2006), Journal of Water Supply: Research and Technology – AQUA, 55 (7–8), 479–491, with permission from the copyright holders, IWA Publishing*).

FIGURE 11-3 CoCoDAFF reactor showing inlet cones during filter backwashing. (Photograph taken at the 210 ML/d water treatment plant at Vaalkop, South Africa.)

It is necessary to precede DAF with a more robust process such as sedimentation to deal with the problem of raw surface water sources that are subject to extreme but short-lived turbidity peaks during flood events. In South Africa, the combination of sedimentation ahead of DAF was first applied at Unit 3 (90 ML/d or 24 MGD) of the Vaalkop water treatment plant. The same process sequence, more recently, was successfully tested in South

Korea under the label SEDAF (Chung et al., 2000; Kwon et al., 2004; Kwak et al., 2005) showing great promise for sources with highly variable quality, provided that the system is carefully operated during the rainy season.

DAF is also placed ahead of sedimentation in a combined process train. This was done, for example, at Unit 2 (60 ML/d) of the Vaalkop water treatment plant where DAF was a retrofit to an existing plant. Space and hydraulic gradient constraints dictated that DAF should precede sedimentation, which turns out to work very efficiently. When turbidity drops and algal concentrations are greater, the flocculated water is diverted with a gate to the DAF units. The water then flows through the sedimentation tanks to the filters, but with no clarification in or desludging of the sedimentation tanks. During high-turbidity periods, the water bypasses the DAF units and the saturation system is shut down to save energy.

11-2 DAF TREATMENT OF VARIOUS WATER QUALITY TYPES

There is considerable experience demonstrating the ability of DAF to treat a variety of water quality cases or source water types. Some sources of information are the various proceedings of the flotation conferences held in Orlando (Ives and Bernhardt, 1995), London (CIWEM, 1997), Helsinki (2001), and Seoul (Han and Edzwald, 2007); the DAF review paper by Edzwald (2010); and the book chapter by Gregory and Edzwald (2010). An older paper by Longhurst and Graham (1987) tracing the development and use of DAF in the United Kingdom is also worthwhile reading.

A question often asked is, how high of a raw water turbidity level can be treated effectively by DAF? This question was addressed in Sec. 10-1. A major factor to consider is the nature of the turbidity. Is it non-mineral (organic-particulate matter from algae) or mineral turbidity (silts and clays)? DAF is effective in treating water supplies for average raw water conditions: mineral turbidity at levels <10 NTU but up to 100 NTU for supplies with high levels of non-mineral turbidity. For maximum turbidity events, DAF is effective for supplies with turbidities that range up to 50 NTU for mineral type particles and 200 NTU for non-mineral particles. These criteria are based on typical design conditions for drinking water clarification of providing delivered air at 10 to 12 mg/L in the contact zone or about 8 mL/L based on volumetric air criteria presented in Chap. 3 or bubble volume concentrations of about 8400 ppm based on the theory of bubble and particle collisions in Chap. 7. If one wishes to treat mineral turbidity supplies with higher turbidities, then additional air may be required. The reader can find additional information on treating turbid waters in the review paper by Edzwald (2010) and the book chapter by Gregory and Edzwald (2010).

In the following presentation, we do not further address treatment of mineral-turbidity water supplies. Rather, we address those types of supplies most amendable to treatment by DAF. DAF is effective in separating low-density particles from waters. It is considered a better particle separation process than sedimentation in treating water supplies with algae, natural color, and low-mineral water turbidities. It is also more effective than sedimentation for treating cold water supplies. Algae are, of course, of low density and following coagulation and flocculation the densities of the flocs containing algae and metal hydroxide precipitated particles are low, not much greater than the density of water. In coagulating waters with natural color and low turbidity, the flocs consist mainly of metal-humate precipitate and metal hydroxide precipitated particles, all of low density. Low-turbidity supplies, after coagulation and flocculation, also yield low-density flocs. What follows are the presentations of DAF treatment for these types of water supplies. Some general comments are presented for each type followed by one or more case studies.

Treatment of Low Turbidity, Low Color, Low TOC Water Supplies

These types of water supplies at one time would have been candidates for direct filtration; however, the additional pathogen barrier provided by DAF makes them, instead, good candidates for treatment by DAF. Often the source waters are reservoirs, but in certain cases rivers may have these water quality characteristics.

Case Study: Fairfield, Connecticut, United States. The William S. Warner (simply, Warner) water treatment plant began operation in 1997. The plant has a capacity of 190 ML/d (50 MGD), and at the time it was commissioned it was the largest DAF facility in North America. It is the largest DAF over filtration (DAFF) plant in North America until the Croton plant for New York City goes on-line in 2012 or 2013. It is located in Fairfield, Connecticut (United States), and it is one of several water treatment plants owned and operated by Aquarion Water Company of Connecticut. The first author has conducted several studies at the plant and has extensive knowledge of the plant and its performance. Additional information was provided by Gary S. Kaminski (2010), a process engineer with Aquarion.

The water supply for the plant is Hemlocks Reservoir. It is a supply of low turbidity (<3 NTU), low color (apparent color ~25 CU), low TOC (2.2 to 4 mg/L), and has seasonal Mn (0.02 to 0.2 mg/L) problems. The Warner plant provides for treatment of turbidity, pathogens (viruses and bacteria and multiple barriers for *Giardia* and *Cryptosporidium*), TOC, and Mn.

A process schematic for the Warner plant is shown in Fig. 11-4. The plant uses a dual-coagulant strategy of alum and a cationic polymer with a high positive-charge density of low molecular weight. Both coagulants are added at the rapid mixing step with the cationic polymer dose constant at 1 mg/L, and alum is varied (range typically of 1 to 2.3 mg/L as Al) to achieve good treatment as described below. Coagulation pH is strictly controlled between pH 6.5 and 6.7 for seasons other than winter. During cold water winter conditions, the coagulation pH is controlled between 6.7 and 6.9. Design and operating values for flocculation, DAF, and filtration are summarized in Table 11-1. Flocculation consists of two stages with a total design hydraulic detention time of 12 min. The plant uses DAFF. Consequently, the nominal hydraulic loading rate is the same for DAF and filtration at 15 m/h (6 gpm/ft^2). The saturator pressure is usually set at 450 kPa, and the recycle ratio varied as needed to maintain treated water quality. Chlorine is applied to the flotation tank above the filter media for Mn removal by oxide-coated media.

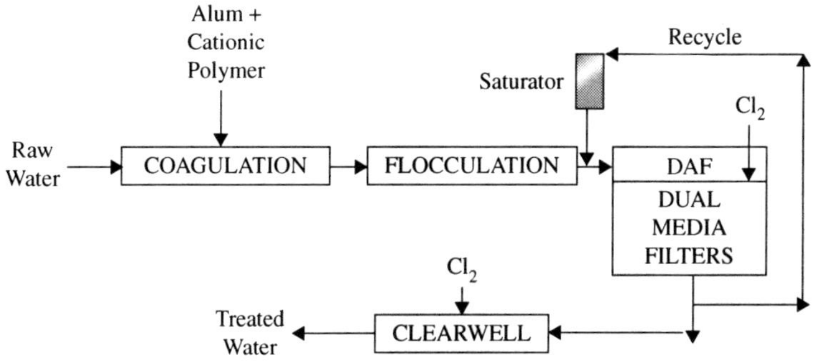

FIGURE 11-4 Process schematic of the Fairfield (Connecticut) flotation over filtration (DAFF) water plant.

TABLE 11-1 Summary Table of Design and Operating Parameters for the Case Studies

	Case studies					
Parameter	Fairfield Connecticut (U.S.)	Saint John's Newfoundland (Canada)	Lysekil (Sweden)	Chingford South (United Kingdom)	Winnipeg (Canada)	Haworth New Jersey (U.S.)
Flow (ML/d)	190	126	22	42.3	400	760
Flocculation						
Number of units	8 + 1 spare	8	3	3 + 1 out-of-service	8	3
Number of stages/unit	2	3	2	2	3	2
Time (minutes)	12	20	28	13	16	10
Type	Propeller mixers	Hydraulic	Propeller mixers	Paddle mixers	Propeller mixers	Propeller mixers
G (s^{-1})	NP	25–60 (tapered)	NP	70 (typical) (20–100)	100 (maximum)	20
Flotation						
Number of units	8 + 1 spare	4	3	3 + 1 spare	8	10
Length in m (total of contact and separation zone)	12.2	12	8.8	9.2	11	13.7
Width in m	5.6	11	4.78	3.8	11	6.1
Depth in m	NP	2.9	3.3 (to top of filter)	3.5	2.5	4.2
Loading (m/h)	15	10	7.3	17	18	39
Sludge removal	Beach drum	Reciprocating scraper	Beach drum	Hydraulic	Reciprocating scraper	Mechanical scrapers

Type of saturator	Packed	Packed	Unpacked eductor	Packed	Packed	Unpacked
Number of saturators	2	2 (1duty + 1 standby)	2	1	4	2/unit
Packing type	90 mm Jaeger Tri-pack®	25 mm Rauschert Polypropylene Hiflow® Rings	NA	Polypropylene Pall Rings	25 mm Rauschert Polypropylene Hiflow® Rings	NA
Saturator pressure (kPa)	415–520 (typically 450)	400–600	500–520	450–560	400–600	585 (typical)
Recycle rate (%)	8–15	10	10 typical and design (12% at low rates due to loss in eductor efficiency)	11 (typical)	8–10 (10 design)	10 (typical)
Injection system (# per tank)	Nozzles (66)	Nozzles (108)	Needle valves (10)	Nozzles (36)	Nozzles (108)	Nozzles (200)
Filtration						
Filter rate (m/h)	Flotation over filtration	10	Flotation over filtration	11	30	15
GAC					2.1 m ES 1.1 mm	NA
Anthracite	0.6 m ES 0.85 mm	0.56 m ES 1.0–1.1 mm	NA	0.6 m	NA	0.69 m ES 1.5 mm
Sand	0.3 m ES 0.45 mm	0.30 m ES 0.45–0.55 mm	See note 1	0.6 m	NA	0.53 m ES 0.6 mm

NA: not applicable.

NP: not provided.

Note 1: Filter 1, 2 m of sand with effective size (ES) of 0.4 to 0.8 mm; Filter 2, 1 m of sand with effective size (ES) of 0.8 to 1.2 mm (original plant design); Filter 3, 1 m of sand with effective size (ES) of 0.8 to 1.0 mm.

The performance of the plant was described in detail by Edzwald and Kaminski (2009). Some of these performance data are summarized here. The DAF process performs well with DAF turbidity generally <0.4 NTU, and average filtered water turbidity of 0.07 NTU as shown in Fig. 11-5. The data in the figure also show that the plant performed well in removing TOC (average treated water TOC of 1.8 mg/L [44 percent removal]) and average treated water UV_{254} of 0.033 cm^{-1}.

Another indicator of DAF performance is its effect on filtration. The water following DAF contains low turbidity and low particle counts, and therefore the plant has longer filter runs compared to sedimentation plants. A way to assess this is to examine the unit filter run volume (UFRV)—that is, the volume of water filtered per unit filter area in a filter run. The Warner plant does exceedingly well with an average UFRV of 640 m^3/m^2/run in 2005 with values up to 730 m^3/m^2/run. Even greater UFRVs could be achieved by the plant if the filters were run longer, but filters are backwashed on a schedule basis at night when power costs are less. Plants with high UFRV consume less energy in backwashing filters, and this is an advantage of DAF plants.

Case Study: Saint John's, Newfoundland, Canada. The Bay Bulls Big Pond Water Treatment Plant (St. John's, Newfoundland, Canada) is scheduled to go on-line in 2012. The plant has a maximum capacity of 126 ML/d (33 MGD). The information and data for this case study were obtained from David J. Pernitsky (2010) of CH2M Hill. What makes this an interesting case study is that the water is extremely cold at 2 to 3°C for eight months of the year making it a challenging water to treat by sedimentation; consequently, DAF was selected for clarification.

The water supply is Bay Bulls Big Pond. It is a lake supply of low turbidity, low color, low to moderate TOC, low alkalinity, seasonal algae problems, and extremely cold water conditions for much of the year. Water quality data follow: turbidity is <1 NTU and averages 0.4 NTU, TOC averages 2.9 mg/L and ranges from 0.5 to 5.5 mg/L, alkalinity is very low, <6 mg/L as $CaCO_3$ and averages only 2.5 mg/L; pH averages 6.1 and is usually

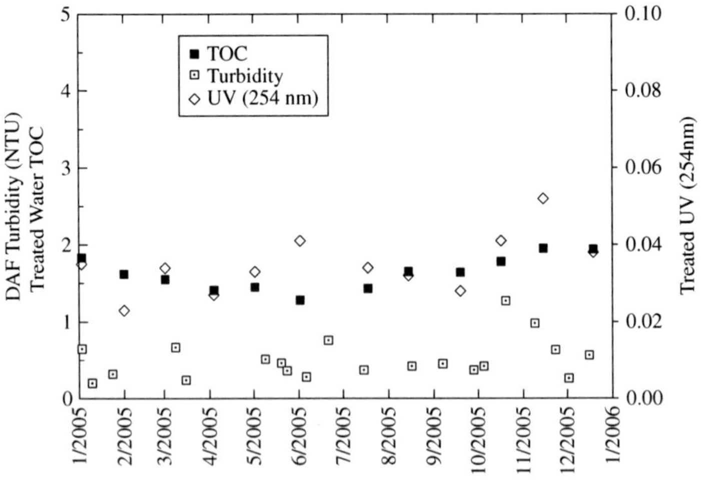

FIGURE 11-5 Fairfield plant performance for DAF turbidity and treated water UV_{254} and TOC *(Source: Gregory and Edzwald (2010), reprinted with permission from Water Quality and Treatment: A Handbook on Drinking Water, 6th ed., Copyright 2010, AWWA).*

between the low 5s and the mid 6s. A major characteristic of the water supply is that the water temperature is between 2 and 3°C for eight months. In the summer the water temperature can reach 22°C and the lake experiences algae problems, especially with diatoms that can clog filters, if not removed by clarification.

The DAF plant replaces a direct filtration plant. To evaluate the feasibility of treating the water with DAF and to develop design criteria, a pilot-plant study was conducted in 2005. The pilot study results showed good DAF and filtration performance when DAF was operated at a hydraulic loading of 10 m/h (4 gpm/ft^2) for cold water conditions and at higher rates during summer conditions.

The full-scale plant is designed for a capacity of 126 ML/d (33 MGD). Polyaluminum chloride (PACl) will be used as the coagulant and will be added to rapid mixing along with lime and carbon dioxide to control coagulation pH at 6.5 to 7.0. Design and operating values for flocculation, DAF, and filtration are summarized in Table 11-1. The flocculation design detention time is 20 min to treat this cold water with three-stage hydraulic flocculation. Tapered flocculation is not often used for DAF plants, but here it will be used with decreasing G through the three stages. DAF is designed at a nominal hydraulic loading rate of 10 m/h (4 gpm/ft^2). The recycle flow to the saturators is taken after DAF. The saturator contains plastic packing and can be operated in the pressure range of 400 to 600 kPa. After dissolving air in the saturators, the pressurized recycle flow will be injected via two manifolds at a rate of about 10 percent with nozzles in the contact zone. Ozone treatment will follow DAF for pathogen inactivation, and this will be followed by dual media filtration with a loading rate of 10 m/h (4 gpm/ft^2). Chloramine disinfection will be applied before the treated water enters the distribution system.

Treatment of Low Turbidity, High Color, High TOC Water Supplies

It is well known that DAF is effective in treating waters of high color containing aquatic humic matter. It was used early on for this purpose in Finland and Sweden—see Chap. 2. Figure 11-6 illustrates the effectiveness of treating highly colored water with a raw water true color of 155 CU. The optimum PACl dose is about 4 mg/L as Al, yielding a high-quality floated water with turbidity <0.5 NTU and true color <10 CU. There are numerous DAF plants around the world treating water supplies with moderate to high levels of natural color. One of the early plants in Sweden is presented below as a case study.

Case Study: Lysekil, Sweden. The Lysekil (Sweden) water treatment plant is of historical significance. It was commissioned in 1974, and it is one of the early DAF plants using the Flofilter™ (see Fig. 11-1) plant concept of DAF over filtration. Purac (Sweden) developed the Flofilter™ process in the late 1960s (see Chap. 2). It is also one of the first DAF plants to use lime and carbon dioxide for control of coagulation pH and addition of alkalinity. Information for this case study was obtained from two sources: first, from an extensive evaluation of the plant conducted in 1994 by Edzwald et al. (1994) and second, from a 2010 site visit by Dahlquist (2010) with Lackeby Water (Purac, Sweden).

The water supply for Lysekil is Karnsjon Reservoir. Raw water quality data are summarized from 1974 to the present. It is a typical Nordic water supply characterized by low turbidity, high color, low water temperatures, and low alkalinity (average of 9 mg/L as $CaCO_3$). Raw water turbidity averages 5.1 NTU (95th percentile of 11 NTU). Raw water true color averages 85 CU (maximum of 200 CU). Some limited SUVA data for the raw water indicate that the NOM is composed largely of aquatic fulvic acid. From mid-October until mid-May, the water temperature is usually <10°C with temperatures in December until March usually at 2 to 4°C. Even during the summer, water temperatures are generally

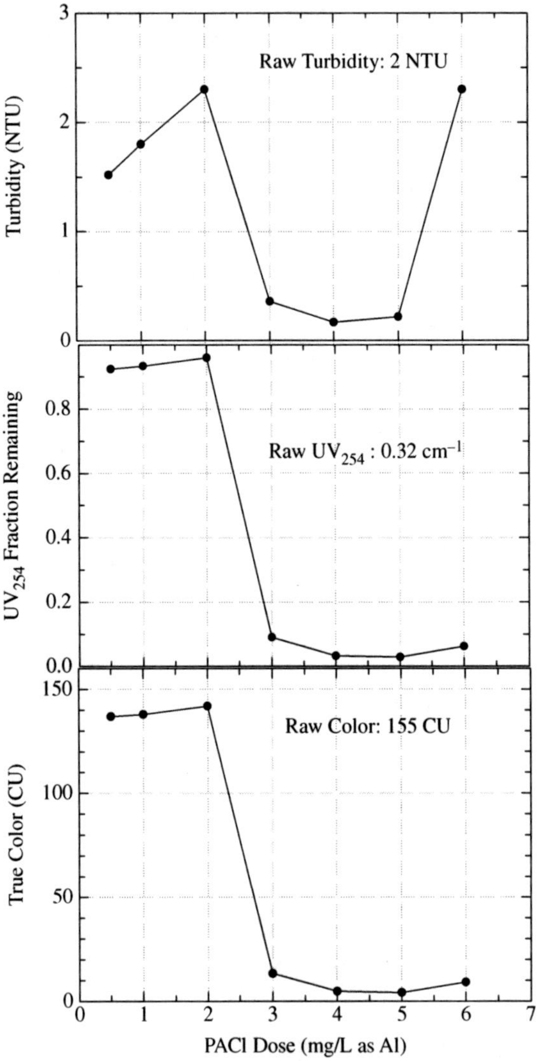

FIGURE 11-6 Flotation of highly-colored water containing humic matter (conditions: high basicity [71%] PACl, coagulation pH of 6.6 ± 0.2, and 5-min pretreatment flocculation).

<18°C. This type of water supply of low turbidity, high color, and cold temperatures is a challenge to treat.

Figure 11-7 shows a process schematic for the plant. The original design capacity was 22 ML/d (5.8 MGD); however, because of a decrease in water demand over the years the plant operates year-round below the design capacity. Table 11-1 summarizes process design and operating parameters. Microscreens (40 μm) are used to remove large particulate

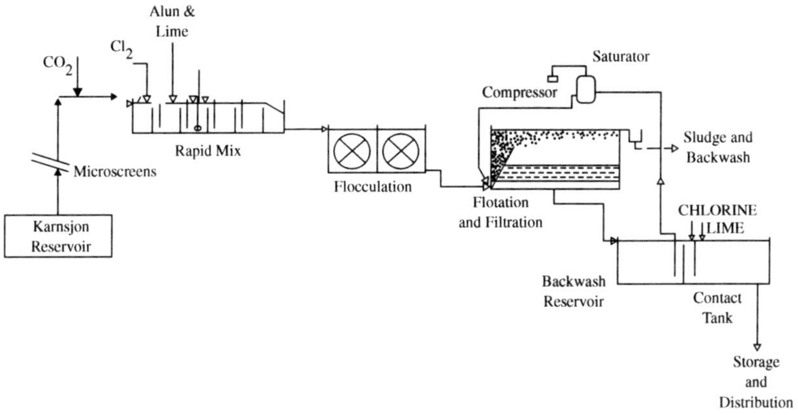

FIGURE 11-7 Schematic process diagram of the Lysekil (Sweden) water treatment plant.

matter from the raw water. Carbon dioxide is added to the water just prior to rapid mixing to increase the inorganic carbon concentration and alkalinity after the pH is adjusted in the clearwell. Rapid mixing is carried out in six stages with a minimum detention time of 30 s each. Chlorine is added in the first stage. Alum and lime (for pH control) are added in the third stage. Coagulation pH is controlled between 6 and 6.5. Flocculation with two stages follows coagulation. The plant was built originally with gate flocculators, but these were replaced with propeller mixers after 1994. There are three Flofilter™ units, and the design nominal loading rate for DAF is 6.4 m/h (2.6 gpm/ft^2). The saturators are unpacked and are operated at 500 to 520 kPa. Typical saturator pressure and recycle rate are 520 kPa and 10 percent, respectively. The pressurized recycle water is injected into the contact zone using needle valves. Sand is used for the filter media. Chlorine and lime are added to the contact tank prior to distribution.

The NOM in the raw water has increased over the years. True color increased from an average value of 60 CU for the period of 1974 to 1992 to an average of 91 CU since that period. The DAF plant must use greater alum dosages sometimes as high as 10 mg/L (as Al) to remove the higher NOM concentration. In spite of the change in raw water quality, the plant's filtered water quality is exceedingly good with turbidity at an average of 0.23 NTU and true color always <5 CU.

Treatment of Water Supplies with Algae

Algae can clog granular media filters causing short filter runs in direct filtration plants and in sedimentation plants when raw water algae counts are high. DAF is a more effective process in removing algae than sedimentation, because of the low densities of algae, and therefore a good choice to treat waters with algae problems. There are several good references summarizing the effectiveness of DAF in removing various types of algae including cyanobacteria (also called blue-green algae) with comparisons to sedimentation (Longhurst and Graham, 1987; Edzwald and Wingler, 1990; Edzwald et al., 1992; van Puffelen et al., 1995; Vlaški et al., 1996; Teixeira and Rosa, 2006 and 2007; Henderson et al., 2008; Gregory and Edzwald, 2010). In general, one can expect 90 to 99 percent removals of algae by DAF except when raw water algae concentrations are low. For sedimentation, algae removals are generally in the range of 60 to 90 percent.

Three case studies are presented dealing with fresh-water supplies containing algae. In Chap. 13, we turn our attention to algae in seawater and the use of DAF pretreatment in desalination plants to protect reverse osmosis membranes.

Case Study: Chingford South Plant, United Kingdom. DAF treatment has had a long history in the United Kingdom going back to the 1970s as described by Longhurst and Graham (1987) and in Chap. 2. For a case study for the United Kingdom, however, we have selected a new plant. The Chingford South Water Treatment Works (greater metropolitan London) DAF plant was commissioned in May 2005 by Thames Water. The plant has a maximum capacity of 42.3 ML/d (11 MGD) and is operated on average near its maximum at 41.7 ML/d. The information for this case study was obtained from Tony Amato (2010), DAF Technology Manager with Enpure Ltd. of Birmingham, England. Enpure designed the treatment plant.

The water supply comes from two reservoirs, King George South and William Girling. In general, the source water quality is characterized as moderate to high alkalinity and hardness with summer blooms of algae and low to moderate levels of turbidity. Reservoir water quality data are summarized for the period of 1998 to 2003. Average chlorophyll *a* concentrations were 13.6 and 23.4 μg/L for the William Girling and King George South Reservoirs, respectively. In the summer there are high algae blooms with 95th percentile chlorophyll *a* concentrations of 49 and 90 μg/L for the William Girling and King George South Reservoirs, respectively, indicating highly eutrophic water supplies. Turbidity for the William Girling Reservoir averaged 2.6 NTU with 95th percentile value of 7 NTU. Turbidity for the King George South Reservoir averaged 2.6 NTU and is generally <10 NTU, but there can be high turbidity events so the plant was designed to handle a maximum turbidity of 50 NTU. TOC averages are low at 2.6 and 1.1 mg/L for the William Girling and King George South Reservoirs, respectively. However, during algae blooms, TOC can be much higher with maximum values at 14.6 and 4.9 mg/L for the two reservoirs. Both reservoirs also contain low concentrations of pesticides.

The Chingford plant has pre-ozone that was originally intended for seasonal use when algae and TOC are high. However, the pre-ozone is used continuously throughout the year because it was shown to improve treatment. Polyaluminum chloride is used for coagulation and dosed into an in-line mixer following ozone. Typical dosage is 2 mg/L as Al, and the coagulation pH is adjusted to about 7 by feeding sulfuric acid. Design and operating values for flocculation, DAF, and filtration are summarized in Table 11-1. Flocculation consists of two stages with a total design hydraulic detention time of 13 min. Paddle mixers are used with an ability to vary the velocity gradient (G) with typical operation at 70 s^{-1}. The plant has four DAF tanks. The plant is designed with one DAF tank out-of-service so that three DAF tanks can provide the maximum plant flow rate of 42.3 ML/d (11 MGD), yielding a nominal DAF loading rate of 17 m/h (7 gpm/ft^2). With all the tanks in service, the loading rate is 12.7 m/h (5.2 gpm/ft^2). Recycle is taken after DAF and pumped to the saturator; one packed saturator is used. Typical operating pressures and recycle rates are 450 to 560 kPa and 11 percent. Filtration is carried out at 10 m/h (4 gpm/ft^2). Following filtration, the water is treated by GAC adsorbers to remove pesticides. Final pH adjustment and chlorination are done before transmission and distribution.

The DAF plant performs exceedingly well: DAF turbidities frequently of 0.2 to 0.4 NTU, filtered water turbidities <0.1 NTU, and <0.05 NTU following GAC.

Case Study: Winnipeg, Canada. The DAF plant for Winnipeg has a capacity of 400 ML/d (106 MGD). It was commissioned in November 2009, and it is the largest DAF plant in Canada and one of the larger ones in the world. Information for this case study was provided by David J. Pernitsky (2010) of CH2M Hill.

The source water is Shoal Lake, which borders the Provinces of Ontario and Manitoba in Canada and Minnesota in the United States. Prior to construction of the DAF plant, the raw water was chlorinated at the intake for control of slimes and zebra mussels, and conveyed by gravity 159 km to Deacon Reservoir at Winnipeg. Water was then re-chlorinated prior to distribution. In 2004, UV disinfection was installed as an additional pathogen barrier.

To provide multiple-barrier drinking water treatment, the City commissioned a series of studies to investigate the most suitable treatment technologies for a future water treatment plant. The first of these studies was a six-week pilot testing program conducted in 1994 using two mobile pilot plants to compare the feasibility of DAF and direct filtration treatment. The City decided, based on the results from the first study phase, to conduct a second phase of pilot work in 1996 and 1997. As part of the second phase pilot testing, the City commissioned the construction of a permanent pilot plant. With the advent of high-rate DAF in the late 1990s and early 2000, the City conducted an additional phase of pilot work in 2003 to compare conventional rate DAF to high-rate DAF. A conventional rate DAF process was eventually selected. Additional information on the pilot studies and plant design are found in a paper by Pernitsky et al. (2007).

The Deacon Reservoir has an unusual water quality. It is one of high TOC (average of 9.3 mg/L, range of 5.0 to 17.0 mg/L), low color (true color <10 CU), and low turbidity (0.3 to 5.3 NTU). The water supply is also characterized as one with high algae (average of 39,700 cells per mL), especially during late spring, summer, and early fall (as high as 666,000 cells per mL). The algae cause taste and odor problems. The DOC is high (4 to 15 mg/L), but natural color is low indicating that the dissolved natural organic matter (NOM) is mainly from algal activity in Deacon Reservoir and other plant matter in Shoal Lake. SUVA data are in the range of 1.5 to 2.7 m^{-1} per mg/L DOC indicating that humic matter is not a main component of the NOM. During the winter months the water is cold, and TOC can be high making it a challenging water to treat.

Figure 11-8 shows a process schematic for the plant. Coagulation is carried out with ferric chloride with pH control between 5.5 and 6.5 to achieve good removals of algae, turbidity, and TOC by coagulation, flocculation, and DAF. Design and operating values for flocculation, DAF, and filtration are summarized in Table 11-1. Flocculation is carried out in three stages with a total detention time of 16 min. The mixers in the flocculation tanks are vertical-axial flow (hydrofoil) with variable speed motors so operators can vary the velocity gradient (G). Eight DAF tanks are designed for a nominal loading rate of 18 m/h (7.4 gpm/ft^2). It is noted that the DAF process is considered conventional rate because of the DAF tank geometry, even though 18 m/h falls into the low end of

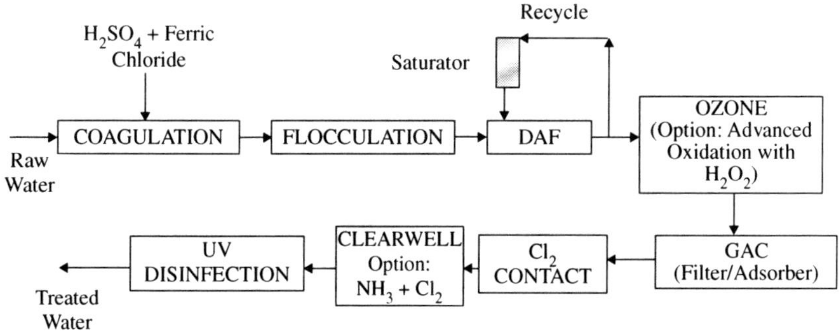

FIGURE 11-8 Process schematic of the 400 ML/d (106 MGD) Winnipeg (Canada) water plant.

loadings for high-rate DAF. The DAF tank is shallow compared to what is typically used in high-rate DAF processes. The recycle flow is taken after DAF. The saturator can be operated at a pressure of 400 to 600 kPa at recycle rates of 8 to 10 percent. Following DAF the clarified water is ozonated for taste and odor destruction, pathogen inactivation, and some conversion of TOC to biodegradable DOC (BDOC). Peroxide may be added for advanced oxidation of taste and odor compounds during severe taste and odor events. The GAC filters/absorbers provide for turbidity reduction, pathogen removal, and biological stability of the water by the removal of BDOC. Some taste and odor reduction is also expected to occur in the GAC filters by adsorption, although the GAC is not intended to be routinely regenerated, and the filters are operated as biologically active carbon (BAC) filters. The GAC beds are designed for a loading rate of 30 m/h (12 gpm/ft^2). The GAC beds are 2.1 m deep mono-media with an effective size of 1.1 mm. A filter-aid polymer is added upstream of the BAC filters. Free chlorine disinfection is provided for virus inactivation. The plant has the option of adding ammonia downstream of the chlorine contact chambers for the formation of chloramines, although chloramination is not currently practiced. UV disinfection provides for inactivation of *Giardia* and *Cryptosporidium*. Its location downstream of chlorine is not common, but resulted from its use and placement in the distribution system pumping area prior to the construction of the DAF plant.

In the first year of operation, post-DAF and filter turbidities have been consistently below 0.8 NTU and 0.1 NTU, respectively, with BAC filter UFRVs consistently in excess of 400 m^3/m^2. The coagulation and DAF processes remove 60 to 70 percent of the influent TOC. Finished water UV_{254} has varied between 0.02 and 0.04 cm^{-1}.

Case Study: Haworth, New Jersey, United States. The Haworth Water Treatment Plant provides water for approximately 800,000 people in Bergen and Hudson counties of northeastern New Jersey. It is owned and operated by United Water New Jersey (UWNJ). Prior to 2009, the Haworth WTP was a direct filtration plant with pre-ozone. Pilot studies took place in 2006 to 2007 to consider a plant renovation and upgrade with DAF for clarification. The full-scale DAF plant went on-line in 2009, and improvements to chlorination were finished in January 2010 to complete the plant upgrade. It is a 760 ML/d (200 MGD) high-rate DAF plant, and it is the largest DAF plant in North America (when the Croton, New York City, plant is completed in 2012 or 2013, it will be the largest at 1100 ML/d [290 MGD]). The information and data for this case study were obtained from Keith Cartnick (2010) (Senior Director of Water Quality and Compliance) and John Dyksen (2010) (Vice President of Capital Investment Planning and Delivery) both with UWNJ.

The supply for the Haworth plant is the Oradell Reservoir, which has moderate to high TOC concentrations and algae blooms, especially in the summer and early autumn. The turbidity varies typically from 0.5 to 14 NTU (average of 5.5 NTU, and a high turbidity event near 50 NTU occurred during a tropical storm). TOC varies from 2 to 10 mg/L (average of 5 mg/L). Algae are the main contributors to the NOM and TOC of the supply. Algal counts vary from 300 to 30,000 cells/mL (average of 2300 counts per mL). A large fraction is blue-green algae (cyanobacteria), which cause tastes and odors. The raw water alkalinity ranges from 55 to 110 mg/L as $CaCO_3$ (average of 84 mg/L $CaCO_3$). The algae activity can increase the pH during the day making control of coagulation problematic. The range in pH is 7 to 9 (average of 7.7). Mn averages 0.1 mg/L.

Pilot studies were conducted over an approximately one-year period during 2006 to 2007. The pilot studies were done because of an aging existing plant, to meet new regulations on *Cryptosporidium* and disinfection by-products, and because of deteriorating raw water quality of higher TOC and algae. The pre-ozone direct filtration plant had short filter runs during algal blooms and was limited in coagulation strategy to remove sufficient

TOC to minimize DBPs. The pilot studies investigated alternative pre-oxidants and disinfectants (ozone, chlorine dioxide) and potassium permanganate. Intermediate placement of ozone and chlorine (following DAF and prior to filtration) were studied. Different filter media were tested: sand, dual media filters, and GAC. An important part of the pilot studies was investigation of high-rate DAF. DAF loading rates were varied from 20 to 44 m/h (8 to 18 gpm/ft^2). The pilot studies demonstrated the feasibility of high-rate DAF. With proper coagulant dosing and pH control, algae removals by DAF were high (average of 93 percent). Improved removals of TOC and reduction in DBPs were also found. The studies also demonstrated that pre-ozone improved particle removals by the filters following DAF and filtration. For the plant renovation, it was decided to use pre-ozone, high-rate DAF, new chlorine contact chambers, and to retain dual media filtration from the existing plant.

Design and operating values for flocculation, DAF, and filtration are summarized in Table 11-1. The plant consists of pre-ozone, coagulation with PACl (typical product dose is 20 to 30 mg/L) and a high charge-density cationic polymer (dosing at 1.5 to 2.5 mg/L), two-stage flocculation with total detention time of 10 min, high-rate DAF (AquaDAF®) at a design hydraulic loading of 39 m/h (16 gpm/ft^2), dual media filtration, and disinfection. Sulfuric acid can be added, prior to ozone for pH control of coagulation, when algal blooms occur causing high pH in the raw water.

Plant performance over the first 12 to 18 months was excellent. DAF effluent turbidities were 0.5 to 0.7 NTU. Filtered water turbidities were 0.03 to 0.07 NTU. Filter run times during summer months are 36 to 60 h, indicative of excellent algae removals by DAF, with maximum run times in the winter of about 185 h, and 90 h in summer during higher demands. THMs were reduced by 40 percent compared to previous years, and chlorine usage has been reduced substantially. Reduction of haloacetic acids (HAAs) has been even greater.

Tastes and Odors

The exceptional ability of DAF to remove algae has been mentioned on numerous occasions. Algal blooms are nearly always associated with real or potential problems with tastes and odors. The challenge of incorporating some process for taste and odor control in conjunction with DAF has therefore been considered by designers for years. With longer, sustained periods of tastes and odors, it is more cost-effective to use fixed-bed GAC that follow DAF. With short, occasional taste and odor events, the economic option is to use powdered activated carbon (PAC).

Conventionally, PAC is added to the raw water, with coagulant addition and flocculation following a PAC contact time of 10 to 20 min. During the subsequent DAF step, the spent PAC is concentrated in the float layer with all the other flocs and particles, and removed by scraping or flushing. Such systems are successfully in use at South African treatment plants. A recent study from South Korea, which used reservoir water spiked with algae, geosmin and 2-MIB, found that the removal of algae stays practically the same whether PAC is used or not (Roh et al., 2008). Their summary findings are shown in Table 11-2.

The PAC-DAF system, first introduced at full scale in South Africa during the 1990s, provides a more sophisticated combination of DAF and PAC (Offringa, 1995). In order to get better utilization of the expensive PAC, the PAC is applied in two stages. After DAF, fresh PAC is dosed and removed by a lamella-assisted sedimentation step. The PAC-enriched sludge from the sedimentation step is then recycled to the inlet of the plant. The PAC is then removed for a second time in the DAF reactor before being disposed of. The process diagram is shown in Fig. 11-9.

TABLE 11-2 Removal Efficiency of Algae and Their Secondary Metabolites by DAF with and without PAC

	Without PAC	With PAC[a]
Anabaena	96%	94%
Microcystis	96%	95%
2-MIB	2%	92%
Geosmin	5%	98%

Source: from Roh et al., 2008.

[a]Conditions: PAC (20 mg/L), adsorption time (50 min), coagulant (polyaluminum chloride, 35 mg/L), initial concentration of *Anabaena* (6.9×10^4 cells/mL), initial concentration of *Microsystis* (5.2×10^5 cells/mL), initial concentration of 2-MIB (26 ng/L), initial concentration of geosmin (31 ng/L).

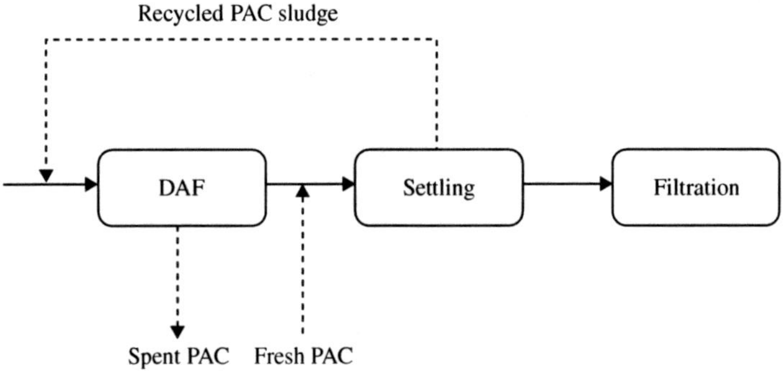

FIGURE 11-9 Process schematic of the PAC-DAF process that allows the countercurrent use of PAC in two treatment stages.

Giardia and Cryptosporidium Removals

Giardia cysts and *Cryptosporidium* oocysts have a low density of about 1070 kg/m^3. Of course, when incorporated into flocs following coagulation the density of the flocs would not be greater except if treating water supplies containing moderate to high amounts of mineral turbidity. Thus, DAF is an ideal process for removing these particles. Several papers have reported on the performance of DAF in removing protozoan cysts from water. Notable ones are those by Hall et al. (1995), Plummer et al. (1995), Edzwald and Kelley (1998), Edzwald et al. (2000), and Edzwald et al. (2003). Some data from the last two are summarized below.

Edzwald et al. (2000) compared DAF and granular media filtration performance against sedimentation and granular media filtration for removals of *Giardia lamblia* cysts and *Cryptosporidium parvum* oocysts under challenge (high) concentrations. The pilot studies were done at the Warner DAF plant in Fairfield, Connecticut (see case study, above). The studies found that DAF achieved *Giardia* removals of 2 to 3-log compared to 0.8 to 1.5-log by plate settling. Likewise, DAF performed much better in removing *Cryptosporidium* with removals of 1.7 to 2.5-log compared to 0.6 to 1.4-log by plate settling. DAF turbidities and particle counts were lower in the DAF effluent compared to plate settling, which is in agreement with the performance in removing *Giardia* and *Cryptosporidium*. Cumulative removals for *Giardia* and *Cryptosporidium* by DAF and filtration or plate settling and filtration were at least 5-log, but an important finding is that the filters had to be relied on to a greater extent for the plate settling plant.

Pilot-plant experiments were run by Edzwald et al. (2003) examining removal of *Cryptosporidium parvum* oocysts that were spiked continuously into the raw water for about 20 to 24 h (filter run duration) at lower raw water concentrations than the work above. The raw water concentrations were about 20 to 40 oocysts per L for summer conditions, and for winter experiments they were about 80 to 160 oocysts per L. The DAF pilot train was run for DAF and filter loadings of 14.6 m/h (6 gpm/ft^2) compared to the plate settling train run at a nominal plate settling loading of 1.7 m/h (0.7 gpm/ft^2) and filter rate of 7.3 m/h (3 gpm/ft^2). Pretreatment flocculation times were 12 min for DAF and 28 min for plate settling. Figure 11-10 shows that DAF achieved 2-log removals of *Cryptosporidium* for summer and winter seasons. Plate settling performed well for summer temperatures at slightly less than 2-log removal, but only achieved 1-log *Cryptosporidium* removal in the winter. DAF turbidities (~0.5 NTU) and particle counts (~500 particles/mL in the 2 to 15 μm size range) were much lower than plate settling turbidities (~1.6 NTU) and particle counts (~2200 particles/mL).

Cumulative *Cryptosporidium* log removals by DAF and dual media filtration or plate settling and filtration were 4 to 5. Plate settling did not perform as well as DAF as stated above so the fate of pathogens and their concentrations in filter backwash water are greatly affected by the efficiency of the clarification process. This is addressed in Chap. 12 under treatment of spent filter backwash water.

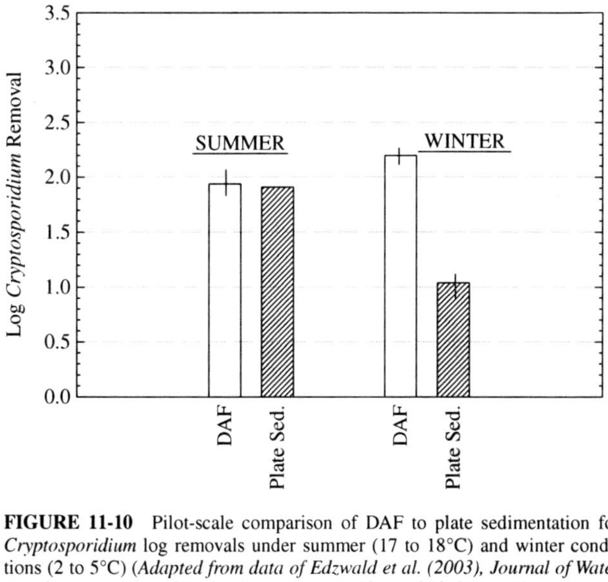

FIGURE 11-10 Pilot-scale comparison of DAF to plate sedimentation for *Cryptosporidium* log removals under summer (17 to 18°C) and winter conditions (2 to 5°C) (*Adapted from data of Edzwald et al. (2003), Journal of Water Supply: Research and Technology – AQUA, 52 (4), 243–248, with permission from the copyright holders, IWA Publishing*).

REFERENCES

Amato, T. (2010), Personal communication, Enpure Ltd., Birmingham, UK.

Amato, T. (2011), Personal communication, Enpure Ltd., Birmingham, UK.

Cartnick, K. (2010), Personal communication, United Water New Jersey, Haworth, New Jersey.

Chung, Y., Kang, H. S., Choi, Y. C., and Kim, J. (2000), Improvement of water purification combining post-dissolved air flotation system in sedimentation, *Proceedings of the 4th International Conference: Flotation in Water and Waste Water Treatment*, September 11–14, Helsinki, Finland.

CIWEM (1997), *Dissolved Air Flotation*, London: Chartered Institution of Water and Environmental Management.

Crossley, I. A., and Valade, M. T. (2006), A review of the technological developments of dissolved air flotation, *Journal of Water Supply: Research and Technology – Aqua*, 55 (7–8), 479–491.

Crossley, I. A., Herzner, J., Bishop, S. L., and Smith, P. D. (2007), Going underground—constructing New York City's first water treatment plant, a 1100 Ml/d dissolved air flotation, filtration and UV facility, *Proceedings of the 5th International Conference on Flotation in Water and Wastewater Systems*, Seoul National University, Seoul, South Korea.

Dahlquist, J. (2010), Personal communication, Lackeby Water, Purac (Sweden).

Dyksen, J. (2010), Personal communication, United Water New Jersey, Haworth, New Jersey.

Eades, A., Jordan, D., and Scheidler, S. (1997), Counter-current dissolved air flotation filtration, in *Dissolved Air Flotation*, London: Chartered Institution of Water and Environmental Management.

Edzwald, J. K. (2010), Dissolved air flotation and me, *Water Research*, 44 (7), 2077–2106.

Edzwald, J. K., and Wingler, B. J. (1990), Chemical and physical aspects of dissolved air flotation for the removal of algae, *Journal of Water Supply Research and Technology – Aqua*, 39 (2), 24–35.

Edzwald, J. K., Walsh, J. P., Kaminski, G. S., and Dunn, H. J. (1992), Flocculation and air requirements for dissolved air flotation, *Journal AWWA*, 84 (3), 92–100.

Edzwald, J. K., Olson, S. C., and Tamulonis, C. W. (1994), *Dissolved Air Flotation: Field Investigations*, AWWA Research Foundation, Denver, CO, pp. 69–77.

Edzwald, J. K., and Kelley, M. B. (1998), Control of *Cryptosporidium*: from reservoirs to clarifiers to filters, *Water Science and Technology*, 37 (2), 1–8.

Edzwald, J. K., Tobiason, J. E., Amato, T., and Maggi, L. J., (1999), Integrating high-rate DAF technology into plant design, *Journal AWWA*, 91 (12), 41–53.

Edzwald, J. K., Tobiason, J. E., Parento, L. M., Kelley, M. B., Kaminski, G. S., Dunn, H. J., and Galant, P. B. (2000), *Giardia* and *Cryptosporidium* removals by clarification and filtration under challenge conditions, *Journal AWWA*, 92 (12), 70–84.

Edzwald, J. K., Tobiason, J. E., Udden, C., Kaminski, G. S., Dunn, H. J., Galant, P. B., and Kelley, M. B. (2003), Evaluation of the effect of recycle of waste filter backwash water on plant removals of *Cryptosporidium*, *Journal of Water Supply: Research and Technology – Aqua*, 52 (4), 243–258.

Edzwald, J. K., and Kaminski, G. S. (2009), A practical method for water plants to select coagulant dosing, *Journal of the New England Water Works Association*, 123 (1) 15–31.

Finlayson, G. M., and Huijbregsen, C. M. (1997), In-filter DAF in Australia: the Rous County Council application, in *Dissolved Air Flotation*, London: Chartered Institution of Water and Environmental Management.

Gregory, R., and Edzwald, J. K. (2010), Sedimentation and flotation, in J. K. Edzwald, ed., *Water Quality and Treatment: A Handbook on Drinking Water*, 6th ed., New York: AWWA and McGraw-Hill.

Hall, T., Pressdee, J., Gregory, R., and Murray, K. (1995), Cryptosporidium removal during water treatment using dissolved air flotation, *Water Science and Technology*, 31 (3–4), 125–135.

Han, M., and Edzwald, J. K. (2007), *Proceedings of the 5th International Conference on Flotation in Water and Wastewater Systems*, Seoul National University, Seoul, South Korea.

Helsinki Conference (2001), Flotation in water and waste water treatment, *Water Science & Technology*, 43 (8), 1–208.

Henderson, R. K., Parsons, S. A., and Jefferson, B. (2008), The impact of algal properties and pre-oxidation on solid-liquid separation of algae, *Water Research*, 42 (8-9), 1827–1845.

Ives, K. J., and Bernhardt, H. J. (eds) (1995), Flotation processes in water and sludge treatment, *Water Science & Technology*, 31 (3-4), 1-346.

Journal AWWA (2010), New water treatment works respects past ... with an eye to the future, *Journal of the American Water Works Association*, 102 (10), 38–39.

Kaminski, G. S. (2010), Personal communication, Aquarion Water Company of Connecticut, Bridgeport, CT.

Kiuru, H. J. (2000), Development of dissolved air flotation technology from the 1st generation to the newest or 3rd one (very thick micro-bubble bed) with high flow-rates (DAF in turbulent conditions), *Proceedings of the 4th International Conference: Flotation in Water and Waste Water Treatment*, September 11–14, Helsinki, Finland.

Kwak, D. H., Jung, H. J., Kim, S. J., Won, C. H., and Lee, J. W. (2005), Separation characteristic of inorganic particles from rainfalls in dissolved air flotation: a Korean perspective, *Separation Science and Technology*, 40 (14), 3001–3015.

Kwon, S. B., Ahn, H. W., Ahn, C. J., and Wang, C. K. (2004), A case study of dissolved air flotation for seasonal high turbidity water in Korea, *Water Science & Technology*, 50 (12), 245–253.

Longhurst, S. J., and Graham, N. J. D. (1987), Dissolved air flotation for potable water treatment: a survey of operational units in Great Britain, *The Public Health Engineer* 14 (6), 71–76.

Officer, J., Ostrowski, J. A., and Woollard, P. J. (2001), The design and operation of conventional and novel flotation systems on a number of impounded water types, *Water Science & Technology: Water Supply*, 1 (1), 63–69.

Offringa, G. (1995), Dissolved air flotation in Southern Africa, *Water Science and Technology*, 31 (3–4), 159–172.

Pernitsky, D. J. (2010), Personal communication, CH2M Hill, Calgary, Canada.

Pernitsky, D. J., Breese, S., Wobma, P., Griffin, G., Kjartansan, K., and Sorokowski, R. (2007), From pilot tests to design on Canada's largest DAF water treatment plant, *Proceedings of the 5th International Conference on Flotation in Water and Wastewater Systems*, Seoul National University, Seoul, South Korea.

Plummer, J. D., Edzwald, J. K., and Kelley, M. B. (1995), Removal of *Cryptosporidium parvum* from drinking water by dissolved air flotation, *Journal AWWA*, 87 (9), 85–95.

Richter, C. A., and Gross, F. (2000), Dissolved air flotation in Latin America, *Proceedings of the 4th International Conference: Flotation in Water and Waste Water Treatment*, September 11–14, Helsinki, Finland.

Roh, S. H., Kwak, D. H., Jung, H. J., Hwang, K. J., Baek, I. H., Chun, Y. N., Kim, S. I., and Lee, J. W. (2008), Simultaneous removal of algae and their secondary algal metabolites from water by hybrid system of DAF and PAC adsorption, *Separation Science and Technology*, 43 (1), 113–131.

Teixeira, M. B., and Rosa, M. J. (2006), Comparing dissolved air flotation and conventional sedimentation to remove cyanobacterial cells of *Microcystis aeruginosa* Part I: The key operating conditions, *Separation and Purification Technology*, 52, 84–94.

Teixeira, M. B., and Rosa, M. J. (2007), Comparing dissolved air flotation and conventional sedimentation to remove cyanobacterial cells of *Microcystis aeruginosa* Part II: The effect of water background organics, *Separation and Purification Technology*, 53, 126–134.

van Puffelen, J., Buijs, P. J., Nuhn, P. N. A. M., and Hijen, W. A. M. (1995), Dissolved air flotation in potable water treatment: the Dutch experience, *Water Science and Technology*, 31 (3–4), 149–157.

Vlaški, A., van Breemen, A. N., and Alaerts, G. J. (1996), Optimisation of coagulation conditions for the removal of cyanobacteria by dissolved air flotation or sedimentation, *Journal of Water Supply: Research and Technology – Aqua*, 45 (5), 252–261.

CHAPTER 12
ADDITIONAL APPLICATIONS

12-1 DAF PRETREATMENT FOR MEMBRANE PROCESSES 12-1
General Features and Operating Conditions of Membrane Processes 12-2
DAF Pretreatment for MF and UF 12-4
Pretreatment for High-Pressure Membrane Systems 12-7
12-2 WATER REUSE 12-7
Possibilities for DAF in Water Reuse Projects 12-7
Case Study: Direct Potable Reuse in Windhoek, Namibia 12-7
12-3 SPENT FILTER BACKWASH WATER 12-9
Model Predictions on Plant Influent Pathogen Concentrations with Recycle of Untreated SFBW 12-10
Effects of SFBW on Clarifier and Filtered Water Performance 12-11
Treatment of SFBW 12-13
REFERENCES 12-14

It was pointed out in Chap. 1 (see Table 1-1) that there are many applications of DAF with the emphasis of the book on drinking water clarification. In Chap. 11, we examined in detail the use of DAF in the conventional mode of placement prior to granular media filtration. In this chapter, we discuss some other applications of DAF. We begin with the use of DAF pretreatment for low-pressure and high-pressure membranes. We follow this with consideration of water reuse and presentation of a case study of direct potable water reuse. We also present one application of DAF for the treatment of a waste stream, specifically in the treatment of spent filter backwash water (SFBW).

12-1 DAF PRETREATMENT FOR MEMBRANE PROCESSES

Membranes have become increasingly important in treating drinking waters for particle removal by low-pressure processes of microfiltration (MF) and ultrafiltration (UF) and for the removal of dissolved matter with high-pressure processes of nanofiltration (NF) and reverse osmosis (RO). We address some general characteristics and operation of membrane processes. This is followed by consideration of DAF pretreatment with low-pressure membranes including a case study. Next, we discuss briefly pretreatment prior to high-pressure membranes noting that the important application of DAF pretreatment in RO desalination is covered in Chap. 13. Our coverage of membranes is brief and so we refer the reader to two recent, primary sources: Chap. 11 by Duranceau and Taylor (2010) in *Water Quality & Treatment: A Handbook on Drinking Water*, and a chapter, *Membrane Filtration*, by MWH (2005). An older, but another good, primary reference is the book, *Water Treatment Membrane Processes* by Mallevialle et al. (1996).

General Features and Operating Conditions of Membrane Processes

The middle of Fig. 12-1 shows different types of particles (solid phase) and their sizes in microns and dissolved matter with respect to molecular weight (MW) and apparent size. The bottom of the figure shows applicable membrane processes. MF and UF are particle separation processes that are alternatives to granular media filtration. They accomplish particle removal by straining; in other words, they remove particles larger than their pore openings. MF has a nominal minimum-rejection size of 0.1 μm, but actual rejection varies around this size because of differences among makes of membranes and distribution of the pore sizes. Without pretreatment of coagulation and flocculation, MF does not remove small colloids, small viruses, and nanoparticles. MF does not remove dissolved NOM (DOC) so coagulation is required to convert the DOC into solid particles. UF has a nominal minimum-rejection size of 0.01 μm, but the actual rejection size varies somewhat. UF can remove, without pretreatment, small colloidal particles and is more effective than MF for virus removal. UF can remove some of the higher MW dissolved NOM, but coagulation and clarification pretreatment is often used for effective integration of UF for removal of particles and NOM—see section below.

NF and RO are used to remove dissolved matter from water. The term nanofiltration was coined because the membrane-cutoff size was about 1 nm (MWH, 2005). NF is used primarily for water softening. There has been some use of NF to remove NOM from water supplies in Norway and Florida (United States), but except for unique situations, it is more effective to use pretreatment to remove NOM. Over time, manufacturers have produced a variety of RO membranes of different formulations and rejection characteristics such that these RO membranes are similar to the original NF membranes. RO is used primarily to remove dissolved salts from water and has important application in treating brackish waters and seawater. DAF pretreatment for desalination of seawater is covered in Chap. 13. NF and RO remove various other dissolved contaminants such as pesticides, other synthetic

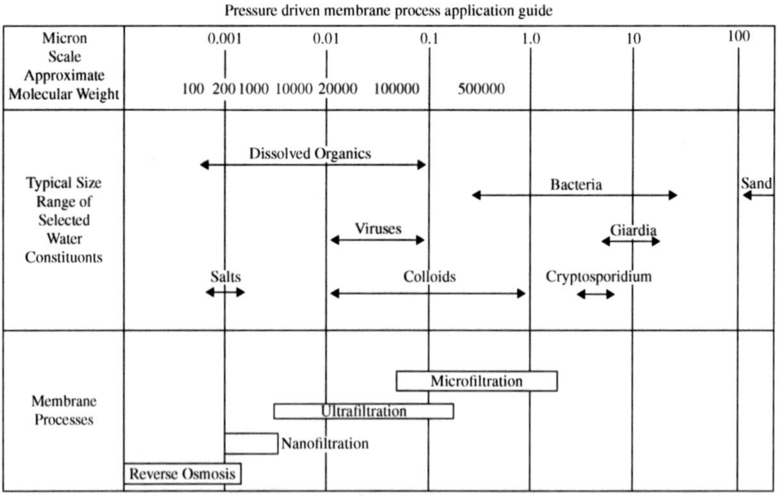

FIGURE 12-1 Sizes of contaminants and membrane processes (*Source: Duranceau and Taylor (2010), reprinted with permission from Water Quality and Treatment: A Handbook on Drinking Water, 6th ed., Copyright 2010, AWWA*).

organic chemicals, and so on (Duranceau and Taylor, 2010). NF and RO are not intended for particle removal so pretreatment to prevent particle fouling is required.

Figure 12-2(a) shows a generic schematic for a membrane process that is used to describe terminology and discuss pressures and flow. The membrane process field has its own unique terminology. The applied flow of water is called the feed (water or stream). The treated water is called permeate, and what is rejected by the membrane is called concentrate or retentate—in the figure we use the common water treatment term, waste stream. What is called the hydraulic loading for other water treatment processes is called the water or permeate flux. It is the flow through per unit area of the membrane; L/m^2-h is the unit used. Typical values are presented for pressure-driven membrane processes in Table 12-1. Note that MF and UF are also available in submerged or immersed vacuum-driven systems—see Duranceau and Taylor (2010) and MWH (2005) for descriptions of these systems. As we move from MF to UF to NF to RO, the pores of the membranes or MW cutoffs decrease; consequently, the pressure increases to produce treated water. Pressure across the membrane is called the transmembrane pressure (TMP). Typical permeate fluxes and TMP as well as other characteristics and operating conditions of the various membrane processes are summarized in Table 12-1.

Common materials used for the membranes in MF and UF are polyvinylidene fluoride (PVDF), polyethersulfone (PES), polysulfone, and cellulose triacetate. The common configuration for these membranes is flow from the outside-in (see Fig. 12-2(b)) or inside-out through hollow fibers. In recent years there has also been development of ceramic membranes (Duranceau and Taylor, 2010). NF and RO membranes are mainly spiral-wound

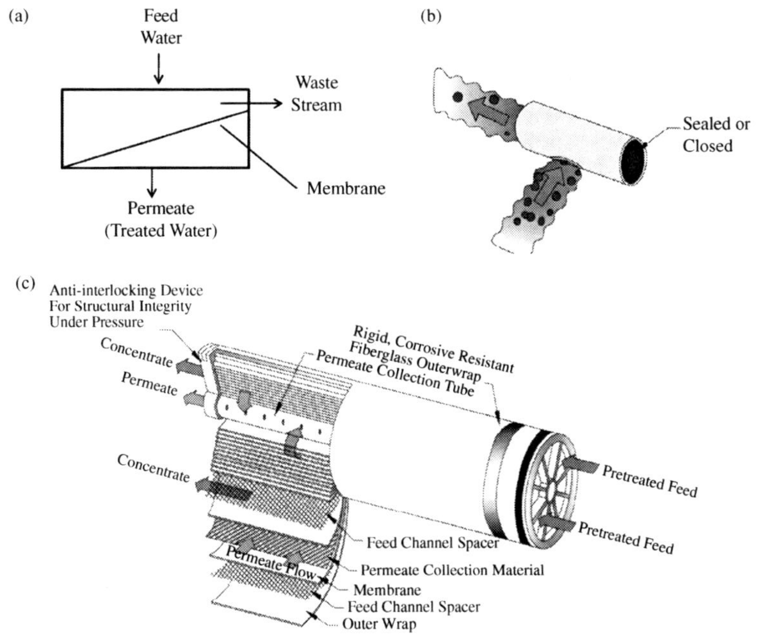

FIGURE 12-2 (a) Schematic of separation by membrane, (b) hollow-fiber membrane with outside-in flow (*Source: Duranceau and Taylor (2010), reprinted with permission from Water Quality and Treatment: A Handbook on Drinking Water, 6th ed., Copyright 2010, AWWA.*), (c) spiral-wound reverse osmosis membrane (*Source: Courtesy of the American Membrane Technology Association, Stuart, FL*).

TABLE 12-1 Typical Operating Characteristics and Conditions for Membrane Processes

Parameter	MF	UF	NF	RO
Permeate flux* (L/m^2-h)	30–170	30–170	1–50	1–50
Transmembrane pressure* (kPa)	20–275	20–275	40–3200	1030–8300†
Recovery (%)	85–98	85–98	70–90	40–90
Typical operation	Single stage, dead-end filtration with backwashing		2 or 3 stage with constant crossflow	
Hydraulic cleaning	Backwash with water (and air); frequency: 10s of minutes to hours		Constant feed-side crossflow to prevent fouling	
Chemical cleaning frequency	<10 to 180 days		~1 month to 1 year	

Source: From MWH (2005), Duranceau and Taylor (2010), Duranceau (2010), Howe (2010), and Tobiason (2010).

*Values for pressure systems; lower values for submerged systems.

†Values depend on the TDS; toward the higher end of the range for seawater.

configurations as illustrated in Fig. 12-2(c). The materials include polyacrylamide and polyamide. Cellulose acetate is used in RO hollow-fiber membranes.

As membranes operate and remove material, fouling occurs that can be caused by several mechanisms: (1) colloidal particle fouling, (2) biofouling in which organic matter of biological origin leads to biofilm growth, (3) organic matter fouling caused by adsorption of NOM or organic compounds, and (4) scaling caused by precipitation of salts on NF and RO membranes. As contaminants are removed or rejected by membranes, the TMP must increase to maintain a constant water flux. MF and UF are backwashed on a frequent basis (see Table 12-2) to remove particles and decrease the operating TMP. Fouling over time occurs requiring chemical cleaning of the membranes. NF and RO membrane plants must have pretreatment to remove particles and organic material that may lead to biofilms.

DAF Pretreatment for MF and UF

For groundwaters with Fe and Mn in reduced oxidation states, the pretreatment prior to MF and UF may be simply an oxidation step in which the oxidized Fe and Mn precipitate and the particles are then removed by MF or UF. For surface waters, several water quality problems can occur that require pretreatment of coagulation, flocculation, and clarification by sedimentation or DAF.

As pointed out above MF is a particle separation process for particles with sizes of about 0.1 μm and larger. Removal of dissolved NOM (DOC, UV_{254}, true color, disinfection byproduct precursors) requires coagulation. Algae can be a cause of biological fouling so pretreatment should be considered. UF is also primarily a particle separation process as pointed out above. UF can remove smaller particles (nominally about 0.01 μm and larger) and a small amount of the large MW dissolved NOM, but for the same reasons as for MF, it is often better to remove as much of the NOM and algae with pretreatment processes. The coagulation and flocculation of algae and NOM produce low-density flocs that are more amenable to removal by DAF than by sedimentation processes. Figure 12-3 shows a treatment process schematic of DAF for either a MF or UF plant. DAF prior to MF or UF can provide an integrated treatment plant with lower energy costs, because

TABLE 12-2 Summary of Design and Operating Parameters for the Case Studies

	Case studies	
Parameter	Scottsdale, AZ (USA) (Membrane Plant)	Windhoek, Namibia (Water Reuse)
Flow (ML/d)	76	21
Flocculation		
Number of units	2	4
Number of stages/unit	2	1
Time (minutes)	10	9
Type	Propeller mixers	Propeller mixers
G (s^{-1})	NP	25–120 (design) 50–70
Flotation		
Number of units	2	4
Length in m (total of contact and separation zone)	12.8	11.2
Width in m	6.1	7
Depth in m	NP	3.5
Loading (m/h)	20	4
Sludge removal	Mechanical scrapers	Hydraulic
Type of saturator	Unpacked	Packed
Number of saturators	2	2 (1.2 m diameter)
Packing type	NA	Pall rings (40 mm)
Saturator pressure (kPa)	620–690	450–550 (480–540 typical)
Recycle rate (%)	10 (typical)	10–11 (typical)
Injection system (# per tank)	Nozzles (136)	Nozzles (96)
Filtration	Note 1	
Filter rate for dual media (m/h)	NA	6.0
BAC upflow filter rate (m/h) (EBCT, min)	NA	6.8 (10)
GAC upflow/down flow rate (m/h) (EBCT, min)	NA	6.8 (20)

NA: not applicable; NP: not provided.

Note 1: ultrafiltration membrane plant; see text.

DAF reduces the fouling (less frequent chemical cleaning) and particle load applied to the membranes.

One example of DAF pretreatment is the 150 ML/d (40 MGD) membrane water treatment plant that opened in 2005 for the South San Joaquin Irrigation District (SSJID) of California. It employs high-rate DAF at 35 m/h (14 gpm/ft^2) prior to immersed hollow-fiber membranes (Crossley and Valade, 2006). Next, we present in some detail, a case study of DAF pretreatment in an ultrafiltration membrane plant.

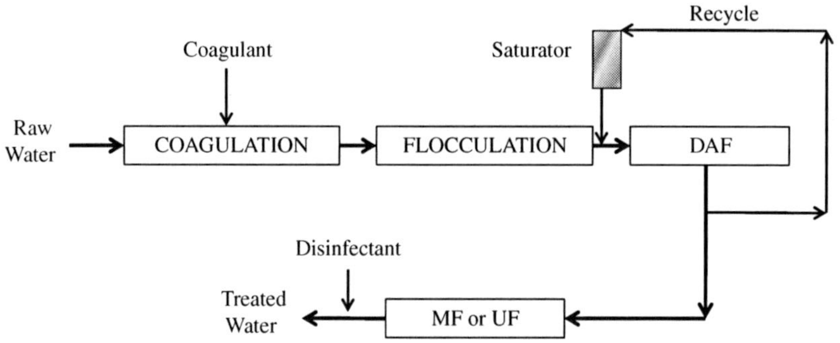

FIGURE 12-3 Treatment process schematic of DAF clarification in a microfiltration or ultrafiltration membrane plant.

Case Study: Scottsdale, Arizona, United States—DAF Pretreatment in an Ultrafiltration Membrane Plant. This case describes a plant expansion for the City of Scottsdale (Arizona, United States) and the Central Arizona Project (CAP) Water Treatment Plant. The City's main supply is the Central Arizona Project Canal (Colorado River water) and is transported by aqueduct in the CAP system to the water treatment plant. The plant expansion of 76 ML/d (20 MGD) consists of high-rate DAF followed by membrane filtration. The information for this case study was provided by Passantino (2010) and Tymkiw (2010) of Malcolm Pirnie, Inc.

To meet increasing water demand, the City embarked on a series of pilot studies in 2006 to examine the feasibility of increasing plant capacity with a small treatment footprint. High-rate flotation was evaluated against ballasted sand sedimentation as pretreatment clarification ahead of membrane filtration. MF and UF were tested. From the pilot study results, DAF clarification was selected based on its ability to (1) provide high quality clarified water to the membranes, (2) produce lower membrane fouling potential, and (3) produce sludge that can be used directly by existing solar drying beds.

The water supply is characterized as low in turbidity and TOC, and high in alkalinity and hardness. Raw water quality data for the period of January 1995 to May 2000 provided average and 99th percentile values (the latter in parenthesis) as follows: turbidity of 3.8 NTU (16 NTU), TOC of 3.0 mg/L (5.2 mg/L), UV_{254} of 0.044 cm^{-1} (0.076 cm^{-1}), alkalinity of 135 mg/L as $CaCO_3$ (154 mg/L), and total hardness of 281 mg/L as $CaCO_3$ (308 mg/L). The water temperature is generally warm with minimum, average, and 99th percentile values of 11, 20, and 29°C, respectively.

The new facility was commissioned in 2010. The plant is presently designed for 76 ML/d (20 MGD) by operating the DAF units at 20 m/h (8 gpm/sf), but it is anticipated that in the future the DAF units can be operated at 30 m/h (12 gpm/sf) increasing the plant capacity to 115 ML/d (30 MGD). Alum coagulation is used. Polymer dosing is not generally needed, but can be used during high turbidity events. Polymer types acceptable to membrane manufactures were evaluated during pilot testing and will be used. Design and operating values for flocculation and DAF are presented in Table 12-2. Two-stage flocculation follows alum coagulation with a hydraulic detention time of 5 min per stage. Two high-rate DAF units are each designed for a hydraulic loading of 20 m/h (8 gpm/sf). The recycle rate is typically 10 percent with a saturator pressure of 620 to 690 kPa (90 to 100 psi). There is one unpacked saturator for each DAF tank. A UF membrane follows DAF. It is a Memcor® hollow-fiber membrane designed for a flux of 74 L/m^2-h (44 gal/ft^2-d) and transmembrane pressure of 55 to 125 kPa (8 to 18 psi).

Pretreatment for High-Pressure Membrane Systems

NF is used mainly for water softening and to treat brackish waters. Water softening is not the subject of this book so we do not address this application. Treatment of seawater by RO and DAF as a pretreatment process is covered in Chap. 13.

There is some use of NF in removing NOM from water supplies and for some groundwater supplies containing both hardness and NOM. Early work on these latter type waters in Florida (United States) was done by Taylor et al. (1987) demonstrating the use of NF and RO membranes. There is some use of NF to remove NOM and color without pretreatment. There are numerous small NF plants in Norway and a few other counties (Thorsen and Fløgstad, 2006) for NOM removal from water supplies of low turbidity and without algae problems. For most water supplies with NOM, especially for larger water plants, pretreatment is essential and even more so with the presence of algae and turbidity. The authors are not aware of any full-scale plants employing DAF pretreatment in NF plants, but DAF would be a good application and should be considered.

There have been some bench-scale studies of DAF pretreatment prior to NF for waters containing *Microcystis aeruginosa* (blue-green alga) and associated microcystins (Teixeira and Rosa, 2006). The benefit here of DAF pretreatment is reduction of the biological fouling potential of the water prolonging NF operation.

12-2 WATER REUSE

Possibilities for DAF in Water Reuse Projects

Reclaimed wastewater may be returned directly to the potable network after extensive treatment (direct potable reuse), or it may also pass through an environmental buffer before returned to the potable network (indirect potable reuse). Alternatively, the reclaimed wastewater may be directed toward non-potable uses. For all these options, DAF may find useful application.

Direct potable reuse was commissioned during 1969 in Windhoek, capital of Namibia. This plant is featured as a case study in the next section. Since that time, the international consensus converged on indirect potable reuse as the preferred option (Drewes and Khan, 2010), leaving Windhoek as the only direct potable reuse system in the world. The number of indirect potable reuse systems is gaining steady acceptance as evidenced by a growing number of full-scale applications (see examples in Drewes and Khan, 2010). DAF could firstly be used for the final polishing of the treated wastewater, whether the water is destined for further treatment (direct reuse) or discharged to the environmental buffer (indirect reuse). Wastewater particles are light and organic in nature, and are ideal for removal by DAF. Secondly, DAF could be used in the treatment train that treats the water to its final quality before distribution. Most modern reuse systems employ membrane processes as the final steps, and the information in the preceding section provided details of DAF as a membrane pretreatment step.

Case Study: Direct Potable Reuse in Windhoek, Namibia

The Goreangab Water Reclamation Plant (GWRP) is part of the direct potable water reuse system that had been serving Windhoek continuously for more than 40 years. Through strict zoning regulations, the domestic wastewater of the city is collected separately from the industrial wastewater. The domestic wastewater is treated at a biological nutrient removal plant and then discharged to a series of shallow maturation ponds, which provide average retention of two to three days. The maturation ponds were introduced for additional viral reduction by viricidal radiation from the sun. From here, the water passes to the potable water reclamation plant.

The different components of the reclamation scheme have been under continuous review and improvement since its inception. By 1995, the fourth different process train had evolved, each an improvement on the previous, while viewing wastewater treatment and advanced water treatment as a continuum (Haarhoff and Van der Merwe, 1996). Before ammonia removal was achieved during wastewater treatment, for example, stripping towers had to be used at the water treatment plant. From a system perspective, it is more effective to have better nitrification at the wastewater treatment plant—this was implemented in 1979 and the stripping towers could be abolished. The GWRP was replaced with a new facility which started production during September 2002, following an extensive process review based on the previous GWRP performance and pilot testing conducted for over two years. After completion, the new GWRP is operated by the Windhoek Goreangab Operating Company (WINGOC), a joint venture which includes Veolia Water, VA TECH Wabag, and Berlinwasser International as partners. The GWRP process train is shown in Fig. 12-4. Design and operating parameters are presented in Table 12-2. Information for this case study was kindly provided by Van der Merwe (2010; Enves Consulting Engineers), Theron-Beukes (2010; WINGOC) and Menge (2011; City of Windhoek).

The DAF plant was constructed as part of a turnkey contract. During commissioning, a number of problems were identified and rectified. Two of these problems warrant discussion. The first relates to the disappointing character of the air suspension which included unexpectedly many coarse bubbles and uneven distribution of white water through the nozzles in the DAF contact zone. This problem was traced to the position of the air saturators (placed 3.5 m below the water level in the DAF reactors) and fairly high frictional and secondary losses in the pipework leading from the saturators to the nozzles in DAF tanks. This, coupled to the high saturation efficiency of the air saturators, led to premature bubble formation upstream of the injection nozzles. The problem with coarse bubbles was rectified by repositioning the saturators at a higher elevation and redesign of the pipework. The problem with the uneven white water distribution was rectified by increasing the nozzles

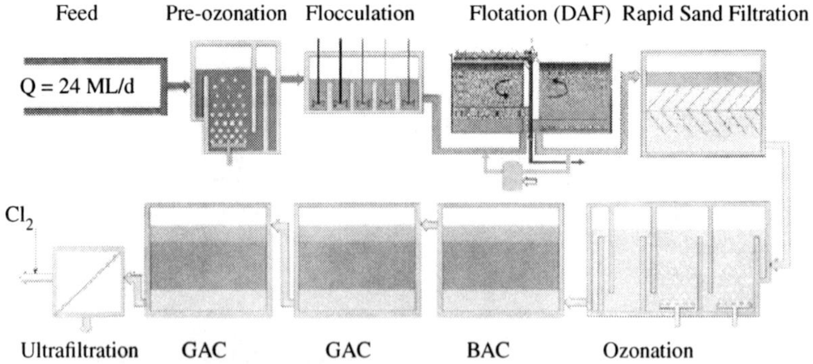

(DAF: dissolved air flotation BAC: biological activated carbon-filter,
GAC: granular activated carton-filter)

FIGURE 12-4 Process Configuration for the Goreangab Water Reclamation Plant (*Courtesy: WINGOC*).

from 6 to 18 nozzles/ML/d. The second problem was related to the incomplete desludging of the sludge hoppers in the floor of the units. Due to undersized desludge pipes, not all the bottom sludge could be removed which led to anaerobic conditions—the sludge had to be manually removed. The operation of the DAF plant improved after the necessary changes were made.

The effluent from the maturation ponds makes up the bulk of the raw water supplied to the GWRP. It is characterized by low turbidity, moderate to high color and alkalinity, and high DOC. The SUVA values, typical of sewage effluents, are low. Algae pose periodic problems and Fe and Mn concentrations are generally low. The main coagulant is $FeCl_3$ which is applied at high doses of 25 to 30 mg Fe/L to improve DOC removal by DAF. During high algal concentrations, a polyacrylamide polymer is added as a floc aid. The properties of the plant influent, as well as the DAF effluent, are summarized in Table 12-3.

12-3 SPENT FILTER BACKWASH WATER

Spent or waste filter backwash water (SFBW) is recycled after or without treatment by many water plants mainly because they do not want to lose the water or they have limited disposal options. The volume of SFBW is typically 2 to 5 percent of a plant's total production. The concerns with recycling SFBW include (1) possible detrimental effects on the plant influent water quality and performance of processes within the plant and (2) possible increases in pathogens, particularly *Cryptosporidium* oocysts in the mixed plant influent flow. This led to regulation of the practice in the United States (USEPA, 2001). The regulations do not necessarily require plants to have treatment before recycling, but they do require that the recycle flow go to the head of the plant and receive full treatment

TABLE 12-3 DAF Performance Data from January 2008 to November 2009 Measured at the Goreangab Water Reclamation Plant (for each parameter, the 10th percentile, median value and 90th percentile are shown)

	Plant Influent			DAF Effluent		
Parameter	P_{10}	Median	P_{90}	P_{10}	Median	P_{90}
pH	7.2	7.8	8.3	7.0	7.4	7.8
Conductivity (mS/m)*	98	110	130	105	120	135
Turbidity (NTU)	1.20	1.69	3.34	0.54	0.73	1.05
Alkalinity (mg/L as $CaCO_3$)	150	177	206	85	107	131
DOC (mg/L)	7.27	9.10	10.69	4.93	5.96	7.00
UV_{254} (cm^{-1})	0.147	0.173	0.196	0.078	0.095	0.113
SUVA (m^{-1} per mg/L)	1.63	1.92	2.26	1.33	1.57	1.85
Chlorophyll *a* (μg/L)	0.92	2.87	8.48	0.23	0.69	2.47

Data provided by Menge (2011).

*Units are milliSiemens per m, in some countries they express as μS/cm—multiply mS/m by 10 to obtain μS/cm)

beginning with coagulation. Many plants, however, do treat the recycle water to minimize any detrimental effects on plant treatment performance and water quality.

In this section, we present a useful model of the fate of pathogens across a conventional water plant and predictions of their concentrations in the plant influent following recycling. Next, we show some data from our work showing that DAF plants are quite effective in removing *Cryptosporidium* oocysts so that DAF plants, if optimized, will not concentrate *Cryptosporidium* within granular media filters. We also show that recycle of SFBW has no adverse impacts on DAF. Finally, we describe briefly treatment options for SFBW and present some information on the use of DAF to treat SFBW.

Model Predictions on Plant Influent Pathogen Concentrations with Recycle of Untreated SFBW

It is instructive to model the fate of pathogens through a conventional water plant in which we make a general assignment for the clarifier to represent sedimentation or DAF and determine whether recycling untreated SFBW increases the plant influent concentration of pathogens. We examine *Cryptosporidium* oocysts as the pathogen of interest but the results apply to other pathogens. We summarize the modelling work presented by Edzwald et al. (2003).

Pathogens can be removed by the clarifier and by granular media filters. A mass balance (actually, number concentration of pathogens) model is used for the following conditions: (1) oocysts not removed by clarification are all removed by the filters, which is the conservative case for evaluating SFBW effects and (2) the SFBW water is collected and mixed uniformly to produce a recycle flow of constant oocyst concentration. The variables in the model are (1) clarifier performance expressed in terms of log removals, (2) the percent of filtered water used for backwashing, and (3) the SFBW recycle rate defined as the recycle flow divided by the raw water flow. The percent water used for backwashing can be varied in the model as desired, but the results are presented for 2.5 percent and we comment on other values. Most conventional water plants use about 2 to 5 percent. In the model, the plant influent flow is maintained constant by reducing the raw water flow to compensate for the recycle flow. Any recycle rate can be used in the model, but we show results for 5, 10, and 20 percent. It is common in the United States to restrict the recycle rate to 10 percent so it serves as a reference value, while 20 percent is a high value case.

The model predictions are presented in Fig. 12-5 showing the plant influent *Cryptosporidium* concentration (C_{inf}/C_{raw}—ratio of oocysts in the mixed influent [raw water flow plus recycle flow] to the raw water concentration) as affected by clarifier performance. A value of C_{inf}/C_{raw} equal to one means there is no effect. The predictions are presented for 5, 10, and 20 percent SFBW recycle rates for 2.5 percent of the filtered water used for backwashing. The predictions show that if the clarifier performance is less than 1.6-log removal (97.5 percent), then oocysts increase in the SFBW to an extent causing the oocysts to increase in the mixed plant influent. This negative impact is greater as the SFBW recycle rate increases. On the other hand, if the clarifier achieves 1.6-log removal, then there is no effect on recycle of SFBW. If the percent of filtered water used for backwashing is greater than the 2.5 percent case for Fig. 12-5, then you are backwashing more frequently and there will be fewer oocysts in the SFBW. Consequently, the required clarifier performance to have no effect on increasing the oocysts concentration in the mixed plant influent flow decreases to 1.3-log removal (95 percent) and 1.0-log removal (90 percent) for 5 and 10 percent, respectively, of filtered water used for backwashing.

It is shown below with performance data that when DAF is operated under good conditions, it can achieve 1.6-log removal and even greater, so for these DAF plants oocysts are concentrated in the DAF floated sludge—not in the filters.

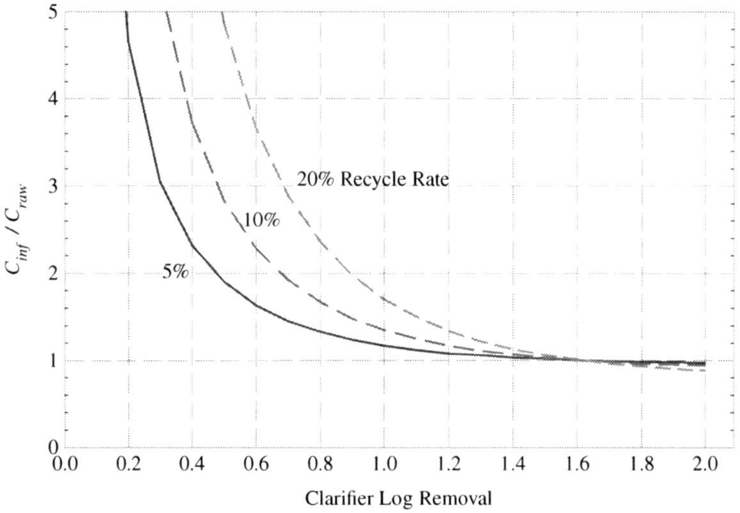

FIGURE 12-5 Effects of clarification performance and recycle rates of 5%, 10%, and 20% for untreated SFBW on the plant influent ratio of pathogens for 2.5% of filtered water used for backwashing (*Source: from Edzwald et al., 2003*).

Effects of SFBW on Clarifier and Filtered Water Performance

We showed above that model predictions indicate that if the clarifier achieves about 1.6-log removal, depending on the percent of filtered water used in backwashing, then recycling untreated SFBW does not increase the plant influent *Cryptosporidium* oocyst concentration. Now, we explore further and more practically the effects of recycling SFBW. SFBW contains suspended solids, organic matter, and other contaminants that can potentially interfere with coagulation and plant performance. We consider the recycle of untreated SFBW and its effect on two types of water plants: both plants employ coagulation, flocculation, clarification, and dual media filtration, but one plant uses DAF for clarification and the other uses plate settling. The reported data are from well-controlled pilot-plant studies in which experiments were run for different seasons to capture any effects of raw water quality and water temperature. The water supply did not have measureable concentrations of *Cryptosporidium*, so oocysts were fed continuously to the raw water entering the pilot plant. SFBW was collected and mixed uniformly, and then used for experiments evaluating the effect of recycling untreated SFBW to the front of the treatment trains. The SFBW quality varied depending on season and had the following characteristics: suspended solids of 70 to 140 mg/L, turbidity of 13 to 29 NTU, UV_{254} (surrogate for DOC) of 0.035 to 0.051 cm^{-1}, pH 7.1 to 7.4, and *Cryptosporidium* concentrations generally of 1100 to 1600 oocysts per 100 L (in one run it was estimated at 30×10^3 oocysts per 100 L). Raw water quality characteristics prior to recycling of SFBW were: turbidity of 0.9 to 1.1 NTU, particle counts (2 to 15 μm size) of 3560 to 5940 particles per mL, and UV_{254} of 0.080 to 0.12 cm^{-1}. The details of the study are reported by Edzwald et al. (2003).

Figure 12-6 shows *Cryptosporidium* oocyst log removal data for DAF and plate sedimentation pilot runs for experiments conducted during the summer (water temperatures of 17 to 18°C) and winter (2 to 5°C) seasons. Comparisons of performance are made for no recycle of SFBW and with recycle. For DAF (bottom of figure), *Cryptosporidium* log

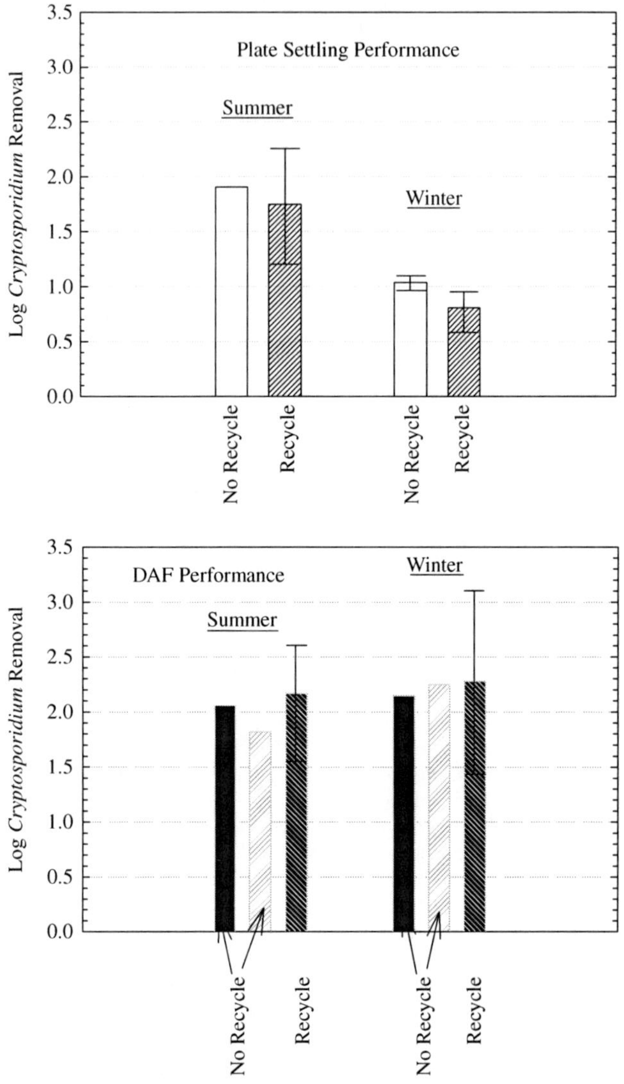

FIGURE 12-6 Comparison of *Cryptosporidium* log removals by plate sedimentation (top) and DAF (bottom) for no recycle and with recycle of waste filter backwash water (*Source: Edzwald et al. (2003), Journal of Water Supply: Research and Technology—AQUA, 52 (4),243–258, with permission from the copyright holders, IWA Publishing*).

removals are 2 or slightly greater, and the data show no adverse effect from recycle and no significant difference in performance for summer and winter seasons. For plate settling, summer performance was good with log removals slightly less than 2 and a small adverse effect from recycle. In the winter, performance was poorer with log removals of 1 or less and a decrease in performance with recycle of SFBW. Other data presented by

Edzwald et al. (2003) showed that DAF produced much lower turbidities and particle counts (2 to 15 μm size) than plate settling with no effect of recycle consistent with the oocyst performance data.

In accordance with the model above, *Cryptosporidium* is concentrated in the floated sludge of the DAF process and does not concentrate in the filters and SFBW as long as the DAF clarifier achieves log removals of about 1.6 depending on the frequency of backwashing. An important plant operation practice, to minimize any effects of SFBW on the quality of the plant influent and on subsequent treatment, concerns the chemistry of the backwash water. The backwash water chemistry should not be altered from that applied to the filters. Tobiason et al. (2003a) have shown for several full-scale plants with good backwash water chemistry and optimum coagulation plant practice that even with an increase in the suspended solids of the mixed plant influent following recycle, there were no effects from recycle on coagulant dosages and often no effects on clarified (DAF or sedimentation) water quality.

Changes in the chemistry of the backwash water from chemicals added for disinfection, acid or base for pH change, and addition of corrosion inhibitors (phosphate-based chemicals) can have negative effects on the SFBW releasing metals and organic matter and cause negative effects on water quality and treatment with recycle (Tobiason et al. 2003b).

Treatment of SFBW

Many water plants have sedimentation and not DAF, some plants are direct filtration, and some water plants alter backwash water chemistry. Sedimentation plants, as presented above, most likely concentrate contaminants in the SFBW. For direct filtration plants that wish to recycle SFBW, then treatment of SFBW must be practiced; otherwise, one is recycling solids with no overall removal from the water streams since the filters are the only particle separation step. As stated above, changes in backwash water chemistry can increase contaminants in the SFBW. Consequently, many water plants choose to treat SFBW because of several concerns with its quality: (1) pathogens, (2) suspended solids (high turbidity), (3) organic matter (DOC, TOC, UV_{254}, and DBP precursors), and (4) metals. We address briefly various options for the treatment of SFBW, and then focus on DAF as a treatment method. Two good references are by Cornwell and Roth (2010) and Cornwell et al. (2010).

Treatment Options SFBW is produced over a short time period (about 15 min) causing high flow rates and widely varying quality. With the possible exception of treatment by discharge to lagoons, equalization through storage is required before treatment so to allow for lower flow rates and a more uniform water quality. It is common to find for equalized and uniformly mixed SFBW fairly high particle concentrations: turbidities of 10 to 100 NTU and TSS of 70 to 300 mg/L (Edzwald et al., 2003; Tobiason et al., 2003a; Cornwell et al., 2010). Organic matter can be high (TOC of several mg/L) and can affect treatment.

There are several treatment options available for the removal of particles (turbidity and TSS). For cases where the TOC is mainly particulate matter, then treatment to remove particles also yields TOC removal. The options are (1) lagoons, (2) conventional sedimentation, (3) tube and plate settling, (4) solids contact sedimentation, (5) ballasted sand settling, and (6) DAF. There have been some bench-scale studies demonstrating the feasibility of low-pressure membranes (Cornwell et al., 2010), but pilot-scale and full-scale demonstrations are needed before consideration of membranes. Some plants also add a disinfection step after these particle separation processes before recycle to the front end of the plant. Of the processes listed, sedimentation is most common with plate settling widely used.

Dissolved Air Flotation Eades et al. (2001) reported on a series of bench-scale and pilot-scale studies assessing the application of DAF for treating SFBW. Important findings from their study follow: (1) effluent DAF turbidities <1 NTU were obtained with feedwater turbidities as high as 50 NTU, (2) no primary coagulants were required but low doses of polymer were required to bind floc particles—type of polymer was site specific, (3) short flocculation times of 10 min or less were feasible; (4) DAF recycle rates as low as 5 percent were adequate and little improvement was found by increasing the recycle rate to 20 percent, and (5) sludge solids of 3.5 to 9.6 percent were obtained with low volumes of sludge (<0.1 percent—based on the raw water volume treated).

An updated and thorough evaluation of the application of DAF at the pilot-scale in treating SFBW was done by Cornwell et al. (2010). The database included studies at several water plants and evaluation of conventional rate and high-rate DAF. The major findings are summarized. A high MW floc-aid polymer was required to achieve good treatment. They report that a cationic polymer of low-charge was often used, but we suggest considering high MW nonionic polymers for SFBW for pH conditions <7. Doses, of course, depend on solids concentrations, but doses of about 1 mg/L and less were effective. A short flocculation period produced good performance compared to no flocculation. In general, DAF recycle rates of 8 to 10 percent provided good treatment, and there was no apparent benefit in providing additional air with greater DAF recycle. Good performance (treated water turbidity < 2 NTU) was achieved with conventional DAF rates of 10 to 15 m/h (4 to 6 gpm/sf) and high-rate DAF up to about 36 m/h (15 gpm/sf). In addition to producing low-turbidity water, the DAF process serves as a thickening process producing sludges of 2 to 4 percent solids.

Boulder (Colorado, United States) employs DAF for treatment of SFBW at the Betasso water plant, a 150 ML/d (40 MGD) conventional treatment facility. The SFBW is collected in a 3.8 ML (1 million gallons) equalization basin. Following the equalization basin, treatment consists of polymer addition to a rapid mixing basin, two-stage flocculation, and DAF. The untreated SFBW water has turbidity <10 NTU but can be >80 NTU. Good performance is achieved with a polymer (type not specified) dose of about 1 mg/L or less. Treated water turbidities are about 1 NTU or slightly greater and <2 NTU 95 percent of the time. Good performance is achieved with a DAF recycle rate of 10 percent with a slight improvement in a test at 15 percent. Floated solids vary with values of 0.3 to 3 percent. The DAF tank is a conventional rate system operated at a nominal hydraulic loading of 5 m/h (2 gpm/sf) or less.

A search of DAF manufacturer web sites revealed that several other water plants have recently built or selected DAF to treat SFBW.

REFERENCES

Cornwell, D. A., and Roth, D. K. (2010), Water treatment plant residuals management, in J. K. Edzwald, ed., *Water Quality and Treatment: A Handbook on Drinking Water*, 6th ed., New York: AWWA and McGraw-Hill.

Cornwell, D. A., Tobiason, J. E., and Brown, R. (2010), *Innovative Applications of Treatment Processes for Spent Filter Backwash*, Denver, CO: Water Research Foundation.

Crossley, I., and Valade, M. T. (2006), A review of the technological developments of dissolved air flotation, *Journal of Water Supply: Research and Technology – Aqua*, 55 (7–8), 479–491.

Drewes, J. E., and Khan, S. J. (2010), Water reuse for drinking water augmentation, in J. K. Edzwald, ed., *Water Quality and Treatment: A Handbook on Drinking Water*, 6th ed., New York: AWWA and McGraw-Hill.

Duranceau, S. J. (2010), Personal communication, University of Central Florida, Orlando, FL.

Duranceau, S. J., and Taylor, J. S. (2010), Membranes, in J. K. Edzwald, ed., *Water Quality and Treatment: A Handbook on Drinking Water*, 6th ed., New York: AWWA and McGraw-Hill.

Eades, A., Bates, B. J., and MacPhee, M. J. (2001), Treatment of spent filter backwash water using dissolved air flotation, *Water Science and Technology*, 43 (8), 59–66.

Edzwald, J. K., Tobiason, J. E., Udden, C., Kaminski, G. S., Dunn, H. J., Galant, P. B., and Kelley, M. B. (2003), Evaluation of the effect of recycle of waste filter backwash water on plant removals of *Cryptosporidium*, *Journal of Water Supply: Research and Technology – Aqua*, 52 (4), 243–258.

Haarhoff, J., and van der Merwe, B. (1996), Twenty-five years of wastewater reclamation in Windhoek, Namibia, *Water Science and Technology*, 33 (10–11), 25–35.

Howe, K. (2010), Personal communication, University of New Mexico, Albuquerque, NM.

Mallevialle, J., Odendall, P. E., and Wiesner, M. R. (1996), *Water Treatment Membrane Processes*, American Water Works Association, Lyonnaise des Eaux, Water Research Commission of South Africa, New York: McGraw-Hill.

Menge, J. G. (2011), Personal communication, City of Windhoek, Namibia.

MWH (2005), *Water Treatment Design: Principles and Design*, 2nd ed., J. C. Crittenden, R. R. Trussell, D. W. Hand, D. W. Howe, G. Tchobanoglous, Hoboken, NJ: John Wiley & Sons.

Passantino, L. (2010), Personal communication, Malcolm Pirnie, Phoenix, AZ.

Taylor, J. S., Thompson, D. M., and Carswell, J. K. (1987), Applying membrane processes to groundwater sources for trihalomethane precursor control, *Journal AWWA*, 79 (8), 72–82.

Teixeira, M. R. and Rosa, M. J. (2006), Integration of dissolved gas flotation and nanofiltration for M. *aeruginosa* and associated microcystins removal, *Water Research*, 40, 3612–3620.

Theron-Beukes, T. (2010), Personal communication, WINGOC, Windhoek, Namibia.

Thorsen, T., and Fløgstad, H. (2006), Nanofiltration in drinking water—literature review, *Techneau Report* D5.3.4B.

Tobiason, J. E. (2010), Personal communication, University of Massachusetts, Amherst, MA.

Tobiason, J. E., Edzwald, J. K., Levesque, B. R., Kaminski, G. S., Dunn, H. J., and Galant, P. B., (2003a), Full-scale assessment of the impacts of recycle of waste filter backwash, *Journal AWWA*, 95 (7), 80–93.

Tobiason, J. E., Edzwald, J. K., Gilani, V., Kaminski, G. S., Dunn, H. J., and Galant, P. B. (2003b), Effects of waste filter backwash recycle operation on clarification and filtration, *Journal of Water Supply: Research and Technology – Aqua* 52 (4), 259–276.

Tymkiw, P. V. (2010), Personal communication, Malcolm Pirnie, Phoenix, AZ.

USEPA (2001), 40 CFR Parts 9, 141, and 142, National Primary Drinking Water Regulations: Filter Backwash Recycling Rule: Final Rule, *Federal Register* 66 (111), 31086–31105.

van der Merwe, B. F. (2010), Personal communication, Environmental Engineering Services, Windhoek, Namibia.

CHAPTER 13

DISSOLVED AIR FLOTATION FOR DESALINATION PRETREATMENT

13-1 SEAWATER CHEMISTRY 13-2
- Salinity and Composition 13-2
- Silica 13-4
- Boron 13-4
- Alkalinity 13-5
- Buffer Intensity 13-6

13-2 CONTAMINANTS 13-8
- Mineral Particles 13-8
- Algae 13-9
- NOM 13-9
- Oil and Grease 13-11

13-3 COAGULANTS AND COAGULATION 13-11
- Organic Polymers 13-11
- Aluminum Coagulants 13-13
- Ferric Coagulants 13-13

13-4 PHYSICAL PROPERTIES OF SEAWATER 13-16
- Effects on Contact Zone Performance 13-16
- Effects on Separation Zone Performance 13-17
- Effects on Dissolving Air in the Saturator 13-17

13-5 AIR SATURATION IN SEAWATER 13-18

13-6 EXAMPLES OF DAF PRETREATMENT 13-19

REFERENCES 13-21

It is essential to have pretreatment prior to reverse osmosis (RO) treatment in desalination water plants. The pretreatment removes contaminants that can foul RO membranes, and the integration of pretreatment with RO makes for far more efficient plant performance with respect to water quality and energy usage. There are numerous pretreatment configurations. All involve coagulation, flocculation, and particle separation. The particle separation processes can consist of (1) sedimentation and granular media filtration, (2) sedimentation and low-pressure membrane filtration, (3) DAF and granular media filtration, or (4) DAF and low-pressure membrane filtration. Many desalination plants are faced with removing algae in pretreatment; some plants face the problem of dealing with harmful algae (red and brown-tide algae). DAF is much more efficient than sedimentation processes in removing algae, and so DAF should be considered as a pretreatment process.

In the first three major sections of this chapter, we cover material that applies to all pretreatment strategies, not just DAF. We start with consideration of the chemistry of seawater and discuss key features that affect coagulation and treatment by RO. Next, we cover the main contaminants in seawater, their properties, concentrations, measurements, and effects on membranes without pretreatment. Coagulation is essential in removing these contaminants in all pretreatment strategies and this is addressed next. In the remaining three major sections of the chapter, we turn our attention to DAF. We begin with the physical properties of seawater that can affect DAF. We examine whether seawater affects the performance of the contact and separation zones compared to freshwater. We next examine air solubility in seawater in saturators. Finally, we present some examples of the use of DAF in full-scale desalination plants.

13-1 SEAWATER CHEMISTRY

Salinity and Composition

The concentration of dissolved matter in seawater is described in terms of the salinity (*S*). It is expressed on a mass basis as grams of dissolved material per kilogram of seawater. It is also expressed as parts per thousand (ppt) in the following way: $S = 35.16‰$, where the symbol, ‰, refers to ppt on a mass basis. In this book, we use both *S* as g/kg and *S*‰. The average salinity of seawater is about 35 g/kg, but varies in the open oceans from about 31 to 45 g/kg. The higher values are found in the Mediterranean Sea, Red Sea (Arabian Gulf), and the Persian Gulf. Lower values are found near the mouths of rivers.

Sometimes salinity of seawater is expressed as chlorinity. Chlorinity is determined by a silver titration measurement of the total Cl^-, Br^-, and I^- in seawater and expressed as total equivalent Cl^-. It is expressed on ppt basis as *Cl*‰. It is related approximately to salinity through Eq. (13-1).

$$S\,{}^{o}\!/_{oo} = 1.81 \times Cl\,{}^{o}\!/_{oo} \tag{13-1}$$

Another measure of the ion content of seawater is the ionic strength (*I*). It is defined by Eq. (13-2). where C_i is normally the molar concentration (mol/L or M) but for seawater it is mol/kg, and z_i is the charge on the ion. The values of *I* for freshwaters range from about 5×10^{-4} to 10^{-2} M corresponding to TDS of 20 to 400 mg/L. Brackish and estuarine waters have values of *I* starting at about 10^{-2} M up to less than that of seawater. The average value of *I* for seawater is approximately 0.7 mol/kg.

$$I = \frac{1}{2}\left(\sum_i C_i z_i^2\right) \tag{13-2}$$

There are numerous processes that affect the chemical composition of seawater (Eby, 2004). These include chemical inputs from rivers, volcanic-seawater reactions, basalt-seawater reactions (basalt magmas at mid-ocean ridges), and the atmosphere. These also include chemical outputs of evaporation and deposition of salts, basalt-seawater reactions, and precipitation and deposition to sea sediments. Various chemical and biological reactions in seawater affect many chemicals such as N, P, inorganic carbon, calcium carbonate, silica, and organic matter.

Many of the major cations and anions in open seawater are at fairly constant concentrations. The concentrations of the major ions are summarized in Table 13-1. For the cations, seawater contains primarily Na^+ with high concentrations of Mg^{2+}, Ca^{2+}, and K^+, and a fairly high concentration of Sr^{2+}. For the anions, seawater contains primarily Cl^- and SO_4^{2-}.

TABLE 13-1 Average Major Ionic Composition of Seawater

Ion	Concentration, g/kg	Concentration, mol/kg
Na^+	10.773	0.467
Mg^{2+}	1.294	5.32×10^{-2}
Ca^{2+}	0.412	1.03×10^{-2}
K^+	0.399	1.02×10^{-2}
Sr^{2+}	7.9×10^{-3}	9.02×10^{-5}
Cl^-	19.344	0.546
SO_4^{2-}	2.717	2.83×10^{-2}
HCO_3^-	0.142	2.33×10^{-3}
Br^-	6.74×10^{-2}	8.44×10^{-4}
B_T	4.45×10^{-3}	4.12×10^{-4}
F^-	1.28×10^{-3}	6.74×10^{-5}

Br^- and B_T (total boron, sum of H_3BO_3 and $B(OH)_4^-$) are greater than found in most freshwaters. Boron is discussed further below. While F^- is greater than found in fresh-surface waters, it can be greater in some groundwaters. The HCO_3^- concentration of seawater is not very much different than of many freshwaters. Its importance with regard to alkalinity, buffer intensity, and inorganic carbon is discussed below.

Seawater contains many other inorganic cations, anions, and dissolved uncharged molecules as well as dissolved organic matter. We discuss some of the important ones. Equilibrium constants for water dissociation and acid-base reactions were adjusted for the ionic strength of seawater and taken from Millero (1995) and Stumm and Morgan (1996). To indicate seawater equilibrium constants, they are designated K^{SW} or pK^{SW} where $pK^{SW} = -\log K^{SW}$. For the solubility of solid SiO_2 (amorphous and quartz) in this section and for amorphous $Al(OH)_3$(am) and $Fe(OH)_3$(am) in Sec. 13-3, thermodynamic data (Stumm and Morgan, 1996; Drever, 1997) were used to calculate equilibrium constants for infinitely dilute solution, and then adjusted for seawater ionic strength using activity coefficients for the ions: monovalent 0.7, divalent 0.25, and trivalent 0.05 (Stumm and Morgan, 1996).

The H^+ concentration is expressed as a molar concentration $[H^+]$ using the pH convention, where $pH = -\log [H^+]$. The pH of seawater varies between 7.5 and 8.3 but is mainly between 8.1 and 8.3. It affects the speciation of many dissolved chemicals, chemical reactions, and is important in coagulation and desalination processes. Because of the ionic strength of seawater, the equilibrium dissociation reaction for water has a different equilibrium constant (K_w^{SW}) than for freshwaters. Expressed as the pK_w^{SW}, it is 13.21 at 25°C compared to the well-known pK_w of 14.0 for freshwaters. Seawater pK_w^{SW} values for 5, 10, 15, and 20°C are 14.06, 13.83, 13.62, and 13.41, respectively.

Dissolved silica in seawater varies from about 2×10^{-6} M (0.12 mg/L as SiO_2) to 10^{-4} M (6 mg/L as SiO_2). It is important because many minerals and aquatic organisms in seawater contain Si, and it has potential effects on alkalinity and buffer intensity. Its chemistry is described in more detail below.

Inorganic carbon is important for several reasons including its relationship to alkalinity and buffer intensity, solubility of carbonate minerals, and its role in algae growth and in the decay of organic matter. Total inorganic carbon (C_T) is defined by Eq. (13-3), where $[H_2CO_3^*]$ is the sum of dissolved carbon dioxide $[CO_2(aq)]$ and $[H_2CO_3]$—$[H_2CO_3^*]$ is mostly $[CO_2(aq)]$ (see Stumm and Morgan, 1996).

$$C_T = \left[H_2CO_3^*\right] + \left[HCO_3^-\right] + \left[CO_3^{2-}\right] \tag{13-3}$$

C_T in seawater averages about 2.33 × 10^{-3} mol/kg and is essentially all HCO_3^- because the concentrations of $H_2CO_3^*$ and CO_3^{2-} are very low relative to HCO_3^- for the pH conditions of seawater. Using a seawater density of 1025 kg/m^3 yields a C_T concentration of 2.36 × 10^{-3} M, which is about the concentration reported in Stumm and Morgan (1996). Stumm and Morgan report an average river C_T of 10^{-3} M with a range of 10^{-4} to 5 × 10^{-3} M. Thus, C_T in seawater is not much different than what is found in many rivers. HCO_3^- is a major component of alkalinity, which is discussed below. The concentration of $[H_2CO_3^*]$ or $[CO_2(aq)]$ in seawater is affected by various chemical reactions (acid–base, solubility of carbonate minerals, and biological reactions), water temperature, and salinity. Reported values are in the range of 10^{-5} to 8 × 10^{-5} M (Berner and Berner, 1996). $CO_2(aq)$ is rarely at equilibrium with its atmospheric partial pressure of about 380 ppm because of the various reactions adding and consuming it, but if it were at equilibrium its concentration would be about 1.2 × 10^{-5} M for temperatures of 20 to 25°C.

P and N are important nutrients in seawater because they affect algae growth. From Berner and Berner (1996) dissolved inorganic P occurs at <0.1 to 3 × 10^{-6} M (or about <30 to 100 μg/L), while dissolved inorganic N occurs at <1 to 50 × 10^{-6} M (or about <14 to 700 μg/L).

Dissolved organic matter in open seawater is low in concentration varying only among the oceans from about 0.72 to 0.96 mg/L as DOC with the higher end for subtropical areas (Ogawa and Tanoue, 2003). The DOC can be much greater in coastal areas and near the mouths of rivers. The TOC (POC [particulate organic carbon] + DOC) is typically at 2 to 5 mg/L. The presence and removal of TOC is an important part of pretreatment in desalination and is discussed in more detail in Sec. 13-2.

Silica

Dissolved Si is of interest because it can potentially affect the alkalinity of seawater. This is discussed below under alkalinity. More importantly, many solids of an inorganic nature are affected by its solubility. Further, Si is incorporated and released by algae (particularly, diatoms). Dissolved Si exists as an acid (H_4SiO_4) and base ($H_3SiO_4^-$). It is a weak acid and for seawater ionic strength of 0.7 mol/kg, the pK_a^{SW} is 9.5 at 25°C. Therefore, 91 to 99 percent of the dissolved silica is H_4SiO_4 for the pH range of seawaters of 7.5 to 8.5.

Si can form amorphous and crystalline solids as $SiO_2(s)$. The prime example of the latter solid is, of course, quartz sand. Figure 13-1 shows solubility curves for seawater for both solids. For pH values of about 8.5 or less, the solubility of the solids is independent of pH. Since dissolved Si concentrations in seawater are in the range of 0.12 to 6 mg/L as SiO_2, Si can be near equilibrium with quartz, but Si is often undersaturated with respect to these two solid phases. However, there are various aluminum silicate solids that can form at lower Si concentrations so the presence of silica solids can be a concern in desalination. This is covered below under Sec. 13-2.

Boron

The chemistry of boron is of interest. First, it can contribute to the alkalinity and buffer intensity of seawater thus affecting pH pretreatment control and coagulation. Second, there is a concern about health effects (developmental and reproductive) so that reduction by desalination is important. The total boron concentration in seawater averages 4.45 × 10^{-3} g/kg (Table 13-1). Using a seawater density of 1027 kg/m^3 at 20°C (from Table D-2) yields

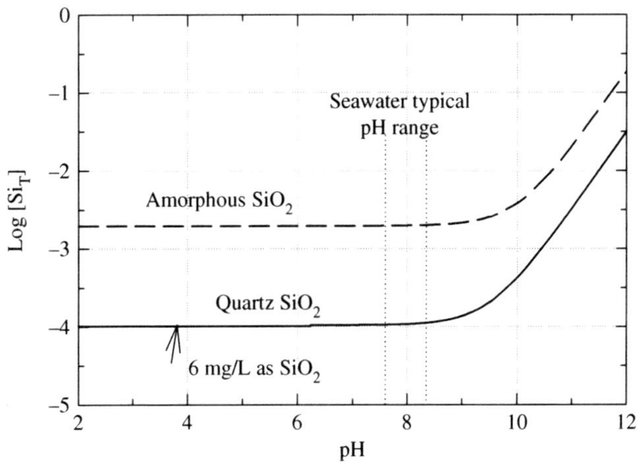

FIGURE 13-1 Solubility of amorphous and quartz in seawater (*S* at 35 g/kg and 25°C) as a function of pH.

a B_T concentration of 4.6 mg/L. The World Health Organization (WHO) had a provisional guideline of 0.5 mg/L, but a revised guideline of 2.4 mg/L may go into effect in 2011 (WHO, 2011). The European Union (EU, 2011) has a standard of 1 mg/L. The USEPA does not have a drinking water standard for B, but it has issued a health advisory document (USEPA, 2008) and notes that several states within the United States have standards ranging from 0.6 to 1 mg/L.

Boron is a weak acid with a pK_a^{SW} of 8.64 for seawater at 20°C. The acid and base forms of B depend on pH as shown in Fig. 13-2. Boric acid (H_3BO_3) dominates at pH below 8.64. If we consider that seawater can range from pH 7.5 to 8.3, then the fraction of boric acid is 69 percent at pH 8.3 and 93 percent at pH 7.5. Borate ($B(OH)_4^-$) only dominates if seawater pH is increased, which is done in some RO plants to enhance the removal of boron—see Sec. 13-6 and discussion of the Tuas Desalination Plant in Singapore.

Alkalinity

Alkalinity (Alk) is the acid neutralizing capacity (ANC) of water. Its role in coagulation of freshwaters was covered in Chap. 6. Weak acids and bases cause Alk and for freshwaters, inorganic carbon, particularly HCO_3^-, is the primary contributor. For seawater, there are several main contributors, these are

- HCO_3^-
- CO_3^{2-}, which becomes more important for pH >8;
- $B(OH)_3^-$, small compared to inorganic carbon and becomes more important for pH >8.

Other potential contributors are small in seawater compared to the above and can be neglected. These are: (1) OH^-, not important unless seawater is adjusted to pH 10 or greater; (2) orthophosphate ($H_2PO_4^-$, HPO_4^{2-}, PO_4^{3-}); (3) base form of dissolved Si, $H_3SiO_4^-$ (this is low for pH <8.5); (4) deprotonated or base forms of humic and

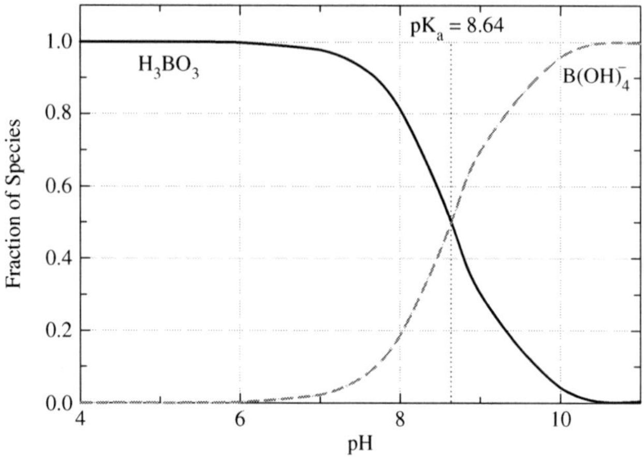

FIGURE 13-2 Speciation of boron in seawater as a function of pH (conditions: *S* at 35 g/kg, 20°C).

fulvic acids, small except at the mouths of some rivers; and (5) others such as NH_3, HS^-, and HF^-.

Considering only the main causes, the Alk of seawater is defined by Eq. (13-4).

$$Alk = \left[HCO_3^-\right] + 2\left[CO_3^{2-}\right] + \left[B(OH)_4^-\right] \tag{13-4}$$

For typical seawater values of inorganic carbon (C_T of 2.2×10^{-3} mol/kg) and boron (B_T of 4.1×10^{-4} mol/kg), we get alkalinities of 2.24 to 2.7 meq/kg for seawater pH between 7.5 and 8.3. Expressing the Alk in a conventional way for a seawater density of 1024 kg/m^3 at 25°C (Table D-2), we obtain alkalinities of about 115 to 150 mg/L as $CaCO_3$. These alkalinities compare to medium and high-alkalinity freshwaters—see Chap. 6.

Buffer Intensity

As pointed out in Chap 6, Alk is a capacity measurement. To properly assess the resistance of unit changes in pH from acid or base addition, we must examine the buffer intensity (β). This is more useful than alkalinity in evaluating changes in seawater pH over the range of interest. Equation (13-5) presents β as a function of pH for seawater.

$$\beta = 2.3\left[\frac{K_B^{SW} B_T\left[H^+\right]}{\left(K_B^{SW}+\left[H^+\right]\right)^2} + \frac{K_1^{SW} C_T\left[H^+\right]}{\left(K_1^{SW}+\left[H^+\right]\right)^2} + \frac{K_2^{SW} C_T\left[H^+\right]}{\left(K_2^{SW}+\left[H^+\right]\right)^2} + \frac{K_w^{SW}}{\left[H^+\right]} + \left[H^+\right]\right] \tag{13-5}$$

Effects: Boron $\quad HCO_3^- \quad CO_3^{2-} \quad OH^- \quad H^+$

Equilibrium constants were adjusted for seawater S of 35 g/kg and 20°C: K_B^{SW} of $10^{-8.64}$ for the boron term, K_1^{SW} of $10^{-5.89}$ for the HCO_3^- term, K_2^{SW} of $10^{-8.99}$ for the CO_3^{2-} term, and K_w^{SW} of $10^{-13.42}$ for the OH^- term. The buffer intensity is plotted as a function of pH in Fig. 13-3 for typical seawater conditions for inorganic carbon (C_T of 2.2×10^{-3} mol/kg) and total boron (B_T of 4.1×10^{-4} mol/kg). Contributions from inorganic carbon and boron are plotted separately, and their sum, total buffer intensity, is also plotted. Buffer intensity due to OH^- and H^+ are not identified in Fig. 13-3, but influence the total buffer intensity at high and low pH. The pH range of most seawater is between 7.5 and 8.3 and is identified by the vertical dotted lines in the figure.

What we learn follows. Buffer intensity is highest at about pH 9, and there is another peak at pH 6. Boron contributes to the buffer intensity between pH 7 and 10, but its contribution is small compared to inorganic carbon. For seawater at pH 8 to 8.3, we see that it has a fairly high buffer intensity so a large amount of acid would be required to decrease the pH to the mid 7s. Then, there is much less buffer intensity so smaller amounts of acid are required to decrease the pH further until the mid 6s. Then at lower pH, buffer intensity increases so larger amounts of acid are required to decrease the pH further to the low 6s and less. Examples follow:

- 0.57 mol of strong acid/kg or 0.58 mol/L is required to decrease the pH from 8.3 to 7.4;
- To then decrease the pH from 7.4 to 6.8 requires only 0.105 mol of strong acid/kg or 0.108 mol/L;
- If we attempt to decrease the pH further from 6.8 to 6, we require an additional 0.57 mol of strong acid/kg or 0.59 mol/L.

The importance of buffer intensity and pH control in coagulation is addressed in Sec. 13-3.

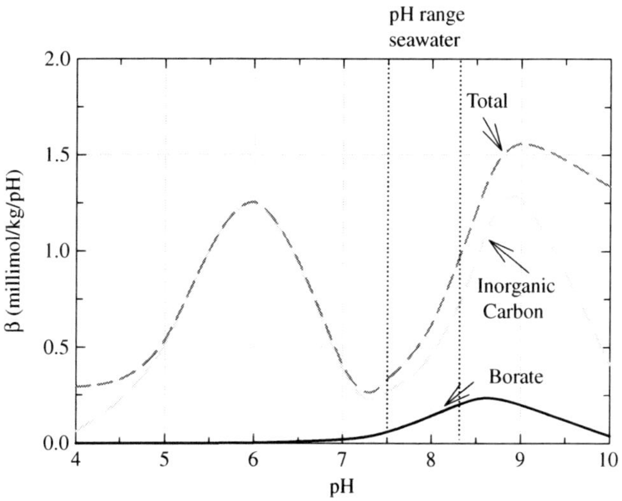

FIGURE 13-3 Buffer intensity as a function of pH for seawater (conditions: S at 35 g/kg, 20°C, total inorganic carbon at 2.2×10^{-3} mol/kg, total boron at 4.1×10^{-4} mol/kg).

13-2 CONTAMINANTS

There are many potential contaminants that must be removed prior to desalination to prevent fouling of RO membranes and to enhance the operational efficiency of RO. The pretreatment processes normally consist of coagulation, flocculation, clarification (our focus is DAF), and either granular media filtration or low-pressure membrane filtration (MF or UF). The occurrence and concentrations of the contaminants depend on location of the desalination plants. Most plants are in coastal areas and the waters in these areas are subject to variation in water quality. Others are near the mouths of rivers, which can adversely affect seawater quality at the plant intake. Many plants are located in areas subject to algae blooms, especially those plants located in warm, shallow seas.

We organize the discussion of contaminants around (1) mineral particles, (2) algae, (3) natural organic matter (NOM), and (4) oil and grease.

Mineral Particles

These particles can be

- Clays (aluminum silicates), carried by rivers and runoff to coastal areas;
- Other aluminum silicate minerals, carried by rivers and runoff or produced in the ocean by precipitation processes;
- $SiO_2(s)$ (sand) particles of small sizes, present in the ocean and coastal areas, and
- $CaCO_3(s)$ solids produced in the ocean by precipitation.

The presence of particles is normally measured as turbidity. Seawater turbidity is usually fairly low at 0.5 to 2 NTU, but can be greater at some coastal locations and when algae blooms occur. Particles can cause colloidal-particle fouling of RO membranes in which a cake of particles forms that reduces the permeate flux (see Chap. 12). Algae can also cause biofouling—see discussion below under algae. Therefore, pretreatment processes to remove particles are essential. One criterion for the water applied to RO membranes is the turbidity of the water following pretreatment. Some membrane companies accept a turbidity of 0.5 NTU, while others desire a turbidity as low as 0.1 NTU (Freeman, 2009; Voutchkov, 2010). A widely used test on the acceptability of the water for RO treatment regarding colloidal-particle fouling is the SDI (Silt Density Index). The details of the test are found in ASTM (2007). The test is a laboratory filtration test using a 0.45-μm-membrane filter using three time intervals to collect filtered water. The SDI is dimensionless. Guidelines for feedwater to RO are: SDI >5 not acceptable, SDI <2 good quality, and SDI maximum of 3 to 4 (Freeman, 2009; Duranceau and Taylor, 2010; Voutchkov, 2010).

Mineral particles are stable (tendency not to flocculate) in freshwaters as discussed in Chap. 5. The primary cause of particle stability is the negative charge on particle surfaces. Other possible causes are hydration and steric effects. As particles approach each other, a repulsive force occurs from electrical double-layer interaction from the particle-surface charge and an attractive force occurs from van der Waals forces at close particle–particle separation distances. In freshwaters, the electrical-double layer thickness and repulsive interaction extends far enough from the particle surfaces that there is a net repulsive force. In seawater, the double-layer thickness is only about 0.4 nm (see Table 5-3). This allows particles to approach close enough to each other so that the attractive van der Waals forces can dominate and particles flocculate. This occurs in brackish or estuarine waters so that colloidal mineral particles carried by rivers flocculate and undergo deposition at the mouths of rivers (Edzwald et al., 1974).

Mineral particles carried by rivers and runoff into the sea therefore should be at relatively low concentration. One may still have some of these particles and other causes of particle stability may apply, particularly adsorbed NOM and hydration effects. Other mineral particles may be present because they were produced in seawater by precipitation such as $CaCO_3$ or some aluminum silicates. In all cases, however, the stability due to charge should be low because of the high ionic strength of seawater and these particles are easily treated by coagulation.

Algae

Algae are often the major problem for desalination plants requiring removal by pretreatment. Almost all groups of phytoplankton-type (floating or suspended) algae are found in seawater: Diatoms, Chlorophyta (green algae), Chrysophyta (golden algae), Dinoflagellates, Phaeophyta (brown algae), and Cyanobacteria (blue-green algae). Unlike mineral particles, algae remain suspended while undergoing growth and do not flocculate until they are in the later stages of their growth phase. With the high ionic strength of seawater, their particle stability is not caused by surface charge (although they do carry a negative charge) but is due to (1) steric effects, (2) motility, and (3) hydration effects. Steric effects are attributed to their surface structure that prevents aggregation. Motility or swimming refers to the fact that many algae have flagella (such as Dinoflagellates and others) or cilia (hair-like structure around the exterior of the cells that provide a means of swimming and steric repulsive interaction). Hydration effects mean the presence of cell-surface groups with bound water that inhibit flocculation.

Removal of algae by pretreatment is essential in preventing algae from accumulating on RO membranes causing particulate fouling. An additional problem is that on the membrane, the transmembrane pressure will cause cell lysis (breaking of the cells) whereby the released cellular matter undergoes decay and causes biofouling of the membranes. Additional discussion of dissolved organic matter is addressed below under NOM.

In many coastal areas and in warm, shallow seas, one can have harmful-algae blooms often referred to as red tides and brown tides depending on the pigments associated with the algae. Some of these harmful algae can also bloom in colder waters. Without proper pretreatment, the algae and associated toxins can stop the operation of desalination plants. Figure 13-4 shows some examples of harmful algae. Many dinoflagellates are associated with red tide, while *Auerococcus anophagefferens* (a chrysophyte) is a cause of brown tide. Some of the algae are large (e.g., dinoflagellates of 10s to 100s of μm) while others are small (e.g., diatoms, green algae, cyanobacteria, and chrysophyte of μm size).

In pretreatment, algae can be coagulated to result in effective separation by DAF compared to sedimentation process. DAF pretreatment is discussed in the last section of the chapter.

NOM

The concentrations and types of NOM present in seawater are highly dependent on the plant intake location. Greater amounts are found at the mouths of rivers carrying NOM and in waters subject to algae blooms. We start with some consideration of collective and surrogate measurements for NOM and then discuss types of organic matter.

TOC is the most widely used measurement for organic matter. It is, of course, a collective measurement. TOC is the sum of particulate organic carbon (POC) and DOC. In

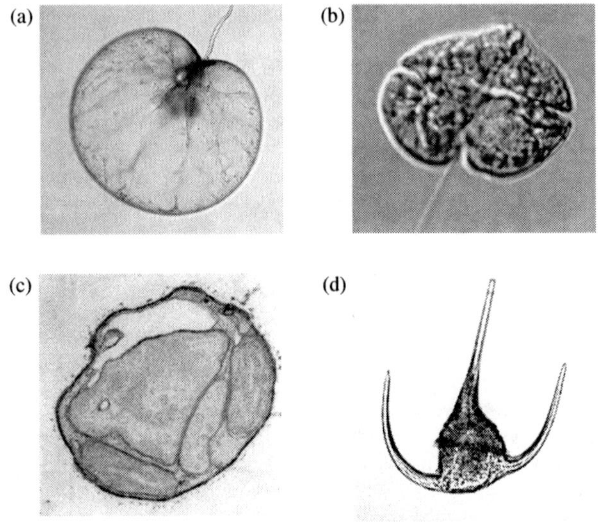

FIGURE 13-4 Photos of examples of algae that are in seawater: (a) *Noctiluca scintillans*, a Dinoflagellate, size 200 to 2000 μm, associated with red tide (*Source: Australian Government, Department of the Environment, Water, Heritage, and the Arts; accessed 12 January 2011, http://www.tafi.org.au/zooplankton/imagekey/dinophyta/*). (b) *Karenia brevis*, a Dinoflagellate, size 25 μm, associated with red tide (*Source: Florida Fish and Wildlife Conservation Commission; accessed 12 January 2011, http://research.myfwc.com/images/articles/23559/23559_5513.jpg&imgrefurl*). (c) *Auerococcus anophagefferens*, a Chrysophyte, size 2 to 3 μm, associated with brown tide, (*Source: U.S. Department of Energy Joint Genome Institute; accessed 12 January 2011, http://www.jgi.doe.gov/*) (d) *Ceratium tripos*, a Dinoflagellate, length 200 to 350 μm, width 65 to 90 μm (*Source: Guiry, M. D. & Guiry, G. M. 2011. AlgaeBase. World-wide electronic publication, National University of Ireland, Galway; accessed 12 January 2011. http://www.algaebase.org*).

the absence of algae blooms, TOC is approximately the DOC. The TOC of seawater used at desalination plants is typically 2 to 5 mg/L (Freeman, 2009; Voutchkov, 2010), but can be greater when algal blooms occur or for seawater affected by aquatic humic matter from rivers—for example, the Tampa Bay (Florida, United States) facility has raw water TOC averaging 6.2 mg/L with a range of 4.1 to 12 mg/L (Schneider, 2011).

A useful surrogate parameter for DOC (or TOC when it is approximately the DOC) is UV_{254}—see Chap. 6 for additional information. Quite often it is low for seawater at 0.01 cm^{-1} or less; consequently, in measuring UV_{254} for raw seawater and especially across the treatment plant, spectrophotometer cells with a path-length greater than the typical 1 cm should be used to increase accuracy. Where seawater is affected by aquatic humic matter and by algal blooms, raw seawater will have higher values than 0.01 cm^{-1}.

A variety of types of organic matter can be present. For seawater that contains aquatic humic matter, the NOM is a mixture of aquatic humic and fulvic acids (especially, fulvic acids). Aquatic humic matter has properties of fairly high MW (100s to 1000s), and while it carries a negative charge it has hydrophobic character, is reactive, and can be removed by coagulation and separation of the precipitated solids (see Chap. 6). Other NOM can come from the decay of plant and algal matter. In addition, algae, while undergoing growth and respiration, impart soluble organic matter to the water called extracellular organic matter (EOM). Another concern is that without good pretreatment, algae cells that are deposited on the membranes undergo cell lysis due to the transmembrane pressure releasing intracellular organic matter (IOM). The compounds composing EOM and IOM consist of amino acids, proteins, simple sugars, and more complex sugars such as polysaccharides. These compounds are soluble, biodegradable, and a major concern regarding biofouling of RO membranes. Measurements for the presence of these compounds are not common, but fluorescence type measurements may have promise.

SUVA was described in Chap. 6 as a simple method for characterizing the nature of NOM. Applying the concept to seawater, we have these guidelines: (1) SUVA of about 4 or greater indicates that the NOM is dominated by aquatic humic matter, (2) SUVA of 2 to 4 indicates that the NOM is composed of a mixture of aquatic humic matter, EOM, and IOM from algae, and (3) SUVA <2 indicates that the NOM is composed largely of algal derived EOM and IOM.

TOC provides a rough parameter for predicting the potential for biofouling. If TOC is <0.5 mg/L biofouling is unlikely and >2 mg/L biofouling is likely (Freeman, 2009; Voutchkov, 2010).

Oil and Grease

Removal of oil and grease (O&G) and oil-based hydrocarbons in pretreatment is essential to prevent fouling of RO membranes. The sources of O&G are from discharges of wastewater or storm drains in the vicinity of the desalination plant intake. Other sources are waste discharge from ships and oil leaks. Voutchkov (2010) reports that O&G >0.02 mg/L can cause membrane fouling. Another good reason to use DAF in pretreatment in desalination plants is because DAF is effective in removing O&G.

13-3 COAGULANTS AND COAGULATION

Table 13-2 is used to guide our presentation. The table lists various coagulants and their use in freshwaters compared to seawater. Chapter 6 contains an extensive discussion of the chemistry of the various coagulants, the reactions with contaminants, and guidance for their use in treatment of freshwaters. Here, we discuss briefly the possible use of polymers in pretreatment of seawater. Then, we discuss PACls and alum and emphasize that their solubility limits their use in desalination pretreatment. Ferric salts are the coagulants of choice for seawater, and we end with discussion of the chemistry of ferric coagulants and conditions for their use.

Organic Polymers

We classify polymers into two categories: (1) low MW, high charge-density cationic polymers and (2) high MW polymers. The former are actually coagulants, while the latter are not used as coagulants but rather as floc-aids, filter-aids, and flotation-aids. The chemistry and use of both categories in the treatment of freshwaters are covered in Chap. 6.

High charge-density cationic polymers are frequently used in a dual-coagulant strategy. Their positive charge can neutralize the negative charge of particles and can satisfy some of the negative charge associated with aquatic humic matter thus reducing the metal coagulant dosage. Their use in seawater coagulation to complement ferric

TABLE 13-2 Comparison of Coagulants for Use in Freshwaters and Seawater

Freshwaters	Seawater
Organic Polymers	*Organic Polymers*
Low MW*, High Charge-Density Cationic Used in direct filtration and in dual-coagulant strategy in conventional treatment	**Low MW, High Charge-Density Cationic** RO membrane manufacturers have concerns about fouling, but authors make case for use in dual-coagulant strategy with ferric chloride
High MW Nonionic and Anionic Used as floc-aid and filter aid polymers in conventional plants; some use as flotation-aids	**High MW Nonionic and Anionic** Concerns about fouling RO membranes
Alum and PACls	*Alum and PACls*
Widely used	Concerns about precipitative scaling
Ferric Salts	*Ferric Salts*
Less use than aluminum coagulants but used extensively, especially ferric chloride	Widely used, especially ferric chloride

*Molecular weight.

coagulation could be advantageous, especially because seawater pH is fairly high limiting the fraction of positively charged Fe species available for charge neutralization—discussed under the section Ferric Coagulants. However, there is concern by RO membrane manufacturers that overdosing of the cationic polymers would lead to carryover to RO membranes where they would adsorb and foul the membrane. It is our opinion that this concern is without sound basis and cationic polymers should be considered in a dual-coagulant strategy with ferric chloride. We do not think they should be a problem for the following reasons.

- First, in a dual-coagulant strategy the cationic polymer dose would be held constant at a low concentration where overdosing cannot occur because ferric chloride would act as the main coagulant. Any change in raw water quality, the ferric chloride dose would be adjusted not the cationic polymer.
- Second, one can use a streaming-current monitor (see Chap. 10) to ensure no overdosing of the cationic polymer.
- Third, even if the cationic polymer dose was too high, which is almost impossible in a dual-coagulant strategy, then flocculation and particle separation by clarification and granular media filtration would perform poorly.
- Fourth, if there were an overdose, then the excess positively charged polymer in solution would then adsorb on the surfaces of granular filter media (sand, anthracite, or GAC), which are negatively charged, and the polymer would not be present in RO feedwater.

For these reasons the authors believe low MW, high charge-density cationic polymers should be considered with ferric coagulation—more on this when ferric coagulation is discussed. Their use and dosage can be easily evaluated in bench-scale and pilot-scale studies.

High MW nonionic and anionic polymers are another matter and are not recommended. These are more easily overdosed and can present a problem. If overdosing of them occurs and granular media filtration is included in pretreatment, then the filters would perform poorly often in the form of plugging near the surface producing high head loss. Overdosing can occur where poor filter performance is not always so obvious and control of the polymer dose can be problematic. There has been some use of these types of polymers as flocaids at dosages of 0.2 to 0.5 mg/L. If these polymers reach RO membranes, they adsorb strongly on the membranes producing scaling and fouling. They should be avoided, but if used then they must be used with caution.

Aluminum Coagulants

The chemistry of PACls and alum and their use as coagulants in freshwaters are covered in detail in Chap. 6. PACls can act by charge neutralization and by precipitation as an aluminum hydroxide solid, and these modes of action are highly dependent on dose and pH. The authors are unaware of full-scale use of PACls in seawater coagulation, and we do not recommend their use for the following reasons.

- If used at pH >7, precipitation of an aluminum hydroxide solid occurs leaving a relatively high soluble Al concentration at the tenths of mg/L for water temperatures <20°C and at 1 mg/L or greater at temperatures of 30°C (high temperature occurs at Middle Eastern desalination plants).
- If used at pH <7, soluble Al can be even greater.

The soluble Al in either pH case would concentrate on the RO membranes leading to precipitative scaling from production of aluminum hydroxide and aluminum silicate solids. The discussion of Al solubility is illustrated in more detail for alum.

Alum is the most widely used coagulant for the treatment of freshwaters. Like PACls, because of its relatively high solubility, it is not recommended for coagulation of seawater. Solubility plots for amorphous aluminum hydroxide, which is produced when alum is used dependent upon dose and pH, are presented in Fig. 13-5. The high ionic strength of seawater makes Al slightly more soluble in seawater compared to freshwaters—note the seawater solubility plot is above that for freshwater. If you were to treat seawater at coagulation pH in the range of 7.5 to 8, a fairly high alum dose is required to produce sweep-floc coagulation—initial dose above the solubility curve. Furthermore, after formation of the sweep-floc aluminum hydroxide solids at equilibrium, soluble Al can be quite high. As the seawater solubility plot in Fig. 13-5 shows, if one practiced coagulation at pH 7.5 to 8, soluble Al at 25°C would be 2.22×10^{-5} to 7.41×10^{-5} M (0.6 to 2 mg/L); a little lower at colder temperatures but greater for higher temperatures. If one practiced coagulation closer to the pH of minimum solubility at pH near 6.3, then the alum dose would be high to overcome the seawater buffer intensity and to decrease the pH, or one would need to add a strong acid such as sulfuric acid with the alum to decrease the pH. Even at this pH condition, the solubility plot (Fig. 13-5) indicates that the residual Al would still be relatively high at about 1.5×10^{-6} M (0.04 mg/L)—lower for colder water but greater for warmer water. In short, Al is too soluble and would be carried to the RO membranes where it can concentrate producing aluminum hydroxide and aluminum silicate solids causing precipitative scaling. The best choice of coagulant is ferric salts.

Ferric Coagulants

Ferric coagulants are the choice for seawater coagulation, with widespread use of ferric chloride. Ferric chloride is very insoluble leaving little residual dissolved Fe in the water

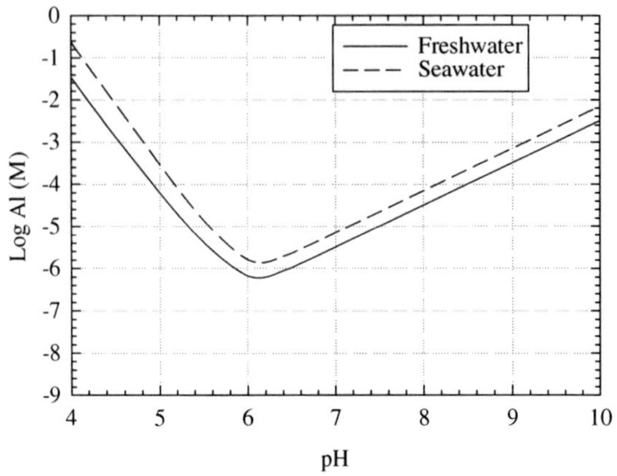

FIGURE 13-5 Solubility of amorphous aluminum hydroxide in freshwater and seawater (*S* at 35 g/kg) when using alum in coagulation (25°C).

after pretreatment and thus avoiding precipitative scaling problems. The reader is referred to Chap. 6 for an extensive presentation of coagulation using ferric coagulants. We discuss here the dissolved speciation of Fe, the solubility of ferric hydroxide, and the effects of seawater buffer intensity on reaching pH coagulation targets.

Figure 13-6 shows the fractional distribution of dissolved Fe in seawater. What this diagram shows is that there is a small fraction of positively charged Fe available for reaction with negatively charged contaminants unless the pH of coagulation is about below 7.5. To maximize positively charged Fe, the coagulation pH must be <7. We address below the desired pH for coagulation after discussing solubility.

Figure 13-7 shows solubility plots for $Fe(OH)_3$(am) in freshwater and seawater. Seawater has a slight effect on Fe solubility increasing it a small amount compared to freshwater, but as we see next its solubility is extremely low. If we coagulate seawater at about pH 7.5 to 8, near the pH of minimum solubility, the residual dissolved Fe after sweep-floc coagulation would be extremely low, $<10^{-9}$ M (<0.1 μg/L). If we coagulate at pH in the 6 to 7 range, we would have immediately after ferric chloride addition positively charged dissolved Fe species (see Fig. 13-6). The dissolved Fe would then quickly precipitate forming ferric hydroxide solids (sweep-floc), and the dissolved residual Fe after coagulation would be low as shown in Fig. 13-7, between 10^{-9} and 10^{-8} M (0.05 and 0.5 μg/L). Fe is insoluble over a wide pH range so we can use as a coagulant and residual dissolved Fe will be low. Therefore, we would carry a low soluble Fe concentration to the RO membranes, and we minimize precipitative scaling compared to Al coagulants. Fe is fairly insoluble over a wide temperature range, again avoiding the problem of warm water effects on Al solubility as discussed above.

Some practical considerations follow regarding dosages and the desired pH. For the coagulation of mineral particles where algae concentrations are low, then ferric chloride doses should be fairly low and coagulation with pH in the range of the low 7s to about 8 should be feasible to accomplish sweep-floc coagulation. To provide some positive charge, a low dosage at about 1 mg/L or less of a low MW, high charge-density cationic polymer may be beneficial as a dual coagulant with the ferric chloride.

If algae blooms are occurring, then the desired pH is lower. High concentrations of algae would also mean the presence of algal by-products (polysaccharides, etc., see prior

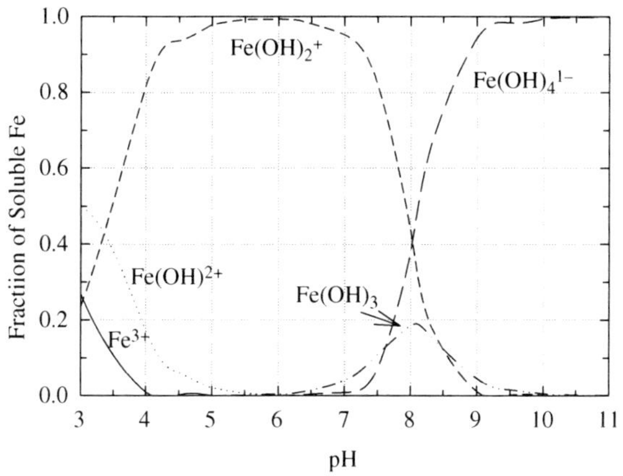

FIGURE 13-6 Fractional distribution of soluble Fe (III) in seawater (conditions: *S* at 35 g/kg, 25°C).

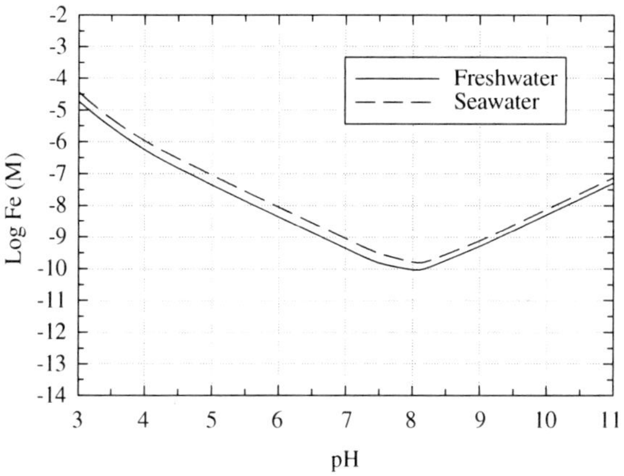

FIGURE 13-7 Solubility of amorphous ferric hydroxide in freshwater and seawater (*S* at 35 g/kg) when using ferric coagulants (25°C).

discussion), so ferric chloride doses will be greater and lower pH conditions are desired to react more favorably with the dissolved organic matter. Coagulation at pH values in the low 6s is desired, but note that the buffer intensity becomes great for pH values in the 6s (see Fig. 13-3 and above discussion on this subject). The effect would be that fairly high ferric chloride dosages are required to decrease the pH and coagulate the contaminants. One alternative would be to add sulfuric acid to decrease the pH in conjunction with ferric chloride dosing.

Generally in treating seawater with algae blooms, one should consider a dual-coagulant strategy with a fixed cationic polymer dose of say 1 mg/L and vary the ferric chloride dose as a function of raw water quality to achieve the desired treatment performance. This dual-coagulant strategy may be feasible without acid addition to decrease seawater pH.

Seawater with aquatic humic matter requires ferric coagulation at certain pH values because of the fact that there is little positively charged Fe available to react with negatively charged humic matter unless the pH is in the low 6s or less (see Fig. 13-6). For seawater with aquatic humic matter at low to moderate TOC concentrations of say 1 to 4 mg/L, coagulation at pH values in the low 6s may be feasible. For seawater with aquatic humic matter at higher TOC (say, 5 mg/L or greater), ferric chloride coagulation at pH close to 6 or a little lower may be desired.

13-4 PHYSICAL PROPERTIES OF SEAWATER

The three physical fluid properties that may affect DAF performance are density, dynamic viscosity, and surface tension. These properties are listed in Tables D-1 (freshwater) and D-2 (seawater). Over the temperature range of 0 to 40°C, the differences between seawater and freshwater are 2.6 to 2.8 percent for density, 6.3 to 8.3 percent for dynamic viscosity, and 0.7 to 1.4 percent for surface tension.

How do these differences affect the performance of DAF?

Effects on Contact Zone Performance

The physical properties of seawater density and viscosity differ from freshwater and potentially affect the contact zone performance model of Eq. (7-15) presented in Chap. 7. Two primary variables in the model depend on these physical properties. One is the rise velocity (v_b) of the air bubbles through the contact zone, which is dependent on both water density and viscosity. The total single collector efficiency (η_T) depends on the particle transport mechanisms of Brownian diffusion (η_{BD}), interception (η_I), and settling (η_S). Interception [Eq. (7-10)] is not dependent on water's physical properties, but Brownian diffusion [Eq. (7-9)] and settling [Eq. (7-11)] are dependent on water density.

We can calculate the ratios of the two single collector terms and rise velocity for seawater versus freshwater, where the subscript, SW, stands for seawater and the subscript, FW, stands for freshwater. The ratios are presented in Eqs. (13-6) through (13-8).

$$\frac{\eta_{BD,SW}}{\eta_{BD,FW}} = \left(\frac{1}{\rho_{FW} - \rho_b}\right)^{-2/3} \left(\frac{1}{\rho_{SW} - \rho_b}\right)^{2/3} \tag{13-6}$$

$$\frac{\eta_{S,SW}}{\eta_{S,FW}} = \left[\frac{(\rho_{FW} - \rho_b)}{(\rho_p - \rho_{FW})}\right]\left[\frac{(\rho_p - \rho_{SW})}{(\rho_{SW} - \rho_b)}\right] \tag{13-7}$$

$$\frac{v_{b,SW}}{v_{b,FW}} = \left[\frac{\mu_{FW}}{(\rho_{FW} - \rho_b)}\right]\left[\frac{(\rho_{SW} - \rho_b)}{\mu_{SW}}\right] \tag{13-8}$$

The results from calculating the ratios follow for seawater at 20°C where the particle or floc density is set at 1050 kg/m^3: (1) $\eta_{BD,SW}$ is 98 percent of what it is in freshwater,

(2) $\eta_{S,SW}$ is 43 percent of the freshwater value, and (3) $v_{b,SW}$ is 96 percent of $v_{b,FW}$. Some seawater facilities are located in hot climates so the calculations were repeated for 35°C. There were no great differences for $\eta_{BD,SW}$ and $v_{b,SW}$, while $\eta_{S,SW}$ increased to 52 percent of the freshwater value.

The effect of seawater on $\eta_{BD,SW}$ is small. Furthermore, after coagulation and flocculation this particle-collision mechanism is not important compared to interception because the floc particles are too large for Brownian diffusion (see Chap. 7). There is an effect of seawater on $\eta_{S,SW}$ reducing it by as much as 57 percent. However, interception is the primary particle-bubble collision mechanism for low-density particles and not settling. Therefore, this effect on $\eta_{S,SW}$ is not important. The effect of seawater on $v_{b,SW}$ (reduction of 4 percent) does not make a practical difference. In summary, there are no practical effects of the physical properties of seawater on the contact zone performance. A final point is that the air bubble volume concentration (Φ_b) can also be affected, but this is discussed in terms of the available air under Sec. 13-5.

Effects on Separation Zone Performance

The theoretical rise rate of the floc-bubble aggregates is given by either Eq. (8-11) (laminar flow) or Eq. (8-13) (transitional flow if the Reynolds number is above 1). From Eq. (8-11), the impact of the fluid properties is expressed as:

$$\frac{v_{fb,SW}}{v_{fb,FW}} = \left(\frac{(\rho_w - \rho_{fb})_{SW}}{(\rho_w - \rho_{fb})_{FW}}\right)\left(\frac{\mu_{SW}}{\mu_{FW}}\right)^{-1.0} \tag{13-9}$$

For transitional flow, a different ratio is obtained from Eq. (8-13):

$$\frac{v_{fb,SW}}{v_{fb,FW}} = \left(\frac{(\rho_w - \rho_{fb})_{SW}}{(\rho_w - \rho_{fb})_{FW}}\right)^{0.8}\left(\frac{\rho_{w,SW}}{\rho_{w,FW}}\right)^{-0.2}\left(\frac{\mu_{SW}}{\mu_{FW}}\right)^{-0.6} \tag{13-10}$$

Equations (13-9) and (13-10) require an estimate of the density of the floc-bubble aggregates. The difference between water density and floc-bubble density should be about the same in seawater and freshwater, and the first bracketed terms in the equations are assumed to be one. For laminar flow, the rise rate of the same floc-bubble aggregate in seawater will be between 94.1 (at 0°C) and 92.2 percent (at 40°C) of the rise rate in freshwater. For transitional flow, the corresponding percentages are 96.3 (at 0°C) and 95.3 percent (at 40°C) of the rise rate in freshwater. For equivalent performance, the hydraulic loading for seawater DAF should be between 4 and 8 percent lower than that of freshwater. Hydraulic loading, however, cannot be determined this precisely and practical design is not likely to be affected by the theoretical difference indicated by Eqs. (13-9) and (13-10).

Effects on Dissolving Air in the Saturator

The wetted area of the plastic packing used in saturators is determined by the density, dynamic viscosity, and surface tension of the water as indicated in Eq. (3-21). For a typical nominal packing size of 50 mm and mass hydraulic loading of 40 kg/m²-s, the wetted packing area in seawater is 99.4 percent of the value in freshwater throughout the temperature range 0 to 40°C.

The mass transfer constant K_L is given by Eq. (3-20). From Eq. (3-20), it follows that the impact of the fluid properties is estimated with:

$$\frac{K_{L,\mathrm{SW}}}{K_{L,\mathrm{FW}}} = \left(\frac{\mu_{\mathrm{SW}}}{\mu_{\mathrm{FW}}}\right)^{-5/6}\left(\frac{\rho_{\mathrm{SW}}}{\rho_{\mathrm{FW}}}\right)^{1/6}\left(\frac{a_{\mathrm{w,SW}}}{a_{\mathrm{w,FW}}}\right)^{-2/3}\left(\frac{D_{\mathrm{SW}}}{D_{\mathrm{FW}}}\right)^{1/3} \tag{13-11}$$

The last factor in Eq. (13-11) shows that the mass transfer rate constant is influenced by the molecular diffusivity of the gas. From Appendix F, the molecular diffusivity of a gas in seawater has conservatively been estimated at 94 percent of that in freshwater. Eq. (13-11) indicates that the mass transfer rate in seawater ranges from 89.2 (0°C) to 87.8 percent (40°C). This number has an effect on the performance of the air saturation system, which is investigated in the next section.

13-5 AIR SATURATION IN SEAWATER

Air saturation at equilibrium is controlled by Henry's law, discussed in Sec. 3-2. Tables F-1 and F-2 provide values for Henry's constant for nitrogen, oxygen, and argon. The Henry's constants are considerably higher, which means that the solubility of the gases is reduced. The Henry's constants for seawater at 0 and 40°C are, as a percentage of those for freshwater, 137.3 and 128.6 percent for nitrogen; 134.7 and 126.2 percent for oxygen; and 134.7 and 126.3 percent for argon. As a result, the solubility of atmospheric air at 20°C and sea level is 0.640 mol/m³ in seawater compared to 0.837 mol/m³ for freshwater—a reduction of 23.5 percent.

The available air concentration in the water from a saturator is a function of both the theoretical maximum (imposed by the Henry's constant) and the transfer efficiency (determined by the design parameters of the saturator). Figure 13-8 shows the available air concentration transferred by a typical saturator, for both freshwater and seawater. For seawater, the available air ranges from only 71 (at 0°C) to 77 percent (at 40°C) of freshwater. This is a large difference. If one assumes that seawater DAF requires the same volumetric air concentration (Φ_b in the contact zone model) as freshwater DAF (and there is no reason at this point to disagree), then the saturation system has to be adapted for seawater. The mass transfer model presented in Sec. 3-4 provides the means for the design of the saturator from first principles, using the physical constants provided in the appendices.

It was shown in Chap. 3 that the nominal packing size and the hydraulic loading have relatively small effects on the available air concentration. This leaves three possible means of adapting the saturation system for seawater. First, the saturator can be designed using the typical freshwater design parameters in Table 3-3, but then the recycle flow must be increased. As a rule of thumb, the recycle rate has to be increased by approximately 100/0.77 = 30 percent to 100/0.71 = 40 percent greater than for freshwater, using the values from Fig. 13-8. Second, the packing depth can be increased. This option can only partly solve the problem. At temperatures below about 10°C, the available air concentration can be increased for seawater by increasing the packing depth from the average 1400 mm in Table 3-3 to about 2500 mm, but at higher temperatures, extra packing depth cannot fully compensate for the lower solubility of air in seawater. Third, the saturator pressure can be increased. This is the most practical option and most likely to be favored by designers.

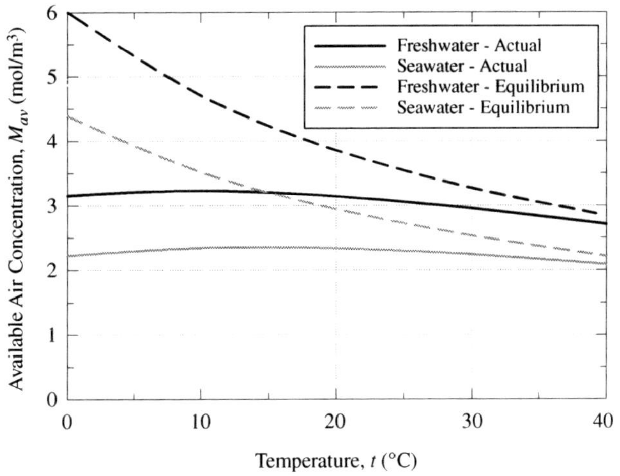

FIGURE 13-8 Comparison of the available air concentration transferred by a typical saturator to freshwater and seawater (dotted lines indicate the maximum concentration for perfect transfer efficiency; the typical saturator is defined in Table 3-3).

Using the typical saturator of Table 3-3 at saturation pressure of 500 kPa, the effect of seawater can be fully compensated for if the saturator pressure is increased to 720 kPa at 0°C, or 650 kPa at 40°C.

13-6 EXAMPLES OF DAF PRETREATMENT

Over the last several years the effectiveness of DAF as a pretreatment process in removing TOC, algae, and oil and grease has been recognized. Several desalination plants now use DAF including plants in (1) Singapore, (2) Chile, (3), Spain, (4) UAE, (5) Kuwait, and (6) a very large plant (1040 ML/d) in Saudi Arabia scheduled to begin operation in 2013. The first three examples are briefly described below. Figure 13-9 shows the pretreatment processes with DAF separate from dual media filtration (Fig. 13-9a) or in an integrated vertical scheme of DAF over filtration or DAFF (Fig. 13-9b).

There are a few DAF-RO plants in Chile. The largest one is the El Coloso plant on the north coast of Chile with a capacity of 45.4 ML/d [11.8 MGD] (Petry et al., 2007). The source water has quality problems with TOC, high algae, and red-tide events. The pretreatment consists of coagulation using ferric chloride and sulfuric acid to control pH (presumably, near pH 7 based on reported preliminary bench-scale studies). Pretreatment continues with flocculation, DAF, two stages in sequence of dual media filters, and cartridge filtration (5 μm nominal cutoff). DAF is high-rate flotation (AquaDAF®) designed for a maximum loading rate of 33 m/h (13.5 gpm/ft^2). Results from pilot studies demonstrated good pretreatment with effluent turbidities following DAF and the first-stage dual media filters of 0.1 to 0.2 NTU and about 0.03 NTU following the second-stage filters. SDI values were 2 to 3 following the second-stage filters.

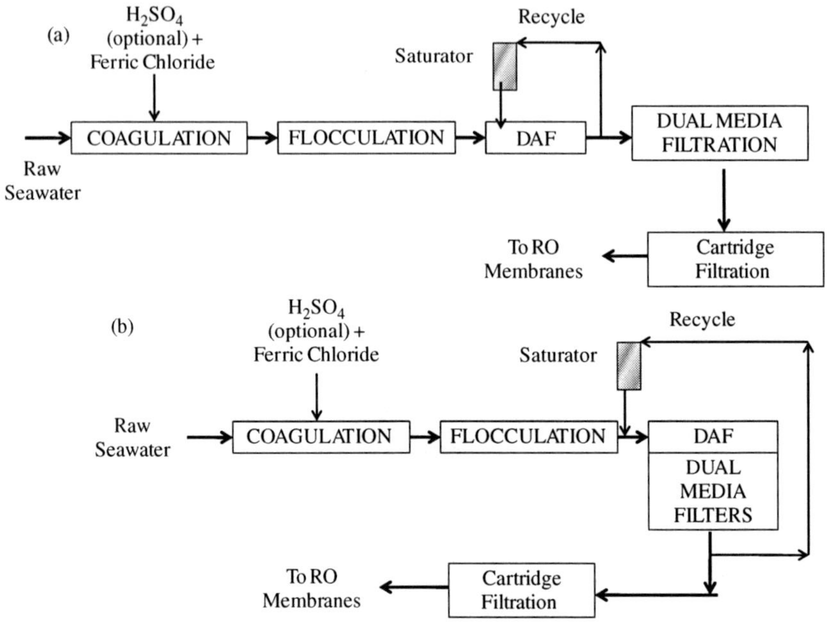

FIGURE 13-9 Schematics of DAF pretreatment RO schemes: (a) DAF with separate dual media filtration and (b) DAF over filtration (DAFF).

A DAF-RO plant in Barcelona (Spain) began operation in 2009. Information on this plant was obtained from Hayword (2009) and Voutchkov (2010). It is a 200 ML/d (53 MGD) plant. The Mediterranean Sea is the source, and the location of the intake is influenced by the discharge of the Llobregat River. Water quality problems consist of algae, TOC, and high turbidity spikes (can reach 100 NTU) because of the influence of the river. Pretreatment consists of coagulation with ferric chloride, flocculation, DAF, two stages of dual media filtration, and cartridge filters (5 μm cutoff).

We end with description of the interesting Tuas Desalination Plant (Singapore) that opened in 2005. Information for this example comes from Kiang et al. (2005) and Voutchkov (2010). Raw water quality problems include SS as high as 50 mg/L, high algae and red-tide events, and O&G up to 10 mg/L

The Tuas Desalination Plant has a capacity of 136.4 ML/d (36 MGD) with extensive pretreatment and two-pass RO membrane treatment. Figure 13-10 shows a schematic of the pretreatment and RO processes. Pretreatment consists of screens followed by coagulation at pH 6 to 6.5 with ferric chloride and sulfuric acid addition. This is followed by flocculation and DAF over filtration (DAFF) at an average loading rate of 10 m/h (4 gpm/ft^2). Sand filters are used for filtration followed by cartridge filtration. The pretreatment is designed to produce acceptable water for RO with the SDI <3 and non-detectable O&G.

The pH is adjusted to 8.2 for the first pass-RO and then raised further for the second-pass RO. The main purposes of the second-pass RO is to reduce boron to <0.5 mg/L (see earlier discussion about boron and health concerns) and produce water with TDS of 50 mg/L.

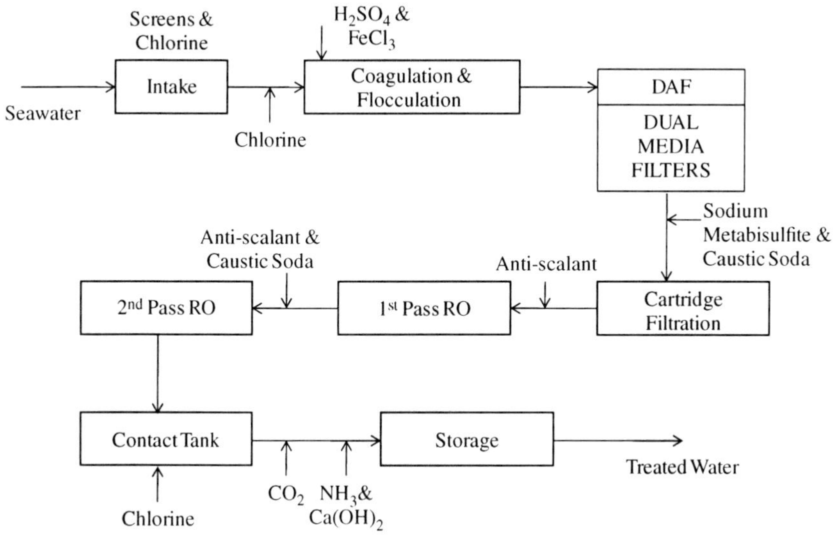

FIGURE 13-10 Process schematic of the Singapore RO plant.

REFERENCES

ASTM D4189-07 (2007), *Standard Test Method for Silt Density Index (SDI) of Water*, Conshohocken, PA, 2007, DOI: 10.1520/D4189-07, www.astm.org

Berner, E. K., and Berner, R. A. (1996), *Global Environment: Water, Air, and Geochemical Cycles*, Saddle River, NJ: Prentice Hall.

Drever, J. I. (1997), *The Geochemistry of Natural Waters*, 3rd ed., Englewood Cliffs, NJ: Prentice Hall.

Duranceau, S. J., and Taylor, J. S. (2010), Membranes, in J. K. Edzwald, ed., *Water Quality and Treatment: A Handbook on Drinking Water*, 6th ed., New York: AWWA and McGraw-Hill.

Eby, G. N. (2004), *Principles of Environmental Geochemistry*, Pacific Grove, CA: Thomson/Brooks Cole.

Edzwald, J. K., Upchurch, J. B., and O'Melia, C. R. (1974), Coagulation in estuaries, *Environmental Science and Technology*, 8 (1), 58–63.

EU (2011), http://ec.europa.eu/environment/water/water-drink/index_en.html (accessed January 2011).

Freeman, S. (2009), Seawater Pretreatment Operations, presentations at the workshop on *Coagulation for Seawater and Reuse Applications*, AWWA, Water Quality Technology Conference, November 15, 2009, Seattle, WA.

Hayword, K. (2009), Desal delivers a supply boost to Barcelona, *Water 21*, December 2009 issue, 40–41.

Kiang, F. H., Arasu, S., Yong, W. W. L., and Ratnayaka, D. D. (2005), Supply of desalinated water by the private sector Singapore's first public-private-partnership, *Presented at the International Desalination Association World Congress*, Singapore, 2005.

Millero, F. J. (1995), Thermodynamics of the carbon dioxide system in the ocean, *Geochimica et Cosmochimica Acta*, 59 (4), 661–677.

Ogawa, H., and Tanoue, E. (2003), Dissolved organic matter in oceanic waters, *Journal of Oceanography*, 59 (3), 129–147.

Petry, M., Sanz, M. A., Langlais, C., Bonnelye, V., Durand, J. P., Guevara, D., Nardes, W. M., and Saemi, C. H. (2007), The El Coloso (Chile) reverse osmosis plant, *Desalination*, 203, 141–152.

Schneider, O. (2011), Personal communication, Senior Environmental Engineer, American Water, Voorhees, NJ.

Stumm, W., and Morgan, J. J. (1996), *Aquatic Chemistry*, 3rd ed., NY: Wiley-Interscience.

USEPA (2008), Drinking Water Health Advisory for Boron, Document Number: 822-R-08-013, Washington, DC.

Voutchkov, N. (2010), *Seawater Pretreatment*, Bangkok, Thailand: Water Treatment Academy.

WHO, (2011), www.who.int/water_sanitation_health/dwq/en (accessed January 2011).

APPENDIX A
ABBREVIATIONS AND EQUATION SYMBOLS

A-1 ABBREVIATIONS

ADV	acoustic Doppler velocimetry
Alk	alkalinity
ANC	acid neutralizing capacity
BAC	biologically active carbon
BDOC	biodegradable dissolved organic carbon
cm	centimeter
CoCoDAFF	counter-current DAFF
C	Celsius
CAP	Central Arizona Project
CFD	computation fluid dynamics
CFSTR	continuous flow stirred tank reactor
CU	color unit
DAF	dissolved air flotation
DAFF	flotation over filtration
DBPs	disinfection by-products
DLVO	Derjaguin-Landau-Verwey-Overbeek
DOC	dissolved organic carbon
EOM	extracellular organic matter
EPM	electrophoretic mobility
EU	European Union
FAD	free air delivery
FTU	formazin turbidity unit
gal	gallons
gpd	gallons per day
gpm/ft^2	gallons per minute per square foot
GAC	granular activated carbon
GWRP	Goreangab Water Reclamation Plant
HAAs	haloacetic acids
IEP	isoelectric point
IOM	intracellular organic matter
kPA	kilopascal
K	Kelvin or equilibrium constant
L	liter
LDV	laser Doppler velocimetry
m	meter
mg	milligram
min	minute
mL	milliliter
M	molar
MF	microfiltration
MGD	million gallons per day
ML/d	megaliters per day
MW	molecular weight
nm	nanometer
NF	nanofiltration
NOM	natural organic matter
NTU	nephelometric turbidity unit
O&G	oil and grease
pH_{iep}	pH of the isoelectric point
pH_{zpc}	pH of zero point of charge
ppm	parts per million

psi	pounds per square inch
PAC	powdered activated carbon
PAC-DAF	powdered activated carbon dissolved air flotation
PACl	polyaluminum chloride
PCBs	polychlorinated biphenols
PES	polyethersulfone
PFR	plug-flow reactor
POC	particulate organic carbon
PVDF	polyvinylidene fluoride
Re	Reynolds number
RO	reverse osmosis
RTD	residence time distribution
S	Siemens or salinity
SEDAF	sedimentation ahead of DAF
SFBW	spent filter backwash water
SS	suspended solids
SUVA	specific ultraviolet absorbance
THMs	trihalomethanes
TMP	transmembrane pressure
TOC	total organic carbon
TDS	total dissolved solids
TSS	total suspended solids
UFRV	unit filter run volume
USEPA	United States Environmental Protection Agency
UV_{254}	ultraviolet absorbance at 254 nm
UWNJ	United Water New Jersey
WHO	World Health Organization
WRc	Water Research Centre
WINGOC	Windhoek Goreangab Operating Company
WTP	water treatment plant
≡XOH	surface of a metal (X) hydroxide or oxide particle
zpc	zero point of charge
μeq	microequivalent
μm	micrometer

A-2 EQUATION SYMBOLS

a	specific packing area
A	area in plan, or cross-sectional area, or Hamaker constant
A_b	projected bubble area
Å	angstrom
C	molar, or mass concentration
C_D	drag coefficient
$Cl‰$	chlorinity in ppt
d	diameter, or nominal size
d_{cb}	critical bubble diameter
d_{dn}	dispersion number
dH	differential reactor depth
D	axial dispersion coefficient, or molecular diffusivity
D_{fr}	fractal dimension
e	overall efficiency of air delivery, or electron charge, and energy
E	efficiency, or unit energy consumption
f	volumetric, or molar air fraction
F_B	buoyant force
F_D	drag force
F_{edl}	electrical double-layer force
F_{hydrop}	hydrophobic force
F_G	gravity force
F_{vdw}	van der Waals force
g	gravitational constant of acceleration
G	root-mean-square velocity gradient, or molar concentration in air

h	altitude
H	Henry's constant
HL	hydraulic loading
HL_{mass}	liquid mass loading rate
h	separation, or interparticle distance
I	ionic strength
J	mass transfer rate per packing volume
k_a	air deficit mass concentration
k_b	Boltzmann's constant
K	mass transfer rate constant, or shape factor, or hydrophobic force constant
L_{cz}	contact zone longitudinal flow length
m	dead volume fraction
mw	molecular weight
M	molar concentration in water
n	ion concentration, or particle number concentration, or bubble number concentration
N	number of bubbles per floc
P	power
p	pressure
Q	volumetric flow rate
r	recycle ratio, or rate of reaction, or degree of neutralization
Re	Reynolds number
S	pressure ratio
$S‰$	salinity in ppt
t	time, or temperature
T	absolute temperature
v_b	rise velocity, or rise rate
v_x	flow velocity in the direction of flow
V	volume
W	reactor width
z	charge on the ion
Z	depth
α	attachment efficiency factor
β	intermediate parameter, or buffer intensity, or collision frequency function
ε	permittivity of water
ε_o	vacuum permittivity
ε_r	dielectric constant of water
ζ	zeta potential
η	single collector efficiency
κ	Debye length parameter
κ_c	conductivity of water
λ	decay length
μ	dynamic viscosity
π	mathematical constant, 3.14
ρ	density
σ	surface tension
τ	theoretical mean detention time
Φ	volume concentration

Superscript

*	indicating equilibrium

Subscripts

abs	indicating absolute
air	indicating air
Ar	indicating argon
av	indicating available
atm	indicating atmosphere, or atmospheric
b	indicating bubble
BD	indicating Brownian diffusion
c	indicating critical, or normalized

CFSTR	indicating continuous flow stirred tank reactor
cz	indicating contact zone
e	indicating effluent
f	indicating floc
fb	indicating floc-bubble
FW	indicating fresh water
i	indicating influent, or *i*-size particles
in	indicating in
I	indicating Interception
IN	indicating Inertia
j	indicating j-size particles
L	indicating liquid
nom	indicating nominal
N	indicating nitrogen
o	indicating initial, or feed
out	indicating out
O	indicating oxygen
p	indicating particle, or packing
pb	indicating particle-bubble
PFR	indicating plug flow reactor
r	indicating recycle
sat	indicating saturator
sc	indicating single collector
standard	indicating standard conditions
sz	indicating separation zone
S	indicating settling
SH	indicating fluid shear
SW	indicating seawater
T	indicating total
vap	indicating vapor
w	indicating water, or wetted
X	indicating gas

APPENDIX B
USEFUL CONVERSIONS

TABLE B-1 Unit Conversion Factors, SI Units to U.S. Customary Units, and U.S. Customary Units to SI Units

SI unit name	Symbol	→	←	Symbol	U.S. customary unit name
		To convert, multiply in direction shown by arrows			
Acceleration					
Meters per square second	m/s^2	3.2808	0.3048	ft/s^2	Feet per square second
Meters per square second	m/s^2	39.3701	0.0254	in/s^2	Inches per square second
Area					
Hectare (10,000 m^2)	ha	2.4711	0.4047	ac	Acre
Square centimeter	cm^2	0.1550	6.4516	in^2	Square inch
Square kilometer	km^2	0.3861	2.5900	mi^2	Square mile
Square kilometer	km^2	247.1054	4.047×10^{-2}	ac	Acre
Square meter	m^2	10.7639	9.2903×10^{-2}	ft^2	Square foot
Square meter	m^2	1.1960	0.8361	yd^2	Square yard
Energy					
Kilojoule	kJ	0.9478	1.0551	Btu	British thermal unit
Joule	J	2.7778×10^{-7}	3.6×10^{6}	kW·h	Kilowatt-hour
Joule	J	0.7376	1.356	$ft \cdot lb_f$	Foot-pound (force)
Joule	J	1.0000	1.0000	W·s	Watt-second
Joule	J	0.2388	4.1876	cal	Calorie
Kilojoule	kJ	2.7778×10^{-4}	3600	kW·h	Kilowatt-hour
Kilojoule	kJ	0.2778	3.600	W·h	Watt-hour
Megajoule	MJ	0.3725	2.6845	hp·h	Horsepower-hour
Force					
Newton	N	0.2248	4.4482	lb_f	Pound force
Flow Rate					
Cubic meters per day	m^3/d	264.2	3.785×10^{-3}	gpd	Gallons per day
Cubic meters per day	m^3/d	2.642×10^{-4}	3.785×10^{3}	MGD	Million gallons per day
Cubic meters per second	m^3/s	35.3147	0.02832	ft^3/s	Cubic feet per second

(continued)

TABLE B-1 Unit Conversion Factors, SI Units to U.S. Customary Units, and U.S. Customary Units to SI Units (*Continued*)

To convert, multiply in direction shown by arrows					
SI unit name	**Symbol**	**⟶**	**⟵**	**Symbol**	**U.S. customary unit name**
Cubic meters per second	m^3/s	22.827	0.0438	MGD	Million gallons per day
Cubic meters per second	m^3/s	15,850.3	6.3090×10^{-5}	gpm	Gallons per minute
Liters per second	L/s	22,827	4.381×10^{-5}	gpd	Gallons per day
Liters per second	L/s	2.2825×10^{-2}	43.8126	MGD	Million gallons per day
Liters per second	L/s	15.852	0.0631	gpm	Gallons per minute
Length					
Centimeter	cm	0.3937	2.540	in	Inch
Kilometer	km	0.6214	1.6093	mi	Mile
Meter	m	39.3701	2.54×10^{-2}	in	Inch
Meter	m	3.2808	0.3048	ft	Foot
Meter	m	1.0936	0.9144	yd	Yard
Millimeter	mm	0.03937	25.4	in	Inch
Mass					
Gram	g	0.0353	28.3495	oz	Ounce
Gram	g	0.0022	4.5359×10^2	lb	Pound
Kilogram	kg	2.2046	0.45359	lb	Pound
Power					
Kilowatt	kW	0.9478	1.0551	Btu/s	British thermal units per second
Kilowatt	kW	1.3410	0.7457	hp	Horsepower
Watt	W	0.7376	1.3558	$ft\text{-}lb_f/s$	Foot-pounds (force) per second
Pressure (force/area)					
Pascal (newtons per square meter)	$Pa(N/m^2)$	1.4504×10^{-4}	6.8948×10^3	lb/in^2	Pounds (force) per square inch
Pascal (newtons per square meter)	$Pa(N/m^2)$	2.0885×10^{-2}	47.8803	lb/ft^2	Pounds (force) per square foot
Pascal (newtons per square meter)	$Pa(N/m^2)$	2.9613×10^{-4}	3.3768×10^3	in Hg	Inches of mercury (60°F)
Pascal (newtons per square meter)	$Pa\ (N/m^2)$	4.0187×10^{-3}	2.4884×10^2	in H_2O	Inches of water (60°F)

(continued)

TABLE B-1 Unit Conversion Factors, SI Units to U.S. Customary Units, and U.S. Customary Units to SI Units (*Continued*)

To convert, multiply in direction shown by arrows

SI unit name	Symbol	⟶	⟵	Symbol	U.S. customary unit name
Kilopascal (kilonewtons per square meter)	kPa (kN/m^2)	0.1450	6.8948	lb/in^2	Pounds (force) per square inch
Kilopascal (kilonewtons per square meter)	kPa (kN/m^2)	9.8688×10^{-3}	1.0133×10^{2}	atm	Atmosphere (standard)
Temperature					
Degree Celsius	°C	1.8(°C) + 32	0.555(°F – 32)	°F	Degree Fahrenheit
Kelvin	K	1.8(K) – 459.67	0.555(°F + 459.67)	°F	Degree Fahrenheit
Velocity					
Meters per second	m/s	2.2369	0.44704	mph	Miles per hour
Meters per second	m/s	3.2808	0.3048	ft/s	Feet per second
Volume					
Cubic centimeter	cm^3	0.0610	16.3781	in^3	Cubic inch
Cubic meter	m^3	35.3147	2.8317×10^{-2}	ft^3	Cubic foot
Cubic meter	m^3	1.3079	0.7646	yd^3	Cubic yard
Cubic meter	m^3	264.1720	3.7854×10^{-3}	gal	Gallon
Cubic meter	m^3	8.1071×10^{-4}	1.2335×10^{3}	acre-ft	Acre-foot
Liter	L	0.2642	3.7854	gal	Gallon
Liter	L	0.0353	28.3168	ft^3	Cubic foot
Liter	L	33.8150	2.9573×10^{-2}	oz	Ounce (U.S. fluid)

Notes: (1) cm is not a SI unit but is included because of its common usage; (2) U.S. gallons are used here and throughout the text and not the imperial (UK) gallon, which is 1.2 × U.S. gallon.

TABLE B-2 Common Water Treatment Conversion Factors, SI units to U.S. Customary Units, and U.S. Customary Units to SI Units

To convert, multiply in direction shown by arrows					
SI unit name	**Symbol**	**⟶**	**⟵**	**Symbol**	**U.S. customary unit name**
Concentration					
Kilogram per cubic meter	kg/m^3	8.34	0.12	lb/MG	Pounds per million gallons
Milligram per liter	mg/L	8.34	0.12	lb/MG	Pounds per million gallons
Flow Rate					
Cubic meter per day	m^3/d	2.642×10^{-4}	3.785×10^3	MGD	Million gallons per day
Megaliter per day	ML/d	0.2642	3.785	MGD	Million gallons per day
Liter per second	L/s	15.852	0.0631	gpm	Gallons per minute
Hydraulic Loading Rate					
Cubic meter per square meter-hour	m^3/m^2-h	0.4098	2.44	gpm/ft^2	Gallons per min per sq ft
Meter per hour	m/h	0.4098	2.44	gpm/ft^2	Gallons per min per sq ft
Mass Loading					
Kilogram per day	kg/d	2.2046	0.45359	lb/d	Pounds per day
Pressure (force/area)					
Kilopascal	kPa	0.1450	6.8948	lb/in^2	Pounds per sq in
Bar	bar	14.504	0.06895	lb/in^2	Pounds per sq in

APPENDIX C
USEFUL CONSTANTS

TABLE C-1 Physical and Chemical Constants

Avogadro's number	6.022×10^{23} mol^{-1}
Boltzmann's constant, k_b	1.3805×10^{-23} J K^{-1}
Earth gravitation, g	9.806 m s^{-2}
Electron charge	1.602×10^{-19} C
Faraday constant, F	96.485 kJ equiv^{-1}V^{-1} = 9.6485×10^{4} C equiv^{-1}
Gas constant, R	1.987 cal mol^{-1} K^{-1} = 8.314 J mol^{-1} K^{-1} = 0.08206 L atm mol^{-1} K^{-1}
Ice point	273.15 K
Molar volume of ideal gas at 273.15 K, 1 atm	22.414 L mol^{-1}
Planck constant, h	6.626×10^{-34} J s
Speed of light in vacuum	2.998×10^{8} m s^{-1}
Vacuum permittivity, ε_o	8.854×10^{-12} C^{2} J^{-1} m^{-1}

APPENDIX D
PROPERTIES OF WATER

TABLE D-1 Physical Properties of Fresh Water* with Salinity S = 0 g/kg

Temperature (°C)	Density, ρ (kg/m³)	Dynamic viscosity† μ × 10³ (N-s/m²)	Kinematic viscosity ν × 10⁶ (m²/s)	Surface tension‡, σ (N/m)	Vapor pressure, p_v (kPa)	Dielectric constant, ε_r (no units)
0	**999.84**	**1.793**	**1.793**	**0.07564**	**0.611**	**88.00**
1	**999.90**	1.733	1.733	0.07550	**0.657**	87.70
2	**999.94**	1.676	1.676	0.07536	**0.706**	87.34
3	**999.97**	1.622	1.622	0.07522	**0.758**	86.97
4	**999.98**	1.570	1.570	0.07508	**0.814**	**86.72**
5	**999.97**	1.521	1.521	0.07494	**0.873**	86.21
6	**999.94**	1.474	1.474	0.07480	**0.935**	85.82
7	**999.90**	1.429	1.429	0.07466	**1.002**	85.43
8	**999.85**	1.386	1.387	0.07452	**1.073**	85.03
9	**999.78**	1.346	1.346	0.07437	**1.148**	84.63
10	**999.70**	**1.307**	**1.307**	**0.07423**	**1.228**	**84.11**
11	**999.61**	1.270	1.270	0.07409	**1.313**	83.83
12	**999.50**	1.235	1.235	0.07394	**1.403**	83.42
13	**999.38**	1.201	1.202	0.07379	**1.498**	83.02
14	**999.25**	1.169	1.170	0.07365	**1.599**	82.63
15	**999.10**	1.138	1.139	0.07350	**1.706**	82.23
16	**998.95**	1.108	1.109	0.07335	**1.819**	81.84
17	**998.78**	1.080	1.081	0.07320	**1.938**	81.45
18	**998.60**	1.053	1.054	0.07305	**2.064**	81.07
19	**998.41**	1.027	1.029	0.07290	**2.198**	80.69
20	**998.21**	**1.002**	**1.004**	**0.07275**	**2.339**	**80.36**
21	**997.99**	0.978	0.980	0.07260	**2.488**	79.94
22	**997.77**	0.955	0.957	0.07245	**2.645**	79.58
23	**997.54**	0.933	0.935	0.07229	**2.810**	79.22
24	**997.30**	0.911	0.914	0.07214	**2.985**	78.86
25	**997.05**	0.891	0.893	0.07198	**3.169**	**78.54**
26	**996.79**	0.871	0.874	0.07183	**3.363**	78.16
27	**996.52**	0.852	0.855	0.07167	**3.567**	77.81
28	**996.24**	0.833	0.836	0.07151	**3.782**	77.47
29	**995.95**	0.815	0.818	0.07136	**4.008**	77.13
30	**995.65**	**0.798**	**0.801**	**0.07120**	**4.246**	**76.75**
31	**995.35**	0.781	0.785	0.07104	**4.495**	76.46
32	**995.03**	0.765	0.768	0.07088	**4.758**	76.12

(continued)

TABLE D-1 Physical Properties of Fresh Water* with Salinity S = 0 g/kg (*Continued*)

Temperature (°C)	Density, ρ (kg/m³)	Dynamic viscosity† μ × 10³ (N-s/m²)	Kinematic viscosity ν × 10⁶ (m²/s)	Surface tension‡, σ (N/m)	Vapor pressure, p_v (kPa)	Dielectric constant, ε_r (no units)
33	**994.71**	0.749	0.753	0.07072	**5.034**	75.78
34	**994.38**	0.734	0.738	0.07056	**5.323**	75.44
35	**994.04**	0.719	0.723	0.07040	**5.627**	75.09
36	**993.69**	0.705	0.709	0.07024	**5.945**	74.74
37	**993.33**	0.691	0.696	0.07008	**6.280**	74.39
38	**992.97**	0.678	0.683	0.06992	**6.630**	74.03
39	**992.60**	0.665	0.670	0.06976	**6.997**	73.65
40	**992.22**	**0.653**	**0.658**	**0.06960**	**7.381**	**73.28**

* The values for fresh water in this and subsequent tables are derived for water with salinity S = 0 g/kg salt. For practical purposes, fresh water values can be assumed for waters with ionic strength <0.01 M; TDS < 400 mg/L or S < 0.4 g/kg.

†Viscosity units of N-s/m² equal to kg/m-s.

‡Surface tension units of N/m equal to kg/s².

TABLE D-2 Physical Properties of Seawater with Salinity S = 35 g/kg

Temperature (°C)	Density, ρ (kg/m³)	Dynamic viscosity† μ × 10³ (N-s/m²)	Kinematic viscosity ν × 10⁶ (m²/s)	Surface tension‡, σ (N/m)	Vapor pressure, p_v (kPa)
0	**1028.00**	**1.906**	**1.854**	**0.07620**	**0.599**
1	1027.93	1.843	1.793	0.07607	0.643
2	1027.86	1.784	1.735	0.07594	0.691
3	1027.77	1.727	1.680	0.07581	0.742
4	1027.68	1.672	1.627	0.07568	0.796
5	1027.58	1.621	1.577	0.07555	0.854
6	1027.48	1.572	1.530	0.07542	0.916
7	1027.36	1.525	1.484	0.07529	0.981
8	1027.23	1.480	1.441	0.07516	1.051
9	1027.10	1.438	1.400	0.07502	1.125
10	**1026.95**	**1.397**	**1.360**	**0.07489**	**1.203**
11	1026.79	1.358	1.323	0.07476	1.286
12	1026.63	1.321	1.287	0.07462	1.374
13	1026.45	1.286	1.253	0.07448	1.468
14	1026.27	1.252	1.220	0.07435	1.566
15	1026.07	1.220	1.189	0.07421	1.671

(*continued*)

TABLE D-2 Physical Properties of Seawater with Salinity S = 35 g/kg (*Continued*)

Temperature (°C)	Density, ρ (kg/m³)	Dynamic viscosity[†] $\mu \times 10^3$ (N-s/m²)	Kinematic viscosity $\nu \times 10^6$ (m²/s)	Surface tension[‡], σ (N/m)	Vapor pressure, p_v (kPa)
16	1025.87	1.189	1.159	0.07407	1.782
17	1025.65	1.159	1.130	0.07393	1.899
18	1025.43	1.131	1.103	0.07379	2.022
19	1025.19	1.103	1.076	0.07366	2.153
20	**1026.95**	**1.077**	**1.049**	**0.07352**	**2.291**
21	1024.70	1.052	1.026	0.07337	2.437
22	1024.43	1.027	1.003	0.07323	2.591
23	1024.16	1.004	0.980	0.07309	2.753
24	1023.88	0.981	0.959	0.07295	2.924
25	1023.58	0.960	0.938	0.07280	3.104
26	1023.28	0.939	0.917	0.07266	3.294
27	1022.97	0.918	0.898	0.07252	3.494
28	1022.66	0.898	0.879	0.07237	3.705
29	1022.33	0.879	0.860	0.07223	3.926
30	**1022.00**	**0.861**	**0.842**	**0.07208**	**4.160**
31	1021.66	0.843	0.825	0.07193	4.405
32	1021.31	0.826	0.809	0.07179	4.662
33	1020.96	0.809	0.793	0.07164	4.932
34	1020.60	0.793	0.777	0.07149	5.216
35	1020.24	0.777	0.762	0.07134	5.514
36	1019.87	0.762	0.747	0.07119	5.827
37	1019.49	0.748	0.733	0.07104	6.154
38	1019.12	0.734	0.720	0.07089	6.497
39	1018.73	0.720	0.707	0.07074	6.857
40	**1018.35**	**0.708**	**0.695**	**0.07059**	**7.233**

[†]Viscosity units of N-s/m² equal to kg/m-s.

[‡]Surface tension units of N/m equal to kg/s².

Notes to Tables D-1 and D-2

1. Values in bold are taken from other sources; other values are interpolated with the correlations provided below.
2. Bold values for kinematic viscosity are calculated from density and dynamic viscosity.
3. Values for fresh water are taken from CRC Handbook (2010).
4. Values for seawater averaged from values provided at S = 30 and S = 40 g/kg in Sharqawy et al. (2010).

Sources

The Chemical Rubber Company (1985), in R. C. Weast, ed., *CRC Handbook of Chemistry and Physics*, 16th ed., Boca Raton, Florida: CRC Press Inc.

Sharqawy, M. H., Lienhard, J. H., and Zubair, S. M. (2010), Thermophysical properties of seawater: a review of the existing correlations and data, *Desalination and Water Treatment*, 16, 354–380.

Mathematical Correlations

To calculate the properties in Tables D-1 and D-2 by mathematical correlation, use the coefficients in Table D-3. All the correlations take the form of a fourth-order polynomial where the water temperature is t (°C).

$$property = a(t)^4 + b(t)^3 + c(t)^2 + d(t) + e \tag{D-1}$$

EXAMPLE The density of seawater (S = 35 g/kg) at 27°C is:

$$\rho(27^{\circ}\text{C}) = 1.04167 \times 10^{-6}(27)^4 - 6.25000 \times 10^{-5}(27)^3 - 3.60417 \times 10^{-3}(27)^2$$
$$- 6.37500 \times 10^{-2}(27) + 1.02800 \times 10^3 = 1022.97 \frac{\text{kg}}{\text{m}^3}$$

TABLE D-3 Coefficients for Water Property Correlations (Temperature in °C)

	a	b	c	d	e
Density (S = 0)	−4.66643E−07	7.70785E−05	−8.75515E−03	6.60883E−02	9.99844E+02
Density (S = 35)	1.04167E−06	−6.25000E−05	−3.60417E−03	−6.37500E−02	1.02800E+03
Dynamic viscosity (S = 0)	1.64167E−10	−2.32333E−08	1.48708E−06	−6.13117E−05	1.79300E−03
Dynamic viscosity (S = 35)	1.81250E−10	−2.50417E−08	1.56937E−06	−6.42708E−05	1.90600E−03
Kinematic viscosity (S = 0)	1.67133E−13	−2.35832E−11	1.50202E−09	−6.14183E−08	1.79328E−06
Kinematic viscosity (S = 35)	1.76318E−13	−2.44044E−11	1.52966E−09	−6.24072E−08	1.85409E−06
Surface tension (S = 0)	8.33333E−11	−5.00000E−09	−2.58333E−07	−1.38000E−04	7.56400E−02
Surface tension (S = 35)	−4.16667E−11	4.16667E−09	−4.45833E−07	−1.26417E−04	7.61950E−02
Vapor pressure (S = 0)	5.46814E−07	1.74419E−05	1.56617E−03	4.36603E−02	6.12128E−01
Vapor pressure (S = 35)	5.33333E−07	1.74167E−05	1.52417E−03	4.28833E−02	5.99000E−01
Dielectric constant (S = 0)	−3.65980E−06	3.00716E−04	−6.92954E−03	−3.39058E−01	8.80465E+01

APPENDIX E
PROPERTIES OF AIR

TABLE E-1 Absolute Pressure of Dry Atmospheric Air p_{atm} as a Function of Altitude

Altitude (m)	Absolute Dry Air Pressure (kPa)
0	101.3
100	100.1
200	98.9
300	97.8
400	96.6
500	95.5
600	94.3
700	93.2
800	92.1
900	91.0
1000	89.9
1100	88.8
1200	87.7
1300	86.7
1400	85.6
1500	84.6
1600	83.5
1700	82.5
1800	81.5
1900	80.5
2000	79.5

TABLE E-2 Moles of Air in 1 L of Atmospheric or Saturator Air

Temperature (°C)	Altitude (m)				
	0	**500**	**1000**	**1500**	**2000**
0	0.0446	0.0420	0.0395	0.0372	0.0350
1	0.0444	0.0419	0.0394	0.0371	0.0348
2	0.0443	0.0417	0.0393	0.0369	0.0347
3	0.0441	0.0416	0.0391	0.0368	0.0346

(*continued*)

TABLE E-2 Moles of Air in 1 L of Atmospheric or Saturator Air (*Continued*)

Temperature (°C)	Altitude (m)				
	0	500	1000	1500	2000
4	0.0439	0.0414	0.0390	0.0367	0.0345
5	0.0438	0.0413	0.0388	0.0365	0.0343
6	0.0436	0.0411	0.0387	0.0364	0.0342
7	0.0435	0.0410	0.0386	0.0363	0.0341
8	0.0433	0.0408	0.0384	0.0361	0.0340
9	0.0432	0.0407	0.0383	0.0360	0.0339
10	0.0430	0.0405	0.0381	0.0359	0.0337
11	0.0429	0.0404	0.0380	0.0358	0.0336
12	0.0427	0.0402	0.0379	0.0356	0.0335
13	0.0426	0.0401	0.0377	0.0355	0.0334
14	0.0424	0.0400	0.0376	0.0354	0.0333
15	0.0423	0.0398	0.0375	0.0353	0.0332
16	0.0421	0.0397	0.0374	0.0351	0.0330
17	0.0420	0.0395	0.0372	0.0350	0.0329
18	0.0418	0.0394	0.0371	0.0349	0.0328
19	0.0417	0.0393	0.0370	0.0348	0.0327
20	0.0416	0.0391	0.0368	0.0347	0.0326
21	0.0414	0.0390	0.0367	0.0345	0.0325
22	0.0413	0.0389	0.0366	0.0344	0.0324
23	0.0411	0.0387	0.0365	0.0343	0.0323
24	0.0410	0.0386	0.0364	0.0342	0.0321
25	0.0409	0.0385	0.0362	0.0341	0.0320
26	0.0407	0.0384	0.0361	0.0340	0.0319
27	0.0406	0.0382	0.0360	0.0339	0.0318
28	0.0404	0.0381	0.0359	0.0337	0.0317
29	0.0403	0.0380	0.0358	0.0336	0.0316
30	0.0402	0.0379	0.0356	0.0335	0.0315
31	0.0400	0.0377	0.0355	0.0334	0.0314
32	0.0399	0.0376	0.0354	0.0333	0.0313
33	0.0398	0.0375	0.0353	0.0332	0.0312
34	0.0397	0.0374	0.0352	0.0331	0.0311
35	0.0395	0.0372	0.0351	0.0330	0.0310
36	0.0394	0.0371	0.0349	0.0329	0.0309
37	0.0393	0.0370	0.0348	0.0328	0.0308
38	0.0391	0.0369	0.0347	0.0327	0.0307
39	0.0390	0.0368	0.0346	0.0326	0.0306
40	0.0389	0.0366	0.0345	0.0324	0.0305

TABLE E-3 Grams of Air in 1 L of Atmospheric and Typical Saturator Air (Composition of Typical Saturator Air Taken at Saturator Gauge Pressure of 500 kPa)

Temperature (°C)	Altitude (m)									
	0	500	1000	1500	2000	0	500	1000	1500	2000
	Atmospheric Air					Typical Saturator Air				
0	1.290	1.216	1.144	1.076	1.012	1.275	1.201	1.130	1.063	0.999
1	1.286	1.211	1.140	1.073	1.008	1.270	1.196	1.126	1.059	0.995
2	1.281	1.207	1.136	1.069	1.005	1.266	1.192	1.122	1.056	0.992
3	1.276	1.202	1.132	1.065	1.001	1.261	1.188	1.118	1.052	0.988
4	1.272	1.198	1.128	1.061	0.997	1.257	1.183	1.114	1.048	0.985
5	1.267	1.194	1.124	1.057	0.994	1.252	1.179	1.110	1.044	0.981
6	1.263	1.189	1.120	1.053	0.990	1.248	1.175	1.106	1.041	0.978
7	1.258	1.185	1.116	1.050	0.987	1.243	1.171	1.102	1.037	0.974
8	1.254	1.181	1.112	1.046	0.983	1.239	1.167	1.098	1.033	0.971
9	1.249	1.177	1.108	1.042	0.980	1.235	1.163	1.094	1.030	0.967
10	1.245	1.173	1.104	1.038	0.976	1.230	1.159	1.090	1.026	0.964
11	1.240	1.169	1.100	1.035	0.973	1.226	1.155	1.087	1.022	0.961
12	1.236	1.164	1.096	1.031	0.969	1.222	1.150	1.083	1.019	0.957
13	1.232	1.160	1.092	1.028	0.966	1.218	1.146	1.079	1.015	0.954
14	1.227	1.156	1.089	1.024	0.963	1.213	1.143	1.075	1.012	0.951
15	1.223	1.152	1.085	1.020	0.959	1.209	1.139	1.072	1.008	0.947
16	1.219	1.148	1.081	1.017	0.956	1.205	1.135	1.068	1.005	0.944
17	1.215	1.144	1.077	1.013	0.953	1.201	1.131	1.064	1.001	0.941
18	1.211	1.140	1.074	1.010	0.949	1.197	1.127	1.061	0.998	0.938
19	1.206	1.137	1.070	1.006	0.946	1.193	1.123	1.057	0.995	0.934
20	1.202	1.133	1.066	1.003	0.943	1.189	1.119	1.054	0.991	0.931
21	1.198	1.129	1.063	1.000	0.940	1.185	1.116	1.050	0.988	0.928
22	1.194	1.125	1.059	0.996	0.937	1.181	1.112	1.046	0.985	0.925
23	1.190	1.121	1.055	0.993	0.933	1.177	1.108	1.043	0.981	0.922
24	1.186	1.117	1.052	0.990	0.930	1.173	1.104	1.039	0.978	0.919
25	1.182	1.114	1.048	0.986	0.927	1.169	1.101	1.036	0.975	0.916
26	1.178	1.110	1.045	0.983	0.924	1.165	1.097	1.033	0.972	0.913
27	1.174	1.106	1.041	0.980	0.921	1.161	1.093	1.029	0.968	0.910
28	1.170	1.103	1.038	0.976	0.918	1.157	1.090	1.026	0.965	0.907
29	1.167	1.099	1.035	0.973	0.915	1.154	1.086	1.022	0.962	0.904
30	1.163	1.095	1.031	0.970	0.912	1.150	1.083	1.019	0.959	0.901
31	1.159	1.092	1.028	0.967	0.909	1.146	1.079	1.016	0.956	0.898
32	1.155	1.088	1.024	0.964	0.906	1.142	1.076	1.012	0.953	0.895
33	1.151	1.085	1.021	0.960	0.903	1.139	1.072	1.009	0.950	0.892

(continued)

TABLE E-3 Grams of Air in 1 L of Atmospheric and Typical Saturator Air (Composition of Typical Saturator Air Taken at Saturator Gauge Pressure of 500 kPa) (*Continued*)

	Altitude (m)									
Temperature (°C)	0	500	1000	1500	2000	0	500	1000	1500	2000
	Atmospheric Air					Typical Saturator Air				
34	1.148	1.081	1.018	0.957	0.900	1.135	1.069	1.006	0.946	0.889
35	1.144	1.078	1.014	0.954	0.897	1.131	1.065	1.003	0.943	0.886
36	1.140	1.074	1.011	0.951	0.894	1.128	1.062	1.000	0.940	0.884
37	1.136	1.071	1.008	0.948	0.891	1.124	1.059	0.996	0.937	0.881
38	1.133	1.067	1.005	0.945	0.888	1.121	1.055	0.993	0.934	0.878
39	1.129	1.064	1.001	0.942	0.886	1.117	1.052	0.990	0.932	0.875
40	1.126	1.060	0.998	0.939	0.883	1.114	1.049	0.987	0.929	0.872

Notes to Tables E-1 to E-3

1. The absolute dry air pressure in Table E-1 is calculated with the correlation below.
2. Typical saturator air in Tables E-2 and E-3 taken at saturator gauge pressure of 500 kPa, and water temperature 15°C.
3. Table E-2 calculated with Eq. (3-3) in Chap. 3 and Table E-3 calculated with Eq. (3-4) in Chap. 3.

Sources

The Chemical Rubber Company (1985), in R. C. Weast, ed., *CRC Handbook of Chemistry and Physics*, 16th ed., Boca Raton, Florida: CRC Press Inc.

Mathematical Correlations

The reader may wish to calculate the absolute dry air pressure in Table E-1 with a correlation adapted from the CRC Handbook (1985). First calculate the temperature t (°C) as a function of the altitude h (m):

$$t = 15 - 0.0065(h) \tag{E-1}$$

The dry atmospheric pressure (kPa) is calculated from:

$$p = 101.325\left(\frac{288.15}{273.15 + t}\right)^{-5.255877} \tag{E-2}$$

TABLE E-4 Moist Air Bubble Densities[a]

Temperature (°C)	ρ_b (kg/m^3)
0	1.29
4	1.27
10	1.24
15	1.22
20	1.19
25	1.17
30	1.15
35	1.12
40	1.10

[a]Calculated from dry air densities and maximum humidity ratio (mass of water vapor in moist air to mass of dry air) for 101.3 kPa absolute pressure.

APPENDIX F
SOLUBILITY OF AIR IN WATER

TABLE F-1 Henry's Constant and Molecular Diffusivity for the Three Principal Air Gases in Fresh Water* with Salinity S = 0 g/kg

Temperature (°C)	$H_{nitrogen}$ (mol/mol)	H_{oxygen} (mol/mol)	H_{argon} (mol/mol)	$D_{nitrogen}$ $\times 10^9$ m^2/s	D_{oxygen} $\times 10^9$ m^2/s	D_{argon} $\times 10^9$ m^2/s
0	41.93	20.46	18.60	0.86	1.14	1.19
1	42.89	20.97	19.05	0.90	1.18	1.23
2	43.85	21.47	19.51	0.93	1.23	1.28
3	44.81	21.98	19.97	0.97	1.27	1.33
4	45.77	22.49	20.43	1.00	1.32	1.38
5	46.72	23.00	20.89	1.04	1.37	1.43
6	47.67	23.51	21.35	1.08	1.41	1.48
7	48.61	24.02	21.80	1.11	1.46	1.53
8	49.55	24.53	22.26	1.15	1.51	1.58
9	50.49	25.04	22.72	1.19	1.57	1.63
10	51.41	25.55	23.18	1.23	1.62	1.69
11	52.33	26.06	23.63	1.27	1.67	1.74
12	53.25	26.56	24.08	1.31	1.72	1.80
13	54.15	27.07	24.54	1.35	1.78	1.86
14	55.05	27.57	24.99	1.39	1.84	1.91
15	55.94	28.07	25.44	1.44	1.89	1.97
16	56.82	28.57	25.89	1.48	1.95	2.03
17	57.70	29.07	26.33	1.52	2.01	2.09
18	58.56	29.57	26.78	1.57	2.07	2.15
19	59.42	30.06	27.22	1.61	2.12	2.22
20	60.27	30.56	27.66	1.66	2.19	2.28
21	61.10	31.05	28.10	1.71	2.25	2.34
22	61.93	31.54	28.53	1.75	2.31	2.41
23	62.75	32.02	28.97	1.80	2.37	2.47
24	63.56	32.51	29.40	1.85	2.44	2.54
25	64.36	32.99	29.83	1.90	2.50	2.61
26	65.15	33.48	30.26	1.95	2.57	2.68
27	65.94	33.95	30.68	2.00	2.63	2.75
28	66.71	34.43	31.11	2.05	2.70	2.82
29	67.47	34.91	31.53	2.10	2.77	2.89
30	68.23	35.38	31.95	2.16	2.84	2.96

(continued)

TABLE F-1 Henry's Constant and Molecular Diffusivity for the Three Principal Air Gases in Fresh Water* with Salinity S = 0 g/kg (*Continued*)

Temperature (°C)	$H_{nitrogen}$ (mol/mol)	H_{oxygen} (mol/mol)	H_{argon} (mol/mol)	$D_{nitrogen}$ ×10⁹ m²/s	D_{oxygen} ×10⁹ m²/s	D_{argon} ×10⁹ m²/s
31	68.97	35.86	32.37	2.21	2.91	3.04
32	69.71	36.33	32.79	2.27	2.98	3.11
33	70.44	36.80	33.21	2.32	3.05	3.19
34	71.16	37.27	33.62	2.38	3.13	3.26
35	71.88	37.74	34.04	2.43	3.20	3.34
36	72.58	38.21	34.45	2.49	3.28	3.42
37	73.28	38.68	34.86	2.55	3.35	3.50
38	73.97	39.15	35.28	2.61	3.43	3.58
39	74.66	39.62	35.69	2.66	3.50	3.66
40	75.34	40.09	36.11	2.72	3.58	3.74

* The values for fresh water in this and subsequent tables are derived for water with salinity S = 0 g/kg salt. For practical purposes, fresh water values can be assumed for waters with ionic strength <0.01 M; TDS < 400 mg/L or S < 0.4 g/kg.

TABLE F-2 Henry's Constant and Molecular Diffusivity for the Three Principal Air Gases in Seawater with Salinity S = 35 g/kg

Temperature (°C)	$H_{nitrogen}$ (mol/mol)	H_{oxygen} (mol/mol)	H_{argon} (mol/mol)	$D_{nitrogen}$ ×10⁹ m²/s	D_{oxygen} ×10⁹ m²/s	D_{argon} ×10⁹ m²/s
0	57.55	27.55	25.05	0.81	1.07	1.12
1	58.72	28.17	25.61	0.84	1.11	1.16
2	59.88	28.79	26.16	0.88	1.15	1.20
3	61.03	29.41	26.71	0.91	1.20	1.25
4	62.17	30.02	27.26	0.94	1.24	1.29
5	63.31	30.64	27.81	0.98	1.28	1.34
6	64.44	31.26	28.35	1.01	1.33	1.39
7	65.56	31.87	28.90	1.05	1.38	1.44
8	66.68	32.48	29.44	1.08	1.42	1.49
9	67.78	33.09	29.98	1.12	1.47	1.54
10	68.88	33.70	30.52	1.16	1.52	1.59
11	69.96	34.30	31.06	1.19	1.57	1.64
12	71.03	34.90	31.59	1.23	1.62	1.69
13	72.10	35.50	32.13	1.27	1.67	1.75
14	73.15	36.10	32.66	1.31	1.73	1.80
15	74.19	36.69	33.18	1.35	1.78	1.86
16	75.22	37.28	33.70	1.39	1.83	1.91

(continued)

TABLE F-2 Henry's Constant and Molecular Diffusivity for the Three Principal Air Gases in Seawater with Salinity S = 35 g/kg (*Continued*)

Temperature (°C)	$H_{nitrogen}$ (mol/mol)	H_{oxygen} (mol/mol)	H_{argon} (mol/mol)	$D_{nitrogen}$ ×10⁹ m²/s	D_{oxygen} ×10⁹ m²/s	D_{argon} ×10⁹ m²/s
17	76.24	37.87	34.23	1.43	1.89	1.97
18	77.25	38.45	34.74	1.48	1.94	2.03
19	78.24	39.03	35.26	1.52	2.00	2.08
20	79.23	39.60	35.77	1.56	2.05	2.14
21	80.20	40.17	36.28	1.60	2.11	2.20
22	81.16	40.74	36.78	1.65	2.17	2.26
23	82.11	41.31	37.29	1.69	2.23	2.33
24	83.05	41.87	37.79	1.74	2.29	2.39
25	83.98	42.43	38.29	1.79	2.35	2.45
26	84.90	42.99	38.78	1.83	2.41	2.52
27	85.81	43.54	39.27	1.88	2.47	2.58
28	86.70	44.09	39.76	1.93	2.54	2.65
29	87.59	44.64	40.25	1.98	2.60	2.72
30	88.47	45.18	40.74	2.03	2.67	2.78
31	89.34	45.73	41.23	2.08	2.73	2.85
32	90.20	46.27	41.71	2.13	2.80	2.92
33	91.06	46.81	42.20	2.18	2.87	2.99
34	91.90	47.35	42.68	2.23	2.94	3.07
35	92.74	47.89	43.16	2.29	3.01	3.14
36	93.58	48.43	43.64	2.34	3.08	3.21
37	94.41	48.97	44.13	2.39	3.15	3.29
38	95.23	49.50	44.61	2.45	3.22	3.36
39	96.06	50.04	45.10	2.50	3.29	3.44
40	96.87	50.58	45.58	2.56	3.37	3.51

TABLE F-3 Saturation Concentrations for the Three Principal Air Gases in Fresh Water with Salinity S = 0 g/kg, with Water Exposed to Atmospheric Air at Sea Level

	Molar concentration				Mass concentration			
Temperature (°C)	Nitrogen (mol/m³)	Oxygen (mol/m³)	Argon (mol/m³)	Total (mol/m³)	Nitrogen (mg/L)	Oxygen (mg/L)	Argon (mg/L)	Total (mg/L)
0	0.831	0.457	0.022	1.310	23.26	14.62	0.89	38.76
1	0.809	0.444	0.022	1.275	22.65	14.21	0.87	37.73
2	0.788	0.432	0.021	1.241	22.07	13.82	0.84	36.74
3	0.769	0.420	0.021	1.210	21.52	13.45	0.82	35.80
4	0.750	0.409	0.020	1.179	21.00	13.10	0.80	34.90
5	0.732	0.399	0.020	1.150	20.49	12.77	0.78	34.04
6	0.715	0.389	0.019	1.123	20.01	12.44	0.76	33.22
7	0.698	0.379	0.019	1.096	19.56	12.14	0.74	32.43
8	0.683	0.370	0.018	1.071	19.12	11.84	0.72	31.68
9	0.668	0.361	0.018	1.047	18.70	11.56	0.71	30.96
10	0.653	0.353	0.017	1.023	18.30	11.29	0.69	30.27
11	0.640	0.345	0.017	1.001	17.91	11.03	0.67	29.61
12	0.626	0.337	0.016	0.980	17.54	10.78	0.66	28.98
13	0.614	0.330	0.016	0.959	17.19	10.54	0.64	28.38
14	0.602	0.322	0.016	0.940	16.85	10.32	0.63	27.79
15	0.590	0.316	0.015	0.921	16.52	10.10	0.62	27.24
16	0.579	0.309	0.015	0.903	16.21	9.89	0.60	26.70
17	0.568	0.303	0.015	0.886	15.91	9.68	0.59	26.18
18	0.558	0.296	0.015	0.869	15.62	9.49	0.58	25.69
19	0.548	0.291	0.014	0.853	15.34	9.30	0.57	25.21
20	0.538	0.285	0.014	0.837	15.08	9.12	0.56	24.75
21	0.529	0.279	0.014	0.822	14.82	8.94	0.55	24.31
22	0.520	0.274	0.013	0.808	14.57	8.77	0.54	23.88
23	0.512	0.269	0.013	0.794	14.33	8.61	0.53	23.47
24	0.504	0.264	0.013	0.781	14.10	8.45	0.52	23.07
25	0.496	0.259	0.013	0.768	13.88	8.30	0.51	22.69
26	0.488	0.255	0.013	0.755	13.66	8.16	0.50	22.32
27	0.481	0.250	0.012	0.743	13.46	8.01	0.49	21.96
28	0.473	0.246	0.012	0.732	13.26	7.88	0.48	21.62
29	0.467	0.242	0.012	0.720	13.06	7.74	0.47	21.28
30	0.460	0.238	0.012	0.710	12.88	7.61	0.47	20.96
31	0.453	0.234	0.012	0.699	12.70	7.49	0.46	20.64
32	0.447	0.230	0.011	0.689	12.52	7.37	0.45	20.34

(continued)

TABLE F-3 Saturation Concentrations for the Three Principal Air Gases in Fresh Water with Salinity S = 0 g/kg, with Water Exposed to Atmospheric Air at Sea Level (*Continued*)

	Molar concentration				Mass concentration			
Temperature (°C)	Nitrogen (mol/m³)	Oxygen (mol/m³)	Argon (mol/m³)	Total (mol/m³)	Nitrogen (mg/L)	Oxygen (mg/L)	Argon (mg/L)	Total (mg/L)
33	0.441	0.227	0.011	0.679	12.35	7.25	0.44	20.04
34	0.435	0.223	0.011	0.669	12.19	7.13	0.44	19.76
35	0.429	0.219	0.011	0.660	12.03	7.02	0.43	19.48
36	0.424	0.216	0.011	0.651	11.87	6.91	0.42	19.21
37	0.419	0.213	0.010	0.642	11.72	6.81	0.42	18.94
38	0.413	0.210	0.010	0.633	11.57	6.70	0.41	18.69
39	0.408	0.206	0.010	0.625	11.43	6.60	0.41	18.44
40	0.403	0.203	0.010	0.616	11.29	6.51	0.40	18.19

TABLE F-4 Saturation Concentrations for the Three Principal Air Gases in Seawater with Salinity S = 35 g/kg, with Water Exposed to Atmospheric Air at Sea Level

	Molar concentration				Mass concentration			
Temperature (°C)	Nitrogen (mol/m³)	Oxygen (mol/m³)	Argon (mol/m³)	Total (mol/m³)	Nitrogen (mg/L)	Oxygen (mg/L)	Argon (mg/L)	Total (mg/L)
0	0.605	0.339	0.017	0.961	16.94	10.85	0.66	28.46
1	0.591	0.331	0.016	0.938	16.55	10.58	0.64	27.77
2	0.577	0.322	0.016	0.915	16.17	10.31	0.63	27.11
3	0.564	0.314	0.015	0.894	15.80	10.06	0.61	26.47
4	0.552	0.307	0.015	0.874	15.46	9.81	0.60	25.87
5	0.540	0.299	0.015	0.854	15.12	9.58	0.58	25.29
6	0.529	0.293	0.014	0.836	14.81	9.36	0.57	24.74
7	0.518	0.286	0.014	0.818	14.50	9.15	0.56	24.21
8	0.507	0.279	0.014	0.801	14.21	8.94	0.55	23.70
9	0.497	0.273	0.013	0.784	13.93	8.75	0.53	23.21
10	0.488	0.267	0.013	0.768	13.66	8.56	0.52	22.74
11	0.478	0.262	0.013	0.753	13.40	8.38	0.51	22.29
12	0.470	0.256	0.013	0.739	13.15	8.21	0.50	21.86
13	0.461	0.251	0.012	0.725	12.91	8.04	0.49	21.44
14	0.453	0.246	0.012	0.711	12.68	7.88	0.48	21.04
15	0.445	0.241	0.012	0.698	12.46	7.73	0.47	20.66
16	0.437	0.237	0.012	0.686	12.25	7.58	0.46	20.29

(*continued*)

TABLE F-4 Saturation Concentrations for the Three Principal Air Gases in Seawater with Salinity S = 35 g/kg, with Water Exposed to Atmospheric Air at Sea Level (*Continued*)

	Molar concentration				Mass concentration			
Temperature (°C)	Nitrogen (mol/m³)	Oxygen (mol/m³)	Argon (mol/m³)	Total (mol/m³)	Nitrogen (mg/L)	Oxygen (mg/L)	Argon (mg/L)	Total (mg/L)
17	0.430	0.232	0.011	0.674	12.04	7.43	0.46	19.93
18	0.423	0.228	0.011	0.662	11.84	7.30	0.45	19.59
19	0.416	0.224	0.011	0.651	11.65	7.16	0.44	19.25
20	0.410	0.220	0.011	0.640	11.47	7.04	0.43	18.93
21	0.403	0.216	0.011	0.630	11.29	6.91	0.42	18.62
22	0.397	0.212	0.010	0.620	11.12	6.79	0.42	18.33
23	0.391	0.209	0.010	0.610	10.95	6.68	0.41	18.04
24	0.385	0.205	0.010	0.601	10.79	6.56	0.40	17.76
25	0.380	0.202	0.010	0.592	10.64	6.46	0.40	17.49
26	0.375	0.198	0.010	0.583	10.49	6.35	0.39	17.23
27	0.369	0.195	0.010	0.574	10.34	6.25	0.38	16.97
28	0.364	0.192	0.009	0.566	10.20	6.15	0.38	16.73
29	0.359	0.189	0.009	0.558	10.06	6.06	0.37	16.49
30	0.355	0.186	0.009	0.550	9.93	5.96	0.37	16.26
31	0.350	0.184	0.009	0.543	9.80	5.87	0.36	16.03
32	0.346	0.181	0.009	0.535	9.68	5.78	0.36	15.82
33	0.341	0.178	0.009	0.528	9.55	5.70	0.35	15.60
34	0.337	0.175	0.009	0.521	9.44	5.62	0.34	15.40
35	0.333	0.173	0.009	0.514	9.32	5.53	0.34	15.19
36	0.329	0.170	0.008	0.508	9.21	5.46	0.34	15.00
37	0.325	0.168	0.008	0.501	9.10	5.38	0.33	14.80
38	0.321	0.166	0.008	0.495	8.99	5.30	0.33	14.62
39	0.317	0.163	0.008	0.489	8.88	5.23	0.32	14.43
40	0.314	0.161	0.008	0.483	8.78	5.16	0.32	14.25

Notes to Tables F-1 to F-4

1. Henry's constants for nitrogen and argon in Tables F-1 and F-2 are calculated with correlations from Hamme and Emerson (2004).
2. Henry's constant for oxygen in Tables F-1 and F-2 is calculated with correlations from García and Gordon (1992).
3. Molecular diffusivity in freshwater in Table F-1 is calculated from Perry and Green (1984).
4. Molecular diffusivity in seawater with S = 35 g/kg in Table F-2 taken as 94 percent of the values in Table F-1, a conservative assumption taken from King et al. (1995).

Sources

García, H. E., and Gordon, L. I. (1992), Oxygen solubility in seawater: better fitting equations, *Limnology and Oceanography*, 37 (6), 1307–1312.

Hamme, R. C., and Emerson, S. R. (2004), The solubility of neon, nitrogen and argon in distilled water and seawater, *Deep Sea Research*, 51 (11), 1517–1528.

King, D. B., De Bryun, W. J., Zheng, M., and Salzman, E. S. (1995), Uncertainties in the molecular diffusion coefficient of gases in water for use in the estimation of air-sea exchange, in B. Jähne and E. Monahan, eds., *Air-Water Gas Transfer*, AEON Verlag.

Perry, R. H., and Green, D. (1984), *Perry's Chemical Engineers' Handbook*, 6th ed., New York: McGraw-Hill.

Mathematical Correlations for Henry's Constants

To calculate the Henry's constants in Tables F-1 and F-2 for any temperature and degree of salinity in Table D-5, first calculate the normalized absolute temperature T_c from the water temperature t (°C):

$$T_c = \ln\left(\frac{298.15 - t}{273.15 + t}\right) \tag{F-1}$$

The concentration of a gas in water is estimated with:

$$\ln C_X^* = A_0 + A_1 \times T_c + A_2 \times T_c^2 + A_3 \times T_c^3 + S\left(B_0 + B_1 \times T_c + B_2 \times T_c^2\right) \tag{F-2}$$

C_X^* is the concentration of the gas in water (mmol/m^3), S is the salinity (g/kg), and A and B are constants.

The Henry's constant is calculated from:

$$H = 44.6 \times \rho_W \left(\frac{f_X}{C_X^*}\right)\left(\frac{273.15}{273.15 + t}\right) \tag{F-3}$$

The volumetric fraction of the gas in atmospheric air is denoted by f_X. The required constants are provided in Table F-5 (García and Gordon, 1992; Hamme and Emerson, 2004).

TABLE F-5 Correlation Coefficients for Gas Concentrations in Water

Constant	Oxygen	Nitrogen	Argon
A_0	5.80818	6.42931	2.79150
A_1	3.20684	2.92704	3.17609
A_2	4.11890	4.32531	4.13116
A_3	4.93845	4.69149	4.90379
B_1	−0.00701	−0.00744	−0.00696
B_2	−0.00726	−0.00803	−0.00767
B_3	−0.00779	−0.01468	−0.01169

EXAMPLE The Henry's constant for argon at 27°C for salinity of 35 g/kg:

The normalized temperature is:

$$T_c = \ln\left(\frac{298.15 - 27}{273.15 + 27}\right) = -0.102$$

The solubility of argon in water is:

$$\ln C^*_{Ar} = 2.79150 + 3.17609(-0.102) + 4.13116(-0.102)^2 + 4.90379(-0.102)^3$$
$$+35(-0.00696 - 0.00767(-0.102) - 0.01169(-0.102)^2) = 2.2856$$

$$C^*_{Ar} = e^{2.2856} = 9.832\frac{\text{mmol}}{\text{m}^3}$$

The density of water at 27°C is 1022.97 kg/m^3 (Table D-2).

The air fraction of argon = 0.0093 (Table 3-1).

The Henry's constant is:

$$H = 44.6(1022.97)\left(\frac{0.0093}{9.832}\right)\left(\frac{273.15}{273.15 + 27}\right) = 39.27$$

INDEX

Note: Page numbers followed by "f" indicate material in figures; by "t", in tables.

A

Absolute dry air pressure and altitude, E-1, E-4
Absolute temperature, 5-17, 7-12, F-7
Acoustic Doppler velocimetry (ADV), 8-11
Actinomycetes, 5-8
Adka vacuum flotation system, 2-3
Adsorption clarification, 10-5
ADV, 8-11
Aeration, 1-2 to 1-3, 3-5
Air concentrations:
  in CFD models, 8-11
  in contact zone, 3-2 to 3-7, 7-7, 11-6
  detention time and, 7-7
  dispersion number and, 7-7
  dissolved, 3-8f
  hydraulic loading and, 8-15
  mass. *See* Mass air concentration
  mixing and, 7-7
  molar, 3-2 to 3-9, 3-6f, 3-26 to 3-28, 3-35, F-4 to F-6
  leaving saturators, 3-20, 3-25
  in separation zone, 8-12
  volumetric. *See* Volumetric air concentration
Air dosing rate, 3-7, 3-10 to 3-11, 3-28 to 3-29, 10-11, 10-16
Air separation tanks, 3-10
Air solubility, 3-23, 13-18
Air transfer efficiency, 3-7
Albert Plant, 4-3t
Algae:
  in Bay Bulls Big Pond, 11-11
  bottom sludge and, 9-6
  carbon and, 13-3
  charge of, 5-12 to 5-13, 6-4
  cilia of, 13-9
  classification of, 6-5
  concentrations of, 5-8
  DAF and, 10-3 to 10-5, 11-6, 11-13,
  in Deacon Reservoir, 11-15
  direct filtration and, 11-13
  EOM from, 13-10 to 13-11
Algae (*Cont.*):
  ferric coagulants and, 13-14 to 13-15
  filtration and, 10-2, 10-3
  flagella of, 13-9
  in float layer, 9-2, 9-5 to 9-8
  flocculation of, 13-9
  granular media filtration and, 11-13
  at Haworth water plant, 11-17
  hydration effects on, 13-9
  IOM from, 13-11
  in King George South Reservoir, 11-14
  in lakes, 5-8, 6-5, 6-6
  membrane filtration of water with, 12-4
  motility of, 13-9
  nitrogen and, 13-4
  odors and, 11-17
  in Oradell Reservoir, 11-16
  ozone for water with, 1-3
  PAC and, 11-17, 11-18t
  pH and, 5-8, 5-13t, 6-4, 11-17
  phosphorus and, 13-4
  photographs of, 13-10f
  pilot-scale testing of, 10-7
  removal of, 10-2
  reverse osmosis and, 13-9
  scraper speed and, 9-7
  in seawater, 13-1, 13-3 to 13-4, 13-8, 13-9
  sedimentation and, 11-6, 11-13, 13-1
  silicon and, 13-4
  size of, 5-9, 5-9f
  steric effects on, 13-9
  surfactants and, 5-6
  tastes and, 11-17
  TOC and, 13-10
  treatment plant type for, 10-3
  turbidity and, 5-10
  types of, 13-9
  in William Girling Reservoir, 11-14
Alkalinity:
  of Bay Bulls Big Pond water, 11-10
  boron and, 13-4

Alkalinity (*Cont.*):
buffer intensity and, 6-11f
carbon and, 13-3 to 13-4
coagulants and, 6-10 to 6-11, 6-34
of Colorado River water, 12-6
definition of, 6-10, 13-5
of Hemlock Reservoir water, 7-29
at Lysekil water plant, 11-11, 11-13
of Oradell Reservoir water, 11-16
pH and, 6-11f
scraper speed and, 9-7 to 9-8
of seawater, 13-3 to 13-6
silicon and, 13-3, 13-4
water hardness and, 6-12
at Windhoek water plant, 12-9t
Allochthonous organic matter, 6-5
Alum. *See also* Aluminum sulfate
alkalinity and, 6-10, 6-34
bench-scale jar testing of, 10-8, 10-9f
cationic polymers and, 6-3
chemical formula for, 6-15t
chemistry of, 6-16 to 6-22
as coagulant, 6-14, 6-17f
dosage, 6-17, 6-34 to 6-35, 6-39 to 6-41
EMP data for, 6-22, 6-34, 6-35f, 10-14, 10-14f
floc volume and, 5-12
in freshwater vs. seawater, 13-12t, 13-13
ligands and, 6-19
at Lysekil water plant, 11-13
pH and, 6-10, 6-17 to 6-23, 6-18f, 6-34 to 6-35, 13-13
properties of liquid, 6-15t
in rapid mixing, 6-13 to 6-14, 6-17 to 6-22
reaction pathways for, 6-22, 6-23f
scaling with, 13-12t, 13-13
specific gravity of, 6-15t
sulfuric acid and, 6-17
at Warner water plant, 6-39, 6-40f, 11-7
water temperature and, 6-12, 6-18 to 6-20, 6-34, 13-13
Aluminum, 1-3, 1-6, 5-8, 5-13, 6-2, 6-22
Aluminum chlorides, 6-27
Aluminum chlorohydrate, 6-28
Aluminum hydroxides:
in coagulation, 5-7, 5-7t, 6-2
freshwater vs. seawater, 13-13, 13-14f
Hamaker constant for, 5-19
hydrogen bubbles and, 5-5, 5-18
particle charge and, 5-12 to 5-13, 6-21 to 6-22
Aluminum hydroxides (*Cont.*):
pH levels and:
freshwater, 5-8, 5-13, 5-13t, 6-19 to 6-23, 6-20f
seawater, 13-13, 13-14f
in rapid mixing, 6-17 to 6-22
residual solids from, 6-20 to 6-21
reverse osmosis and, 13-13
in seawater pretreatment, 13-13
solubility of:
freshwater, 6-12, 6-19, 6-20f, 6-34
seawater, 13-3, 13-13, 13-14f
water temperature and, 6-12, 6-19, 6-20f, 13-13
Aluminum silicates, 5-7t, 13-4, 13-8 to 13-9, 13-13. *See also* Clays
Aluminum sulfate, 2-2, 6-2, 6-15t, 6-17
Amato, Tony, 11-14
American Water Works Association (AWWA):
on streaming current, 10-14
Water Quality and Treatment. See *Water Quality and Treatment*
Amino acids, 6-5, 13-11
Ammonia, 11-16, 12-8, 13-6
Anabaena sp., 5-8, 11-18t
Anthracite, 11-9t
Antwerp Conference, 2-8
AquaDAF®, 11-3, 11-17, 13-19
Aquarion Water Company. *See* Warner water plant
Argon:
air solubility and, 3-23, 3-24f
air temperature and, F-1 to F-6
in atmospheric air, 3-2, 3-2t, 3-17
correlation coefficients for, F-7
Henry's constant for, F-1 to F-3, F-8
mass concentration of, F-4 to F-6
mass transfer constant for, 3-19 to 3-20, 3-24f
molar concentration of, F-4 to F-6
molecular diffusivity for, F-1 to F-3
in saturator air, 3-2t, 3-15, 3-17
in seawater, 13-18
Arizona, 12-5t, 12-6
Arvika water plant, 2-6 to 2-7
Atmospheric air, 3-2 to 3-3, 3-2t, 3-3t, 3-17, E-1 to E-4
Attachment efficiency:
in contact zone:
bubble size and, 7-25f
coagulants and, 7-31 to 7-32, 7-31t
description of, 7-16 to 7-17
diffusion vs. interception, 7-18f

Attachment efficiency, in contact zone (*Cont.*):
experimental data on, 7-30f
forces and, 7-17t
particle charge and, 7-31t
particle size and, 7-17, 7-18f, 7-29 to 7-31, 7-30f
pH and, 7-31, 7-31t
removal efficiency and, 7-12
in flocculation, 6-41, 6-44, 6-45
Auerococcus anophagefferens, 13-9, 13-10f
Australia, 11-3
Autochthonous organic matter, 6-5
Available air:
altitude and, 3-17
dead- vs. open-end saturation, 3-28
definition of, 3-16
end effects and, 3-25
energy consumption and, 3-31
in equilibrium conditions, 3-16 to 3-17, 3-21f, 3-22f, 3-23, 3-23f
freshwater vs. seawater, 13-18, 13-19f
hydraulic loading and, 3-22f, 3-23
macro-bubbles and, 4-5
mass transfer model for, 3-20 to 3-23
measurement of, 3-26, 3-27f, 10-16
packing material and, 3-22f, 3-23, 3-23f
pressure and, 3-16f, 3-21f, 3-23
recycle flow and, 10-16
saturation pressure and, 3-23
volume delivered, 10-16
water temperature and, 3-16f, 3-21f, 3-22f, 3-23, 3-23f, 13-19f
wetted packing area and, 3-23
Avogadro's number, C-1
AWWA. *See* American Water Works Association

B

BAC filters, 11-16, 12-5t, 12-8f
Bacteria. *See also specific types of*
charge of, 5-12 to 5-13, 6-4
classification of, 6-5
inactivation of, 5-8
in lakes, 5-8
microbiological assays of, 5-11
pH and, 5-13t, 6-4
size of, 5-9, 5-9f, 12-2f
sources of, 5-7t, 5-8
Baffles, 6-46, 7-2, 7-7, 8-12 to 8-14
Ballast rings, 3-19t
Ballast saddles, 3-19t
Ballasted-sand sedimentation, 10-5, 12-6, 12-13
Barcelona desalination plant, 13-20
Bay Bulls Big Pond, 11-10 to 11-11
BDOC, 11-16
Beach drum, 11-8t
Beach plate, 9-2, 9-3 to 9-4, 9-6 to 9-7, 9-9, 10-15
Beach scraper, 9-6 to 9-7, 9-7f
Belgium, 2-7
Bench-scale jar testing:
about, 10-5 to 10-8
available air measurements in, 3-26
benefits of, 10-7
conditions for, 10-7 to 10-8
flocculation in, 10-6, 10-7
of nanofiltration, 12-7
PACl dose for, 6-29
photograph of setup, 10-5f
saturator in, 3-25
of SFBW, 12-14
turbidity in, 10-6, 10-9f
UV light absorbance in, 10-6, 10-9f
Bessel, Adolph and August, 2-1
Betasso water plant, 12-14
Bicarbonate, 13-3, 13-3t, 13-5 to 13-7
Biodegradable DOC (BDOC), 11-16
Biofilms, 12-4
Biologically active carbon (BAC) filters, 11-16, 12-5t, 12-8f
Blue-green algae. *See* Cyanobacteria
Bogs, 6-5, 6-6, 6-6t
Boltzmann's constant, C-1
Borate, 13-5, 13-6f, 13-7f
Borescopes, 4-2
Boric acid, 13-5, 13-6f
Boron, 13-3 to 13-7, 13-3t, 13-6f, 13-20
Boulder Betasso water plant, 12-14
Bound water, 5-14
Bromide, 13-3, 13-3t
Brown algae, 13-9
Brown-tide algae, 13-1, 13-9, 13-10f
Brownian diffusion, 6-41 to 6-45, 6-42t, 7-11 to 7-15, 7-11f, 13-16 to 13-17
Bubble bed, 8-11 to 8-17
Bubble clusters, 8-5. *See also* Floc-bubble aggregates
Bubble concentration:
bubble-to-floc concentration ratio, 5-2 to 5-5
CFD assumptions, 7-7
contact zone efficiency and, 7-19, 7-25

Bubble concentration (*Cont.*):
design considerations, 5-2, 7-26t
mass, 5-2, 5-3f, 7-19, 7-25 to 7-26, 7-26t
in nozzles, 4-13
number. *See* Bubble number concentration
particle concentrations and, 5-2 to 5-5
recycle rate and, 7-23t
recycle ratio and, 8-17
saturation pressure and, 7-23t, 8-16 to 8-17
in separation zone, 8-16 to 8-17
volume. *See* Bubble volume concentration
water temperature and, 5-2, 5-3f
in white water, 4-2 to 4-3
in white water bubble-blanket model, 7-10
Bubble mass concentration, 5-2, 5-3f, 7-19, 7-25 to 7-26, 7-26t. *See also* Mass air concentration
Bubble number concentration:
about, 7-26
bubble mass concentration and, 5-3f
bubble size and, 5-2, 5-4
bubble volume concentration and, 5-2, 5-3f
calculating, 5-2
contact zone efficiency and, 7-16
designing for, 7-26, 7-26t
particle number concentration and, 5-2 to 5-3, 7-26
water temperature and, 5-2, 5-3f
in white water blanket model, 7-10
Bubble rise velocity:
bubble coalescence and, 4-8 to 4-9
bubble density and, 7-19, 8-3
bubble size and, 7-19, 7-23, 8-3, 8-4f
bubble volume concentration and, 3-3
calculating, 7-19 to 7-20, 8-2 to 8-5
contact zone efficiency and, 7-16, 7-23
DAF rate coefficient and, 7-10
definition of, 7-11
forces and, 5-5, 8-2
freshwater vs. seawater, 13-16 to 13-17
gravity and, 7-19, 8-3
illustrations of, 7-11f
particle size and, 7-22
Reynolds number and, 8-2 to 8-3
shape of bubbles and, 4-4
slipping conditions and, 8-3 to 8-4, 8-4f
Vrablik's study of, 2-4
water density and, 7-19, 8-2, 8-3
water temperature and, 7-15, 7-23, 8-3 to 8-4, 8-4f
water viscosity and, 7-19, 7-23t, 8-3
Bubble size:
bubble number concentration and, 5-2, 5-4
bubble rise velocity and, 7-19, 7-23, 8-3, 8-4f
CFD assumptions, 7-7
contact vs. separation zone, 8-2
in contact zone:
attachment efficiency and, 7-25f
bubble number concentration and, 7-16
depth and, 4-3
vs. filter size, 7-9t
needle valves and, 4-3t, 4-12f
nozzles and, 4-3t, 7-23t
performance efficiency and, 7-24 to 7-25, 7-31t, 7-33
pressure and, 4-3, 4-3t, 7-23t
vs. separation zone, 4-3, 8-2
single collector efficiency and, 7-14 to 7-15, 7-14f
in dispersed air flotation, 4-4, 5-15, 7-25
distribution of, 4-2 to 4-4, 4-4f, 4-5f, 10-15 to 10-16
float layer and, 4-5
for floc-bubble aggregates, 5-20, 5-24, 7-7 to 7-8
forces and, 5-16
growth patterns, 4-8
impinging surface and, 4-13 to 4-14
macro, 4-5 to 4-6, 4-9, 4-12 to 4-14
needle valves and, 4-3t, 4-6, 4-12f
nozzles and:
air saturation method and, 3-11
basics of, 1-8
contact zone efficiency and, 7-23t, 7-25, 7-31t, 7-33
particles and, 4-8
pilot-scale testing of, 10-16
studies of, 4-3t
particle density and, 1-3
particle size and, 7-25, 7-25f
pumps and, 3-11
Reynolds number and, 8-2
saturation pressure and, 3-11, 4-3t, 4-6 to 4-7, 7-33, 10-16
in separation zone, 4-3, 8-2
shape and, 4-4
single collector efficiency and, 7-12 to 7-15, 7-13t, 7-14f, 7-20 to 7-21, 7-23t, 7-24
zeta potential and, 5-6
Bubble volume concentration. *See also* Volumetric air concentration
about, 7-26

Bubble volume concentration (*Cont.*):
bubble density and, 5-2
bubble mass concentration and, 5-2, 5-3f, 7-19
bubble number concentration and, 5-2, 5-3f
bubble rise velocity and, 3-3
calculating, 5-2
contact zone efficiency and, 5-2, 7-16, 7-19, 7-26 to 7-29, 7-28f, 7-33
designing for, 5-2, 7-26, 7-26t, 7-33
detention time and, 7-28 to 7-29, 7-28f
factors affecting, 7-33
freshwater vs. seawater, 13-17, 13-18
particle density and, 5-4
particle mass concentration and, 5-4
particle size and, 7-28 to 7-29, 7-28f
particle volume concentration and, 5-3 to 5-4
recycle rate and, 7-23t, 7-31t, 7-33
rise velocity of aggregates and, 7-26
saturation pressure and, 7-23t, 7-31t, 7-33
separation zone efficiency and, 5-2
turbidity and, 11-6
water temperature and, 5-2, 5-3f
in white water bubble-blanket model, 7-26
Bubbles:
area per bubble, 8-2
buoyancy of, 8-2
coalescence of, 2-4, 4-3 to 4-9, 4-7f, 4-13 to 4-14, 5-5 to 5-6, 9-2
concentration of. *See* Bubble concentration
density of, 5-2, 7-12, 7-19, 8-2, 8-3, E-5
in dispersed air flotation, 4-4, 7-15
under float layer, 9-1 to 9-2
forces on, 5-5 to 5-6, 8-2 to 8-3
formation of, 1-1, 4-6 to 4-8, 4-7f, 12-8
interactions between, 5-5
particles and. *See* Floc-bubble aggregates
Reynolds number of, 8-2 to 8-3
rise rate of. *See* Bubble rise velocity
rotation of, 8-3
separation distance between, 7-9
shape of, 4-4, 8-2
size of. *See* Bubble size
stability of, 5-5
streamlines around rising, 7-11, 7-11f
surface tension of, 5-16f
velocity of, 8-11
volume per bubble, 5-2, 8-2
zeta potential of, 5-5 to 5-6, 5-16, 5-19, 5-20 to 5-22, 5-24, 7-18
Buffer intensity:
coagulants and, 6-10 to 6-11, 13-13, 13-15
definition of, 6-10
pH and, 6-11f, 13-6 to 13-7, 13-7f
of seawater, 13-3, 13-4, 13-6 to 13-7, 13-7f
Bulk oil flotation, 2-1

C

Calcite, 5-7t. *See also* Lime
Calcium, 5-13, 6-12, 13-3t
California, 10-12t, 12-5
Cambridge DAF pilot study, 10-12t
Camp's velocity gradient, 6-14, 6-43
Canada:
Manitoba. *See* Manitoba
Newfoundland. *See* Newfoundland
Shoal Lake, 11-15
CAP, 12-6
Carbohydrates, 6-5
Carbon, 6-6
Carbon dioxide, 2-6, 6-11, 6-13, 11-11, 11-13, 13-3 to 13-4
Carbonates, 2-1 to 2-2, 5-8, 5-9, 13-5 to 13-7
Cartnick, Keith, 11-16
Cartridge filtration, 13-19 to 13-20
Caustic, 6-13
Cellulose acetate, 12-4
Cellulose triacetate, 12-3
Central Arizona Project (CAP), 12-6
Centrifugal pumps, 1-4, 1-4f, 3-10 to 3-11
Ceramic membranes, 12-3
Ceratium tripos, 13-10f
CFD, 7-7, 8-11, 8-15
CFSTR, 6-47, 7-4 to 7-6, 7-4f
Chain-and-flight scraper, 9-6 to 9-7, 9-7f
Chile, desalination plants in, 13-19
Chingford South water plant, 11-8t to 11-9t, 11-14
Chitosan, 6-32
Chloramine, 11-11, 11-16
Chloride, 13-3t
Chlorine, 11-7, 11-13, 11-16 to 11-17
Chlorine dioxide, 11-17
Chlorophyll *a* concentrations, 10-3, 11-14, 12-9t. *See also* Algae
Chrysophytes, 13-9, 13-10f
Clari-DAF™, 11-3
Clays:
charge of, 5-12 to 5-13, 6-4, 6-8
composition of, 5-7t
flocculation rate and, 6-3
particle size in, 5-9

Clays (*Cont.*):
pH range for, 6-8
in seawater, 13-8
Coagulant aid, 6-3
Coagulants:
alkalinity and, 6-10 to 6-12, 6-28, 6-34
attachment efficiency and, 7-31 to 7-32, 7-31t
bench-scale jar testing of, 10-6 to 10-8
buffer intensity and, 6-10 to 6-11
definition of, 6-2
for direct filtration, 6-2, 13-12t
dissolved residual, 10-6
DOC and dosage of, 7-31, 7-32f
dosage, 5-17, 6-33 to 6-34
float layer and, 9-8f
for freshwater vs. seawater, 13-12t
NOM demand for, 6-8
particle charge and, 5-17, 5-24
pH and, 6-10
pilot-scale testing of, 10-7
in rapid mixing, 6-14, 10-7
in seawater pretreatment, 13-11 to 13-16
for sedimentation, 6-2, 6-33 to 6-34
streaming current and, 10-15
for sweep floc coagulation, 5-8
turbidity and, 6-8, 7-31, 7-32f, 10-6
water temperature and, 6-12
Coagulation:
bench-scale jar testing for, 10-6, 10-7
buffer intensity and, 6-10 to 6-11
contact zone efficiency and, 7-16
definition of, 6-2
electrostatic forces and, 5-16 to 5-17
EMP data for, 10-14
enhanced, 6-3
factors affecting, 6-4
filtration and, 5-24
flocculation and, 6-2, 6-3, 6-44
fulvic acids and, 5-14
goal of, 1-7
granular media filtration and, 6-2
hydroxide production in, 5-7
before membrane filtration, 6-2, 12-4
of metal particles, 5-7t, 5-8, 5-12
microfiltration and, 12-2
NOM and, 5-8, 6-2, 6-5, 6-8
optimum, 6-3, 6-8, 6-33
particle charge and, 5-13, 5-24
particle size and, 5-21
particle zeta potential for, 10-14
pesticides and, 5-8
Coagulation (*Cont.*):
pH of, 6-9
purpose of, 1-6, 5-12, 6-33, 6-41
before reverse osmosis, 13-1 to 13-2
role of, 7-31
SFBW and, 12-11
SUVA and, 6-7
sweep floc. *See* Sweep-floc coagulation
ultrafiltration and, 12-2
van der Waals force and, 5-22, 5-23f
at Warner water plant, 6-39
water hardness and, 6-12
at Winnipeg water plant, 11-15, 11-16
CoCoDAFF, 8-2, 11-4, 11-5f
Coefficient of variation, 6-14
Collision efficiency functions, 6-41 to 6-45, 6-42t. *See also* Single collector efficiency
Colloids, 5-6, 5-9f, 12-2, 12-2f, 12-4
Colorado, 12-14
Colorado River, 12-6
Computational fluid dynamics (CFD), 7-7, 8-11, 8-15
Concentrate, 12-3, 12-3f
Conductivity, 5-17, 5-18t, 12-9t
Connecticut:
Fairfield. *See* Warner water plant
Hemlocks Reservoir, 7-29, 10-12t, 11-7
Contact zone:
air concentration in, 3-2 to 3-7, 7-7, 11-6
attachment efficiency in:
bubble size and, 7-25f
coagulants and, 7-31 to 7-32, 7-31t
description of, 7-16 to 7-17
diffusion vs. interception, 7-18f
experimental data on, 7-30f
forces and, 7-17t
particle charge and, 7-31t
particle size and, 7-17, 7-18f, 7-29 to 7-31, 7-30f
pH and, 7-31, 7-31t
removal efficiency and, 7-12
boundary with separation zone, 8-1 to 8-2
bubble concentration in, 7-9t
bubble size in:
attachment efficiency and, 7-25f
bubble number concentration and, 7-16
depth and, 4-3
vs. filter size, 7-9t
needle valves and, 4-3t, 4-12f
nozzles and, 4-3t, 7-23t
performance efficiency and, 7-24 to 7-25, 7-31t, 7-33

Contact zone, bubble size in (*Cont.*):
pressure and, 4-3, 4-3t, 7-23t
vs. separation zone, 4-3, 8-2
single collector efficiency and, 7-14 to 7-15, 7-14f
in circular tanks, 11-2
counter-current flow and, 11-4
depth of, 7-9t, 7-23t
description of, 1-8
detention time in:
air concentration and, 7-7
depth and, 7-23t, 7-37
for disinfection requirements, 7-5
dispersion number and, 7-7
in efficiency modeling, 7-27 to 7-29, 7-28f, 7-33 to 7-34
flow rate and, 7-31t, 7-33, 7-37
particle size and, 7-28 to 7-29, 7-28f, 7-32 to 7-33
in PFR, 7-3 to 7-4, 7-5
water velocity and, 7-23t
diffusion in, 7-6
dispersion number for, 7-6 to 7-7
eddies in, 7-6
efficiency of, 3-3, 5-2, 7-16 to 7-29
features of, 7-9t
floc-bubble aggregates in, 1-8, 6-33, 7-1
flow in, 7-6 to 7-7
granular media filtration and, 7-9, 7-9t
hydraulic loading of, 7-9, 7-9t, 7-29
illustrations of, 4-12f, 7-2f, 7-36f
modeling of:
assessment of, 7-8
background of, 7-7 to 7-8
basics of, 7-9
batch reactor, 7-37
with CFD, 7-7
considerations, 7-8
derivation of, 7-35 to 7-38
design variables for, 7-30 to 7-34, 7-31t
dispersion and, 7-5
filtration models and, 7-9 to 7-10
operating variables for, 7-30 to 7-34, 7-31t
performance efficiency, 7-16 to 7-29, 13-16 to 13-17
as PFR, 7-2, 7-5 to 7-7
single collector efficiency, 7-10 to 7-15
verification of, 7-29 to 7-30, 7-30f
particle size in, 5-9
RTD of freshwater in, 7-2
seawater in, 13-16 to 13-17
separation distance between bubble collectors in, 7-9t

Contact zone (*Cont.*):
steady state conditions in, 7-37
tracer tests of, 7-6
white water bubble-blanket model of, 7-10 to 7-39
Continuous-flow stirred-tank reactor (CFSTR), 6-47, 7-4 to 7-6, 7-4f
Conventional drinking water treatment plants, 1-6 to 1-8, 1-6f, 1-7f
Corrosion control, 6-11, 12-13
Counter-Current DAFF (CoCoDAFF), 8-2, 11-4, 11-5f
Croton water plant, 6-46, 10-12t, 11-2, 11-7, 11-16
Cryptosporidium:
DAF and:
seasonal variations in, 11-19f
vs. sedimentation, 10-2, 11-18 to 11-19
in SFBW, 12-10 to 12-13, 12-11f, 12-12f
study on, 10-12t
density of, 11-18
disinfection and, 5-8
granular media filtration and, 11-18 to 11-19, 12-10
in Hemlock Reservoir water, 10-12t, 11-7
mass balance model for, 12-10
number concentrations of, 5-11
pH of zero point charge of, 5-13t
plate settling and, 11-18 to 11-19, 11-19f, 12-11 to 12-12, 12-12f
sedimentation and, 10-2, 11-18 to 11-19
in SFBW, 12-9 to 12-13, 12-11f, 12-12f
size of, 5-9, 5-9f, 12-2f
at Warner water plant, 11-18 to 11-19
Cryptosporidium parvum, 11-18
Cyanobacteria, 5-8, 11-13, 11-16, 12-7, 13-9

D

DAF over filtration (DAFF):
advantages of, 11-3
CoCoDAFF, 8-2, 11-4, 11-5f
float layer in, 9-2
Flofilter, 2-7, 11-3, 11-4f, 11-11, 11-13
flotation aids for, 6-48
flow patterns in, 8-12
illustration of, 11-4f
at Lysekil water plant, 2-7
in seawater pretreatment, 13-19 to 13-20, 13-20f
separation zone outlets in, 8-12
tanks for, 11-2
at Warner water plant, 11-7

DAFRapide®, 11-3
DBP, 6-2, 6-5, 6-7, 11-16 to 11-17
Deacon Reservoir, 11-15
Dead-end filtration, 12-4t
Dead-end saturation systems:
  air flow into, 3-28, 3-29 to 3-30
  description of, 3-11 to 3-13
  end effects of, 10-16
  energy requirements for, 3-31 to 3-32, 3-31f
  gas fractions in, 3-2
  transfer efficiency of, 10-16
Debye length, 5-14, 5-16 to 5-17, 5-20, 5-23, 5-25
"deep shaft" flotation, 1-5
Derjaguin-Landau-Verwey-Overbeek (DLVO) theory, 5-14 to 5-20, 5-23 to 5-24
Detention time:
  bubble volume concentration and, 7-28 to 7-29, 7-28f
  in CFSTRs, 7-5
  at Chingford South water plant, 11-8t, 11-14
  in contact zone:
    air concentration and, 7-7
    depth and, 7-23t, 7-37
    for disinfection requirements, 7-5
    dispersion number and, 7-7
    in efficiency modeling, 7-27 to 7-29, 7-28f, 7-33 to 7-34
    flow rate and, 7-31t, 7-33, 7-37
    particle size and, 7-28 to 7-29, 7-28f, 7-32 to 7-33
    in PFR, 7-3 to 7-4, 7-5
    water velocity and, 7-23t
  in flocculation:
    at Chingford South water plant, 11-8t, 11-14
    at Haworth water plant, 11-8t, 11-17
    at Lysekil water plant, 11-8t
    in orthokinetic formula, 6-46
    particle size and, 7-31t
    at St. John's water plant, 11-8t, 11-11
    at Scottsdale water plant, 12-5t, 12-6
    stages and, 6-43
    tank sizing for, 10-11
    at Warner water plant, 11-7, 11-8t
    at Windhoek water plant, 12-5t
    at Winnipeg water plant, 11-8t, 11-15
  flotation, 10-6
  at Haworth water plant, 11-8t, 11-17
  at Lysekil water plant, 11-8t, 11-13
  for metal coagulants, 6-14
  in nozzles, 4-13
Detention time (*Cont.*):
  particle size and, 7-28 to 7-29, 7-28f, 7-31t, 7-32 to 7-33
  in PFR, 7-3 to 7-4, 7-5
  in pipes, 7-5
  in rapid mixing, 11-13
  reaction time and, 6-13
  at St. John's water plant, 11-8t, 11-11
  at Scottsdale water plant, 12-5t, 12-6
  at Warner water plant, 11-7, 11-8t, 11-19
  water velocity and, 7-23t
  at Windhoek water plant, 12-5t
  at Winnipeg water plant, 11-8t, 11-15
Dew point, 5-2
Diatoms, 11-11, 13-4, 13-9
Dielectric constant, D-1 to D-4
Diffuse double-layer thickness, 5-15, 5-17, 5-20
Dinoflagellates, 13-9, 13-10f
Dipole forces, 5-5, 5-14
Direct filtration:
  algae and, 11-13
  chlorophyll and, 10-3
  coagulants for, 6-2, 13-12t
  vs. DAF, 10-12
  disadvantage of, 10-3
  for disinfection, 10-3
  at Haworth water plant, 11-16
  at St. John's water plant, 11-11
  SFBW and, 12-13
  TOC and, 10-3, 10-4f, 11-7
  turbidity and, 10-3, 10-4f, 11-7
  water color and, 10-3, 11-7
  at Winnipeg water plant, 10-12t, 11-15
Disinfectant residual concentration, 7-5
Disinfection:
  backwash water chemistry and, 12-13
  bacteria and, 5-8
  chloramine for, 11-11
  chlorine for, 11-16
  contact zone detention time for, 7-5
  direct filtration and, 10-3
  of freshwater, 7-5
  NOM and, 6-5
  protozoa and, 5-8
  residence time for, 7-5
  at St. John's water plant, 11-11
  UV light for, 6-5, 10-3, 11-15, 11-16
  viruses and, 5-8
  at Winnipeg water plant, 11-15, 11-16
Disinfection by-product (DBP) precursors, 6-2, 6-5, 6-7, 11-16 to 11-17

Dispersed air flotation:
- aeration in, 1-3
- applications for, 1-2
- bubble-particle interactions in, 5-15
- bubble size in, 4-4, 5-15, 7-25
- vs. DAF, 1-2 to 1-3, 7-25
- description of, 1-2 to 1-3
- history of, 2-2
- ozone with, 1-3
- single collector efficiency and, 7-10
- surfactants for, 1-3

Dispersion coefficient, 7-6
Dispersion number, 7-6
Dissolved air concentration, 3-8f
Dissolved air flotation (DAF):
- acceptance of, 2-9 to 2-10, 2-10f
- advantages of, 10-2
- algae and, 10-3 to 10-5
- applications for, 1-1, 1-2t
- aspect ratio of DAF units, 11-2 to 11-3
- conventional vs. high-rate, 11-3
- description of, 1-1
- vs. direct filtration, 10-12
- vs. dispersed air flotation, 1-2 to 1-3, 7-25
- energy requirements for, 10-2
- feasibility study for, 10-1 to 10-5
- high-rate, 11-2 to 11-3
- history of, 2-1 to 2-10, 11-2, 11-6
- installed capacity by year, 2-9f
- plants for, 2-8
- preliminary design proposal for, 3-33 to 3-35
- pressure. *See* Pressure DAF
- sedimentation and, 10-2, 10-12, 10-12t, 11-1 to 11-2, 11-4 to 11-6, 11-13, 12-6
- vacuum, 1-4

Dissolved organic carbon (DOC):
- bench-scale jar testing of, 10-6, 10-8
- biodegradable, 11-16
- cationic polymers and, 6-2
- coagulant dose and, 7-31, 7-32f
- composition of, 6-5
- in Deacon Reservoir water, 11-15
- membrane fouling by, 12-4
- microfiltration and, 12-2
- in pretreatment, 1-6
- in seawater, 13-4
- size of particles, 12-2f
- TOC and, 6-6
- UV light absorbance and, 6-7
- at Windhoek water plant, 12-9t

Dissolved residual coagulant, 10-6
DLVO theory, 5-14 to 5-20, 5-23 to 5-24
DOC. *See* Dissolved organic carbon
Dolomite, 5-7t
Dorr Company, 2-3
Drag coefficient, 8-2 to 8-3
Dual coagulants, 6-3, 6-39, 6-40f, 11-7, 13-11 to 13-12, 13-12t
Dual media filtration:
- *Cryptosporidium* removal with, 11-19
- at Haworth water plant, 11-17
- at St. John's water plant, 11-11
- in seawater pretreatment, 13-19 to 13-20, 13-20f
- SFBW and, 12-11
- turbidity and, 6-35f
- at Warner water plant, 6-39, 11-7f
- at Windhoek water plant, 12-5t

DWL nozzles, 4-10f
Dyes, 7-4
Dyksen, John, 11-16

E

Efficient Application of Flotation with Very Small Bubbles to the Clarification of Waste Water, The (Hanisch), 2-4
El Coloso desalination plant, 13-19
Electrical double layer:
- attachment efficiency and, 7-17t
- bubble size and, 5-16
- coagulation and, 5-22, 5-23f
- conductivity and, 5-18t
- description of, 5-14, 5-16 to 5-18
- in freshwater, 5-20 to 5-21, 5-22f, 13-8
- ionic strength and, 5-15, 5-17, 5-18t, 6-12
- particle size and, 5-16
- in seawater, 13-8
- separation distance and, 5-16
- in streaming-current measurements, 10-14 to 10-15
- TDS and, 5-18t
- thickness of, 5-17, 13-8
- van der Waals force and, 13-8
- water permittivity and, 5-16
- zeta potentials and, 5-16, 5-19, 5-24, 7-18

Electrolytic flotation, 1-3
Electron charge, C-1
Electrophoretic mobility (EPM) data:
- for alum, 6-22, 6-34, 6-35f, 10-14, 10-14f
- for PACls, 7-32f, 10-14, 10-14f
- particle charge indicator, 7-31, 7-32f
- process for gathering, 10-13 to 10-14
- zeta potential from, 10-13

Elmore, Francis, 2-2
End effects, 3-24 to 3-25, 3-27, 10-16
Enflo-vite®, 11-3
England, 9-2
Enpure, 11-3, 11-14
EOM, 13-10 to 13-11
Epichlorohydrin dimethylamine (Epi-DMA), 6-15t, 6-32, 6-39
EPM data. *See* Electrophoretic mobility
Equalization basin, 12-14
Estuarine water, 5-15, 5-18t, 13-2, 13-8
European Union (EU), 13-5
Extracellular organic matter (EOM), 13-10 to 13-11

F

FAD rate, 3-29
Fairfield, Warner water plant. *See* Warner water plant
Faraday constant, C-1
Fatty acids, 6-5
Feed water, 12-3, 12-3f, 13-8
Felixstowe Conference, 2-8, 2-10
Ferric chloride. *See also* Polyferric chlorides
  chemical formula for, 6-15t
  as coagulant, 6-2
  dosage, 6-26, 6-36, 6-37f, 13-14 to 13-15
  freshwater vs. seawater, 13-12t
  pH range for, 6-26, 6-36, 6-37f
  properties of liquid, 6-15t, 6-24
  residual from, 13-13 to 13-14
  in seawater pretreatment, 13-12, 13-13 to 13-15, 13-19 to 13-20, 13-20f
  solubility of, 13-13
  TOC and, 6-36
  at Windhoek water plant, 12-9
  at Winnipeg water plant, 11-15
Ferric coagulants. *See also* Iron coagulants
  algae and, 13-14 to 13-15
  alum and, 6-14
  buffer intensity and, 13-15
  chemistry of, 6-23 to 6-27
  dosage, 6-36, 6-37f
  floc volume and, 5-12
  freshwater vs. seawater, 13-12t, 13-15f
  manganese in, 6-24
  pH and, 5-8, 6-10, 6-24 to 6-27, 6-36, 13-14 to 13-16
  in rapid mixing, 6-13 to 6-14
  reaction pathways for, 6-24f
  scaling with, 13-14
Ferric coagulants (*Cont.*):
  in seawater pretreatment, 13-12 to 13-16, 13-15f
  water temperature and, 6-12, 6-24 to 6-25
Ferric hydroxides:
  ligands and, 6-26
  particle charge and, 5-12 to 5-13, 6-27
  pH and, 5-13t, 6-24 to 6-27, 6-25f, 6-26f, 6-36
  production of, 5-7t, 6-2
  in rapid mixing, 6-24
  residual solids from, 6-25 to 6-27
  in seawater pretreatment, 13-14, 13-15f
  solubility of, 6-25, 6-26f, 13-3, 13-14
Ferric sulfate, 6-2, 6-15t, 6-24, 6-26. *See also* Polyferric sulfates
Filter aids:
  definition of, 6-3
  HPAM, 6-16t
  PAM, 6-16t, 6-32, 12-4, 12-9
  polymers as, 6-47 to 6-48, 13-11
  at Winnipeg water plant, 11-16
Filter bed, 7-9t
Filter rate, 11-9t
Filters:
  BAC, 11-16, 12-5t, 12-8f
  bubbles and, 11-3
  clogging of, 11-11
  flotation and performance of, 10-6
  GAC. *See* Granular activated carbon (GAC) filters
  gravity, 1-4
  pilot-scale testing of, 10-7
  plug flow behavior in, 7-5 to 7-6
  pore opening between collectors, 7-9t
  pressure belt, 1-4
  SFBW, 1-2t, 10-12t, 12-9 to 12-14, 12-11f, 12-12f
  size of, 7-9t
  UFRV and backwashing, 11-10
Filtration:
  alum and, 6-34
  at Chingford South water plant, 11-9t, 11-14
  DAF integration with, 10-2, 10-3, 11-3 to 11-4
  direct. *See* Direct filtration
  dispersion in, 7-6
  dual media. *See* Dual media filtration
  granular media. *See* Granular media filtration

Filtration (*Cont.*):
at Haworth water plant, 11-9t
hydraulic loading of, 7-9t
at Lysekil water plant, 11-9t
membrane. *See* Membrane filtration
particle size and, 5-9
pilot-scale testing of, 10-10
at St. John's water plant, 11-9t
sand. *See* Sand
in seawater pretreatment, 13-19
sedimentation and, 10-3
single collector efficiency and, 7-10
two-stage, 10-3
at Warner water plant, 11-9t, 11-10
water temperature and, 6-12
at Windhoek water plant, 12-5t
at Winnipeg water plant, 11-9t
Finland:
float layer studies in, 9-2, 9-4
Helsinki Conference, 2-8, 2-9, 11-6
history of water treatment in, 2-3, 2-5 to 2-8
Kemijärvi water plant, 2-6 to 2-7
Kokkola water plant, 2-6
Kristinestad water plant, 2-7
Kusaankoski water plant, 2-7
Fittings, location of, 4-1
Fixed-orifice nozzles, 4-2, 4-9 to 4-10
Flen water plant, 2-6
Flexirings, 3-19t
Float layer:
about, 1-8
algae in, 9-2, 9-5 to 9-8
alum and, 2-6
breakup of, 9-2
bubbles and, 4-5, 9-1 to 9-2
coagulants and, 9-8f
Cryptosporidium in, 12-10, 12-13
environment and, 9-4, 9-6
floc-bubble aggregates in, 8-1, 8-14
flocculant aids and, 9-2
flotation aids and, 6-3
in froth flotation, 2-2
grid stabilization of, 9-9
photographs of, 9-3f
removal of, 9-1, 9-2 to 9-9, 10-15, 11-2
SFBW and, 12-14
turbidity and, 9-5f, 9-8, 9-8f
water color and, 9-2, 9-7 to 9-8
Floc-blanket clarification, 10-5
Floc-bubble aggregates:
about, 8-5
behavior of, 8-6 to 8-9
Floc-bubble aggregates (*Cont.*):
from Brownian diffusion, 7-11 to 7-12, 7-11f
bubble formation and, 4-8
bubble size for, 5-20, 5-24, 7-7 to 7-8
bubble-to-floc concentration ratio, 5-2 to 5-5
coagulation and, 5-22 to 5-23, 5-23f
contact angle of, 5-15, 5-16f
in contact zone, 1-8, 6-33, 7-1, 7-8 to 7-33
density of, 5-3 to 5-4, 7-26, 8-7 to 8-8, 13-17
description of, 1-7
diameter of, 8-7
in dispersed air flotation, 5-15
DLVO theory on, 5-15 to 5-20
drag force on, 8-7 to 8-8
electrical double layer and, 5-16 to 5-18, 7-18
in float layer, 8-1, 8-14
in freshwater, 5-20 to 5-21, 5-22f
in freshwater vs. seawater, 13-17
Hamaker constant for, 5-18 to 5-20
hydrodynamic forces and, 5-6, 5-19, 7-18
hydrophobic forces and, 5-19 to 5-20, 7-18
number of bubble per floc, 8-6 to 8-7
particle concentration for, 7-7 to 7-8
particle size and, 5-9, 5-20, 5-24 to 5-25, 7-7 to 7-8
rise velocity of, 7-22, 7-26, 8-6 to 8-9, 8-8f, 13-17
in separation zone, 1-8, 1-9, 7-1
single collector efficiency and, 7-11 to 7-12, 7-11f
stability of, 5-19
stratified flow structure and, 8-15
van der Waals force and, 5-18 to 5-19, 7-18
Flocculant aids:
definition of, 6-3
float layer and, 9-2
polymers as, 6-34, 6-47 to 6-48, 13-11, 13-13
for sedimentation, 6-48
for SFBW, 12-14
Flocculants:
colloids, 5-6, 5-9f, 12-2, 12-2f, 12-4
definition of, 6-3
NOM and, 6-3
PAM, 6-16t, 6-32, 12-4, 12-9
Flocculation:
air dosing rate and, 10-11
of algae, 13-9

Flocculation (*Cont.*):
attachment efficiency in, 6-41, 6-44, 6-45
bench-scale jar testing of, 10-6, 10-7
Brownian diffusion in, 6-41 to 6-45, 6-42t, 7-17t, 7-19, 7-22
bubble coalescence and collisions in, 6-2
bubble concentration and, 7-7 to 7-8
at Chingford South water plant, 11-8t, 11-14
coagulant selection and, 6-34
coagulation and, 6-2, 6-3, 6-44
collision efficiency functions and, 6-43 to 6-44, 6-44f
contact zone efficiency and, 7-16, 7-17, 7-18f, 7-19 to 7-22, 7-31t, 7-32 to 7-33
definition of, 6-2
detention time in:
at Chingford South water plant, 11-8t, 11-14
at Haworth water plant, 11-8t, 11-17
at Lysekil water plant, 11-8t
in orthokinetic formula, 6-46
particle size and, 7-31t
at St. John's water plant, 11-8t, 11-11
at Scottsdale water plant, 12-5t, 12-6
stages and, 6-43
tank sizing for, 10-11
at Warner water plant, 11-7, 11-8t
at Windhoek water plant, 12-5t
at Winnipeg water plant, 11-8t, 11-15
fluid shear in, 6-41 to 6-45, 6-42t
at Haworth water plant, 11-8t, 11-17
hydraulic loading and, 10-10 to 10-11
interceptions and, 7-17t, 7-19, 7-22
at Lysekil water plant, 11-8t, 11-13
before membrane filtration, 12-4
microfiltration and, 12-2
monodispersed particle suspensions, 6-45 to 6-46
objective of, 6-48
orthokinetic equation, 6-45 to 6-46
particle charge and, 5-13, 6-33
particle concentration and, 5-11, 6-41, 6-44, 7-7 to 7-8, 7-19
particle size and:
Brownian diffusion in, 6-42t, 6-45, 7-17t, 7-19, 7-22
coagulant selection and, 6-34
coalescence and collisions in, 6-2
collision efficiency functions and, 6-43 to 6-44, 6-44f
contact zone efficiency and, 7-17, 7-18f, 7-19 to 7-22, 7-31t, 7-32 to 7-33

Flocculation, particle size and (*Cont.*):
description of, 6-41 to 6-45
detention time and, 7-31t
distribution of sizes, 6-43
fluid shear in, 6-42t, 6-45
forces and, 5-21 to 5-22, 5-23f
interceptions and, 7-17t, 7-19, 7-22
modeling of, 7-7 to 7-8
particle concentration and, 6-44, 7-19
in sedimentation, 6-48
settling and, 7-17t, 7-19
stages of, 7-31t
perikinetic, 6-42
pH during, 6-9
pilot-scale testing of, 10-10 to 10-11
in plug flow conditions, 6-47, 6-48f
in pretreatment, 1-7
purpose of, 6-41
residence time in, 6-47, 6-48f
before reverse osmosis, 13-1
in rivers, 13-9
at St. John's water plant, 11-8t, 11-11
at Scottsdale water plant, 12-5t, 12-6
in seawater pretreatment, 13-19 to 13-20
in sedimentation, 1-7, 6-34, 6-41, 6-46, 6-48
settling and, 7-17t, 7-19
in settling tanks, 6-41
SFBW and, 12-11, 12-14
stability of suspensions and, 5-12
stages of, 6-43, 6-46 to 6-47, 6-48f, 7-31t, 11-8t
time for, 6-46 to 6-47, 6-48
turbidity and, 5-10, 10-6
velocity gradient and, 6-42t, 6-43, 6-45, 6-46, 6-48, 7-7 to 7-8
at Warner water plant, 11-7, 11-8t
water density and, 6-46
water temperature and, 6-12, 6-45
water velocity and, 6-46
at Windhoek water plant, 12-5t, 12-8f
at Winnipeg water plant, 11-8t, 11-15
Flocculation tanks, 6-46, 10-2, 10-6, 10-11, 10-15, 11-2
Flocculators:
gate, 6-46, 11-13
head loss through, 6-46
hydraulic, 6-46, 6-47f, 11-8t
impellers, 6-14, 6-46, 6-47f, 6-48, 11-8t
paddle, 6-46, 6-47f, 11-8t, 11-14
propeller, 6-46, 11-8t, 11-13, 12-5t

Flocs, 6-2, 6-41, 6-48, 7-22. *See also* Floc-bubble aggregates
Flofilter, 2-7, 11-3, 11-4f, 11-11, 11-13
Florida, 6-6, 12-2, 12-7, 13-10
Flotation:
  Adka vacuum system, 2-3
  for algae, 1-3
  bench-scale jar testing of, 10-7 to 10-8
  bulk oil, 2-1
  at Chingford South water plant, 11-8t
  "deep shaft," 1-5
  detention time, 10-6
  dispersed air. *See* Dispersed air flotation
  dissolved air. *See* Dissolved air flotation
  electrolytic, 1-3
  filter performance, 10-6
  froth, 2-2, 5-6, 7-10
  at Haworth water plant, 11-8t
  at Lysekil water plant, 11-8t
  at St. John's water plant, 11-8t
  at Scottsdale water plant, 12-5t
  turbidity and, 10-6
  vacuum, 1-4, 2-2 to 2-3, 2-5
  at Warner water plant, 11-8t
  at Windhoek water plant, 12-5t, 12-8f
  at Winnipeg water plant, 11-8t
Flotation aids:
  definition of, 6-3
  float layer and, 6-3
  PAM, 6-16t, 6-32, 12-4, 12-9
  Photometric Dispersion Analyzer and, 10-15
  polymers as, 6-48, 13-11
Flotation jar tests, 7-32
Flotation tanks, 7-16, 7-23, 11-7
Flottazone, 1-3
Flow meter location, 4-1
Fluoride, 6-22, 7-4, 13-3, 13-3t
Formaldehyde, 2-2
Formazin, 5-10
Fraction, recycle. *See* Recycle ratio
France, 1-3
Free air delivery (FAD) rate, 3-29
Froth flotation, 2-2, 5-6, 7-10
Full-flow pressurization, 1-5, 1-5f, 2-2
Fulvic acids:
  charge of, 5-13, 6-6, 6-8
  coagulation and, 5-14
  complexation constants for, 6-22
  composition of, 6-6
  vs. humic, 6-6
  in Karnsjon Reservoir water, 11-11
  pH and, 6-6, 6-8
Fulvic acids (*Cont.*):
  in seawater, 13-5 to 13-6, 13-10
Fundamental Principles of Dissolved-Air Flotation for Industrial Wastes (Vrablik), 2-4

G

GAC filters. *See* Granular activated carbon (GAC) filters
Gas constant, C-1
Gate flocculators, 6-46, 11-13. *See also* Paddle flocculators
Geosmin, 11-17, 11-18t
Germany, 2-3 to 2-5
Giardia:
  DAF and, 10-2, 11-18 to 11-19
  density of, 11-18
  disinfection and, 5-8
  granular media filtration and, 11-18 to 11-19
  in Hemlock Reservoir water, 11-7
  number concentrations of, 5-11
  plate settling and, 11-18 to 11-19
  sedimentation and, 10-2, 11-18
  size of, 5-9, 5-9f, 12-2f
  at Warner water plant, 11-18 to 11-19
Giardia lamblia, 11-18 to 11-19
Gibbs Company, 2-3
Glencorse water plant, 11-4
Golden algae, 13-9
Goreangab Water Reclamation Plant (GWRP), 12-7 to 12-9, 12-8f, 12-9t
Graincliffe Plant, 4-3t
Granular activated carbon (GAC) filters:
  at Chingford South water plant, 11-14
  fouling of, 6-5
  odors and, 11-17
  ozone with, 10-12
  tastes and, 11-17
  at Windhoek water plant, 12-5t, 12-8f
  at Winnipeg water plant, 11-9t, 11-16
Granular media filtration:
  algae and, 11-13
  coagulation and, 6-2
  contact zone and, 7-9, 7-9t
  Cryptosporidium and, 11-18 to 11-19, 12-10
  after DAF, 10-2
  EMP data for, 10-14
  features of, 7-9t
  Giardia lamblia and, 11-18 to 11-19
  microfiltration and, 12-2
  particle charge in, 13-12
  particle concentration after, 5-11

Granular media filtration (*Cont.*):
particle size and, 5-9
pH during, 6-9
polymers and, 6-47 to 6-48, 13-12
before reverse osmosis, 13-1
of SFBW, 12-10
single collector efficiency equation for, 7-13
turbidity and, 5-11
ultrafiltration and, 12-2
Graphite, 2-1
Gravitational constant, C-1, 7-12
Green algae, 13-9
Gregory, John, 10-15
Groundwater, 6-6, 6-6t, 12-4, 12-7, 13-3
Gullspång water plant, 2-7
GWRP, 12-7 to 12-9, 12-8f, 12-9t

H

Hague nozzles, 4-3t
Haloacetic acids, 11-17
Hamaker constant, 5-18 to 5-20, 5-25
Hanisch, Baldefrid, 2-4, 2-5
Haworth water plant, 11-3, 11-8t to 11-9t, 11-16 to 11-17
Haynes, William, 2-1
Hazen's sedimentation theory, 7-6, 8-9
Helsinki Conference, 2-8, 2-9, 11-6
Hemlocks Reservoir, 7-29, 10-12t, 11-7
Henry's constant:
correlation coefficients for, F-7
freshwater, 13-18, F-1 to F-2
ionic strength and, 3-7
seawater, 13-18, F-2 to F-3, F-8
solubility and, 3-13, 3-23
water temperature and, 3-29, F-1 to F-3
Henry's law, 3-7, 13-18
Hercules Powder, 2-6 to 2-7
Hiflow rings, 3-19t, 11-9t
Hollow balls, 3-19t
Hollow-fiber membrane, 12-3 to 12-4, 12-3f
HPAM, 6-16t
Humic acids, 5-13 to 5-14, 6-5 to 6-6, 13-5 to 13-6, 13-10
Hyde, Bob, 2-6
Hydraulic flocculators, 6-46, 6-47f, 11-8t
Hydrocarbons, 13-11
Hydrochloric acid, 2-1
Hydrodynamic forces, 5-5 to 5-6, 5-14 to 5-15, 5-19, 7-17t, 7-18
Hydrofoil, 11-15
Hydrogen, 1-3, 5-5 to 5-6, 6-6, 13-3
Hydrolyzed polyacrylamide (HPAM), 6-16t
Hydrophilic forces, 5-5 to 5-7
Hydrophilic fractions, 6-5
Hydrophobic forces:
attachment efficiency and, 7-17t
between bubbles, 5-5, 5-6
coagulation and, 5-22, 5-23f
floc-bubble aggregates and, 5-19 to 5-21, 7-18
in freshwater, 5-21, 5-22f
between particles, 5-5 to 5-7, 5-15, 5-16f
surfactants and, 5-15
Hydrophobic fractions, 6-5

I

Ice point, C-1
Illite, 5-7t
Impellers, 6-14, 6-46, 6-47f, 6-48, 11-8t
"In-Filter" DAF, 2-7. *See also* DAF over filtration
Induced air saturation, 1-4 to 1-5, 1-4f, 3-9
Industrial water/wastewater treatment, 1-2t, 2-3
Inertia, 7-11 to 7-13
Infilco Degremont, 11-3
Interceptions:
description of, 7-11
vs. diffusion, 7-18f
flocculation and, 7-17t, 7-19, 7-22
illustration of, 7-11f
particle size and, 7-15, 7-17t, 7-18f, 7-19, 7-21
single collector efficiency equation for, 7-12 to 7-15
Intermolecular & Surface Forces (Israelachvili), 5-5
Intracellular organic matter (IOM), 13-11
Introduction to the Theory of Flotation, An (Klassen & Mokrousov), 2-4
IOM, 13-11
Iron, 1-4, 5-8, 5-13, 6-2, 12-4
Iron coagulants, 1-6, 6-3, 6-10 to 6-11, 6-12. *See also* Ferric coagulants
ITT WWW, 11-3

J

Jaeger Tri-Pack, 3-12f, 11-9t
Joule, 7-20

K

Kaminski, Gary S., 11-7
Kaolinite, 5-7t, 5-13t

Karenia brevis, 13-10f
Karlström, Adolf, 2-3, 2-5
Karlström and Rausing process, 2-5
Karlström, Bengt, 2-5
Karnsjon Reservoir, 11-11 to 11-12
Kemijärvi water plant, 2-6 to 2-7
King George South Reservoir, 11-14
Klassen, V. I., 2-4
Kokkola water plant, 2-6
Kristinestad water plant, 2-7
Kusaankoski water plant, 2-7
Kuwait, desalination plant in, 13-19

L

Lagoons, 12-13
Lakes, 5-8, 5-10 to 5-12, 6-5 to 6-6, 6-6t, 7-15
Laser Doppler velocimetry (LDV), 8-11, 8-15
Lead oxide, 1-3
Leopold, USA, 11-3
Lewis-Whitman two-film approach, 3-18
Ligands, 6-19, 6-22
Lime, 6-13, 11-11, 11-13, 13-8 to 13-9. *See also* Calcite
Llobregat River, 13-20
London Conference, 2-8, 2-9, 11-6
London dispersion forces, 5-5, 5-14
Lower Lliw Reservoir, 10-12t
Lysekil water plant, 2-7, 11-2, 11-8t to 11-9t, 11-11 to 11-13, 11-13f

M

Magnesium, 13-3t
Magnesium hydroxide, 5-6, 5-18
Manganese:
  in ferric coagulants, 6-15t, 6-24
  in Hemlock Reservoir water, 11-7
  from lakes, 5-8
  membrane filtration of water with, 12-4
  in Oradell Reservoir water, 11-16
  in vacuum DAF, 1-4
Manganese oxides, 5-7t, 5-8, 5-13t, 6-2
Manitoba:
  Deacon Reservoir, 11-15
  Winnipeg water plant, 10-12t, 11-2, 11-8t to 11-9t, 11-14 to 11-16, 11-15f
Marshes, 6-5, 6-6, 6-6t
Mass air concentration. *See also* Bubble mass concentration
  air temperature and, 3-3t, 3-6f, F-4 to F-6
  altitude and, 3-3t, 3-6f
  atmospheric vs. saturator air, 3-3t
Mass air concentration (*Cont.*):
  contact zone efficiency and, 7-19, 7-25 to 7-26, 7-33
  designing for, 3-35
  dry, equation for, 3-3
  factors affecting, 7-33
  volumetric air concentration and, 3-4, 3-6f
Mass precipitation model, 4-6
Mass transfer constant, 3-19 to 3-20, 3-23, 13-18
Mass transfer model, 3-17 to 3-18, 13-18
Massachusetts, 10-12t
Mediterranean Sea, 13-2, 13-20
Membrane filtration:
  backwashing after, 12-4
  in bench-scale testing, 10-6
  coagulation and, 6-2, 12-4
  flocculation and, 12-4
  fouling during, 12-4
  microfiltration, 12-2 to 12-6, 12-2f, 12-4t, 12-6f
  nanofiltration, 12-2 to 12-4, 12-2f, 12-4t, 12-7
  NOM and, 12-4
  particle size and, 5-9, 12-2, 12-2f
  before reverse osmosis, 13-1
  RO. *See* Reverse osmosis
  scaling during, 12-4
  for SDI, 13-8
  transmembrane pressure in, 12-3 to 12-4, 12-4t, 12-6, 13-9, 13-11
  ultrafiltration, 12-2 to 12-6, 12-2f, 12-4t, 12-5t, 12-6f, 12-8f
Membrane plants, 1-2t, 10-12, 10-12t, 12-3 to 12-5, 12-3f, 12-6f
Memcor membrane, 12-6
MF, 12-2 to 12-6, 12-2f, 12-4t, 12-6f
Microcystis, 11-18t
Microcystis aeruginosa, 5-8, 12-7
Microfiltration (MF), 12-2 to 12-6, 12-2f, 12-4t, 12-6f
Microscreens, 11-12 to 11-13
Mokrousov, V. A., 2-4
Molar air concentration, 3-2 to 3-9, 3-6f, 3-26 to 3-28, 3-35, F-4 to F-6
Molar volume of ideal gas, C-1
Molecular diffusivity, 3-23, 13-18, F-1 to F-3
Monodispersed particle suspensions, 6-45 to 6-46
Montmorillonite, 5-7t, 5-13t
Moringa oleifera, 6-32

N

Namibia:
- GWRP, 12-7 to 12-9, 12-8f, 12-9t
- history of water treatment in, 2-6
- photographs of saturators in, 3-12f
- Windhoek water plant, 2-6, 12-5t, 12-7 to 12-9, 12-9t

Nanofiltration (NF), 12-2 to 12-4, 12-2f, 12-4t, 12-7

Nanoparticles, 5-8, 5-9f, 12-2

Natural organic matter (NOM):
- allochthonous, 6-5
- autochthonous, 6-5
- bench-scale jar testing of, 10-8
- charge of, 5-13 to 5-14, 6-8 to 6-10, 6-21 to 6-22, 6-27
- coagulation and, 5-8, 6-2, 6-5, 6-8
- color from, 6-5
- disinfectants and, 6-5
- flocculants and, 6-3
- fractionation method for, 6-5
- hydrophilic vs. hydrophobic, 5-6
- at Lysekil water plant, 11-13
- membrane fouling by, 12-4
- microfiltration and, 12-2
- nanofiltration and, 12-2, 12-7
- oxidants and, 6-5
- in pretreatment, 1-6
- in rivers, 5-7
- in seawater, 13-9 to 13-11
- in SFBW, 12-11
- SUVA range and composition of, 6-7, 6-8t, 13-11
- treatment plant type for, 10-3
- ultrafiltration and, 12-2

Needle valves, 4-3t, 4-6, 4-10 to 4-12, 4-12f, 11-9t

Nephelometric turbidity units (NTU), 5-10

Netherlands, 2-7, 7-31, 9-2, 9-3, 9-6

New Jersey:
- Haworth water plant, 11-3, 11-8t to 11-9t, 11-16 to 11-17
- Oradell Reservoir, 11-16

New York City, Croton water plant, 6-46, 10-12t, 11-2, 11-7, 11-16

Newfoundland:
- Bay Bulls Big Pond, 11-10
- St. John's water plant, 11-2, 11-8t to 11-9t, 11-10 to 11-11

NF, 12-2 to 12-4, 12-2f, 12-4t, 12-7

Nitrogen:
- air solubility and, 3-23, 3-24f
- air temperature and, F-1 to F-6
- in atmospheric air, 3-2, 3-2t, 3-15, 3-17
- correlation coefficients for, F-7
- Henry's constant for, F-1 to F-3
- in humic and fluvic acids, 6-6
- mass concentration of, F-4 to F-6
- mass transfer constant for, 3-19 to 3-20, 3-24f
- molar concentration of, F-4 to F-6
- molecular diffusivity for, F-1 to F-3
- pressure and, 3-15f
- in saturator air, 3-2t, 3-14 to 3-17, 3-15f
- in seawater, 13-4, 13-18

NIWR nozzles, 4-10f

Noctiluca scintillans, 13-10f

NOM. *See* Natural organic matter

Nor-Pac, 3-19t

Norway, 2-2, 12-2, 12-7

Novalox, 3-19t

Nozzles:
- air precipitation efficiency of, 3-26 to 3-28, 3-27f, 4-8, 4-14
- bubble size and:
  - air saturation method and, 3-11
  - basics of, 1-8
  - contact zone efficiency and, 7-23t, 7-25, 7-31t, 7-33
  - particles and, 4-8
  - pilot-scale testing of, 10-16
  - studies of, 4-3t
- at Chingford South water plant, 11-9t
- concurrent vs. countercurrent blending, 4-9
- cross-section of, 4-11f
- dead volumes within, 4-13
- design objectives, 4-13
- detention time in, 4-13
- diameter of, 4-6
- DWL, 4-10f
- with fixed orifices, 4-2, 4-9 to 4-10
- flow velocity through, 4-8, 4-13
- Hague, 4-3t
- at Haworth water plant, 11-9t
- impinging surface in flow path of, 4-13 to 4-14
- lab testing of, 10-11
- NIWR, 4-10f
- number of, 4-9
- photographs of, 4-11f
- pilot-scale testing of, 10-16

Nozzles (*Cont.*):
pneumatically-controlled, 4-11f, 4-12 to 4-13
pressure and, 4-2
Rictor, 4-10f
at St. John's water plant, 11-9t
saturation pressure and, 4-2
at Scottsdale water plant, 12-5t
spacing of, 4-9
at Warner water plant, 11-9t
water flow direction and, 4-9
at Windhoek water plant, 12-5t, 12-8 to 12-9
at Winnipeg water plant, 11-9t
WRC, 4-3t, 4-10f

O

Oil and grease, 13-11, 13-20
Onda correlations, 3-18
Open channel static mixers, 6-13f
Open-end saturation systems:
air flow into, 3-28 to 3-29
air flow rate in, 10-16
description of, 3-9 to 3-10
efficiency of, 3-26
energy requirements for, 3-31 to 3-32, 3-31f
gas fractions in, 3-2
saturator pump capacity of, 10-16
specifications for, 3-10 to 3-11
Oradell Reservoir, 11-16
Orlando Conference, 2-8, 2-10, 11-6
Orthokinetic-flocculation equation, 6-45 to 6-46
Orthophosphate, 6-22, 13-5
Östana paper mill, 2-6
Oxygen:
air solubility and, 3-23, 3-24f
air temperature and, F-1 to F-6
in atmospheric air, 3-2, 3-2t, 3-17
correlation coefficients for, F-7
demand of freshwater, 3-5 to 3-7
in electroflotation, 1-3
Henry's constant for, F-1 to F-3
in humic and fluvic acids, 6-6
mass concentration of, F-4 to F-6
mass transfer constant for, 3-19 to 3-20, 3-24f
molar concentration of, F-4 to F-6
molecular diffusivity for, F-1 to F-3
in saturator air, 3-2t, 3-15 to 3-17
in seawater, 13-18
in vacuum DAF, 1-4
Ozoflot, 1-3
Ozone:
algae and, 1-3
at Chingford South water plant, 11-14
with dispersed air flotation, 1-3
with GAC filters, 10-12
at Haworth water plant, 11-16 to 11-17
at St. John's water plant, 11-11
at Windhoek water plant, 12-8f
at Winnipeg water plant, 11-16

P

PAC, 5-7 to 5-9, 5-7t, 6-2, 10-6, 11-17, 11-18f
Packed saturators:
available air. *See* Available air
at Chingford South water plant, 11-9t
end effects, 3-24 to 3-25, 3-27, 10-16
hydraulic loading of, 3-14t, 3-25
mass transfer constant for, 3-19
packing in. *See* Packing material
photographs of, 3-12f
pressure in, 3-14t, 3-25
at St. John's water plant, 11-9t
schematic cross-section of, 3-12 to 3-13, 3-13f
vs. unpacked, 3-11 to 3-13
at Warner water plant, 11-9t
water level in, 3-25
at Windhoek water plant, 12-5t
at Winnipeg water plant, 11-9t
Packham, Ron, 2-6
Packing material:
at Chingford South water plant, 11-9t
clogging of, 3-11, 10-16
depth of, 3-13f, 3-14t, 3-22f, 3-23, 13-18
dry packing area, 3-18
mass transfer and, 3-9, 3-17 to 3-19
photographs of, 3-12f
at St. John's water plant, 11-9t
size of, 3-14t, 3-18, 3-23, 3-23f
specific area of, 3-19, 3-19t
support plate for, 3-13, 3-13f
surface tension of, 3-19
at Warner water plant, 11-9t
water flow through, 3-12
wetted packing area, 3-18, 3-23, 13-17 to 13-18
at Winnipeg water plant, 11-9t
PACls. *See* Polyaluminum chlorides
Paddle flocculators, 6-46, 6-47f, 11-8t, 11-14. *See also* Gate flocculators

Pall rings, 3-12f, 3-19t, 11-9t, 12-5t
PAM, 6-16t, 6-32, 12-4, 12-9
Paper and pulp industry, 1-2t, 2-2 to 2-3, 2-5
Particle charge:
  attachment efficiency and, 7-31t
  cationic polymers and, 5-17 to 5-18, 5-24, 6-32 to 6-33, 13-11
  coagulant and, 5-17 to 5-18, 6-5, 7-31, 10-14
  electrical double layer and, 5-14, 5-16 to 5-17, 13-8
  flocculation and, 5-13, 6-33
  functional group ionization, 5-12 to 5-13, 5-15
  in granular media filtration, 13-12
  hydrophobic forces and, 7-18, 7-31
  ion adsorption, 5-12, 5-13
  isomorphic substitution, 5-12
  NOM charge and, 5-13 to 5-14, 6-8 to 6-9
  pH and, 5-12 to 5-13, 5-13t, 6-4, 6-10, 6-21 to 6-23, 10-14
  stability and, 13-8 to 13-9
  streaming-current measurement and, 10-14 to 10-15
  zeta potential and, 10-13
Particle concentration:
  coagulation and, 6-4 to 6-5, 10-15
  for floc-bubble aggregates, 7-7 to 7-8
  flocculation and, 5-11, 6-41, 7-7 to 7-8, 7-19
  after granular media filtration, 5-11
  in lakes, 5-11 to 5-12
  in mondispersed suspensions, 6-45
  particle density and, 5-4, 5-11
  particle size and, 6-43
  pressure and, 2-3
  in rivers, 13-9
  in SFBW, 12-13
  suspended solids. *See* Suspended solids
  TSS, 5-10, 12-13
  turbidity and, 5-10 to 5-11
  in white water bubble-blanket model, 7-9 to 7-10
Particle counters, 5-11, 10-15 to 10-16
Particle density:
  bubble size and, 1-3
  bubble volume concentration and, 5-4
  in dispersed air flotation, 1-2
  floc settling rate and, 8-6
  in freshwater, 7-14
Particle density (*Cont.*):
  in lakes, 7-15
  particle concentration and, 5-4, 5-11
  particle size and, 7-14f
  in reservoirs, 7-15
  in rivers, 7-15
  single collector efficiency and, 7-11, 7-12 to 7-13, 7-13t, 7-14f, 7-15
  streamline flow and, 7-11
  turbidity and, 7-15
  water color and, 7-15
  water density and, 5-3
Particle number concentration:
  bubble number concentration and, 5-2 to 5-3, 7-26
  contact zone efficiency and, 7-8, 7-10, 7-16, 7-26
  description of, 5-11
Particle size:
  aggregation rate and, 5-9
  attachment efficiency and, 7-17, 7-18f, 7-29 to 7-31, 7-30f
  Brownian diffusion and, 6-43, 7-11, 7-17t, 7-18f, 7-19, 7-20
  bubble attachment and, 5-9
  bubble rise velocity and, 7-22
  bubble size and, 7-25, 7-25f
  bubble volume concentration and, 7-28 to 7-29, 7-28f
  coagulation and, 5-21
  collision efficiency function and, 6-42t, 6-43, 6-44f
  concentration and, 5-11, 6-44
  in contact zone, 5-9, 5-21, 7-7 to 7-8
  contact zone efficiency and:
    attachment efficiency and, 7-29 to 7-31, 7-30f
    detention time and, 7-28 to 7-29, 7-28f, 7-32 to 7-33
    flocculation and, 7-17, 7-18f, 7-19 to 7-22, 7-31t, 7-32 to 7-33
    mixing and, 7-32 to 7-33
    single collector efficiency and, 7-31t
    water temperature and, 7-24f
  detention time and, 7-28 to 7-29, 7-28f, 7-31t, 7-32 to 7-33
  distribution of, 6-43, 10-15
  electrostatic forces and, 5-16
  filtration and, 5-9
  floc-bubble aggregate rise rate and, 8-9

Particle size (*Cont.*):
flocculation and:
Brownian diffusion in, 6-42t, 6-45, 7-17t, 7-19, 7-22
coagulant selection and, 6-34
coalescence and collisions in, 6-2
collision efficiency functions and, 6-43 to 6-44, 6-44f
contact zone efficiency and, 7-17, 7-18f, 7-19 to 7-22, 7-31t, 7-32 to 7-33
description of, 6-41 to 6-45
detention time and, 7-31t
distribution of sizes, 6-43
fluid shear in, 6-42t, 6-45
forces and, 5-21 to 5-22, 5-23f
interceptions and, 7-17t, 7-19, 7-22
modeling of, 7-7 to 7-8
particle concentration and, 6-44, 7-19
in sedimentation, 6-48
settling and, 7-17t, 7-19
stages of, 7-31t
granular media filtration and, 5-9
gravity and, 7-11
interceptions and, 7-15, 7-17t, 7-18f, 7-19, 7-21
microfiltration and, 12-2, 12-2f
nanofiltration and, 12-2, 12-2f
of PAC, 5-9
particle concentration and, 5-11
range, in water, 5-9f
reverse osmosis and, 12-2, 12-2f
scrapers and, 10-15
sedimentation and, 5-9, 6-48
in separation zone, 5-9
settling and, 7-17t, 7-19, 7-21
single collector efficiency and, 7-12 to 7-15, 7-13t, 7-14f, 7-19 to 7-24, 7-23t
transportation mechanisms and, 7-11
turbidity and, 5-9
ultrafiltration and, 12-2, 12-2f
water temperature and, 7-14f, 7-15
Particle zeta potential:
calculation of, 5-24, 6-22, 10-13
coagulation and, 5-22, 7-31, 10-14
electrical double layer and, 5-16, 5-19, 5-24, 7-18
equipment for calculating, 10-13f
in freshwater, 5-20 to 5-21, 7-18
streaming-current measurement and, 10-15
Particles:
bubbles and. *See* Floc-bubble aggregates
charge of. *See* Particle charge
Particles (*Cont.*):
coagulation and, 5-7 to 5-8, 5-7t, 6-4 to 6-5, 6-8
DAF vs. dispersed air flotation, 5-15
density of. *See* Particle density
drag force on, 8-6
electrostatic force between, 5-14 to 5-16
flocculation rate of, 6-4
Hazen theory on, 7-6, 8-9
hydrodynamic forces between, 5-5 to 5-7, 5-14 to 5-15
hydrophilic portions of, 5-5 to 5-7
hydrophobic portions of, 5-5 to 5-7, 5-15, 5-16f
mass balance statement for, 7-35
mass concentration of, 5-3 to 5-4, 5-10
number concentrations of. *See* Particle number concentration
from pretreatment, 5-7 to 5-8, 5-7t
restabilization of, 6-33
separation processes for, 5-8
settling rate of, 8-6
size of. *See* Particle size
stability of, 5-12 to 5-15, 6-4, 6-5, 6-12, 6-41, 13-8 to 13-9
steric effects on, 5-14
structure of, 8-6
transportation mechanisms for, 7-10
van der Waals force between, 5-14 to 5-16
volume concentration of, 5-3 to 5-5, 5-11 to 5-12
water and, 5-6, 5-7, 5-14, 5-16f, 8-6
Particulate organic carbon (POC), 6-5 to 6-7
Patch model, 6-33
PAX. *See* Polyaluminum chlorides
Peat, 6-6
Peclet number, 7-6
Pennsylvania, 10-12t
Perikinetic flocculation, 6-42
Permeate, 12-3, 12-3f
Permeate flux, 12-3, 12-4t, 12-6, 13-8
Pernitsky, David J., 11-10, 11-14
Peroxide, 11-16
Persian Gulf, 13-2
PES, 12-3
Pesticides, 5-8, 11-14, 12-2
Petersen, Nils, 2-2
PFR, 6-47, 6-48f, 7-2 to 7-7, 7-3f
pH:
algae and, 5-8, 5-13t, 6-4, 11-17
alkalinity and, 6-11f

pH (*Cont.*):
alum and, 6-10, 6-17 to 6-23, 6-18f, 6-34 to 6-35, 13-13
attachment efficiency and, 7-31, 7-31t
of backwash water, 12-13
bacteria and, 5-13t, 6-4
of Bay Bulls Big Pond water, 11-10 to 11-11
bench-scale jar testing of, 10-6 to 10-8
buffer intensity and, 6-11f, 13-6 to 13-7, 13-7f
at Chingford South water plant, 11-14
during DAF, 6-9
ferric coagulants and:
effects of, 6-10
fraction of dissolved iron and, 6-25f
freshwater, 6-24 to 6-27, 6-36
seawater, 13-14 to 13-16
solubility of, 5-8, 6-26f
zero point of charge, 5-13t
fulvic acids and, 6-6, 6-8
hydrogen concentration and, 13-3
of the isoelectric point, 5-13
at Lysekil water plant, 11-11, 11-13
of Oradell Reservoir water, 11-16
PACls and, 6-10 to 6-12, 6-28 to 6-32, 6-30f, 6-36 to 6-37, 13-13
particle charge and, 5-12 to 5-13, 5-13t, 6-4, 6-10, 6-21 to 6-23, 10-14
recarbonation and, 2-6
at St. John's water plant, 11-11
sand and, 5-12 to 5-13, 5-13t
of seawater, 13-3 to 13-7, 13-5f, 13-6f, 13-7f
of SFBW, 12-11, 12-14
streaming current and, 10-15
sweep-floc coagulation and, 6-10, 6-19 to 6-22, 6-25, 6-27, 6-31 to 6-32, 6-34 to 6-37
viruses and, 6-4
at Warner water plant, 11-7
at Windhoek water plant, 12-9t
at Winnipeg water plant, 11-15
of zero point of charge, 5-13
zeta potential and, 5-5 to 5-6
Phenols, 6-5
Phosphates, 12-13
Phosphorus, 6-22, 13-4
Photometric Dispersion Analyzer, 10-15
Pilot-scale testing:
available air measurements in, 3-26, 10-16
of CoCoDAFF, 11-4
DAF footprint for, 11-2
Pilot-scale testing (*Cont.*):
design variables for, 10-10 to 10-11
focus of, 10-7
at Haworth water plant, 11-16 to 11-17
on hydraulic loading, 11-2
need for, 10-9 to 10-10
photographs of, 10-10f, 10-11f
at St. John's water plant, 11-11
saturator in, 3-25
at Scottsdale water plant, 12-6
of SFBW, 12-11, 12-12f, 12-14
studies of, 10-11 to 10-12, 10-12t
at Warner water plant, 11-18 to 11-19
at Winnipeg water plant, 11-15
Pipe injection tanks, 6-13f
Pipe static mixers, 6-13f
Pipes:
detention time in, 7-5
dispersion number for, 7-6
losses in, 3-30, 12-8
plug flow behavior in, 7-5, 7-6
premature air precipitation in, 4-1
sizing of, 12-9
turbulent flow in, 7-5, 7-6
Planck constant, C-1
Plant throughput flow, 1-8 to 1-9
Plate settling, 11-18 to 11-19, 11-19f, 12-11 to 12-13, 12-12f
Plug flow behavior, 7-5 to 7-6
Plug-flow reactor (PFR), 6-47, 6-48f, 7-2 to 7-7, 7-3f
POC, 6-5 to 6-7
Poly-DADMAC, 6-15t, 6-32
Polyacrylamide (PAM), 6-16t, 6-32, 12-4, 12-9
Polyaluminum chlorides (PACls):
alkalinity and, 6-11 to 6-12, 6-28
basicity of, 6-28 to 6-32
chemistry of, 6-27 to 6-32
at Chingford South water plant, 11-14
as coagulant, 6-2, 6-14, 13-12t
degree of neutralization of, 6-28 to 6-29
DOC and, 7-32f
dosage, 6-31 to 6-32, 6-36 to 6-38
EMP data for, 7-32f, 10-14, 10-14f
at Haworth water plant, 11-17
PAC and, 11-18t
pH and, 6-10 to 6-12, 6-28 to 6-32, 6-30f, 6-36 to 6-37, 13-13
properties of liquid, 6-15t
at St. John's water plant, 11-11
scaling with, 13-12t
in seawater pretreatment, 13-13

Polyaluminum chlorides (PACls) (*Cont.*):
solubility of, 6-30, 6-31f
specific gravity of, 6-28 to 6-29
sulfate with, 6-28
turbidity and, 6-31, 6-37, 6-38f, 7-32f, 11-12f
UV light absorbance and, 6-31 to 6-32, 6-36 to 6-38, 6-38f, 7-32f
water color and, 11-11, 11-12f
water temperature and, 6-12, 6-29 to 6-30, 6-31f, 6-36, 13-13
Polyaluminum sulfates, 6-27
Polyamide, 12-4
Polydiallyldimethyl ammonium chloride (Poly-DADMAC), 6-15t, 6-32
Polyelectrolytes, cationic, 5-6, 5-13, 6-14, 6-28, 6-32 to 6-33. *See also* Polymers, cationic
Polyethersulfone (PES), 12-3
Polyferric chlorides, 6-14, 6-15t, 6-27
Polyferric sulfates, 6-14, 6-15t, 6-27
Polymers, anionic, 6-3, 13-12 to 13-13
Polymers, cationic:
alum, iron, and, 6-3, 13-11 to 13-12
branching agents for, 6-32
chemistry of, 6-32 to 6-33
chitosan, 6-32
as coagulants, 6-2 to 6-3, 13-11 to 13-13, 13-12t
DOC and, 6-2
dosage, 6-33, 6-38 to 6-39
Epi-DMA, 6-15t, 6-32, 6-39
freshwater vs. seawater, 13-12t
in froth flotation, 5-6
at Haworth water plant, 11-17
HPAM, 6-16t
molecular weight of, 6-15t to 6-16t, 6-32
Moringa oleifera, 6-32
PAM, 6-16t, 6-32, 12-4, 12-9
particle charge and, 5-17 to 5-18, 5-24, 6-32 to 6-33, 13-11
the patch model for, 6-33
pH levels and, 5-24, 6-32, 6-38
Poly-DADMAC, 6-15t, 6-32
in rapid mixing, 6-14
scaling with, 13-13
in seawater pretreatment, 13-11 to 13-15
for SFBW, 12-14
structure of, 6-15t to 6-16t
SUVA and, 6-33
TOC removal with, 6-33
at Warner water plant, 6-39, 11-7
Polysaccharides, 6-5, 13-11
Polysulfone, 12-3
Polyvinylidene fluoride (PVDF), 12-3
Potassium, 13-3t
Potassium permanganate, 11-17
Potter-Delprat method, 2-1 to 2-2
Powdered activated carbon (PAC), 5-7 to 5-9, 5-7t, 6-2, 10-6, 11-17, 11-18f
Power number, 6-14
Premature air precipitation, 4-1 to 4-2
Pressure DAF:
about, 1-1, 1-4 to 1-5
history of, 2-2 to 2-4
photographs of, 1-7f
process schematics for, 1-4f, 1-5f
recycle stream in, 3-11
Pressure ratio, 3-15, 3-17
Propeller flocculators, 6-46, 11-8t, 11-13, 12-5t. *See also* Impellers
Proteins, 6-5, 13-11
Protozoa, 5-7t, 5-8, 5-12 to 5-13, 6-4, 11-18 to 11-19, 12-10. *See also specific types of*
Purac, 2-5, 2-6 to 2-7, 11-3, 11-11
PVDF, 12-3

Q

Quartz, 13-3, 13-4

R

Rapid mixing:
acids in, 6-14
addition of chemicals in, 6-14
alum in, 6-13 to 6-14, 6-17 to 6-22
in CFSTRs, 7-4
coagulants in, 6-3
detention time in, 11-13
for EMP measurements, 10-14
ferric hydroxides in, 6-24
intensity of, 6-14
at Lysekil water plant, 11-13
open channel static mixers for, 6-13f
pH level in, 6-9, 6-13
pipe static mixers for, 6-13f
in process schematic, 1-6
purpose of, 6-12
at St. John's water plant, 11-11
of SFBW, 12-14
tanks for, 6-13f
at Warner water plant, 11-7
water density and, 6-12
water temperature and, 6-12
water viscosity and, 6-12

Rauschert Polypropylene Hiflow Rings, 11-9t
Rausing, Ruben, 2-5
Reaction time, 6-13
Reaction zone, 1-8, 7-1. *See also* Contact zone
Reciprocating scraper, 9-6 to 9-7, 9-7f, 11-8t
Reclamation/reuse, 1-2t
Recycle flow:
  air dosing rate and, 3-7
  air flow rate and, 3-10 to 3-11, 3-28 to 3-29
  available air and, 10-16
  in contact zone, 1-8
  freshwater vs. seawater, 13-18
  for oxygen deficient freshwater, 3-5 to 3-7
  pumping power for, 3-32 to 3-33
  recycle ratio and, 1-8
  at St. John's water plant, 11-11
  in saturators, 1-8, 3-13 to 3-16
  in separation zone, 1-8 to 1-9
  SFBW in, 12-10 to 12-13, 12-12f
Recycle-flow pressurization, 1-1, 1-5, 1-5f, 3-11
Recycle rate:
  bubble volume concentration and, 7-23t, 7-31t, 7-33
  at Chingford South water plant, 11-9t, 11-14
  contact zone efficiency and, 7-29, 7-33
  designing for, 3-35
  freshwater vs. seawater, 13-18
  at Haworth water plant, 11-9t
  at Lysekil water plant, 11-9t
  at St. John's water plant, 11-9t
  at Scottsdale water plant, 12-5t, 12-6
  of SFBW, 12-10, 12-11f, 12-14
  in United States, 12-10
  at Warner water plant, 11-9t
  at Windhoek water plant, 12-5t
  at Winnipeg water plant, 11-9t, 11-16
Recycle ratio:
  bench-scale jar testing of, 10-6 to 10-8
  bubble bed depth and, 8-17
  bubble concentration and, 8-17
  definition of, 1-8
  designing for, 3-35
  in drinking water, 1-8
  energy requirements and, 3-32
  molar air concentration and, 3-5
  plant throughput flow and, 1-8
  saturator air and, 3-7
  turbidity and, 10-7
Red Sea, 13-2
Red-tide algae, 13-1, 13-9, 13-10f
Removal efficiency, 7-12, 7-16
Reservoirs. *See also specific bodies of water*
  algae in, 6-5, 6-6, 11-4
  particle density in, 7-15
  suspended solids in, 5-8
  TOC of, 6-6
  treatment plant type for, 10-4f, 10-5
  turbidity of, 11-4
Residence time, 6-47, 6-48f, 7-3 to 7-5, 8-14
Residence time distribution (RTD) of water, 6-47, 7-2, 7-4 to 7-5, 7-4f
Retentate, 12-3
Retention time, 8-13. *See also* Residence time
Return flow, 8-14 to 8-15
Reverse osmosis (RO):
  algae and, 13-9
  applications for, 12-2 to 12-3
  cleaning after, 12-4t
  DAF before, 13-1
  for desalination of seawater, 1-2t, 12-2, 13-1
  fouling during, 13-8
  material for, 12-3 to 12-4
  particle size and, 12-2, 12-2f
  permeate flux, 12-4t
  pretreatment before, 13-1, 13-20f
  recovery percentage, 12-4t
  scaling during, 12-4
  TMP, 12-4t
  turbidity before, 13-8
Reynolds number, 7-11, 7-20, 8-2 to 8-3, 13-17
Rhodamine WT, 7-4
Rhode Island, 10-12t
Richardson number, 8-12
Rictor, 2-5 to 2-6, 4-10f, 11-2 to 11-3
Rivers:
  algae in, 6-5
  bacteria in, 5-8
  flocculation in, 13-9
  NOM in, 5-7
  paper industry and, 2-2
  particle concentration in, 5-11 to 5-12, 13-9
  particle density in, 7-15
  suspended solids in, 5-8, 5-10
  TOC of, 6-6, 6-6t
  total inorganic carbon in, 13-4
  treatment plant type for, 10-4f, 10-5
  turbidity of, 5-10, 11-4
RO. *See* Reverse osmosis

S

St. John's water plant, 11-2, 11-8t to 11-9t, 11-10 to 11-11
Salinity, 8-12

Salt, 4-9, 7-4, 12-2, 12-2f, 12-4
San Luis Obispo, 10-12t
Sand:
ballasted-sand sedimentation, 10-5, 12-6, 12-13
cationic polymers and, 13-12
charge of, 5-12 to 5-13
at Chingford South water plant, 11-9t
chlorophyll and, 10-3
at Haworth water plant, 11-9t
at Lysekil water plant, 11-9t
particle size, 5-9, 12-2f
pH and, 5-12 to 5-13, 5-13t
quartz, 13-4
at St. John's water plant, 11-9t
at Scottsdale water plant, 12-6
in seawater, 13-8
for SFBW, 12-13
slow sand filtration, 10-3
at Tuas Desalination Plant, 13-20
turbidity and, 10-3
at Warner water plant, 11-9t
at Windhoek water plant, 12-8f
Saturation concentration, 7-25 to 7-26
Saturation pressure:
adjustment of, 4-13
available air and, 3-23
bench-scale jar testing of, 10-6 to 10-8
bubble bed depth and, 8-17
bubble concentration and, 7-23t, 8-16 to 8-17
bubble nucleation and, 4-6
bubble size and, 3-11, 10-16
bubble volume concentration and, 7-23t, 7-31t, 7-33
at Chingford South water plant, 11-9t, 11-14
in drinking water, 1-8
energy to generate, 3-30 to 3-31, 10-2
freshwater vs. seawater, 13-18 to 13-19
at Haworth water plant, 11-9t
at Lysekil water plant, 11-9t, 11-13
macro-bubbles and, 4-5
measurement of, 3-11
nitrogen and, 3-15f
nozzles and, 4-2
pilot-scale testing of, 10-7
at St. John's water plant, 11-9t, 11-11
at Scottsdale water plant, 12-5t, 12-6
in vacuum DAF, 1-4
at Warner water plant, 11-7, 11-9t
at Windhoek water plant, 12-5t
at Winnipeg water plant, 11-9t
Saturator air:
absolute pressure of, 3-2
air temperature and, 3-3t, E-1 to E-4
altitude and, 3-3t, E-1 to E-4
bubble mass concentration and, 7-25 to 7-26
designing for, 3-4, 7-25 to 7-26
freshwater vs. seawater, 13-17 to 13-18
gas fractions in, 3-2 to 3-3, 3-2t, 3-13 to 3-16
grams of air in one liter of, E-2 to E-4
moles of air in one liter of, E-1 to E-2
pressure of, 1-8
recycle ratio and, 3-5 to 3-7, 3-7
solubility of, 3-9
water temperature and, 3-6 to 3-7
Saturators:
air concentration leaving, 3-20, 3-25
air flow test for, 3-26
available air. *See* Available air
at Chingford South water plant, 11-9t
efficiency of, 3-25 to 3-26
energy requirements for, 3-30 to 3-32
flow control, 3-25
at Haworth water plant, 11-9t
location of, 4-1
at Lysekil water plant, 11-9t, 11-13
mass transfer in, 3-17 to 3-24
packed. *See* Packed saturators
photographs of, 3-12f
recycle flow in, 1-8, 3-5 to 3-7
at St. John's water plant, 11-9t
at Scottsdale water plant, 12-5t
unpacked, 3-11, 3-12f, 11-9t, 12-5t
at Warner water plant, 11-9t
water flow test for, 3-26
at Windhoek water plant, 12-5t, 12-8
at Winnipeg water plant, 11-9t, 11-16
Saudi Arabia, desalination plant in, 13-19
Scotland, Glencorse water plant, 11-4
Scottsdale water plant, 12-5t, 12-6
Scrapers:
beach, 9-6 to 9-7, 9-7f
chain-and-flight, 9-6 to 9-7, 9-7f
continuous use of, 9-8
depth of penetration of, 9-7
grid stabilization and, 9-9
at Haworth water plant, 11-8t
interval between scrapes, 9-8, 9-8f
knockdown and, 9-2, 10-15
particle size and, 10-15
reciprocating, 9-6 to 9-7, 9-7f, 11-8t
residual coagulant and, 9-8f

Scrapers (*Cont.*):
at St. John's water plant, 11-8t
at Scottsdale water plant, 12-5t
speed of, 9-7 to 9-8, 9-8f
tank shape and, 11-2
turbidity and, 9-8 to 9-9, 9-8f
at Winnipeg water plant, 11-8t
SDI, 13-8, 13-19, 13-20
Seawater:
algae in, 13-1, 13-3 to 13-4, 13-8, 13-9
alkalinity of, 13-3 to 13-6
chlorinity of, 13-2
coagulants for pretreatment of, 13-11 to 13-16
composition of, 13-2 to 13-7, 13-3t
conductivity of, 5-18t
contact zone efficiency and, 13-16 to 13-17
contaminants in, 13-8 to 13-11
density of, 13-4, 13-6, 13-16 to 13-18, D-2 to D-4
desalination of, 1-2t, 12-2, 13-1
electrical double layer thickness in, 5-18t
equilibrium constants for, 13-3
Henry's constant for, F-2 to F-3, F-8
ionic strength of, 5-15, 5-18t, 13-2 to 13-4, 13-9
mass concentration of gases in, F-5 to F-6
molar concentration of gases in, F-5 to F-6
molecular diffusivity for, F-2 to F-3
pH of, 13-3 to 13-7, 13-5f, 13-6f, 13-7f
salinity of, 13-2, 13-4
separation zone efficiency and, 13-17
surface tension of, 13-16 to 13-18, D-2 to D-4
TDS in, 5-18t
temperature of, 13-16 to 13-18
TMP, 12-4t
turbidity, 13-8
vapor pressure of, D-2 to D-3
viscosity of, 13-16 to 13-18, D-2 to D-3
SEDAF, 11-5 to 11-6
Sedimentation:
algae and, 11-6, 11-13, 13-1
ballasted-sand, 10-5, 12-6, 12-13
coagulants for, 6-2, 6-33 to 6-34
Cryptosporidium and, 10-2, 11-18 to 11-19
DAF and:
algae, NOM, and, 11-13, 12-6
coagulants for, 6-2, 6-33 to 6-34
combinations, 11-4 to 11-6
cost and efficiency, 2-6, 10-2
floc size, 6-41, 6-48
Sedimentation, DAF and (*Cont.*):
footprint for, 10-2
performance efficiency, 10-12
studies of, 10-12t
tank flow patterns, 8-12, 11-1 to 11-2
flocculant aids for, 6-48
flocculation and, 1-7, 6-34, 6-41, 6-46, 6-48
flow patterns in, 8-12
Giardia lamblia and, 11-18
Hazen's sedimentation theory, 7-6, 8-9
PAC removal by, 11-17
particle concentration in, 6-34
particle size in, 5-9, 6-48
plate settling, 11-18 to 11-19, 11-19f, 12-11 to 12-13, 12-12f
before reverse osmosis, 13-1
at Scottsdale water plant, 12-6
SFBW and, 12-13
TOC and, 10-5
turbidity and, 8-12, 10-5, 11-5 to 11-6
water color and, 2-7, 10-5, 11-6
water temperature and, 2-7, 8-12, 11-6, 11-10
Seoul Conference, 2-8, 11-6
Separation zone:
ADV studies of, 8-11
air concentration in, 3-2 to 3 7
boundary with control zone, 8-1 to 8-2
breakthrough conditions in, 8-13f, 8-14 to 8-15
bubble bed in, 8-11 to 8-17, 11-3
bubble concentration in, 8-16 to 8-17
bubble sizes in, 4-3, 4-3t
CFD modeling of, 8-11
counter-current flow in, 11-4
description of, 1-8 to 1-9, 8-1
efficiency of, 3-3, 5-2
float layer in. *See* Float layer
flow patterns in, 8-11 to 8-15, 8-16f, 9-7, 11-1
function of, 7-1
hydraulic loading of, 1-9, 8-13 to 8-17, 8-16f, 11-2 to 11-3
illustrations of, 4-12f, 7-2f, 7-36f, 8-10f, 8-15f
inlet to, 8-13 to 8-14
layers in, 8-14
LDV studies of, 8-11
outlets from, 8-12
particle size in, 5-9
removal efficiency of, 8-17

Separation zone (*Cont.*):
seawater and performance efficiency, 13-17
stratified flow structure in, 8-14 to 8-15, 8-15f, 9-7, 11-3
tracer studies of, 8-10, 8-12 to 8-13, 8-13f
water velocity in, 8-16f, 9-7
Settling, 7-11 to 7-15, 7-11f, 10-4f, 13-16 to 13-17
Settling tanks, 6-41, 7-5 to 7-6
SFBW, 1-2t, 10-12t, 12-9 to 12-14, 12-11f, 12-12f
Shoal Lake, 11-15
Silicon, 13-4
Silicon dioxide, 5-7t, 5-13t, 13-3, 13-4, 13-5f, 13-8. *See also* Sand
Silt Density Index (SDI), 13-8, 13-19, 13-20
Silts, 5-9
Singapore, desalination plant in, 13-19, 13-20, 13-21f
Single collector efficiency:
about, 7-10 to 7-15
Brownian diffusion equation, 7-12 to 7-15
bubble size and, 7-12 to 7-15, 7-13t, 7-14f, 7-23t
contact zone efficiency and, 7-16
dispersed air flotation and, 7-10
filtration and, 7-10
freshwater vs. seawater, 13-16 to 13-17
granular media filtration equation, 7-13
inertia equation, 7-12 to 7-13
interception equation, 7-12 to 7-15
particle density and, 7-13t, 7-14f, 7-15
particle size and, 7-12 to 7-15, 7-13t, 7-14f, 7-17t, 7-19 to 7-24, 7-23t
Reynolds number for, 7-11, 7-20
settling equation, 7-12 to 7-15
water density and, 7-12
water temperature and, 7-13t, 7-14f, 7-15, 7-23
water viscosity and, 7-12 to 7-13
Slimes, 11-15
Slow sand filtration, 10-3
Sludge:
algae and, 9-6
anaerobic conditions in, 12-9
bench-scale jar testing of, 12-14
in circular tanks, 11-2
DAFF and, 6-48
dry solids in, 9-3 to 9-4, 12-14
dual-coagulant strategy for, 7-29
electroflotation for thickening, 1-3
floating. *See* Float layer
measurement of. *See* Total suspended solids
PAC-enriched, 11-17, 11-18f
Sludge (*Cont.*):
Photometric Dispersion Analyzer for, 10-15
pipe sizing for, 12-9
polymers and, 6-3, 6-34, 6-38, 6-39, 6-48, 10-15
pressure DAF for thickening, 2-3
removal of, 9-2 to 9-9, 11-2, 11-8t, 12-5t
retention of, 10-15
from Scottsdale water plant, 12-6
vacuum flotation for thickening, 1-4
Smoluchowski equation, 10-13
Sodium, 13-3t
Sodium aluminate, 6-14, 6-15t
Sodium bisulphate, 2-2
Soil, 6-6
Solar radiation, 12-8
Solubility limit, 7-25
South Africa:
CoCoDAFF plants in, 11-4, 11-5f
draining tanks in, 9-6
float layer studies in, 9-2, 9-5
history of water treatment in, 2-6, 2-8
nozzles at plants in, 4-9
PAC use in, 11-17
process selection in, 10-3
Vaalkop water plant, 11-5 to 11-6, 11-5f
South America, DAFF in, 11-3
South Korea, 2-8, 8-11, 10-15, 11-5 to 11-6, 11-17
South San Joaquin Irrigation District (SSJID), 12-5
Spain, desalination plant in, 13-19, 13-20
Specific UV absorbance (SUVA):
alum and, 6-22 to 6-23, 6-34 to 6-35, 6-38f
cationic polymers and, 6-33
coagulant dose and, 6-8
in Deacon Reservoir water, 11-15
description of, 6-7
ferric hydroxides and, 6-27, 6-36, 6-37f
of freshwater, 6-39
in Karnsjon Reservoir water, 11-11
PACls and, 6-31 to 6-32, 6-36 to 6-38, 6-38f
of seawater, 13-11
water composition and, 6-8t
at Windhoek water plant, 12-9t
Spectrophotometer cells, 13-10
Spent filter backwash water (SFBW), 1-2t, 10-12t, 12-9 to 12-14, 12-11f, 12-12f
Spiral-wound membrane, 12-3 to 12-4, 12-3f
Split-flow pressurization, 1-5, 1-5f, 3-11
SS. *See* Suspended solids:
Stainless steel, 1-3

Static mixing devices, 3-10
Steel, 1-3
Step input tracer test, 7-4
Stokes' flow conditions, 7-11
Stokes' law, 7-19, 8-3
Strängnås water plant, 2-6
Streaming-current measurements, 6-22, 10-14 to 10-15, 13-12
Streams, 6-5, 6-6, 6-6t, 12-3
Strontium, 13-3t
Subnatant, 1-8, 8-1
Sugars, 6-5, 13-11
Sulfate, 13-3t
Sulfuric acid:
- alkalinity and, 6-11
- alum and, 6-17
- at Chingford South water plant, 11-14
- Delprat on, 2-2
- at Haworth water plant, 11-17
- for pH control, 6-13, 13-13, 13-15
- Potter on, 2-1 to 2-2
- in seawater pretreatment, 13-19 to 13-20, 13-20f

Super Intalox, 3-19t
Surfactants, 1-3, 5-6, 5-15
Suspended solids (SS):
- backwash water chemistry and, 12-13
- coagulation and, 12-13
- concentration of, 8-15
- in lakes, 5-8, 5-10
- in reservoirs, 5-8
- in rivers, 5-8, 5-10
- in seawater, 13-20
- in SFBW, 12-11, 12-13
- water density and, 8-12

Suutarinen, Oiva, 2-5
SUVA. *See* Specific UV absorbance
Sveen, Karl, 2-2
Sveen-Petersen system, 2-2 to 2-3
Swamps, 6-5, 6-6, 6-6t
Sweden:
- Arvika water plant, 2-6 to 2-7
- Flen water plant, 2-6
- Gullspång water plant, 2-7
- history of water treatment in, 2-2 to 2-3, 2-5, 2-7
- Karnsjon Reservoir, 11-11 to 11-13
- Lysekil water plant, 2-7, 11-2, 11-8t to 11-9t, 11-11 to 11-13, 11-13f
- Östana paper mill, 2-6
- Strängnås water plant, 2-6

Sweep-floc coagulation:
- with alum, 6-13 to 6-14, 6-19 to 6-22, 6-34
- description of, 5-8
- with ferric coagulants, 6-25 to 6-27, 6-36
- with PACls, 6-31 to 6-32, 6-36 to 6-37
- particle concentration in, 6-4
- pH and, 6-10, 6-19 to 6-22, 6-25, 6-27, 6-31 to 6-32, 6-34 to 6-37
- in seawater pretreatment, 13-13, 13-14
- water temperature and, 6-12, 6-19 to 6-20, 6-25, 6-34, 6-36

T

Tampa Bay desalination plant, 13-10
Tanks:
- air concentration in, 3-2 to 3-7
- air separation, 3-10
- baffles in, 6-46, 7-2, 7-7, 8-12 to 8-14
- bottom sludge in, 9-6
- CFD for designing, 7-7
- at Chingford South water plant, 11-8t
- circular vs. rectangular, 11-2
- contact zone of. *See* Contact zone
- flocculation tanks, 6-46, 10-2, 10-6, 10-11, 10-15, 11-2
- floor plates in, 8-12, 9-7, 11-2 to 11-3
- flotation, 7-16, 7-23, 11-7
- flushing of, 9-5 to 9-6
- footprint for, 11-2
- at Haworth water plant, 11-8t
- hydraulic loading of, 1-8 to 1-9, 8-9 to 8-10, 8-13 to 8-17
- lab testing of, 10-11
- at Lysekil water plant, 11-8t
- manifolds in, 4-2, 4-10
- photographs of, 1-7f, 10-11f
- pipe injection, 6-13f
- in process schematic, 1-6f
- for rapid mixing, 6-13f
- at St. John's water plant, 11-8t
- saturator, 1-7f, 3-12f
- at Scottsdale water plant, 12-5t
- sections of, 1-8
- sedimentation, 8-12, 11-2
- separation zone of. *See* Separation zone
- settling tanks, 6-41, 7-5 to 7-6
- shapes and sizes of, 11-1 to 11-2
- size of, 11-3
- at Warner water plant, 11-8t
- water flowing on inside edges of, 9-4

Tanks (*Cont.*):
at Windhoek water plant, 12-5t
at Winnipeg water plant, 11-8t
TDS, 5-17, 12-4t, 13-20
Teflon, 5-6
Tellerettes, 3-19t
Tetrapak, 2-5
THM, 6-7, 11-17
Titanium, 1-3
TMP, 12-3 to 12-4, 12-4t, 12-6, 13-9, 13-11
TOC. *See* Total organic carbon
Total dissolved solids (TDS), 5-17, 12-4t, 13-20
Total inorganic carbon, 6-11, 11-13, 13-3 to 13-4, 13-6 to 13-7, 13-7f
Total organic carbon (TOC):
algae and, 13-10
alum and, 6-22 to 6-23, 6-34 to 6-35
in Bay Bulls Big Pond, 11-10
bench-scale jar testing of, 10-8
cationic polymers and, 6-33
coagulant dose and, 6-8
in Colorado River water, 12-6
DAF and, 10-3 to 10-4, 10-4f
in Deacon Reservoir water, 11-15
direct filtration and, 10-3, 10-4f, 11-7
ferric hydroxides and, 6-27, 6-36, 13-16
in Hemlock Reservoir water, 7-29, 11-7
in King George South Reservoir water, 11-14
in lakes, 6-6
in Oradell Reservoir water, 11-16
PACls and, 6-31 to 6-32, 6-36 to 6-38
pilot-scale testing of, 10-7
POC and, 6-6
in seawater, 13-4, 13-10, 13-11, 13-16
sedimentation and, 10-5
in SFBW, 12-13
streaming current and, 10-15
at Tampa Bay desalination plant, 13-10
treatment plant type for, 10-3 to 10-5, 10-4f
USEPA on, 6-3
UV light absorbance and, 6-7, 6-8t
at Warner water plant, 6-39, 6-41, 11-10
of water supplies, 6-6, 6-6t
in William Girling Reservoir water, 11-14
at Winnipeg water plant, 11-16
Total suspended solids (TSS), 5-10, 12-13
Tracer studies, 7-3f, 7-4 to 7-6, 7-4f, 8-10
Tracers, definition of, 7-4
Transmembrane pressure (TMP), 12-3 to 12-4, 12-4t, 12-6, 13-9, 13-11
Trihalomethane (THM), 6-7, 11-17
Tripacks, 3-19t
TSS, 5-10, 12-13
Tuas Desalination Plant, 13-20, 13-21f
Turbidity:
algae and, 5-10
alum and, 6-22, 6-34, 6-35f, 6-38f, 6-40f
in Bay Bulls Big Pond, 11-10
bench-scale jar testing of, 10-6 to 10-8, 10-9f
bottom sludge and, 9-6
bubble volume concentration and, 11-6
at Chingford South water plant, 11-14
coagulant dose and, 6-8, 7-31, 7-32f
of Colorado River water, 12-6
DAF and, 5-11, 10-3 to 10-5, 10-4f, 11-4 to 11-6
in Deacon Reservoir water, 11-15
definition of, 5-10
direct filtration and, 10-3, 10-4f, 11-7
dual media filtration and, 6-35f
ferric hydroxides and, 6-27, 6-36, 6-37f
float layer and, 9-5f, 9-8, 9-8f
flocculation and, 5-10, 10-6
flotation and, 10-6
granular media filtration and, 5-11
at Haworth water plant, 11-17
in Hemlock Reservoir water, 11-7
in Karnsjon Reservoir water, 11-11
in King George South Reservoir water, 11-14
of lakes, 5-10 to 5-11
at Lysekil water plant, 11-13
in Oradell Reservoir water, 11-16
PACls and, 6-31, 6-37, 6-38f, 7-32f, 11-12f
particle concentration and, 5-10 to 5-11
particle density and, 7-15
particle size and, 5-9, 5-10
pathogens and, 11-18
pilot-scale testing of, 10-7, 10-10
plate settling and, 11-19
recycle ratio and, 10-7
of reservoirs, 11-4
before reverse osmosis, 13-8, 13-19
of rivers, 5-10, 11-4
sand and, 10-3
scrapers and, 9-8 to 9-9, 9-8f
seawater, 13-8
sedimentation and, 8-12, 10-5, 11-5 to 11-6
in SFBW, 12-11, 12-13, 12-14
streaming current and, 10-15
treatment plant type for, 10-3 to 10-5, 10-4f
at Warner water plant, 6-39, 6-40f, 11-10
Water Quality and Treatment on, 5-10

Turbidity (*Cont.*):
  in William Girling Reservoir water, 11-14
  at Windhoek water plant, 12-9t
  at Winnipeg water plant, 11-16
2-MIB, 11-17, 11-18t
Two-stage filtration, 10-3

U

UF, 12-2 to 12-6, 12-2f, 12-4t, 12-5t, 12-6f, 12-8f
UFRV, 10-10 to 10-12, 11-10, 11-16
Ultrafiltration (UF), 12-2 to 12-6, 12-2f, 12-4t, 12-5t, 12-6f, 12-8f
Ultraviolet (UV) light:
  absorbance of:
    alum dose and, 6-38f, 6-39
    bench-scale jar testing of, 10-6, 10-8, 10-9f
    coagulant dose and, 7-31, 7-32f
    by Colorado River water, 12-6
    DOC/TOC levels and, 6-7
    PACls and, 6-31 to 6-32, 6-36 to 6-38, 6-38f, 7-32f, 11-12f
    in seawater, 13-10
    in SFBW, 12-11
    SUVA. *See* Specific UV absorbance
    treatment plant type for, 10-3
    at Warner water plant, 6-39, 6-40f, 11-10
    at Windhoek water plant, 12-9t
    at Winnipeg water plant, 11-16
  for disinfection, 6-5, 10-3, 11-15, 11-16
Unit filter run volume (UFRV), 10-10 to 10-12, 11-10, 11-16
United Arab Emirates (UAE), desalination plant in, 13-19
United Kingdom:
  Chingford South water plant, 11-8t to 11-9t, 11-14
  CoCoDAFF plants in, 11-4
  England, 9-2
  history of water treatment in, 2-6 to 2-8, 11-6, 11-14
  King George South Reservoir, 11-14
  London Conference, 2-8, 2-9, 11-6
  Lower Lliw Reservoir, 10-12t
  Scotland, 11-4
  Wales, 10-12t
  Wessex Water plant, 2-7
  William Girling Reservoir, 11-14
United States. *See also specific states*
  on boron in drinking water, 13-5
United States (*Cont.*):
  history of water treatment in, 2-3 to 2-4
  Orlando Conference, 2-8, 2-10, 11-6
  recycle rate in, 12-10
United States Environmental Protection Agency (USEPA):
  on boron, 13-5
  on Enhanced Coagulation, 6-3
  on SFBW, 12-9 to 12-10
United Water New Jersey, 11-16
Upflow-filtration through plastic, 10-5
USSR, 2-4

V

Vaalkop water plant, 11-5 to 11-6, 11-5f
"Vacuator," 2-3
Vacuum flotation, 1-4, 2-2 to 2-3, 2-5
Vacuum permittivity, C-1
Valves:
  location of, 4-1
  needle, 4-3t, 4-6, 4-10 to 4-12, 4-12f, 11-9t
van der Waals force:
  attachment efficiency and, 7-17t
  between bubbles, 5-5 to 5-6
  coagulation and, 5-22, 5-23f
  electrical double layer and, 13-8
  floc-bubble aggregates and, 5-18 to 5-19, 7-18
  in freshwater, 5-21, 5-22f, 13-8
  between particles, 5-14 to 5-16
  in seawater, 13-8
Vapor pressure, 3-2, D-1 to D-4
Variation coefficient, 6-14
Velocity gradient, RMS:
  bench-scale jar testing of, 10-6, 10-7
  Camp's, 6-14, 6-43
  at Chingford South water plant, 11-8t, 11-14
  dispersion coefficient and, 7-6
  flocculation and, 6-45, 6-46, 7-7 to 7-8
  fluid shear and, 6-42 to 6-43, 6-42t
  at Haworth water plant, 11-8t
  particle volume concentration and, 5-11
  at St. John's water plant, 11-8t, 11-11
  water temperature and, 6-45
  water viscosity and, 5-11, 6-45
  water volume and, 5-11
  at Windhoek water plant, 12-5t
  at Winnipeg water plant, 11-8t
Veolia Water, 1-3
Vertical-axial flow mixers, 11-15

Viruses. *See also specific types of*
charge of, 6-4
inactivation of, 5-8
microbiological assays of, 5-11
microfiltration and, 12-2
pH and, 6-4
size of, 5-9, 5-9f, 12-2f
solar radiation and, 12-8
sources of, 5-7t
ultrafiltration and, 12-2
Viscous effect, 5-19. *See also* Hydrodynamic force
Volumetric air concentration, 3-2 to 3-7, 3-6f, 3-35, 13-18. *See also* Bubble volume concentration
Volumetric air flow rate, 3-14, 3-28 to 3-29
Vrablik, Edward, 2-4

W

Wachusett Reservoir, 10-12t
Wales, 10-12t
Warner water plant:
case study of, 11-7 to 11-10
Cryptosporidium at, 11-18 to 11-19
DAFF at, 11-2, 11-7
dual-coagulant strategy at, 6-39, 6-40f, 11-7
Giardia at, 11-18 to 11-19
parameters of, 11-8t to 11-9t
pilot-scale testing at, 7-29
process schematic for, 11-7f
TOC at, 6-39, 6-41, 11-10, 11-10f
turbidity at, 6-40f, 11-10, 11-10f
UFRV at, 11-10
UV light absorbance at, 6-39, 6-40f, 11-10, 11-10f
Wastewater treatment:
dispersed air flotation for, 1-2 to 1-3
electroflotation for, 1-3
induced air saturation for, 1-5
oil and grease in discharge after, 13-11
packed vs. unpacked saturators for, 3-11
pressurization flow for, 1-5
reuse after, 12-7
of SFBW, 12-9 to 12-14
at Windhoek water plant, 12-7 to 12-9, 12-8f, 12-9t
Wastewater treatment plants, 7-4, 12-7 to 12-9, 12-8f, 12-9t
Water:
color of. *See* Water color
concentration of gas in, F-7
conductivity of, 5-17, 5-18t
Water (*Cont.*):
density of. *See* Water density
electrical double layer thickness in, 5-18t
equilibrium constants for, 13-3
feed, 12-3, 12-3f, 13-8
fulvic acid in, 6-6
hardness of, 6-8, 6-12, 12-6
ionic strength of, 5-17, 5-25, 6-12, 13-2
permeate, 12-3, 12-3f
permittivity of, 5-20 to 5-21, 10-13
salt water. *See* Seawater
spent filter backwash, 1-2t, 10-12t
TDS in, 5-17, 5-18t, 13-2
temperature of. *See* Water temperature
velocity of, 6-46, 7-23t, 8-11, 8-13 to 8-14, 8-16f, 9-7
viscosity of. *See* Water viscosity
volume of, in system, 6-14
Water color:
bench-scale jar testing of, 10-8
coagulation and, 6-2, 6-8
DAF and, 10-3 to 10-4, 10-4f, 11-6, 11-11
in Deacon Reservoir water, 11-15
direct filtration and, 10-3, 11-7
float layer and, 9-2, 9-7 to 9-8
of Hemlock Reservoir water, 11-7
at Karnsjon Reservoir, 11-11, 11-13
at Lysekil water plant, 11-13
nanofiltration and, 12-7
NOM and, 6-5
PACls and, 11-11, 11-12f
particle density and, 7-15
scraper speed and, 9-7 to 9-8
sedimentation and, 2-7, 10-5, 11-6
treatment plant type for, 10-3 to 10-5, 10-4f
Water density:
above baffles, 8-12
Brownian diffusion and, 7-12
bubble density and, 5-2
bubble rise velocity and, 7-19, 8-2, 8-3
correlation coefficients for, D-4
for design, 3-33
floc-bubble aggregate density and, 5-3 to 5-4
flocculation and, 6-46
inertia and, 7-12
in mass transfer model, 3-18
rapid mixing and, 6-12
Reynolds number and, 8-2
settling and, 7-12
single collector efficiency and, 7-12
suspended solids and, 8-12

Water density (*Cont.*):
TDS and, 8-12
volumetric loading and, 3-13
water temperature and, 6-12, 8-12, D-1 to D-3
Water Management, 2-6 to 2-7
Water Quality and Treatment (AWWA):
on algae, 5-8
assessment of, 2-9
on membrane processes, 12-2f, 12-3f
on particle sizes, 5-9f
on particle types and sources, 5-7t
on SUVA, 6-8t
on turbidity, 5-10
on Warner plant performance, 11-10f
Water Research Centre, 2-6 to 2-8
Water Science and Technology, 2-8
Water softening, 12-2, 12-7
Water temperature:
air density and, 5-2
aluminum coagulants and, 6-12, 6-18 to 6-20, 6-20f, 6-34
available air and, 3-16f, 3-21f, 3-22f, 3-23, 3-23f, 13-19f
in Bay Bulls Big Pond, 11-11
bench-scale jar testing of, 10-8
Brownian diffusion and, 6-42t, 7-15, 7-23
bubble concentration and, 5-2, 5-3f
bubble density and, 5-2
bubble rise velocity and, 7-15, 7-23, 8-3 to 8-4, 8-4f
of Colorado River water, 12-6
contact zone efficiency and, 7-15, 7-23 to 7-24, 7-24f
DAF and, 6-12, 11-6
dielectric constant and, D-1 to D-3
equilibrium constants and, 13-3 to 13-4
ferric coagulants and, 6-12, 6-24 to 6-25
filtration and, 6-12
flocculation and, 6-12, 6-45
Henry's constant and, 3-29, F-1 to F-3
hydraulic loading and, 11-2
at Karnsjon Reservoir, 11-11 to 11-12
PACls and, 6-12, 6-29 to 6-30, 6-31f, 6-36, 13-13
particle size and, 7-14f, 7-15, 7-24f
pilot-scale testing of, 10-7
rapid mixing and, 6-12
saturator air and, 3-6 to 3-7
sedimentation and, 2-7, 8-12, 11-6, 11-10
single collector efficiency and, 7-13t, 7-14f, 7-15, 7-23
Water temperature (*Cont.*):
surface tension and, D-1 to D-3
sweep-floc coagulation and, 6-12, 6-19 to 6-20, 6-25, 6-34, 6-36
vapor pressure and, D-1 to D-3
velocity gradient and, 6-45
water density and, 6-12, 8-12, D-1 to D-3
water viscosity and, 6-12, 6-45, 7-24, D-1 to D-3
Water Treatment Membrane Processes (Mallevialle et al.), 12-1
Water viscosity:
Brownian diffusion and, 6-42t
bubble formation and, 4-6
bubble rise velocity and, 7-19, 7-23t, 8-3
Camp's velocity gradient and, 6-14
correlation coefficients for, D-4
for design, 3-33
mass transfer rate constant and, 3-18
Reynolds number and, 8-2 to 8-3
single collector efficiency and, 7-12 to 7-13
velocity gradient and, 5-11, 6-45
water temperature and, 6-12, 6-45, 7-24, D-1 to D-3
zeta potential and, 10-13
Wessex Water plant, 2-7
West Chester pilot study, 10-12t
Wetted packing area, 3-18, 3-23, 13-17 to 13-18
White water, 1-8, 4-2 to 4-3, 8-2, 12-8 to 12-9
White water blanket, 7-9
White water bubble-blanket model:
contact zone efficiency, 7-16 to 7-29
derivation of, 7-35 to 7-38
description of, 7-10
single collector efficiency, 7-10 to 7-15
verification of, 7-29 to 7-34
WHO, 13-5
William Girling Reservoir, 11-14
Windhoek Goreangab Operating Company (WINGOC), 12-8
Windhoek water plant, 2-6, 12-5t, 12-7 to 12-9, 12-9t
Winnipeg water plant, 10-12t, 11-2, 11-8t to 11-9t, 11-14 to 11-16, 11-15f
World Health Organization (WHO), 13-5
WRC nozzles, 4-3t, 4-10f

Z

Zebra mussels, 11-15
Zevenbergen water plant, 2-7

CPSIA information can be obtained at www.ICGtesting.com
Printed in the USA
BVOW021501020112

279526BV00004B/1/P

9 780071 745628